Interference Mitigation and Energy Management in 5G Heterogeneous Cellular Networks

Chungang Yang
Xidian University, China

Jiandong Li
Xidian University, China

A volume in the Advances in Wireless Technologies and Telecommunication (AWTT) Book Series

IGI GLOBAL
DISSEMINATOR OF KNOWLEDGE

www.igi-global.com

Published in the United States of America by
IGI Global
Information Science Reference (an imprint of IGI Global)
701 E. Chocolate Avenue
Hershey PA, USA 17033
Tel: 717-533-8845
Fax: 717-533-8661
E-mail: cust@igi-global.com
Web site: http://www.igi-global.com

Library of Congress Cataloging-in-Publication Data

Names: Yang, Chungang, 1982- editor. | Li, Jiandong, 1962- editor.
Title: Interference mitigation and energy management in 5G heterogeneous cellular networks / Chungang Yang and Jiandong Li, editors.
Description: Hershey, PA : Information Science Reference, [2017] | Includes bibliographical references and index.
Identifiers: LCCN 2016044349| ISBN 9781522517122 (hardcover) | ISBN 9781522517139 (ebook)
Subjects: LCSH: Radio resource management (Wireless communications) | Cell phone systems--Protection. | Wireless communication systems--Energy conservation. | Wireless communication systems--Energy consumption. | Radio--Interference--Prevention.
Classification: LCC TK5103.4873 .I58 2017 | DDC 621.3845/6--dc23 LC record available at https://lccn.loc.gov/2016044349

This book is published in the IGI Global book series Advances in Wireless Technologies and Telecommunication (AWTT) (ISSN: 2327-3305; eISSN: 2327-3313)

British Cataloguing in Publication Data
A Cataloguing in Publication record for this book is available from the British Library.

All work contributed to this book is new, previously-unpublished material. The views expressed in this book are those of the authors, but not necessarily of the publisher.

For electronic access to this publication, please contact: eresources@igi-global.com.

Advances in Wireless Technologies and Telecommunication (AWTT) Book Series

Xiaoge Xu
The University of Nottingham Ningbo China, China

ISSN:2327-3305
EISSN:2327-3313

Mission

The wireless computing industry is constantly evolving, redesigning the ways in which individuals share information. Wireless technology and telecommunication remain one of the most important technologies in business organizations. The utilization of these technologies has enhanced business efficiency by enabling dynamic resources in all aspects of society.

The **Advances in Wireless Technologies and Telecommunication Book Series** aims to provide researchers and academic communities with quality research on the concepts and developments in the wireless technology fields. Developers, engineers, students, research strategists, and IT managers will find this series useful to gain insight into next generation wireless technologies and telecommunication.

Coverage

- Telecommunications
- Grid Communications
- Digital Communication
- Wireless sensor networks
- Wireless Broadband
- Virtual Network Operations
- Cellular Networks
- Radio Communication
- Network Management
- Broadcasting

IGI Global is currently accepting manuscripts for publication within this series. To submit a proposal for a volume in this series, please contact our Acquisition Editors at Acquisitions@igi-global.com or visit: http://www.igi-global.com/publish/.

The Advances in Wireless Technologies and Telecommunication (AWTT) Book Series (ISSN 2327-3305) is published by IGI Global, 701 E. Chocolate Avenue, Hershey, PA 17033-1240, USA, www.igi-global.com. This series is composed of titles available for purchase individually; each title is edited to be contextually exclusive from any other title within the series. For pricing and ordering information please visit http://www.igi-global.com/book-series/advances-wireless-technologies-telecommunication-awtt/73684. Postmaster: Send all address changes to above address.

Titles in this Series

For a list of additional titles in this series, please visit: www.igi-global.com

Handbook of Research on Advanced Trends in Microwave and Communication Engineering
Ahmed El Oualkadi (Abdelmalek Essaadi University, Morocco) and Jamal Zbitou (Hassan 1st University, Morocco)
Information Science Reference • copyright 2017 • 716pp • H/C (ISBN: 9781522507734) • US $315.00 (our price)

Handbook of Research on Wireless Sensor Network Trends, Technologies, and Applications
Narendra Kumar Kamila (C. V. Raman College of Engineering, India)
Information Science Reference • copyright 2017 • 589pp • H/C (ISBN: 9781522505013) • US $310.00 (our price)

Handbook of Research on Advanced Wireless Sensor Network Applications, Protocols, and Architectures
Niranjan K. Ray (Silicon Institute of Technology, India) and Ashok Kumar Turuk (National Institute of Technology Rourkela, India)
Information Science Reference • copyright 2017 • 502pp • H/C (ISBN: 9781522504863) • US $285.00 (our price)

Self-Organized Mobile Communication Technologies and Techniques for Network Optimization
Ali Diab (Al-Baath University, Syria)
Information Science Reference • copyright 2016 • 416pp • H/C (ISBN: 9781522502395) • US $200.00 (our price)

Advanced Methods for Complex Network Analysis
Natarajan Meghanathan (Jackson State University, USA)
Information Science Reference • copyright 2016 • 461pp • H/C (ISBN: 9781466699649) • US $215.00 (our price)

Emerging Innovations in Wireless Networks and Broadband Technologies
Naveen Chilamkurti (La Trobe University, Australia)
Information Science Reference • copyright 2016 • 292pp • H/C (ISBN: 9781466699410) • US $165.00 (our price)

Critical Socio-Technical Issues Surrounding Mobile Computing
Norshidah Mohamed (Prince Sultan University, Kingdom of Saudi Arabia and Universiti Teknologi Malaysia, Malaysia) Teddy Mantoro (USBI-Sampoerna University, Indonesia) Media Ayu (USBI-Sampoerna University, Indonesia) and Murni Mahmud (International Islamic University Malaysia, Malaysia)
Information Science Reference • copyright 2016 • 357pp • H/C (ISBN: 9781466694385) • US $210.00 (our price)

Handbook of Research on Next Generation Mobile Communication Systems
Athanasios D. Panagopoulos (National Technical University of Athens, Greece)
Information Science Reference • copyright 2016 • 604pp • H/C (ISBN: 9781466687325) • US $370.00 (our price)

Table of Contents

Detailed Table of Contents

capacity. With the rapidly increasing energy consumption, in this chapter, the authors investigate the NP-hard energy-efficient context-aware resource allocation problem and formulate it as a one-to-one matching problem. The preference lists in the matching are modeled based on the optimum energy efficiency (EE) under specified matching, which can be obtained by using an iterative power allocation algorithm based on nonlinear fractional programming and Lagrange dual decomposition. Next, on account of the Gale-Shapley algorithm, an energy-efficient matching algorithm is proposed. Some properties of the proposed algorithm are discussed and analyzed in detail. Moreover, the authors extend the algorithm to the matching with indifferent and incomplete preference lists. Finally, the significant performance gain of the proposed algorithm is demonstrated through simulation results.

Chapter 3
Challenges in Energy-Efficient Communications as Enablers for Green Solutions on the 5G Heterogeneous Networks 58

Irma Uriarte-Ramírez, Universidad Autónoma de Baja California, Mexico
Norma A. Barboza-Tello, Universidad Autónoma de Baja California, Mexico
Paul Medina-Castro, Universidad Autónoma de Baja California, Mexico

One of the most representative challenges of multi-tier heterogeneous network architectures is the interference management among different Radio Access Technologies. In this sense, in order to achieve the potential increase on network capacity forecasted for 5G, the issue of interference must be addressed. Intelligent energy management, as proposed in the growing area of Green Communications could serve as an appealing solution to the interference issue. However developing an optimal energy management plan requires the understanding of the characteristics of 5G cellular networks that could be exploited to improve energy and resources efficiency. In this work we present a study about challenges in energy-efficient communications and technologies as enablers for green solutions and how this challenges can be extended to meet those of the 5G heterogeneous networks, in order to identify possible solutions to address the energy efficiency and interference mitigation issues.

Section 2
Enhanced Interference Management Technology with Featured Characteristics

Chapter 4
Interference Management for Full-Duplex Massive MIMO Relaying System with Hardware Impairments 78

Kui Xu, PLA University of Science and Technology, China
Xiaochen Xia, PLA University of Science and Technology, China
Youyun Xu, PLA University of Science and Technology, China
Dongmei Zhang, PLA University of Science and Technology, China

Massive MIMO full-duplex relaying (MM-FDR), where multiple source-destination pairs communicate simultaneously with the help of a common full-duplex relay equipped with very large antenna arrays, is studied in this chapter. Different from the traditional MM-FDR protocol, a general model where sources/destinations are allowed to equip with multiple antennas is considered. The effect of hardware impairments is taken into consideration, and is modeled using transmit/receive distortion noises. We propose a low complexity hardware impairments aware transceiver scheme (named as HIA scheme) to

mitigate the distortion noises by exploiting the statistical knowledge of channels and antenna arrays at sources and destinations. A joint degree of freedom and power optimization algorithm is presented to further optimize the spectral efficiency of HIA based MM-FDR. The results show that the HIA scheme can mitigate the ``ceiling effect" appears in traditional MM-FDR protocol, if the numbers of antennas at sources and destinations can scale with that at the relay.

Yong Liao, Chongqing University, China
Yufeng Li, Chongqing University, China
Shumin Zhang, Chongqing University, China
Ming Zhao, Chongqing University, China
Xin Zhou, Chongqing University, China
Ling Chen, Chongqing University, China
Xuanfan Shen, Chongqing University, China
Yi Hu, Chongqing University, China

For the coexistence and increasing interference of satellite-terrestrial network and terrestrial wireless network, we analyze a typical scenario where the GEO satellite-terrestrial network and the 4G mobile communication network coexist heterogeneously. Besides, a multi-user cognitive system model that secondary satellite terminals interfere the primary MIMO 4G base stations is also proposed, with whose general signal processing is deduced. Meanwhile, DBF technology in 4G base station system is adopted to minimize the cognitive interference caused by multi-antennas and multi-users. And we propose an OBW-FAI. Weight vector is only related to the azimuth of the interferences, thus the proposed algorithm does not need real-time and repeat calculations, and has small complexity. Finally, the numerical simulation results verify that the proposed system and algorithm can effectively reduce interference between satellite-terrestrial network and terrestrial wireless network to a certain extent.

Yun Meng, Chang'an University, China
Yuan Dong, Chang'an University, China
Song Shi, Chang'an University, China

Different with the traditional resource allocation problems, the consistency of resource occupation and the diversity of interference of participants in interference alignment (IA) clusters bring about the complexity in the combination of resource allocation and IA. Therefore, this chapter gives a transformed conflict graph-based solution framework which considers the low complexity of chordal graph, where the selection criteria of IA clusters are determined by the influence of IA on the resource occupation and the interference. The simulation results show that the proposed schemes can improve the network performance.

Small-cell technologies are seen as one of the most promising solutions for the rapid growth of wireless data services and 5G requirements. However, because most SBSs are deployed with minimum intervention from the end users and the service providers, it is hard to mitigate co-tiered interference. It is significant to study the self-organized distributed co-tiered interference mitigation and resource allocation. Game theory is an effective distributed approach towards handling the distributed co-tiered interference mitigation problem without a central controller. This chapter is to address the application of game theory and distributed learning solutions for distributed co-tiered interference mitigation. Two potential game models for static and dynamic co-tiered interference mitigation are presented and discussed for small-cell networks with fixed loads and dynamic loads separately. In addition, two distributed learning algorithms are presented and results are discussed. Finally, some future research directions are given.

In evolving fifth generation (5G), some demands that to be addressed are improved data rate, increased capacity, decreased latency and better quality of service. Heterogeneous networks (HetNets) is considered as an effective way to meet these prime objectives. Deployment of HetNets faces a number of challenges, among which interference management is of much importance. This chapter has summarized the major challenges and solutions of interference management. Interference cancellation, avoidance and coordination are discussed in detail, including enhanced intercell Interference Coordination (eICIC), Coordinated Multi-point Transmission (CoMP) et al. In addition, the future challenges of interference management have also been presented.

The high-speed railway (HSR) is a typical application case in 5G systems. Mobile relay stations (MRSs) which are mounted in a high-speed train (HST) is popular system architecture for high-speed mobile communications. However, sharing spectrums between the macro cell and the MRS cell, interference exists in this hybrid system. In this chapter, we investigate the downlink of a multi-cellular decode and forward (DF) relayed orthogonal frequency division multiple access (OFDMA) system and formulate the problem to maximize the system sum rate of all cells subject to a total power constraint and a new proposed time delay constraint. An effective resource allocation scheme combined by greedy sub-carriers allocation and geometric programming (GP) based power allocation algorithm is proposed to optimize subcarrier allocation and power allocation. Numerical experiments verify that the proposed resource allocation scheme outperforms the other traditional approaches and the necessity of introducing the time delay constraint.

Driven by the continuous growth of wireless and mobile communication users, Long Term Evolution (LTE) and Long Term evolution-Advanced (LTE-A) have taken up densification of network as a new paradigm to meet the growing capacity demands. Small cells come with the advantage of enhanced coverage in indoor and hard reach areas and offer traffic offloading capacity in hotspots. However, there are challenges of interference management and self-adaptability with the overlaying macro cellular network since most small cell base stations are user deployed and do not have centralized control on their configuration and operation. The purpose of this chapter is to elaborate the concept of heterogeneous network and Self Organization Network (SON) in LTE-A. The various use cases of SON that can benefit the heterogeneous network have been discussed laying emphasis on interference management use case. Further the current trend of research in this field has been highlighted. It provides a holistic picture of the heterogeneous network and SON in LTE-A and upcoming mobile communication generations.

Game theory has found an extensive application in wireless communication networks including cognitive radio networks, heterogeneous cellular networks, cooperative relay networks. Also, cognitive radio networks, green communications and heterogeneous cellular networks have attracted a wide attention on improve the spectrum efficiency and energy efficiency; therefore, the capacity, the coverage and the energy consumption. However, interference problem and energy consumption are critical for these networks. Introducing hierarchy among different decision-making players in cognitive, heterogeneous, green, cooperative cellular networks can both save energy and mitigate interference, thus enhance throughput. Stackelberg game suits to model, analyze and design the distributed algorithms in these hierarchical decision-making networking scenarios. In this chapter, we introduce basics of Stackelberg game and survey the extensive applications of Stackelberg game in cognitive, heterogeneous, cooperative cellular networks with the emphasis on resource management, green commutations design and interference management. This chapter highlights the potentials and applications with the promising vision of Stackelberg game theoretic framework for future cognitive green heterogeneous cellular networks.

Foreword

To meet the recent wireless traffic dramatic increase, 5G heterogeneous cellular networks have been investigated as the most powerful network architectures to enhance both spectrum and energy efficiency in a cost-effective way. In such an architecture, various kinds of small cells with different downlink powers coexist in the coverage of macrocells. However, these good opportunities provided by the 5G heterogeneous cellular networks always come with various technical challenges. For instance, both the inter- and the intra-tier interference damages the potential performance improvement. This is mainly due to the limited spectrum resources, thus leading to these underlying small cells fully sharing the licensed spectrum to the macrocells. On the other hand, denser deployment of small cells means higher energy consumption, and backhaul and fronthaul expenditures. This book investigates interference mitigation and energy management, with the aim to release the potential capabilities of the 5G heterogeneous cellular networks with the following sections:

In Section 1, Chapters 1 and 2 concentrate on specific energy efficient technologies from different perspectives, e.g., sectorisation and densification (Chapter 1) and matching framework-based catching (Chapter 2). Chapter 3 surveys the current research status and the ongoing projects in the field being carried out around the world. Finally, future research directions are presented with some detailed discussion. In Section 2, Chapters 4, 5, 6, and 7 investigate the interference management schemes with featured characteristics of the full-duplex massive MIMO relay (Chapter 4), the cognitive MIMO system (Chapter 5), the combination of resource allocation and interference alignment (Chapter 6), and game theory (Chapter 7). Finally, Chapter 8 summarizes the major challenges and solutions of interference management. Interference cancellation, avoidance and coordination are also discussed in detail in Chapter 8. In Section 3, novel mathematical frameworks are provided. In section 3, a geometric programming formulation is first presented (Chapter 9), thus leading to the discussion of the necessity of self-organizing networks in the 5G era (Chapter 10). Following that, we introduce two surveys of Stackelberg game (Chapter 11) and pricing methodology (Chapter 12) to meet the hierarchical architecture in 5G heterogeneous cellular networks.

The topic is timely and interesting. The editors carefully chosen a promising topic, and selected excellent up-to-date material addressing these technical aspects. The targeted audience of the book are either students, researchers or industry managers to get valuable insights into their own research or future applications. I am sure it will be an important reference for many years in the field of future wireless communications.

Mohsen Guizani
University of Idaho, USA
September 2016

Preface

To meet the recent wireless traffic volume increment, 5G heterogeneous cellular networks have been investigated as the most powerful network architecture to enhance both spectrum and energy efficiency in a cost-effective way, where various kinds of small cells with different downlink powers coexist in the coverage of macrocells. These small cells can be pico cells, femtocells, remote radio heads, relays, and device-to-device communication pairs. These small cells can offload the traffic, extend the coverage, and better meet the subscribers' quality of experience (QoE). In summary, 5G heterogeneous cellular networks will counter the challenge of cost-effectively supporting a 1000-fold increase in traffic demand over the next decades.

However, these good opportunities provided by the 5G heterogeneous cellular networks always come with various technical challenges, for instance, both the inter- and the intra-tier interference damages the potential performance improvement. This is mainly due to the limited spectrum resources, thus leading to these underlying small cells fully sharing the licensed spectrum to the macrocells.

On the other hand, denser deployment of small cells means higher energy consumption, and backhaul and fronthaul expenditures. Therefore, the spectrum, energy, and cost efficiency problems are becoming more serious, in particular, in the ultra-dense deployment scenarios. This edited book will concentrate on the promising interference mitigation and energy management, with the aim to release the potential capabilities of the 5G heterogeneous cellular networks.

By now, extensive research has been poured on this topic, which can be viewed from the following aspects: first, theoretic performance limits of interference-limited systems have been analyzed in the framework of stochastic geometry and optimization theory. However, researchers prefer the advanced mathematical tools with features of characterizing the interaction and rational behaviors and distributed control schemes. Second, although there exist different interference mitigation technology, they should be further studied and applied to the 5G heterogeneous cellular networks. Meanwhile, there should exist a survey on the interference alignment, enhance inter-cell interference coordination, and interference cancellation. In addition, different energy management methodology should be jointly considered including the cell-zooming and cell range expansion to enhance the energy efficiency. Last but not the least, new network architectures and typical features are involved in the heterogeneous cellular networks, including cloud computing, cognitive radio, and software-defined networks and network virtualization. These will largely enhance the capabilities of 5G heterogeneous cellular networks.

The promising 5G featured technology has been widely studied to mitigate interference and save energy, thus improving both spectrum and energy efficiency. Researchers from academia, industry, and standardization communities are still working hard on these promising topics. Therefore, it is necessary to call for them to propose their novel ideas to contribute this book.

It is known that the energy efficient, green, and interference management should be explored and exploited for better understanding and design wireless networks. This book looks into discuss and address interference and energy management for the emerging 5G heterogeneous cellular networks, where we concentrate on building a overall picture on the topic, surveying the current research status, and providing novel mathematical frameworks, e.g., matching, game, and pricing.

This book is therefore both timely and apposite. The book will provide the novel and potential way to solve the encountered problems for wireless network researcher and engineer. It also will help various kinds of networking engineers for understanding and designing the suitable wireless networks.

TARGET AUDIENCE

Researchers, advanced-level students, technology developers, and engineers will find this text useful in furthering their research on Interference Mitigation and Energy Management in 5G Heterogeneous Cellular Networks. This text will assist wireless engineers in furthering their own research efforts and interests, where they will find useful and valuable guidance from the surveyed techniques, the specific technology applied to specific scenarios, and the mathematical frameworks.

CHAPTER OVERVIEW

After careful and rigorous review, finally we accepted 12 chapters according to the review results, and content, concentration, and objective of this book; and then we divide them into three sections according to their internal logic.

Section 1: Energy-Efficient Communications and Energy Management Techniques

In the section, we have three chapters, where the first two chapters (Chapter 1 and Chapter 2) concentrate on specific energy efficient technology from different perspectives, and Chapter 3 surveyed the current research status and the ongoing projects all over the world, finally the authors looked into the future research directions. In detail, Chapter 1 studied two kind of green communication techniques, including adding more sectors per site and deploying an overlay of small cells, which are key methods to enhance the RAN coverage and capacity. The impact of these two techniques on cellular network energy consumption was also investigated in Chapter 1 with the aim to find an energy efficient deployment strategy when trading-off the order of sectorisation with the intensity of small cell densification. In addition, Chapter 1 evaluated the amount of network energy efficiency improvement when various adaptive sectorisation schemes were implemented, where the conclusion was the strategy of adding more sectors was less energy efficient than directly deploying an overlay of small cells, even when adaptive sectorisation was implemented.

In Chapter 2, the authors studied caching to ultra-densely deployed small cells to save the energy consumption, where the authors investigated a NP-hard energy-efficient context-aware resource allocation problem and formulated it as a one-to-one matching. The preference lists in the matching were modeled based on the optimum energy efficiency under specified matching and established by using an iterative power allocation algorithm based on nonlinear fractional programming and Lagrange dual decomposition. Some properties of the proposed Gale-Shapley algorithm were discussed and analyzed in detail. Moreover, the authors extended the algorithm to the matching with indifferent and incomplete preference lists in Chapter 2. In green communications, the network operator makes efforts to reduce energy consumption., where different research schemes about energy efficient communication techniques, the design of energy-aware network architectures and protocols, energy-friendly software, and renewable energy sources have been well studied recently. However developing an optimal energy management plan requires the understanding of the characteristics of 5G heterogeneous cellular networks that could be exploited to improve energy and resources efficiency. In Chapter 3, the authors presented a study about challenges in energy-efficient communications and communications technologies as enablers for green solutions and how this challenges could be extended to meet those of the 5G heterogeneous networks, in order to identify possible solutions to address the energy efficiency and interference mitigation issues. Chapter 3 surveyed the current research status and the ongoing projects allover the world, finally the authors looked into the future research directions.

Section 2: Enhanced Interference Management Technology with Featured Characteristics

In this section, we have five chapters, where Chapters 4, 5, 6, and 7 investigated the interference management schemes with featured characteristics of the full-duplex massive MIMO relay, the cognitive MIMO system, the combination of resource allocation and interference alignment, and game theory. Finally, Chapter 8 summarized the major challenges and solutions of interference management. Interference cancellation, avoidance and coordination were discussed in detail in Chapter 8.

Massive MIMO full-duplex relaying (MM-FDR) was studied in Chapter 4, where multiple source-destination pairs simultaneously communicated with the help of a common full-duplex relay. Different from the traditional MM-FDR protocol, a general model was formulated where sources and destinations were allowed to equip with multiple antennas. A low complexity hardware impairments aware transceiver scheme (named as HIA scheme) was presented to mitigate the distortion noises by exploiting the statistical knowledge of channels and antenna arrays at sources and destinations. The results showed that the HIA scheme can mitigate the ceiling effect appearing in traditional MM-FDR protocol when the numbers of antennas at sources and destinations can scale with that at the relay. In Chapter 5, the authors analyzed a typical scenario where the GEO satellite-terrestrial network and the 4G mobile communication network coexist heterogeneously. Besides, a multi-user cognitive system model was also proposed, where secondary satellite terminals interfered the primary MIMO 4G base stations. Meanwhile, DBF in 4G base station system was adopted to minimize the cognitive interference caused by multi-antennas and multi-users, where weight vector was only related to the azimuth of the interference, thus the proposed algorithm did not need real-time and repeat calculations, and had small complexity. Finally, the numerical simulation results verified that the proposed system and algorithm can effectively reduce interference between satellite-terrestrial network and terrestrial wireless network to a certain extent.

Chapter 6 and Chapter 7 are from the resource management perspective to manage the interference. However, different with the traditional resource allocation problems, the consistency of resource occupation and the diversity of interference of participants in interference alignment (IA) clusters bring about the complexity in the combination of resource allocation and IA. Therefore, Chapter 6 gave a transformed conflict graph-based solution framework which considered the low complexity of chordal graph, where the selection criteria of IA clusters were determined by the influence of IA on the resource occupation and the interference. The simulation results showed that the proposed schemes can improve the network performance. Meanwhile, it is significant to study the self-organized distributed co-tiered interference mitigation and resource allocation. Game theory is an effective distributed approach towards handling the distributed co-tiered interference mitigation problem without a central controller. Chapter 7 was to address the application of game theory and distributed learning solutions for distributed co-tiered interference mitigation. Two potential game models for static and dynamic co-tiered interference mitigation were presented and discussed for small-cell networks with fixed loads and dynamic loads separately. In addition, two distributed learning algorithms were presented and results are discussed. Finally, some future research directions were given.

In addition to the listed interference management schemes, Chapter 8 summarized the major challenges and solutions of interference management. Interference cancellation, avoidance and coordination were discussed in detail, including enhanced intercell Interference Coordination(eICIC), Coordinated Multi-point Transmission (CoMP) et al. In addition, the future challenges of interference management had also been presented.

Section 3: Novel Mathematical Frameworks for Interference and Energy Management

We have four chapters in this section, where we provided novel mathematical frameworks, although there have exited the specific applications of matching in Chapter 2 and game in Chapter 7. In this section, a geometric programming formulation was first given, thus leading to the discussion of the necessity of self-organizing networks in the 5G era. Following that, we introduced two surveys of Stackelberg game and pricing methodology to meat the hierarchical architecture of Interference Mitigation and Energy Management in 5G Heterogeneous Cellular Networks.

In Chapter 9, the authors studied the high-speed moving scenario, such as high-speed railway (HSR). Which was a typical application scenario of 5G communication systems. Recently, the HSR has been rapidly developed in the world. Mobile relay stations (MRSs) which are mounted in a high-speed train (HST) is the popular system architecture for high-speed mobile communications. However, sharing spectrum between the macro cell and the MRS cell, interference exists in this hybrid system. Chapter 9 investigated the downlink of a multi-cellular decode and forward (DF) relayed OFDMA)system and formulated the problem to maximize the system sum rate of all cells subject to a total power constraint and a new proposed time delay constraint. An effective resource allocation scheme combined by a greedy sub-carriers allocation and geometric programming (GP) based power allocation algorithm was proposed to optimize subcarrier allocation and power allocation. Particularly, a rigid time delay constraint was introduced to accommodate the high speed mobile environment except for the common power constraint. Numerical experiments verified that the proposed resource allocation scheme outperforms the other traditional approaches and the necessity of introducing the time delay constraint. Although optimal solutions can be found in the convex optimization frameworks, there comes the novel challenges of distributed interference management and self adaptability with the overlaying macro cellular network since most small cell base stations are user deployed and do not have centralized control on their configuration and operation. The purpose of Chapter 10 is to elaborate the concept of heterogeneous network and Self Organization Network (SON) in LTE-A. The various use cases of SON that can benefit the heterogeneous network has been discussed laying emphasis on interference management use case. Further the current trend of research in this field has been highlighted. It provides a holistic picture of the heterogeneous network and SON in LTE-A and upcoming mobile communication generations.

Chapters 11 and 12 considered the joint interference mitigation and energy management, where the concentration is to introduce the novel game and pricing techniques to facilitate the design of distributed schemes. Game theory has found an extensive application in wireless communication networks including cognitive radio networks, heterogeneous cellular networks, cooperative relay networks. Also, cognitive radio networks, green communications and heterogeneous cellular networks have attracted a wide attention on improve the spectrum efficiency and energy efficiency; therefore, the capacity, the coverage and the energy consumption. However, interference problem and energy consumption are critical for these networks. Introducing hierarchy among different decision-making players in cognitive, heterogeneous, green, cooperative cellular networks can both save energy and mitigate interference, thus enhance throughput. Stackelberg game suits to model, analyze and design the distributed algorithms in these hierarchical decision-making networking scenarios. In this chapter, we introduce basics of Stackelberg game and survey the extensive applications of Stackelberg game in cognitive, heterogeneous, cooperative cellular networks with the emphasis on resource management, green commutations design and interference management. Chapter 11 highlighted the potentials and applications with the promising vision of Stackelberg game theoretic framework for future cognitive green heterogeneous cellular networks.

Pricing theory and methodology from economic fields have found extensive applications in cognitive radio and multi-tier heterogeneous cellular networks. Conventional economic pricing applications in wireless communications are always related to the monetary revenue of network operators. In this chapter, the authors concentrate on different technical applications and research directions of pricing theory and methodology, where the authors investigate following technical applications and functions of pricing including cooperative incentive mechanism design, Pareto- and social optimality improvement, distributed algorithm design with the low signaling overhead. the authors first clarify different concepts of pricing, summarize the motivation, present a taxonomy according to these different technical applications. Then, the authors survey applications of pricing theory and methodology with understandings and observations in cognitive radio and multi-tier heterogeneous cellular networks. the authors emphasize some of the recent critical problems, such as the cooperation incentive, resource and interference management and economics of small cells. Finally, the authors conclude this chapter with the possible research directions and more potential network applications of pricing theory and methodology.

CONCLUSION

This comprehensive and timely publication aims to be an essential reference source, building on the available literature in the field of interference mitigation and energy management for wireless communication networks, and in particular, the readers will find useful and valuable guidance from the surveyed techniques, the specific technology applied to specific scenarios, and the mathematical frameworks.

The book will provide the novel and potential way to solve the encountered problems for wireless network researcher and engineer. It also will help various kinds of networking engineers for understanding and designing the suitable wireless networks.

Chungang Yang
Xidian University, China

Jiandong Li
Xidian University, China

Section 1

Energy-Efficient Communications and Energy Management Techniques

Chapter 1
Toward Green Evolution of Cellular Networks by High Order Sectorisation and Small Cell Densification

Abdelrahman Arbi
University of Sheffield, UK

Timothy O'Farrell
University of Sheffield, UK

Fu-Chun Zheng
Harbin Institute of Technology, Shenzhen, China

Simon C. Fletcher
Real Wireless Ltd, UK

ABSTRACT

Network densification by adding either more sectors per site or by deploying an overlay of small cells is always considered to be a key method for enhancing the RAN coverage and capacity. The impact of these two techniques on cellular network energy consumption is investigated in this chapter. The aim is to find an energy efficient deployment strategy when trading-off the order of sectorisation with the intensity of small cell densification. A new enhanced base station power consumption model is presented, followed by a novel metric framework for the evaluation of the RAN energy efficiency. The use of the power model and the proposed metrics is demonstrated by applying them to a RAN case study when the two techniques are used to improve the network capacity. In addition, the chapter evaluates the amount of network energy efficiency improvement when various adaptive sectorisation schemes are implemented. The results show that the strategy of adding more sectors is less energy efficient than directly deploying an overlay of small cells, even when adaptive sectorisation is implemented.

INTRODUCTION

The unprecedented exponential growth in the demand for mobile data services is expected to continue over the next five years. The latest Cisco visual networking index reported that global mobile data traffic grew by 69 percent in 2014, and predicted about a tenfold increase in traffic from 2.5 EB to 24.3 EB between 2014 and 2019 (Cisco Visual Networking, 2015). This trend is mainly due to the proliferation

DOI: 10.4018/978-1-5225-1712-2.ch001

of the use of smart devices and the expansion of mobile services in developing societies. In response, operators need to enhance the capacity of their networks to meet this growth in demand. There exist three main technical approaches for the operators to improve their radio access network (RAN) capacity. The first is to improve the link spectral efficiency by implementing more advanced signal processing techniques such as multiple-input multiple-output (MIMO) antennas and efficient modulation and coding schemes. The second approach is the addition of new radio spectrum while the third is to increase the geographic spectrum reuse by cell densification of the RAN.

Historically, the reuse of radio spectrum was the main reason for the growth of capacity in cellular mobile radio systems. According to Martin Cooper (Arraycomm.com, 2015), a 1600 fold increase during the last 45 years has come from the reuse of spectrum, whereas a 625 fold increase has resulted from adding more spectrum and implementing more advanced signal processing techniques to enhance the link capacity. The wireless channel link capacity is very close to its maximum Shannon limit. To significantly increase capacity by adding more spectrum, technologies such as millimetre wave radio (Rappaport et al., 2013) and visible light communications are being considered (Singh, O'Farrell, & David, 2014) and (Singh, O'Farrell, & David, 2013). Also, technology based on large antenna arrays such as Massive MIMO has the potential to significantly increase spectral efficiency without using more spectrum (Larsson, Edfors, Tufvesson, & Marzetta, 2014). However, the deployment of more cells (i.e. base station densification) will continue to be the key enabling technique to achieve high target capacities in the next generation of cellular mobile networks. Therefore, the trend of base station densification in cellular networks is expected to intensify in 5G RANs. However, such growth in the number of deployed base stations will be accompanied by an increase in the total network energy consumption, which represents a major concern for network operators for the following reasons:

1. **Economic Reasons:** Reducing the network operating cost is an important target for the operators as the network energy consumption bill represents a significant cost for the operators (Li, Sheng, Yang, & Wang, 2013). For example, the cost of energy represents 30% of the total operating expenses of the networks in the urban areas of India, and around 50% in rural areas (Rakesh Kumar Bhatnagar & Panigrahi, 2013).
2. **Environmental Reasons:** The information and communication technology (ICT) sector was responsible for about 2% of the global CO_2 emissions in 2009 and such emissions are predicted to grow to about 2.8% in 2020 (Forster, Dickie, Maile, Smith, & Crisp, 2009).

Network densification can be accomplished by adding either more macro base station sites, more sectors per site or by deploying an overlay of small cells to form what is known as a heterogeneous network. Deploying more macro sites and adding more sectors have been the traditional methods that operators have used to expand the coverage and capacity of their existing networks. However, reducing the site to site distance between the macro sites cannot continue without limit due to the technical issues of inter cell interference and, more importantly, due to the cost of macro base stations coupled with the difficulty of finding new sites, especially in urban and dense urban areas.

This chapter focuses on RAN densification by considering the deployment of more sectors per site (i.e. high order sectorisation) and the deployment of a second layer of small cells when needed, as two approaches to improve the RAN capacity. We address how the RAN energy consumption is influenced either when one, the other or both of these approaches is implemented. The objective is to find an energy efficient deployment strategy for mobile network operators seeking to deploy small cells alongside choosing the order of sectorisation in a RAN. Small cell densification is considered to be a significant method for enhancing the capacity and coverage of a 5G RAN. However, using these techniques come with the price of increasing the overall RAN energy consumption. This is due to the fact that densification increases the number of deployed base stations, which adds more power consumption to the RAN. Similarly, increasing the number of sectors at the base station site amounts to increasing the number of radio frequency (RF) chains in the base station, which in turn increases the overall power consumption by a factor approximately equal to the number of sectors used.

In this chapter an evaluation framework for the energy efficiency of these two deployment approaches is presented. The impact of small cell densification versus high order sectorisation on the energy efficiency of a heterogeneous LTE based RAN is considered in the context of enhancing the capacity of a 5G RAN. Energy consumption is compared for two deployment options. The first option involves the densification of a 3-sector macro-cell RAN by deploying an overlay of small cell base stations. The second option involves first upgrading the 3-sector sites to 6-sectors and then deploying the small cell overlay. A new detailed model to estimate the power consumption of the radio base station is presented. The RAN energy consumption gain (ECG) is presented as the main figure of merit to quantify any gains or losses in the RAN energy efficiency. The impact of implementing an adaptive sectorisation technique on the RAN energy efficiency for the two options of 3 and 6 sector base stations is also investigated.

In addition to a background and review of the literature about sectorisation and heterogeneous networks, the chapter presents a radio base station power consumption model for both the sectorised macro-cell and small-cell deployments. An energy efficiency metric framework is also presented followed by a detailed methodology to evaluate the energy efficiency in the sectorised and heterogeneous cellular networks. At the end of the chapter, the proposed energy efficiency metrics and the base station power consumption model are used to evaluate the energy efficiency of a RAN case study, and the case study analysis includes investigating the impact of applying various adaptive sectorisation schemes on the RAN energy efficiency.

BACKGROUND AND LITERATURE REVIEW

In this section, a background and non-exhaustive review of the literature on the impact of sectorisation, adaptive sectorisation, and heterogeneous networks on the RAN spectral and energy efficiency is presented.

High Order Sectorisation

Present day 4G cellular network deployments consist mainly of three sector macro-cell base stations, and as the traffic demand grows, operators need to evolve their cellular networks to provide the required capacity. Adding more sectors to the existing macro sites is considered to be the most cost effective evolution path by the operators as it increases the network capacity without needing to add more base station sites to the network with backhaul capabilities. However, due to the interference leakage between the adjacent sectors, the gain from adding more sectors does not scale up linearly with the number of sectors added to a site. The careful design of the antenna beamwidths can keep this effect small. On the other hand, a linear increase in the power consumption of macro sites is observed when more sectors are deployed.

The literature has reported the impact of high order sectorisation on the RAN capacity. In (Huang et al., 2010), the authors evaluated the mean throughput per macro site against the antenna beamwidth for 3-, 6- and 12-sectors circular antenna array deployments. Their results showed that when the optimum beamwidth is used, a 12-sectors deployment is able to improve the site's mean throughput by a factor of 3.2 compared to a 3-sectors deployment. It has been shown by simulation and field trials (Joyce, Morris, Brown, Vyas, & Zhang, 2015) that high order sectorisation can offer potential capacity gains when utilised in HSPA+ cellular networks. Cooperation between the sectors of the same macro site was shown to mitigate the interference leakage between the sectors, leading to improvements in both the site throughput and the fairness between users (Riedel & Fettweis, 2011).

A combination of horizontal and vertical sectorisation can also be used. For example, an antenna array of 10 vertical sectors and 16 horizontal sectors can offer up to 1.5 Gbit/s throughput when deployed inside a stadium environment (Maslennikov et al., 2013). The performance of horizontal and vertical high order sectorisation is evaluated in (Athley, Johansson, & Nilsson, 2013). The authors concluded that both the horizontal and vertical sectorisation techniques give similar results in the case of small urban macro-cells, whereas the horizontal sectorisation techniques outperform the vertical case in a large macro-cell scenario. In addition to the evaluation of the throughput gain when more sectors are deployed, the impact of using high order sectorisation techniques on the RAN energy consumption is also evaluated by (Arbi & O'Farrell, 2015), where it was shown that adding more sectors is beneficial to network operators only if the RAN energy efficiency is not a significant concern.

Adaptive Sectorisation

The previous section highlighted that using high order sectorisation incurs an increase in the overall RAN energy consumption. A number of studies in the literature have investigated the impact of sectorisation on the RAN energy efficiency and have proposed using an adaptive sectorisation technique to reduce the overall energy consumption. Also, adaptive sectorisation is proposed to address the limitations of designing the RAN to meet the peak traffic demand only. When the traffic intensity is low, the fixed sectorisation solution is no longer energy efficient as radio infrastructure unnecessarily remains switched on. In adaptive sectorisation, the radio base station site is reconfigured to have fewer and larger beamwidth sectors during periods of low traffic load. Results reported in the literature show a 30% power saving using adaptive sectorisation compared to a fixed tri-sectored site with only a 7.6% decrease in throughput (Hevizi & Godor, 2011). The authors in (Yinan, Imran, & Tafazolli, 2011) have proposed an energy-aware adaptive sectorisation strategy, to improve the energy efficiency of LTE networks, where each base station is capable of adapting to the temporal traffic variation by switching off one of its sectors, and changing the beamwidth of the two remaining sectors to maintain the required coverage level. Their results show a 21% energy saving when one sector is switched off during the low traffic load period.

An electrically reconfigurable antenna concept has been proposed for adaptive sectorisation (Weisi Guo, Rigelsford, Ford, & O'Farrell, 2012). The concept allows the number of active sectors and the antenna beamwidth to be changed by using a tuneable reflector. The authors have compared the energy consumption between an always active 6-sector base station, and a dynamic 6-sector base station, which can dynamically switch off its sectors while adapting the antenna beamwidth. The work reports peak energy savings of 75% and mean energy savings of 38%. The authors in (Hossain, Munasinghe, & Jamalipour, 2013) proposed a traffic-sensitive dynamic sectorisation technique based on a low complexity heuristic algorithm. The proposed dynamic algorithm outperforms semi-static sectorisation schemes as well as base station on/off schemes. Adaptive sectorisation has also been proposed to enhance the network throughput in unbalanced traffic scenarios (Zhang & Cuthbert, 2011). A dynamic sectorisation technique based on a genetic algorithm was used in an OFDMA network to adapt the antenna pattern when an unbalanced traffic distribution is encountered. An enhancement of up to 50% was shown in the network throughput compared to a fixed sectorisation scenario.

Heterogeneous Networks

The concept behind Heterogeneous Networks (Het Nets) is to overlay an existing homogenous macro network with additional smaller low power base stations, and these small base stations can be micro, pico or femto base station, or a combination thereof. Densifying the macro-cell RAN by deploying small-cell base stations as an overlay has the advantage of immediately targeting traffic hot spots in a cost effective way. Several published works have shown that Het Net deployments improve the RAN throughput by increasing the spectrum reuse factor while helping to alleviate the network energy consumption by deploying low power base stations where and when they are needed. The results of (Badic, O'Farrell, Loskot, & He, 2009) demonstrated that a homogeneous small-cell deployment is theoretically always more energy efficient than a large-cell deployment if the base station overhead power tends to zero. In practice, this is not possible and some limit to densification is reached for realistic overhead powers (He et al., 2010).

The impact of a Het Net deployment strategy on the power consumption of the mobile network has been evaluated by varying the number of micro sites per macro cell as well as the macro cell size while aiming to achieve the same target RAN area spectral efficiency. It has been shown that the area power consumption can be decreased under full load conditions without affecting the target area throughput (Fehske, Richter, & Fettweis, 2009). The authors in (Tombaz, Usman, & Zander, 2011) have also analysed the impact of different heterogeneous deployment strategies on the area power consumption for uniform and non-uniform user distributions. The results demonstrated that in WCDMA networks, the area power consumption can be decreased by up to 50% when small, low power base stations are deployed at the edges of the macro-cells. In (Quek, Cheung, & Kountouris, 2011), an analytical and numerical evaluation of the energy efficiency of a network consisting of macro-cell base stations overlaid by randomly distributed pico-cells, proved that an optimal pico-cell to macro-cell density ratio exists, which maximises the energy efficiency of the network. The authors have used a bit/s/m^2/joule metric for the energy efficiency. In the same paper, the authors also showed that this optimum ratio decreases when the macro density increases.

(Micallef, Mogensen, Scheck, & Lang, 2011) studied a long term energy efficient strategy for the evolution of cellular mobile networks. The authors evaluated the possible energy gains obtained when considering the upgrading of mobile network infrastructure over a period of 8 years. They compared the strategy of upgrading the macro-cells only with the strategy of joint macro/pico-cell deployment. High, medium and low pico-cell densities were considered. The results showed that deploying a high density of pico-cells consumes the least RAN energy, while the carried traffic volume increased on average by 16%. The disadvantage is that a large number of backhaul links are needed for the pico-sites. The authors also showed that the strategy of deploying a low density of pico-cells is cheaper than a macro-cell only upgrade path while a reduction in the total energy consumption by 10.5% is achieved. This energy saving can be increased by switching off the idle pico-cells during periods of low traffic volume.

The authors of (Soh, Quek, Kountouris, & Shin, 2013) showed that the energy efficiency gains obtained by adding an overlay of small-cells to a macro-cell RAN saturates as the density of small-cells increases. The use of discontinuous transmission (DTX) to save energy in Het Nets was considered in (Nie, Yi, Zheng, Zhang, & Farrell, 2015), which showed that there is an optimal mute probability that maximises the network energy efficiency when random DTX is implemented, and that this probability is an increasing function of the small-cell density. The authors of (Yang, Guo, Sheng, Li, & Yue, 2014) have showed that offloading portion of the macro-cell traffic to an overlay of small-cells, and reducing the macro-cell transmitted power can lead to an optimal trade-off between the energy and spectral efficiency of the RAN.

More recently, the strategy of ultra-dense deployment of small-cells has gained significant attention and is regarded as one evolutionary path towards the achievement of the target capacities in 5G. However, deploying a large number of small-cells poses several concerns including the energy consumption. A reduction in the overall RAN RF transmit power is achieved when the density of small-cell base stations increases (i.e. the inter-site distance is reduced). This leads to a saving in RAN RF energy consumption which is enhanced if an efficient sleep mode capability is used (Lopez-Perez, Ming, Claussen, & Jafari, 2015). However, this observation may not hold when the total power consumption of each small cell is taken into account, and the reason is that the base station still consumes a non-negligible amount of energy even in the sleep mode status. Another important issue that arises when the deployment density of small-cells increases is the energy consumed by the backhaul links, which becomes significant, and can consume up to 50% of the RAN total power consumption (Tombaz et al., 2014) and (O'Farrell & Fletcher, 2015), which may limit the advantages of heterogeneous networks.

POWER CONSUMPTION MODELLING OF RADIO BASE STATION SITES

The radio base station (RBS) represents the main energy consuming portion of the cellular network. RBSs consume about 60% of the total RAN energy (Lister, 2009). For this reason, an accurate estimate of the power consumption of the base station is an important step in the evaluation of the RAN energy efficiency. Base station sites can vary in design (i.e. integrated or distributed) and type (macro, micro, pico etc) according to the shape and size of the serving cells. However, each base station site consists of seven common key units: the antennas and feeder cables system, the power amplifier (PA), the RF transceiver, the baseband processing unit, the backhaul unit, the cooling system; and the AC/DC power supply unit.

Modelling the base station power consumption usually involves an intensive phase of measurements of power consumption of the radio base stations in different operating environments and conditions. A number of base station power consumption models are available in the open literature (Soh et al., 2013), (Debaillie, Desset, & Louagie, 2015). These models provide an estimate of the base station operational power and its dependence on the RF transmitted power. The base station power consumption in most of the existing models is characterised by the sum of two parts, a static overhead part and a dynamic part. The static part consists of an overhead power that depends on the base station type and design, but it does not depend on the RF transmitted power. It includes power consumed in the baseband processing unit, the RF transceiver, the AC/DC power supply unit and the cooling unit. Whereas the dynamic part consists of power that depends on the average offered traffic load and includes the losses in the power amplifier as well as part of the power supply and cooling units. A revised version of the Green Radio (GR) base station power consumption model is developed in this chapter (O'Farrell & Fletcher, 2015). The GR model was developed by the MVCE using manufacturers' and operators' data to model the power consumption of a single antenna base station. Figure 1 depicts a schematic block diagram of the units consuming power in a macro-cell base station.

Estimation of the Base Station Power Consumption

The separate base station units are classified into: per antenna units, which include the PA and the RF transceiver units; per sector units, which include the baseband processing unit and power supply unit; and per base station site units, which include the cooling and backhaul units. The overall base station site average power consumption is modelled as the sum of the power consumption of the seven units, as expressed in Equations (1) and (2).

$$P_{site} = P_{cool} + P_{rect} + P_{bh} + P_{bts} \tag{1}$$

$$P_{bts} = n_s \cdot n_t \left(P_{bb} + P_{trx} + P_{pa}\right) \tag{2}$$

In Equation (1), the term P_{site} represents the total base station site power consumption, P_{cool} denotes the cooling system power consumption, P_{rect} is the power consumed by the power supply unit, P_{bh} is the backhaul power consumption, and P_{bts} is the base transceiver station power consumption. In Equation (2) the term P_{bb} is the baseband signal processing power consumption, P_{trx} is the RF transceiver power consumption, P_{pa} is the power consumed in the power amplifier, n_s refers to the number of sectors per site and n_t is the number of transmit antennas per sector. The cooling and power unit power consumption is calculated as a percentage of P_{bts}. The method for estimating the power consumption of each unit is explained in the following subsections.

Figure 1. Base station site block diagram

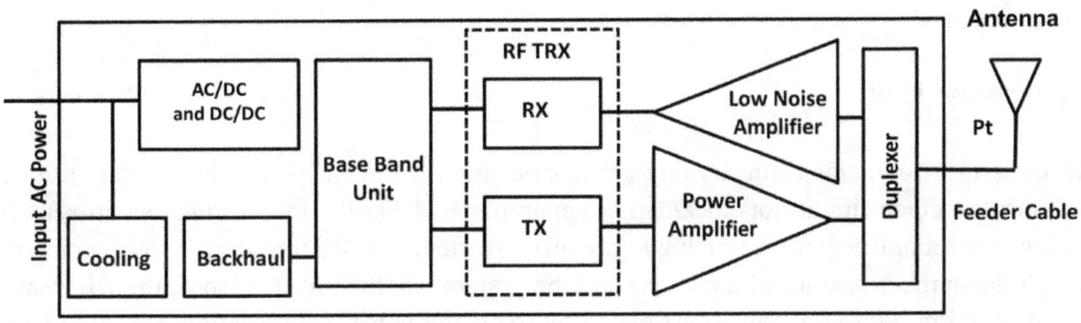

Power Amplifier

The power amplifier (PA) usually represents the main power consuming unit in the base station and may consume over 50% of the total base station power (Guelzim & Obaidat, 2013). The power consumption of the power amplifier can be expressed as a function of the average transmitted RF power per antenna, the power amplifier efficiency η_{pa}, and the feeder cable losses σ_{feed} as shown in (3).

$$P_{pa} = \frac{P_t}{n_t \cdot \eta_{pa} \cdot \left(1 - \sigma_{feed}\right)} \tag{3}$$

where the transmitted RF power P_t is expressed as a function of thc normalised offered average traffic load α as shown in (4).

$$P_t = \alpha \cdot P_{t,\max} \tag{4}$$

In (4) $P_{t,\max}$ is the maximum transmitted RF power per sector whereas α denotes the average offered load during a suitable observation time and has a value between 0 and 1. When α is equal to unity, all of the RF resources are assigned all of the time, and the base station transmits at maximum power $P_{t,\max}$. Due to the peak to average power ratio (PAPR) in multicarrier signals like OFDMA, the power amplifier is required to operate at a relatively small power level, compared to the maximum saturation power level $P_{pa,\max}$. This helps to prevent distortion of the signal peaks by the saturation region of the PA, which then leads to degradation in the power amplifier efficiency. The value of the power amplifier efficiency η_{pa} at different load values needs to be measured experimentally to be accurate, it can be also approximated by Equation (5) according to the results of (Persson, Eriksson, & Larsson, 2013).

$$\eta_{pa} = \sqrt{\frac{P_t}{P_{pa,\max}}} \cdot \eta_{pa,\max} = \sqrt{\frac{\alpha \cdot P_{t,\max}}{P_{pa,\max}}} \cdot \eta_{pa,\max} \tag{5}$$

In (5), the term $\eta_{pa,\max}$ is the maximum efficiency of the PA, which is reached when the power amplifier operates at $P_{pa,\max}$.

RF Transceiver Unit

The RF transceiver contains elements that are responsible for the RF processing in the base station. These elements include circuits for clock/carrier generation and distribution, modulators, mixers, filters, buffers, low noise amplifier and the analogue/digital converters. The RF transceiver power consumption scales with the number of transmit antennas and the system bandwidth B. Hence, the RF transceiver power consumption per sector can be estimated by (6) if the power consumption of a reference base station configuration is available (Holtkamp, Auer, Giannini, & Haas, 2013).

$$P_{trx} = \frac{n_t \cdot B \cdot P_{trx,ref}}{B_{ref}} \tag{6}$$

In (6) the term $P_{trx,ref}$ is the RF transceiver power consumption at a reference bandwidth B_{ref} with one transmit antenna per sector.

Baseband Processing Unit

The baseband (BB) processing unit is responsible for signal processing, radio resource management and site control functionality. Similar to the RF transceiver, the BB unit power consumption is usually modelled as being dependent on the number of transmit antennas and the system bandwidth. The power consumption of this unit has relatively small dependence on the average traffic load, and its power consumption can be considered load independent (Holtkamp et al., 2013). An expression for the BB power consumption is provided in (7), where $P_{bb,ref}$ is the BB unit power consumption at a reference bandwidth B_{ref} with one transmit antenna per sector.

$$P_{bb} = \frac{n_t \cdot B \cdot P_{bb,ref}}{B_{ref}} \tag{7}$$

Power Supply Unit

The Power supply unit includes the AC/DC and DC/DC converters within the base station site. The power losses in this unit P_{rect} are usually modelled as a fixed percentage of the overall power consumption of the base station. The common default values of these losses typically range from 10% to 15% (Desset et al., 2012).

Cooling Unit

Site cooling is required at macro-cell base station sites in order to maintain an appropriate operating temperature inside the base station cabinet. The power consumption of the cooling unit depends on the base station power consumption, the internal and ambient temperatures of the base station cabinet and on the power efficiency of the cooling unit. Typical estimates of the cooling unit power consumption range from 10% to 30% of the total power consumption of the base station (O'Farrell & Fletcher, 2015). A theoretical model to estimate the cooling unit power consumption is presented in this chapter. Previous studies mostly used empirical data, which may differ from one base station site to another.

When estimating the power consumption of the cooling unit, an ambient temperature of 25°C is usually considered. The amount of heat generated inside the cabinet is obtained by assuming that the overall power consumed by the equipment inside the cabinet, denoted by P_c, is dissipated as heat and can be calculated in British Thermal Units (BTU) per hour by (8).

$$BTU\ Per\ Hour = 3.4121 \cdot P_c \tag{8}$$

The value of P_c is then given by (9).

$$P_c = P_{bts} - \frac{n_s \cdot P_t}{1 - \sigma_{feed}} \tag{9}$$

When calculating P_c in Equation (9), the losses in the feeder cable (σ_{feed}) and the number of sectors (n_s) are taken into account when subtracting the transmitted RF power P_t per sector from the base transceiver station power consumption. The energy efficiency of a cooling unit is determined by its energy efficiency rating which indicates the number of BTUs per hour removed for each consumed watt of power. The Energy Efficiency Ratio (*EER*) is usually indicated on the unit by an energy guide label. The larger the EER value, the greater the energy efficiency of the cooling unit. Hence, the cooling unit power consumption can be estimated by (10).

$$P_{cool} = \frac{BTU\ Per\ Hour}{EER} \tag{10}$$

Table 1 tabulates the estimated power consumption of the cooling unit when the base station total power consumption is 1000 W. The numerical values suggest that the cooling unit power consumption accounts for about 20% of the total BTS power consumption. This figure is consistent with values found in the open literature. Less power is consumed when cooling units with higher *EER* values are considered. In small-cell base stations (e.g. micro- and pico-cells), there is no requirement for active cooling. Equally, there is no need for cooling when distributed macro-cell base stations using remote radio heads are deployed.

Backhaul Unit

The backhaul network encompasses the intermediate links that connect the base station sites to the core network. Different communications technologies are used in the backhaul such as optical fibre, copper wire or microwave links. Most studies of power consumption in macro-cellular networks tend to omit the power contributed by the backhaul, P_{bh}. However, the backhaul power consumption is an important

Table 1. Cooling unit power consumption

$P_{bts}(W)$	$P_t = P_{t,\max}(W)$	$\sigma_{feed}(dB)$	n_s	EER	$P_{cool}(W)$
1000	40	-3	3	11	215

component of the overall power consumption in a heterogeneous or small-cell network. The value of the ratio of the backhaul power consumption to the total RAN power consumption depends on both the technology used and the network topology deployed. For example, (Skubic, De Betou, Ayhan, & Dahlfort, 2012) have estimated the power consumption of a fibre backhaul as lying in the range 8 to 17 W. Often, a rule-of-thumb value of 10 W is found in the open literature (Blume, Ambrosy, Wilhelm, & Barth, 2013). In contrast, when microwave links are ultilised in the backhaul, the power consumption per link can be as high as ~93 W (Monti, Tombaz, Wosinska, & Zander, 2012), which is considerably greater than the fibre case.

Base Station Power Consumption vs. Traffic Load

The total base station site power consumption is the sum of the power consumption of different base station subsystems which requires the estimation of the power consumption of each subsystem individually before obtaining the overall power consumption. A simpler approach is usually used by writing the power model as the sum of a load independent static part and a load dependent dynamic part, as given by Equation (11), where P_{oh} is the load independent overhead part and P_{rh} is the load dependent "radio head" part. The term α denotes the base station normalised average load ranging from 0 to 1. In cellular networks like LTE, the load includes both traffic and control data and while the traffic data maybe zero at times, there is always some level of signalling and control data present.

$$P = P_{oh} + P_{rh} \cdot \alpha \tag{11}$$

Table 2 lists the parameters of the model that have been used to estimate the total power consumption of either macro-, micro- or pico-cell base station sites when using Equations (1) and (2). These values are obtained from (O'Farrell & Fletcher, 2015). They typically differ between manufacturer's equipment and depend on the manufacturing date as well as the technology used. A common model that is independent of any particular manufacturer is usually preferred and this is obtained by averaging characteristics from several manufactures' power consumption data sets. The PA efficiency in Table 2 is the PA efficiency when the transmit power per sector equals $P_{t,\max}$, the PA efficiency at $P_{pa,\max}$ can be used, if its value is known.

Figure 2 shows the total site power consumption when the new power model (referred to as the Enhanced Green Radio base station power consumption model) is used to estimate the consumption of a 3-sector, macro-cell base station that uses a single antenna per sector, and at bandwidth of 20 MHz (the blue solid line). The total power consumption is plotted as a function of the normalised average load α. On the same graph, the total power consumption which is estimated by a theoretical linear model is shown (the dotted red line), which is obtained from Equation (12).

$$P = P_{oh} + \frac{\Delta P}{\Delta \alpha} \cdot \alpha \tag{12}$$

Table 2. Power model parameters

Parameter	Macro-Cell	Micro-Cell	Pico-Cell
Transmit Power Per Sector ($P_{t,\max}$, W)	40	6.3	0.13
Bandwidth $\left(B_{ref}, MHz\right)$	20	20	20
Transceiver Unit ($P_{trx,ref}$, W)	13	6.5	1.0
Processing Unit ($P_{bb,ref}$, W)	30	27	3
Fibre Backhaul (P_{bh} , W)	10	10	10
PA Efficiency at $P_{t,\max}$	30%	23%	7%
Antenna Feed Cable Loss (dB)	-3	-1	0

Figure 2. Power consumption for a macro-cell base station site using 3-sectors and 1 antenna/sector

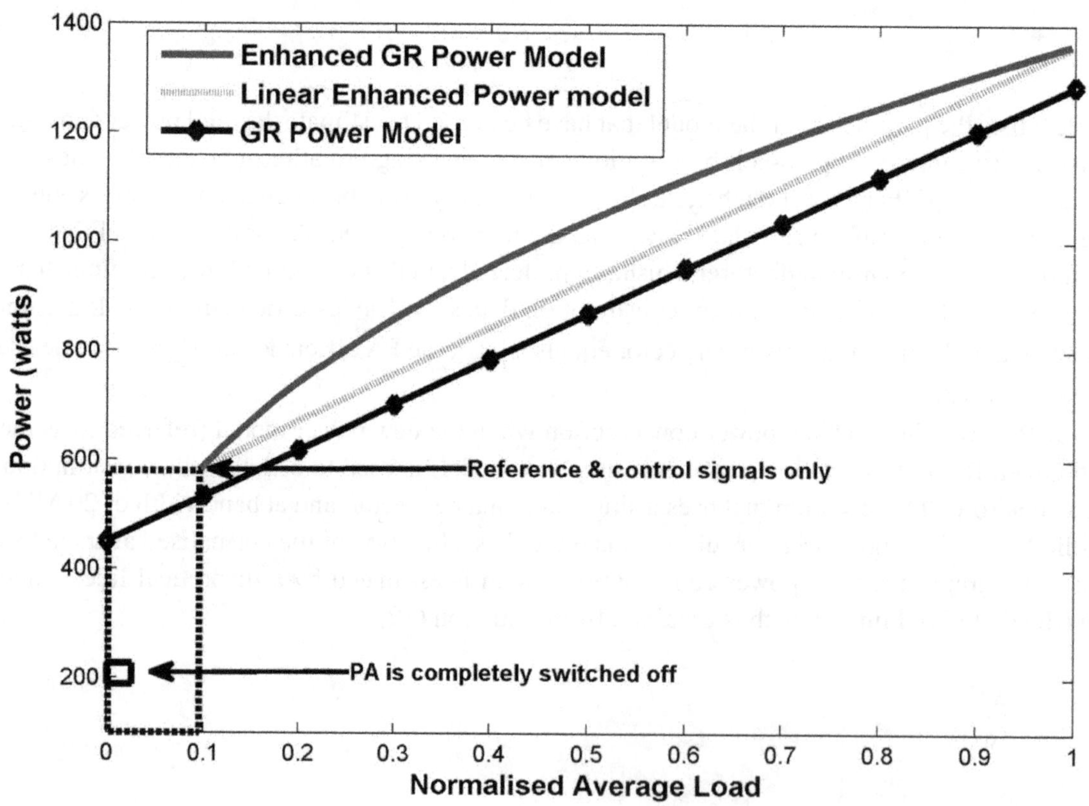

In (12) the gradient $\frac{\Delta P}{\Delta \alpha}$ is equivalent to the load dependent part P_{rh} in (11) and P_{oh} is equivalent to the load independent part in (11). A value of $\frac{\Delta P}{\Delta \alpha} = 864$ was used in Figure 2. The black curve shows the base station power consumption when the previous version of the Green Radio model is used.

In Figure 2, the base station power consumption remains at 480 watts when the teletraffic component of the load is zero. The solid black box shown encloses a region where the power consumption is attributable to the load independent part P_{oh} and the load dependent part P_{rh} corresponding to the transmission of control signalling only (i.e. reference and synchronisation signals). In an LTE single antenna base station with 1.4 MHz bandwidth, the reference signals and signalling data amount to approximately 10% of the total load, which is appreciable. At instances of zero traffic data, it may be possible to switch off completely the PA. For the macro-cell parameters considered, this would reduce the instantaneous power consumption to approximately 200 W, which gives a considerable energy saving. Even lower levels of power consumption can be achieved by switching off more parts of the base station subsystems, which are not required at low traffic load. Such sleep-mode techniques offer useful power savings to the network operator.

ENERGY EFFICIENCY METRICS

The capacity and coverage are two significant key performance indicators (KPIs) in cellular mobile networks. Well-defined and accurate metrics already exist for both of these KPIs. In contrast, the energy efficiency of a cellular network has only recently been established as a KPI. A number of energy efficiency metrics have been proposed in the open literature for this purpose (Zhao, Zhao, & O'Farrell, 2013) and (Thompson, Grant, Fletcher, & O'Farrell, 2012), such metrics, in general, are classified into two categories (Hamdoun, Loskot, O'Farrell, & He, 2012). In the first category, absolute energy metrics are used, which calculate the amount of energy consumed to support a certain network performance. Examples of absolute metrics include the Energy Consumption Rate (ECR) measured in J/bit, the ETSI energy efficiency (EE) measured in bit/J and the area power density measured in W/m^2. In the second category, relative metrics are used, which compare the energy consumption of two distinct networks. An examples of a relative energy metric is the Energy Consumption Gain (ECG) (O'Farrell & Fletcher, 2015).

Absolute EE metrics such as the J/bit and bit/J ratios exhibit an inherent ambiguity when comparing the energy consumption of networks as it is unclear whether or not the gains indicated are attributable to true energy savings or to increases in the data throughput. This can result in misleading claims when comparing the EE of two networks if the throughput or data volume is not defined. The requirement for an unambiguous RAN energy efficiency evaluation framework was addressed by (O'Farrell & Fletcher, 2015). The framework developed separately quantifiable energy and throughput efficiencies as well as a new joint metric formed from the product of both. That is, the ECG was used as a relative energy measure, with respect to a reference network, whereas the "data volume gain" (DVG) was used as a relative throughput measure. Their product, ECG times DVG, forms a new, joint, relative EE metric termed the "energy throughput gain" (ETG).

The ECG between two networks (RAN_1 and RAN_2) is given by equation (13). A reference network RAN_1, consisting of N_1 cells each transmitting with average power P_1, is compared to a test network

RAN_2, consisting of N_2 cells each transmitting with average power P_2. The total energy consumption of each RAN is denoted $\{E_i : i = 1,2\}$. Implicit in (13) is the condition that the observation time T (i.e. the sufficiently long period over which the assessment is made), the RAN coverage area A and the RAN offered traffic load have the same respective values in both RANs.

$$ECG = \frac{E_1}{E_2} = \frac{N_1 \cdot P_1 \cdot T}{N_2 \cdot P_2 \cdot T} \tag{13}$$

Similarly, the DVG between two networks is given by Equation (14). The reference network RAN_1 transports a data volume M_1 bits during the observation time T, whereas test network RAN_2 transports a data volume M_2 bits during the same observation period.

$$DVG = \frac{M_2}{M_1} \tag{14}$$

Then an expression for the ETG is obtained by taking the product of (13) and (14) to give Equation (15). The expressions in (15) demonstrate how the ratio of the RAN_1 and RAN_2 energy efficiencies EE_1 and EE_2, respectively, with units of bit/J, can be partitioned into an energy ratio component and a data volume ratio component. This partition of the ETG metrics allows energy savings and throughput gains to be assessed separately, in particular to see if a true energy saving has been obtained.

$$ETG = ECG \cdot DVG = \frac{E_1}{E_2} \cdot \frac{M_2}{M_1} = \frac{M_2 / E_2}{M_1 / E_1} = \frac{EE_2}{EE_1} \tag{15}$$

For homogeneous RANs providing coverage to a fixed geographic area A, we note that the ratio N_1 / N_2 is approximated by the ratio A_2 / A_1, where A_1 and A_2 are the cell coverage areas in RAN_1 and RAN_2, respectively. Then from (13), (14) and (15) we can obtain estimates of the ECG, DVG and the ETG from the individual base station site area power and area throughput densities, respectively, as shown in equation (16).

$$ECG \equiv \frac{P_1 / A_1}{P_2 / A_2}, DVG \equiv \frac{S_2 / A_2}{S_1 / A_1}, ETG \equiv \frac{P_1 / S_1}{P_2 / S_2} \tag{16}$$

The advantage of Equation (16) is that the cell power consumptions P_i, throughputs S_i and coverage areas A_i are readily calculable or measurable, thereby providing an accurate, straightforward means of comparing the energy consumption and data volume capacity of two RANs. It is also worth noting that when the DVG equals unity, the ECG and ETG amount to the same measurement.

ENERGY EFFICIENCY ANALYSIS IN SECTORISED AND HETEROGENEOUS NETWORKS

In this section, we compare the energy efficiencies of two up-grade paths for a cellular RAN: one based on introducing high order sectorisation and another based on introducing small-cell overlays. The evaluation also considers the amalgamation of both techniques. The base station site power consumption model and the EE metric framework introduced in the previous sections are used in the analysis.

High Order Sectorisation

Deploying sectorised base station sites improves a RAN's capacity, and in principle offers a capacity gain almost equal to the number of sectors deployed. In practice, the interference leakage between the adjacent sectors limits the capacity gain. The sectorisation capacity gain G, shown in Equation (17), for different sectorisation orders, needs to be known in order to evaluate a RAN's energy efficiency.

$$G = \frac{sectored\ site\ average\ capacity}{omni\ site\ average\ capacity} \tag{17}$$

In this chapter, a MATLAB based static system level simulator was used to estimate the sectorisation gain G for different orders of sectorisation ranging from 1 to 6. The network model considered consists of a homogeneous macro-cell, multi-user RAN covering a flat, square geographic area of 6-by-6 km. Table 3 tabulates the salient simulation parameters used. The mobile users are deployed uniformly in the $\left(x, y\right)$ coordinates of the target service area and the downlink SINR for each user is calculated by considering the interference from all the base stations in the service area. Performance data is obtained for the central cells only in order to avoid underestimating the amount of intercell interference experienced by the users located at the edge of the service area. Further, all the mobile users are assumed to be located indoors. The base station antennas have azimuth $A\left(\theta\right)$ and elevation $A\left(\varphi\right)$ beam pattern gains in dBi given by Equations (18) and (19), respectively, while the 3D antenna gain is approximated by (20).

$$A\left(\theta\right) = -\min\left[12\left(\frac{\theta}{\theta_{3dB}}\right)^2, A_m\right] \tag{18}$$

$$A(\varphi) = -\min\left[12\left(\frac{\varphi - \varphi_{tilt}}{\varphi_{3dB}}\right)^2, A_m\right] \tag{19}$$

$$A_G(\theta,\varphi) = A_{GB} - \min\left[-\left(A(\theta) + A(\varphi)\right), A_m\right] \tag{20}$$

The terms in θ_{3dB} and φ_{3dB} denote the 3 dB beamwidth in the azimuth and elevation directions, respectively, A_m is the minimum antenna attenuation (a value of 25 dB was used in the model), φ_{tilt} is the boresight tilt angle and A_{GB} is the antenna boresight gain. Using the simulation model developed, the RAN sectorisation gain with respect to an omnidirectional base station site is plotted in Figure 3 for sectorisation orders from 1 to 6. A key penalty associated with this gain in capacity is an increase in the overall RAN power consumption. When increasing the number of sectors between RANs from $n_{s1} = 1$ to $n_{s2} = 6$, the ECG monotonically decreases with increasing number of sectors according to the ratio $ECG \propto n_{s1} / n_{s2}$ as plotted in Figure 4. When sectorisation orders higher than 6 are considered, the sectorisation gain tends to reduce due to the increase of inter sector interference, whereas the *ECG* continues to reduce.

Table 3. System level simulation parameters

Parameter	Value
Frequency (MHz)	2000
Bandwidth (MHz)	20
RF Power (dBm)	Macro/Micro/Pico: 46/ 38/ 21
Macro Site ISD (km)	0.866
BS antenna height (m)	Macro/small-cell: 30/6
BS antenna tilt angle (degrees)	Macro/small-cell: -5/0
Antenna boresight gain (dBi)	6/3/1-sector macro:17.5/15/12
Small-cell omni antenna gain (dBi)	Micro/Pico-cell: 6/2
Shadowing standard deviation (dB)	8
Outdoor/Indoor wall loss (dB)	20
Pathloss model (dB)	$128 + 37 \cdot \log_{10}(d)$
Traffic Model	Full buffer
User location distribution	(x,y) Uniform

Figure 3. Sectorisation capacity gain

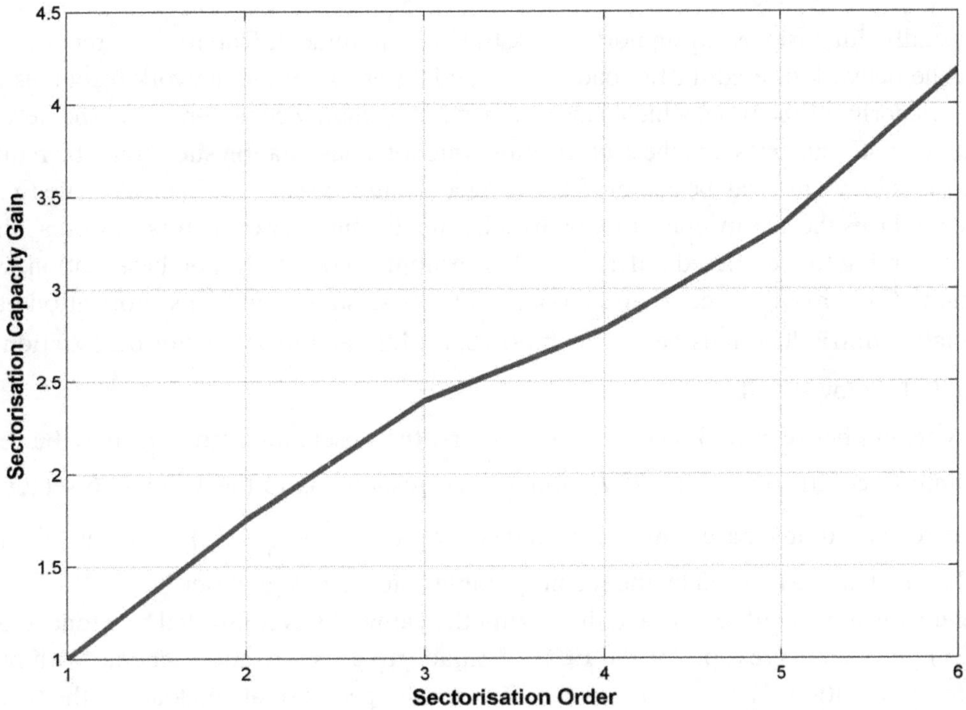

Figure 4. ECG vs. sectorisation order

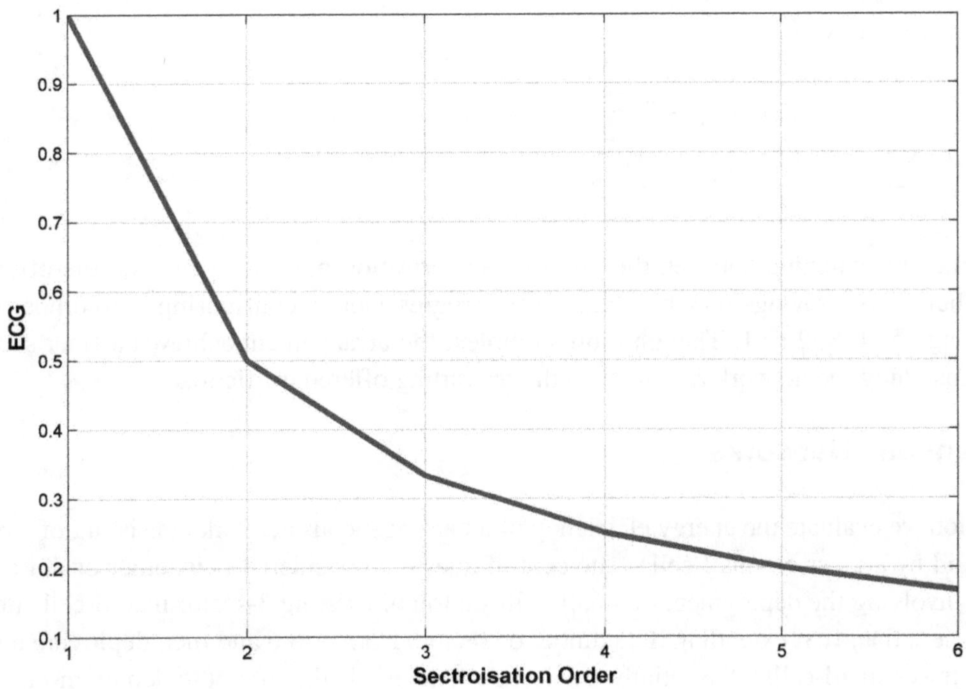

Adaptive Sectorisation

A RAN is usually dimensioned to support a peak traffic requirement. Due to the temporal and spatial variation of the network offered traffic load, a substantial portion of the network resources are underutilised for a majority of the time, which causes unnecessary energy consumption in the network. Such a traffic characteristic suggests that the sectorisation order of a base station site should be reduced when the traffic load is low. This may be achieved by using an adaptive sectorisation technique (Yinan et al., 2011), which reduces the energy consumption by adapting the number of sectors as well as the antenna beamwidth according to the offered traffic load. For example, a 6-sector radio base station, with a sectorisation gain of 4.2, may be reconfigured to a 3-sector base station, with a sectorisation gain of 2.4, when the relative traffic load falls below the ratio $\left(2.4 / 4.2\right)$ and to a 1-sector base station when the relative load falls below $\left(1 / 4.2\right)$.

When switching between 6, 3, and 1 active sectors, the observation time T may be regarded as partitioned into three time intervals $\left\{T_1, T_3, and\ T_6\right\}$ corresponding to the 1, 3 and 6 sector operating modes in the reconfigurable base station constrained by $\left\{T = T_1 + T_3 + T_6\right\}$. Such an adaptive sectorisation technique may be realised by the reconfigurable antenna design described in (Weisi Guo et al., 2012), whereby the number of sectors and their azimuth beamwidth is controlled by a tuneable reflector. Equation (21) gives an expression for the ECG of an adaptive sectorisation scheme with respect to a fixed 6-sector base station. The term in α is the diurnal averaged normalized load for the 6-sector case, whereas α_i represents the averaged normalised load for the i-sector case. Similarly, the terms in $P_{oh,6}$ and $P_{rh,6}$ represent the overhead and radiohead base station power consumptions for 6-sectors, respectively, whereas $P_{oh,i}$ and $P_{rh,i}$ represent the overhead and radiohead base station power consumptions for i-sectors, respectively.

$$ECG_{adaptive} = \frac{\left(P_{oh,6} + \alpha \cdot P_{rh,6}\right) \cdot T}{\sum_{i=1,3,6} \left(P_{oh,i} + \alpha_i \cdot P_{th,i}\right) \cdot T_i} \tag{21}$$

Automatically switching between the different sectorisation modes depends on the offered traffic value. Further energy savings may be obtained by progressively reconfiguring the number of sectors from 6 through 5, 4, 3, 2 or 1. Though more complex, the arrangement achieves a finer sectorisation resolution matching the network resources to the prevailing offered traffic load.

Heterogeneous Networks

In this section we evaluate the energy efficiency of a heterogeneous network consisting of a macro-cell layer overlaid by a layer of small-cells. The central question considered is whether or not a RAN upgrade path involving the deployment of small-cells on top of existing 3-sector macro-cell sites is more energy efficient than first extending the number of sectors from 3 to 6 and then deploying a small-cell overlay (of fewer small-cells). Essentially, the issue is one of whether or not to deploy more small-cells (the first deployment option) or to deploy more sectors first (the second deployment option).

Network Model

We evaluate a multi-cell, multi-user heterogeneous RAN using a static system level simulator as a modelling tool. The evaluation is done for the same RAN coverage area with a high target capacity. In a first deployment option (Option-1), small-cells are deployed as an overlay on top of a 3-sector macro-cell layer. In a second deployment option (option-2), small-cells are deployed as an overlay on top of a 6-sector macro-cell layer. The 6-sector RAN offers a sectorised capacity gain G over the 3-sector RAN (see Figure 3), therefore fewer small-cells are needed on top of 6-sector macro-cells in order to support the same offered traffic load. The same macro-cell RAN is used as a starting point for the up-grade, hence, the difference between the two RANs is the number of deployed small-cells and the number of sectors per base station site. The small-cells deployed can be either micro- or pico-cells in the scenarios considered. The small-cells are deployed uniformly in a regular layout around the 3-sector and 6-sector sites as depicted in Figure 5.

Denoting the first (3-sector macro-cell) and second (6-sector macro-cell) deployment options by subscripts 1 and 2, respectively, the RAN ECG can be expressed by (22). The terms P_m and P_s denote the average power consumption of the 3-sector macro-cell and omnidirectional small-cells, respectively. The terms N_1 and N_2 denote the numbers of small-cells deployed per macro-cell in options-1 and 2, respectively. The power consumption for the 6-sector macro-cell sites is then expressed in terms of the consumption for the 3-sector sites as approximately equal to $2 \cdot P_m$.

$$ECG = \frac{(P_m + N_1 \cdot P_s) \cdot T}{(2P_m + N_2 \cdot P_s) \cdot T} \tag{22}$$

Figure 5. 6-sector and 3-sector deployments

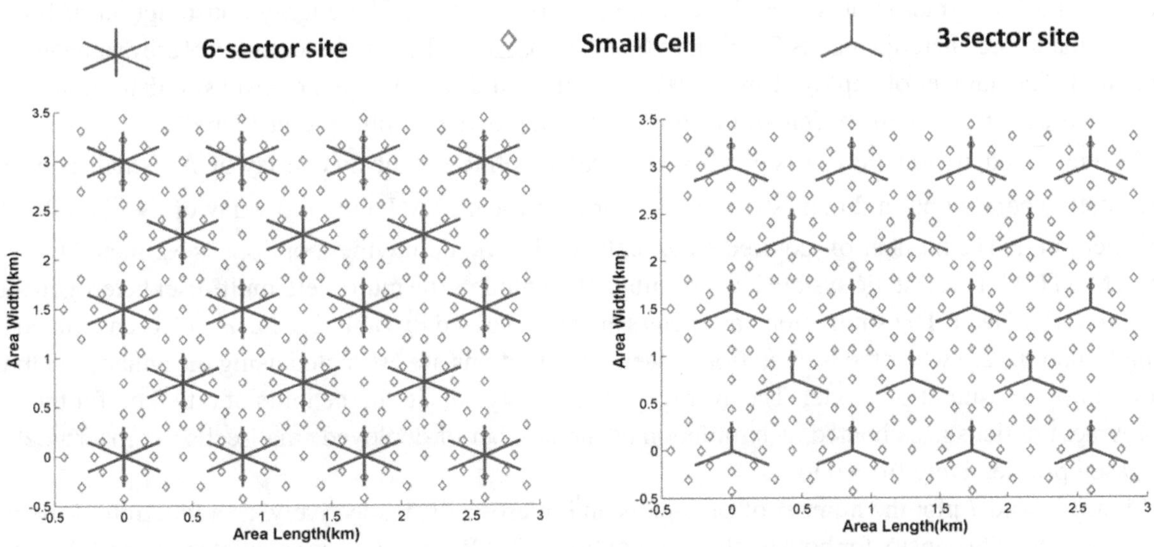

Typically, P_s is much smaller than P_m, and is given by $P_s = m \cdot P_m$, where $0 \leq m < 1$ and its numeric value is obtained from the ratio P_s / P_m. Both N_1 *and* N_2 are estimated by running the system level simulator for the two deployment options while increasing progressively the number of deployed small-cells per base station site. The number of small-cells required to achieve a certain RAN capacity per unit area can then be interpolated. Since N_1 is always greater than N_2 for the same target capacity, the RAN ECG may be expressed by (23).

$$ECG = \left(\frac{1 + m \cdot N_1}{2 + m \cdot N_2} \right) \tag{23}$$

From Equation (23), when the ECG is less than unity, the heterogeneous RAN with 3-sector macro-cells is more energy efficient than the heterogeneous RAN with 6-sector macro-cells. If the small-cell power consumption is very low, then the value of m tends to zero and the heterogeneous RAN ECG approaches 0.5 in the limit, which demonstrates that deployment option-1 is always more energy efficient than option-2, providing that the small-cell power consumption P_s is sufficiently small. The value of m is obtained using the power consumption model of macro- and small-cell base station described earlier, and yields typical values for m of approximately 0.099 for micro-cells and 0.016 for pico-cells.

CASE STUDY

Small Cell Densification

In this section, the energy efficiency of a heterogeneous network case study is considered based on the option-1 and 2 up-grade routes identified in the previous section. The energy consumption of both a 3-sector macro-cell heterogeneous RAN and a 6-sector macro-cell heterogeneous RAN are evaluated and compared. The number of deployed small-cells is estimated through a system level simulator to achieve the same area capacity density (bit/s/km^2) for the two deployment options considered.

As mentioned, option-1 consists of a 3-sector macro-cell RAN and an overlay of N_1 small-cells per macrocell, whereas option-2 consists of a 6-sector macro-cell RAN and an overlay of N_2 small-cells per macro-cell. The number of required small-cells in the two deployment options is estimated by running a MATLAB based static system level simulator that models the multi-cell, multi-user heterogeneous networks (Het Nets). The simulation parameters are summarised in Table 2. The ECG is calculated by using Equation (22) while the base station power consumptions are estimated using the enhanced Green Radio power consumption model. The average area capacity density in the units of bit/s/km^2 for the two deployment options is estimated by simulation as the number of deployed small-cells per macro-cell is increased progressively from 0 to 24.

Figures 6 and 7 plot the number of pico-cells and micro-cells, respectively, as a function of the targeted area capacity density for both deployment options. The RAN ECG is calculated from (22) at each target area capacity density and plotted in Figure 8 for both the 3 sectors and 6 sectors of macro/pico-cell

Figure 6. Number of pico-cells

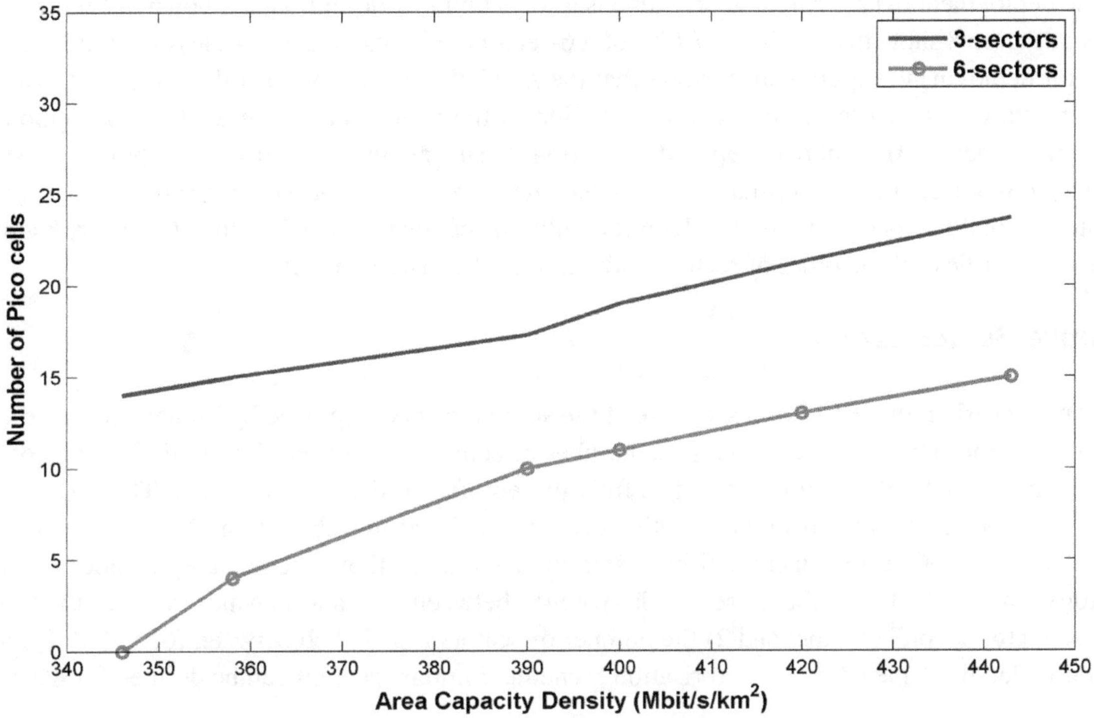

Figure 7. Number of micro-cells

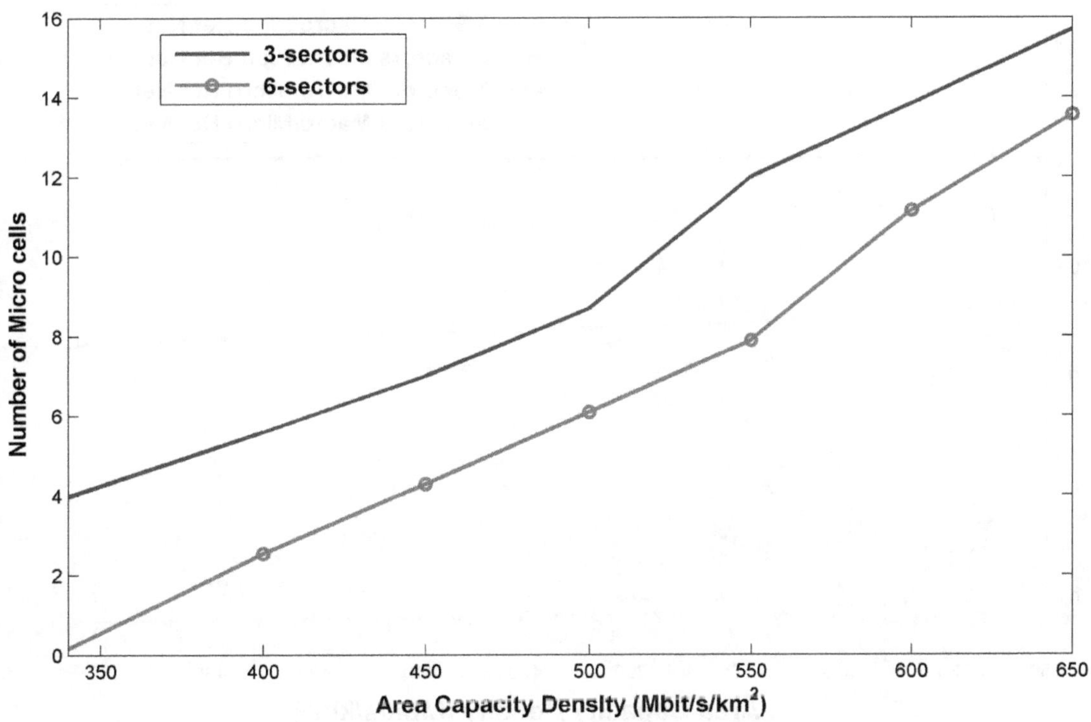

and macro/micro-cell Het Net deployments. The 3 sectors macro/pico-cell RAN is used as the reference baseline deployment in (22) and its *ECG* value is always unity by definition, as shown in Figure 8. The curves in Figure 8 show that the RAN ECG of a 6-sector deployment is always less than the ECG of a 3-sector deployment. Figure 8 also shows that the *ECG* of a macro/micro-cell deployment is always less than that of a macro/pico-cell deployment. Hence, the results show that deployment option-1 is always more energy efficient than deployment option-2. This result is only true if the power consumption of a pico- or micro-cell is substantially less than the power consumption of a macro-cell. The results indicate also that the macro/micro-cell Het Nets is always less energy efficient than the macro/pico-cell Het Nets, regardless of the order of sectorisation used in the macro-cell site.

Adaptive Sectorisation

As discussed earlier, the ECG of a sectorised base station may be improved by implementing an adaptive sectorisation scheme, which selects the number of active sectors according to the average offered traffic load. Unused radio equipment is then fully turned off in order to save energy. The ECG for an adaptive 3- or 6-sector macro-cell base station can be calculated from Equation (21), with respect to a fixed 3-sector and 6-sector macro-cell base station. Two adaptation schemes are considered for the 6-sector solutions: 1) the number of sectors is switched between (6, 3 and 1) which is called the "coarse adaptive sectorisation" scheme; and 2) the number of sectors is switched between (6, 5, 4, 3, 2 and 1), which is called the "fine adaptive sectorisation" scheme. Similar schemes can be defined for the 3-sec-

Figure 8. ECG for macro/pico and macro/micro het nets

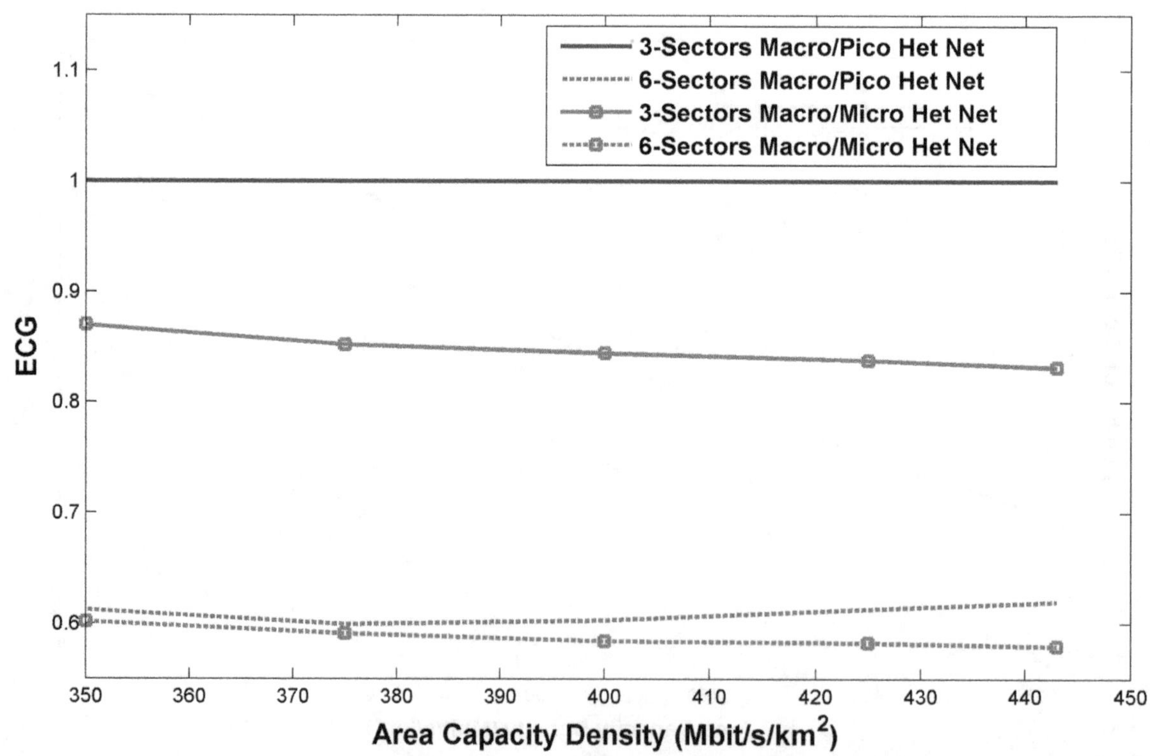

tor solutions: coarse adaptation corresponding to switching between 3 and 1 sectors; and fine adaptation corresponding to switching between 3, 2 and 1 sectors.

Table 4 shows the percentage improvements in ECG of the coarse and fine adaptive schemes over the fixed schemes. The 6-sector macro-cell base station benefits most with a 25% improvement in ECG for a fine adaptive scheme over the fixed scheme. However, the percentage ECG improvement of the fine adaptive scheme over the coarse scheme is marginal for the temporal traffic profile shown in Figure 9. Greater improvements may be possible for other types of temporal traffic distributions or when the spatial distribution of traffic is taken into account.

The ECG improvements in Table 4 for adaptive sectorisation may be implemented in the Het Net scenarios considered earlier in order to improve the energy consumption of an adaptive 6-sector macro-cell Het Net (i.e. adaptive 6-sector macro-cells plus omni small-cells) with respect to a fixed or an adaptive 3-sector macro-cell Het Net (i.e. adaptive 3-sector macro-cells plus omni small-cells). Figure 10 plots ECG versus the area capacity density for macro/pico and macro/micro Het Nets when the two

Table 4. ECG improvements for adaptive sectorisation

	Percentage ECG Improvement	
Macro-cell Sectorisation Type	3-Sector	6-Sector
Coarse Adaptive Sectorisation	10%	19%
Fine Adaptive Sectorisation	14%	25%

Figure 9. Daily traffic profile

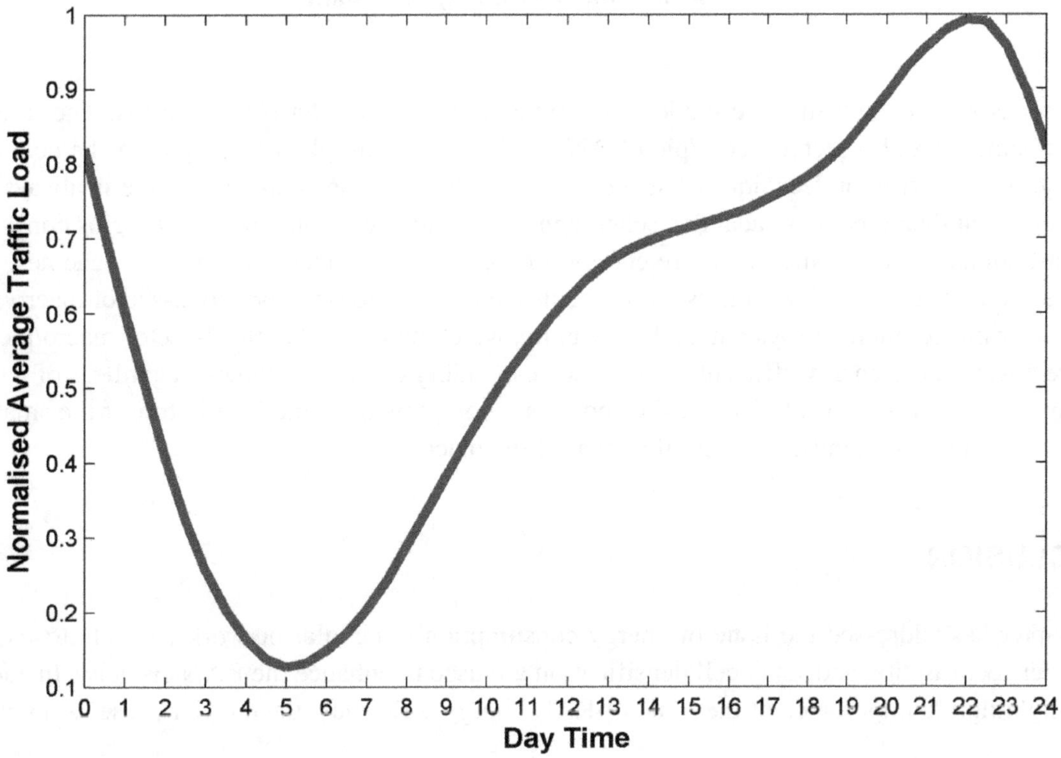

Figure 10. ECG of macro/pico and macro/micro het nets with adaptive sectorisation

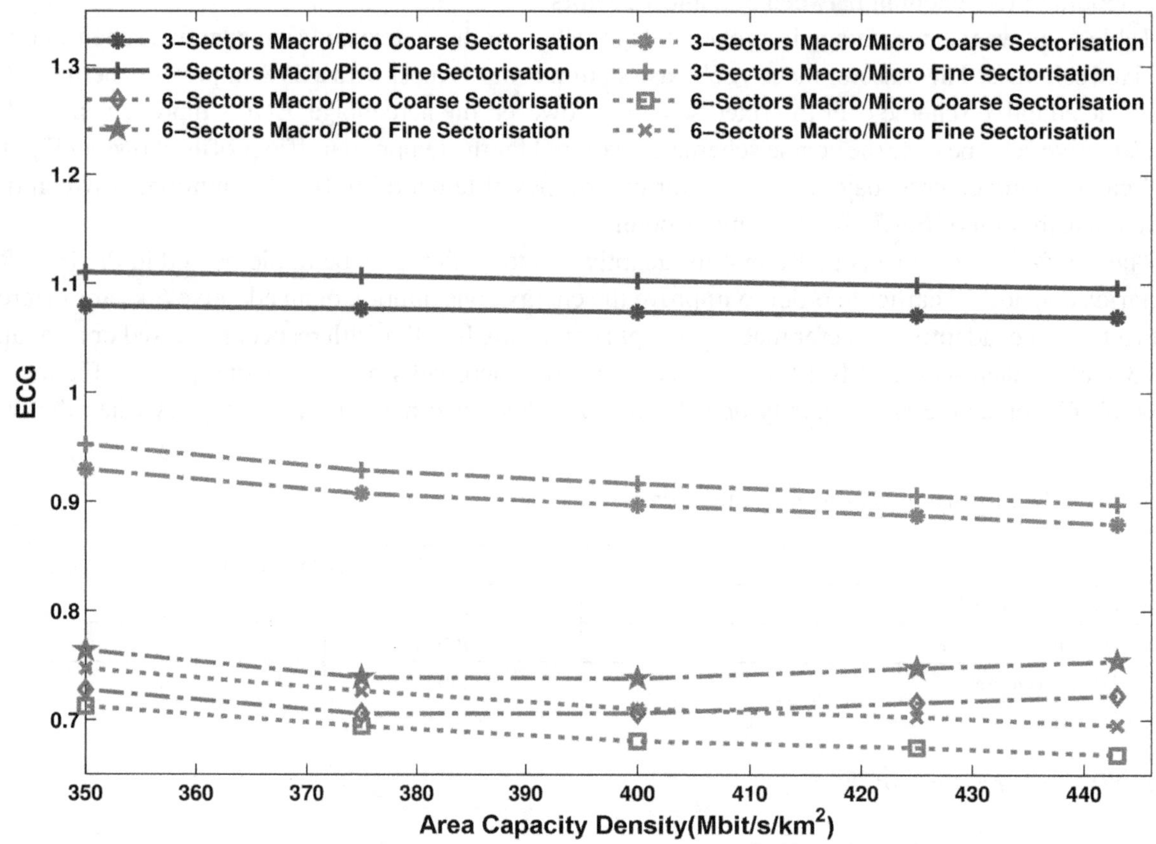

adaptive sectorisation schemes are implemented in 3- and 6-sector Het Net scenarios. The reference network being a fixed 3-sector macro/pico RAN. ECG curves are plotted comparing the coarse and fine adaptive sectorisation techniques. The curves show that the fine adaptive scheme is always more energy efficient than the coarse adaptive scheme in all the Het Net deployments scenarios considered, and that both adaptive schemes are more energy efficient than the fixed sectorisation schemes. However, the ECG of the adaptive sectorisation schemes in the 6-sector macro/pico, 3-sector macro/micro and 6-sector macro/micro always remain below unity, which indicates that the 3-sector macro/pico Het Nets are always more energy efficient than the corresponding 6-sector schemes, regardless of whether the adaptive sectorisation is implemented or not. As before, this outcome is only true if the small-cell power consumption is significantly less than that of the macro-cell.

CONCLUSION

The chapter has addressed the issue of energy consumption in cellular networks, which arises when high order sectorisation and small cell densification are used to enhance the RAN capacity. In addition to highlighting the importance of the issue of RAN energy efficiency, the chapter presents an evaluation methodology to assess the RAN energy consumption by proposing an enhanced base station power consumption model, as well as a novel energy efficiency metric framework. A RAN case study is used to illustrate how the proposed model and metrics can be used to evaluate the RAN energy efficiency. The chapter has investigated the energy efficiency of a heterogeneous RAN comparing 3- and 6-sector macro-cell underlays. The aim was to ascertain if network operators can obtain greater energy efficiency by increasing the sectorisation order before densifying with a small-cell overlay, or should an overlay of small-cells be added without increasing the sectorisation order. The analysis was extended by considering various adaptive sectorisation schemes to quantify how much energy is saved when sectors are switched off in low traffic conditions. The results showed that the range of energy efficiency improvements due to adaptive sectorisation lies between 10% to 25% depending on the implemented scheme, and on whether a 3-sector or 6-sector macro-cell deployment is used. For the traffic profile considered, and for all adaptive schemes, the 3-sector macro-cell heterogeneous network was found to be always the most energy efficient. In summary, the results show that while increasing sectorisation order to meet the capacity demands allows fewer small-cells to be deployed, the strategy is less energy efficient than using fewer sectors and more small cells to meet the same RAN capacity requirements which is in contrast to the practice of most network operators in which more sectors are deployed before small cells. A cost-benefit analysis over the operational lifetime of the network would add further insight in respect of which upgrade route to consider(W. Guo & O'Farrell, 2012).

REFERENCES

Arbi, A., & O'Farrell, T. (2015). *Energy efficiency in 5G access networks: Small cell densification and high order sectorisation.* Paper presented at the Communication Workshop (ICCW), 2015 IEEE International Conference on.

Arraycomm.com. (2015). *Cooper's Law*. Retrieved 17 December, 2015, from http://www.arraycomm.com/technology/coopers-law/

Athley, F., Johansson, M. N., & Nilsson, A. (2013). *Increased Sectorization: Horizontal or Vertical?* Paper presented at the Vehicular Technology Conference (VTC Fall), 2013 IEEE 78th.

Badic, B., O'Farrrell, T., Loskot, P., & He, J. (2009). *Energy Efficient Radio Access Architectures for Green Radio: Large versus Small Cell Size Deployment*. Paper presented at the Vehicular Technology Conference Fall (VTC 2009-Fall), 2009 IEEE 70th.

Bhatnagar, R. K., & Panigrahi, P. K. (2013). *Energy Efficiency, Clean Power as Engine for Sustainable Growth*. Retrieved from http://www.itu.int/en/ITU-T/climatechange/symposia/201305/Documents/info/Presentations/s1p3-India.pdf

Blume, O., Ambrosy, A., Wilhelm, M., & Barth, U. (2013). *Energy Efficiency of LTE networks under traffic loads of 2020*. Paper presented at the Wireless Communication Systems (ISWCS 2013).

Cisco Visual Networking. (2015). *Global Mobile Data Traffic Forecast Update 2014–2019*. White Paper c11-520862. Available on http://www.cisco.com/c/en/us/solutions/collateral/serviceprovider/visual-networking-index-vni/white_paper_c11-520862. html

Debaillie, B., Desset, C., & Louagie, F. (2015). *A Flexible and Future-Proof Power Model for Cellular Base Stations*. Paper presented at the Vehicular Technology Conference (VTC Spring), 2015 IEEE 81st.

Desset, C., Debaillie, B., Giannini, V., Fehske, A., Auer, G., Holtkamp, H., et al. (2012). *Flexible power modeling of LTE base stations*. Paper presented at the Wireless Communications and Networking Conference (WCNC).

Fehske, A. J., Richter, F., & Fettweis, G. P. (2009). *Energy Efficiency Improvements through Micro Sites in Cellular Mobile Radio Networks*. Paper presented at the GLOBECOM Workshops, 2009 IEEE.

Forster, C., Dickie, I., Maile, G., Smith, H., & Crisp, M. (2009). *Understanding the environmental impact of communication systems*. Ofcom Final Report.

Guelzim, T., & Obaidat, M. S. (2013). Green Computing and Communication Architecture. In M. S. Obaidat, A. Anpalagan, & I. Woungang (Eds.), *Handbook of green information and communication systems* (pp. 209–227). Academic Press. doi:10.1016/B978-0-12-415844-3.00008-5

Guo, W., & O'Farrell, T. (2012). *Capacity-Energy-Cost Tradeoff in Small Cell Networks*. Paper presented at the Vehicular Technology Conference (VTC Spring), 2012 IEEE 75th.

Guo, W., Rigelsford, J. M., Ford, L., & O'Farrell, T. (2012). Dynamic basestation antenna design for low energy networks. *Progress in Electromagnetics Research, C*, 31, 153–168.

Hamdoun, H., Loskot, P., O'Farrell, T., & He, J. (2012). Survey and applications of standardized energy metrics to mobile networks. *Annals of Telecommunications, 67*(3-4), 113-123.

He, J., Loskot, P., O'Farrell, T., Friderikos, V., Armour, S., & Thompson, J. (2010). *Energy efficient architectures and techniques for Green Radio access networks*. Paper presented at the Communications and Networking in China (CHINACOM), 2010 5th International ICST Conference on. doi:10.4108/chinacom.2010.136

Hevizi, L. G., & Godor, I. (2011). *Power savings in mobile networks by dynamic base station sectorization*. Paper presented at the Personal Indoor and Mobile Radio Communications (PIMRC), 2011 IEEE 22nd International Symposium on.

Holtkamp, H., Auer, G., Giannini, V., & Haas, H. (2013). A Parameterized Base Station Power Model. *Communications Letters, IEEE, 17*(11), 2033–2035. doi:10.1109/LCOMM.2013.091213.131042

Hossain, M. F., Munasinghe, K. S., & Jamalipour, A. (2013). Energy-Aware Dynamic Sectorization of Base Stations in Multi-Cell OFDMA Networks. *Wireless Communications Letters, IEEE, 2*(6), 587–590. doi:10.1109/WCL.2013.080813.130481

Huang, H., Alrabadi, O., Daly, J., Samardzija, D., Tran, C., Valenzuela, R., & Walker, S. (2010). *Increasing throughput in cellular networks with higher-order sectorization*. Paper presented at the Signals, Systems and Computers (ASILOMAR), 2010 Conference Record of the Forty Fourth Asilomar Conference on. doi:10.1109/ACSSC.2010.5757638

Joyce, R., Morris, D., Brown, S., Vyas, D., & Zhang, L. (2015). *Higher Order Horizontal Sectorisation Gains for 6, 9, 12 and 15 Sectored Cell Sites in a 3GPP/HSPA+ Network*. Academic Press.

Larsson, E. G., Edfors, O., Tufvesson, F., & Marzetta, T. L. (2014). Massive MIMO for next generation wireless systems. *IEEE Communications Magazine*, *52*(2), 186–195. doi:10.1109/MCOM.2014.6736761

Li, Y., Sheng, M., Yang, C., & Wang, X. (2013). Energy Efficiency and Spectral Efficiency Tradeoff in Interference-Limited Wireless Networks. *IEEE Communications Letters*, *17*(10), 1924–1927. doi:10.1109/LCOMM.2013.082613.131286

Lister, D. (2009). *An operator's view on green radio*. Keynote Speech, GreenComm.

Lopez-Perez, D., Ming, D., Claussen, H., & Jafari, A. H. (2015). Towards 1 Gbps/UE in Cellular Systems: Understanding Ultra-Dense Small Cell Deployments. *IEEE Communications Surveys and Tutorials*, *17*(4), 2078–2101. doi:10.1109/COMST.2015.2439636

Maslennikov, R., Trushanin, A., Testov, O., Vechkanov, M., Antipova, A., Thomas, T. A., et al. (2013). *Azimuth and elevation sectorization for the stadium environment*. Paper presented at the Global Communications Conference (GLOBECOM). doi:10.1109/GLOCOM.2013.6831694

Micallef, G., Mogensen, P., Scheck, H. O., & Lang, E. (2011). *Energy Efficient Evolution of Mobile Networks: Macro-Only Upgrades vs. a Joint-Pico Deployment Strategy*. Paper presented at the Vehicular Technology Conference (VTC Spring), 2011 IEEE 73rd.

Monti, P., Tombaz, S., Wosinska, L., & Zander, J. (2012). *Mobile backhaul in heterogeneous network deployments: Technology options and power consumption*. Paper presented at the Transparent Optical Networks (ICTON), 2012 14th International Conference on.

Nie, W., Yi, Z., Zheng, F., Zhang, W., & Farrell, T. O. (2015). HetNets with Random DTX Scheme: Local Delay and Energy Efficiency. *IEEE Transactions on Vehicular Technology*, (99), 1-1. doi: 10.1109/TVT.2015.2477374

O'Farrell, T., & Fletcher, S. (2015). Green Communication Concepts, Energy Metrics and Throughput Efficiency for Wireless Systems. In K. Samdanis, P. Rost, A. Maeder, M. Meo, & C. Verikoukis (Eds.), *Green Communications: Principles, Concepts and Practice* (pp. 19–42). Chichester, UK: John Wiley & Sons, Ltd. doi:10.1002/9781118759257.ch2

Persson, D., Eriksson, T., & Larsson, E. G. (2013). Amplifier-Aware Multiple-Input Multiple-Output Power Allocation. *Communications Letters, IEEE*, *17*(6), 1112–1115. doi:10.1109/LCOMM.2013.043013.130050

Quek, T. Q. S., Cheung, W. C., & Kountouris, M. (2011). *Energy Efficiency Analysis of Two-Tier Heterogeneous Networks*. Paper presented at the Wireless Conference 2011 - Sustainable Wireless Technologies (European Wireless), 11th European.

Rappaport, T. S., Sun, S., Mayzus, R., Zhao, H., Azar, Y., & Wang, K. et al. (2013). Millimeter Wave Mobile Communications for 5G Cellular: It Will Work! *IEEE Access*, *1*, 335–349. doi:10.1109/ACCESS.2013.2260813

Riedel, I., & Fettweis, G. (2011). *Increasing throughput and fairness in the downlink of cellular systems with N-fold sectorization*. Paper presented at the GLOBECOM Workshops (GC Wkshps), 2011 IEEE. doi:10.1109/GLOCOMW.2011.6162429

Singh, R., O'Farrell, T., & David, J. P. R. (2013). *Performance evaluation of IEEE 802.15.7 CSK physical layer*. Paper presented at the 2013 IEEE Globecom Workshops (GC Wkshps).

Singh, R., O'Farrell, T., & David, J. P. R. (2014). An Enhanced Color Shift Keying Modulation Scheme for High-Speed Wireless Visible Light Communications. *Journal of Lightwave Technology*, *32*(14), 2582–2592. doi:10.1109/JLT.2014.2328866

Skubic, B., De Betou, E. I., Ayhan, T., & Dahlfort, S. (2012). Energy-efficient next-generation optical access networks. *Communications Magazine, IEEE*, *50*(1), 122–127. doi:10.1109/MCOM.2012.6122542

Soh, Y. S., Quek, T. Q. S., Kountouris, M., & Shin, H. (2013). Energy Efficient Heterogeneous Cellular Networks. *IEEE Journal on Selected Areas in Communications*, *31*(5), 840–850. doi:10.1109/JSAC.2013.130503

Thompson, J. S., Grant, P. M., Fletcher, S., & O'Farrell, T. (2012). Minimising Power Consumption to Achieve More Efficient Green Cellular Radio Base Station Designs. In H. Zhang (Ed.), *Green Communications: Theoretical Fundamentals, Algorithms and Applications* (pp. 555–578). CRC Press. doi:10.1201/b13083-22

Tombaz, S., Monti, P., Farias, F., Fiorani, M., Wosinska, L., & Zander, J. (2014). *Is backhaul becoming a bottleneck for green wireless access networks?* Paper presented at the Communications (ICC), 2014 IEEE International Conference on.

Tombaz, S., Usman, M., & Zander, J. (2011). *Energy efficiency improvements through heterogeneous networks in diverse traffic distribution scenarios*. Paper presented at the Communications and Networking in China (CHINACOM), 2011 6th International ICST Conference on.

Yang, C., Guo, K., Sheng, M., Li, J., & Yue, J. (2014). *Energy-efficient capacity offload to smallcells with interference compensation*. Paper presented at the 2014 IEEE Wireless Communications and Networking Conference (WCNC). doi:10.1109/WCNC.2014.6952513

Yinan, Q., Imran, M. A., & Tafazolli, R. (2011). *Energy-aware adaptive sectorisation in LTE systems*. Paper presented at the Personal Indoor and Mobile Radio Communications (PIMRC), 2011 IEEE 22nd International Symposium on.

Zhang, D., & Cuthbert, L. (2011). *Dynamic sectorisation based on genetic algorithms for OFDMA networks*. Academic Press.

Zhao, L., Zhao, G., & O'Farrell, T. (2013). *Efficiency metrics for wireless communications*. Paper presented at the 2013 IEEE 24th Annual International Symposium on Personal, Indoor, and Mobile Radio Communications (PIMRC).

Chapter 2
Stable-Matching-Based Energy-Efficient Context-Aware Resource Allocation for Ultra-Dense Small Cells

Zhenyu Zhou
North China Electric Power University, China

Chen Xu
North China Electric Power University, China

Zheng Chang
University of Jyvaskyla, Finland

Tapani Ristaniemi
University of Jyvaskyla, Finland

ABSTRACT

Implementing caching to ultra-densely deployed small cells provides a promising solution for satisfying the stringent quality of service (QoS) requirements of delay-sensitive applications with limited backhaul capacity. With the rapidly increasing energy consumption, in this chapter, the authors investigate the NP-hard energy-efficient context-aware resource allocation problem and formulate it as a one-to-one matching problem. The preference lists in the matching are modeled based on the optimum energy efficiency (EE) under specified matching, which can be obtained by using an iterative power allocation algorithm based on nonlinear fractional programming and Lagrange dual decomposition. Next, on account of the Gale-Shapley algorithm, an energy-efficient matching algorithm is proposed. Some properties of the proposed algorithm are discussed and analyzed in detail. Moreover, the authors extend the algorithm to the matching with indifferent and incomplete preference lists. Finally, the significant performance gain of the proposed algorithm is demonstrated through simulation results.

I. INTRODUCTION

It is predicted that the connected devices will reach almost 50 billion by 2020, and the corresponding mobile data traffic is expected to grow more than 1000 times (Boccardi et al., 2014; Bello & Zeadally, 2014). Since acquiring new spectrum is becoming increasingly difficult due to the physical limits of the

DOI: 10.4018/978-1-5225-1712-2.ch002

spectral efficiency (SE), it is extremely challenging for the current long term evolution (LTE) system to meet the demands of increasing capacity by another 1000 times (Li, Niu, Papathanassiou, & Wu, 2014).

One promising approach when solving the above challenge is to deploy a large number of small cells (SCs) within the legendary macro cell network for further SE enhancement by *reducing the distance between users and contents* (Shanmugam, Golrezaei, Dimakis, Molisch, & Caire, 2013). Compared with traditional large-scale, high-cost macro cell base stations (MBSs), small cell base stations (SCBSs) with the advantage of lower power consumption and operational cost enable high-density spectrum reusing and short-range communications (Shanmugam et al., 2013). However, several resource allocation challenges remain to be addressed for the ultra-dense deployment of SBSs along with the centralized MBS. For delay-sensitive applications, high-speed backhaul links are an essential requirement for reliable quality of service (QoS) provision. When the cell density and user density are of the same order of magnitude, the cost to equip the ultra-densely deployed SBSs with high-speed backhaul links will be enormously high (Shanmugam et al., 2013; Rusek et al., 2013). To address the above challenges, caching popular contents at the edge of the network provides an efficient solution for SBSs with limited capacity backhaul links to accommodate highly overloaded traffic. In this way, the lost-cost storage capacity can be considered as an effective substitute for the high-cost backhaul capacity.

In this chapter, we will firstly introduce the related works and the state-of-the-art progress in the direction of edge-caching based resource allocation. There has been extensive research on the resource allocation problems in edge-caching based SC networks. These works mainly aimed at maximizing the SE in the process of resource allocation regardless of the energy efficiency (EE). The solutions proposed in these works are not easy to be extended for optimizing EE since it is not always possible to realize the maximum EE and SE simultaneously (Chan, Zhang, Xu, & Li, 2011). In addition, EE has already become a crucial factor for the performance of cellular networks with rapidly increasing energy consumption (Chan et al., 2011; Bu, Yu, Cai, & Liu, 2012). The annual operational expenditure (OPEX) of mobile network in terms of electricity has already reached 10 billion US dollars, and the electricity expenditure of BSs accounts for almost 60%-80% (Oh, Krishnamachari, Liu, & Niu, 2011). The amount of CO_2 generated by cellular networks is in the same order of magnitude of 8 million vehicles (Bu et al., 2012). In the future, we can predict that the energy consumption and CO_2 emission will experience a significant increase with the ultra-dense deployment of a large number of SBSs and their relevant supporting infrastructures. Therefore, the focus of this chapter is placed on how to design an energy-efficient context-aware matching approach for resource allocation by exploiting properties of matching theory, nonlinear fractional programming, and Lagrange dual decomposition (Z. Zhou, Dong, Ota, & Chang, 2015). Matching theory provides a low-complexity decentralized self-organizing solution to the *two-sided matching problem* in college admissions (Gale & Shapley, 1962), marriage stability (Roth & Sotomayor, 1991), labor markets (O'Malley, 2007), etc., and has been widely applied for solving resource allocation problems in cellular networks (EI-Hajj, Dawy, & Saad, 2012; Pantiano, Bennis, Saad, & Debbah, 2015; Pantisano, Bennis, Saad, & Debbah, 2014a), cognitive radios (Feng et al., 2014), social networks (Gu, Zhang, Pan, & Han, 2015), D2D communications (Gu, Zhang, Pan, & Han, 2014), and mobile energy-harvesting networks (Niyato, Wang, Pink, Saad, & Kim, 2014), etc.

The contributions of this chapter are summarized in the following four main aspects:

- We provide thorough introduction and summary of the related works and the state-of-the-art progress in the research direction of edge-caching based resource allocation. We have categorized the existing literatures based on research motivations and application scenarios. We provide in-depth analysis and discussion on the contributions of the surveyed works, common assumptions, application scenarios, advantages, disadvantages, and possible future directions. The extensive review of available works sheds new insights to the under-explored open issues of energy-efficient context-aware resource allocation design in ultra-dense SCs.

- An energy-efficient context-aware resource allocation algorithm is proposed for solving the two-sided matching problem. Both SBSs and UEs select each other to form a channel-reusing partnership to optimize EE while guaranteeing some practical constraints such as QoS, transmission power, backhaul and storage capacity, etc. The formulated problem with both binary and continuous optimization variables is actually a NP-hard mixed integer nonlinear programming (MINLP) problem, which is reformulated as a one-to-one matching problem based on the preferences of SBSs and UEs.
- Before the matching operation, UEs and SBSs build preference lists regarding the candidates in the opposite set according to the achievable maximum EE. For any SBS, the nonconvex optimization function in fractional form is transformed into a convex function in subtractive form with the same optimum value. Then, an iterative power allocation algorithm is proposed to obtain the maximum EE by employing nonlinear fractional programming (Dinkelbach, 1967) and Lagrange dual decomposition (Boyd & Vandenberghe, 2004).
- Inspired by the Gale-Shapley (GS) algorithm (Gale & Shapley, 1962), we propose an energy-efficient algorithm to match SBSs and UEs with complete and strict preference lists. The eventual matching result is proved to be stable for both of the two sets and also weak Pareto optimal for SBSs. We then propose some preference deletion and tie-breaking rules to solve the matching problem with incomplete preference lists and ties. The properties of the proposed algorithm are discussed and analyzed in detail. Finally, we evaluate the proposed algorithm through simulation results to validate its efficiency and superiority.

The rest of this chapter is outlined as follows: related works are presented in Section II. System model and problem formulation are described in Section III. Section IV introduces the proposed energy-efficient context-aware matching algorithm and analyzes the stability, optimality, implementation issues and algorithmic complexity. Section V presents simulation results and performance evaluations. Conclusions are given in Section VI.

II. RELATED WORKS

The concept of content caching has been studied intensively for supporting various applications in wireless networks. There are several works investigating how to support mobility using caching (Pantisano et al., 2014a; Xylomenos, Vasilakos, Tsilopoulos, Siris, & Polyzos, 2012; Wang, Song, & Han, 2015; Guan, Xiao, Feng, Shen, & Cimini, 2014; Kwon, Oh, Shin, & Chung, 2015). The primary focus of these works is to cache contents locally within its access region to reduce the delay caused by moving from cell to cell. A one-to-one matching game based SC-UE association algorithm was proposed in (Pantisano et al., 2014a) considering practical constraints such as content availability, QoS requirement, and backhaul capacity, etc. In (Xylomenos et al., 2012), the authors studied the publish-subscribe Internet (PSI) architecture and its ability to support caching and mobility. The PSI architecture supports seamless mobility of UEs by caching identified information items at the network layer. In (Wang et al., 2015), the authors studied the influence of changing network topology caused by UEs' mobility on the content refreshing rate, and proposed both centralized and distributed caching algorithm for SCs. A mobility-aware content caching algorithm was proposed in (Guan et al., 2014) to ease the stress on the backhaul networks. By assuming that both the content preferences and mobility patterns of UEs were known *a priori*, a sub-optimal polynomial-time heuristic solution was derived with bounded approximation ratio. Specifically, the authors combined both the caching strategies and proactive content pushing policies together to improve network performance for moving UEs. In (Bai, Sadagopan, & Helmy, 2003), the authors proposed a framework to investigate the influence of the mobility patterns on content caching policy, which were modeled as a Manhattan grid map.

Another genre of works seeks to maximize the hit ratio of content proposals through proactive caching. Predicting requested files properly and precisely based on *content popularity* is pivotal to determine which contents to be cached at SCs with limited storage capacity. The Zipf distribution was used as a common simplified model for the global content popularity (Shanmugam et al., 2013; Breslau, Cao, Fan, Phillips, & Shenker, 1999; S. Zhou, Gong, Z. Zhou, Chen, & Niu, 2015). More complex models were studied for future content request predictions in (Pantiano et al., 2015; Bastug, Bennis, & Debbah, 2014; Su & Xu, 2015), which were based on user-specific content popularity and social information, respectively. In (Hamidouche, Saad, & Debbah, 2014), the content placement problem was formulated as a *many-to-many matching game* between service provider servers (SPSs) and SCs by exploiting matching theory. A joint user clustering and caching optimization algorithm was proposed in (ElBamby, Bennis, Saad, & Latvaaho, 2014) by exploiting content requesting similarity, which employed a reinforcement learning algorithm to learn the content popularity distribution in each cluster. In (Chen & Kountouris, 2015), the authors modeled the local content popularity as a Zipf-like distribution with variables reflecting the correlation level of user interests. In (Zhang et al., 2015), both long-term and short-term content access patterns were exploited to improve the cache hit ratio by developing separate buffers for popularity-based content caching and mobility prediction-based content prefetching. In (Perabathini, Bastug, Kountouris, Debbah, & Conte, 2015), the Poisson point process (PPP) model was employed to evaluate the impacts of edge caching on area power consumption (APC) and EE. Content ranking and content popularity based caching algorithm was proposed to update contents cached at SCs in (Pingyod & Somchit, 2015). In (Blasco & Gunduz, 2014), the content popularity distribution was assumed as unknown and was obtained by using the multi-arm bandit (MAB) framework. Three heuristic proactive caching algorithms were proposed to study the exploitation-exploration trade-offs. There are also some works that exploit the associations among caching strategies, service characteristics, encoding schemes, and user behaviors. For example, in video streaming, while full caching would incur tremendous cost, partial caching that only stores part of the requested video stream based on viewing patterns provides a promising way for efficient content splitting (Gao, Zhang, Wen, Wang, & Zhu, 2015; Bao, Wang, Zhou, & Niu, 2012). Interactions of UEs with the YouTube video distribution infrastructure was monitored to obtain key parameters that are relevant to improve the reliability and effectiveness of in-network caching (Braun, Klein, Carle, Reiser, & Eisl, 2012). For multimedia contents, a service is often interrupted and aborted before being finished. Thus, forepart chunks with higher popularity was cached in the edge router and a cache pipelining was established in relaying routers to reduce service delay (Lim, Ko, Jung, Kim, & Jang, 2014).

Instead of focusing on individual client, collaborative frameworks for coordinating content retrieval have also attracted intensive research interests. In (Pantisano, Bennis, Saad, & Debbah, 2014b), in-network caching that allows SCs to implement cooperative caching was studied to develop decentralized

SC-specific caching strategies. Each SC was allowed to make individual optimal decisions on how to cooperate with other SCs and how to update the cached contents. Similarly, a greedy algorithm based caching placement strategy was proposed for SCs with many caching helpers in (J. Song, H. Song, & Choi, 2015). More practical scenarios such as overlapped coverage of neighboring helpers were considered in (Shim, Min, K. Kim, Jang, & Kim, 2014). Cooperative caching strategies that take advantage of social ties and service workloads were developed in (Nikolaou, Renesse, & Schiper, 2015). In (Hu, Zheng, Wang, & Song, 2015), the scenarios that contents are from different service providers were considered. A multi-step multi-objective auction based algorithm was proposed to enable content providers to evaluate the price of contents and bid for caching space at SCs at each step. A Stackelberg game based video pricing and caching algorithm was developed for heterogeneous networks consisting of video retailer, network service providers, and UEs in (Li, Chen, Xiao, Shu, & Liu, 2015). Cooperative caching algorithms based on device-to-device (D2D) communications were developed in (Golrezaei, Molisch, Dimakis, & Caire, 2012, 2013). With D2D communications, a variety of contents were firstly cached in the libraries of UEs and then were exchanged directly among UEs that are close enough to each other without going through the BS.

Some works have also considered the joint optimization of caching and other key design factors. In (Shanmugam et al., 2013; Gerami, Ming, & Skoglund, 2014), the authors proposed joint caching and coding schemes to enable robust and fast content dissemination in large-scale dynamic networks at the cost of increased computation complexity. In (Kantarci, 2014; Li, Wang, & Hu, 2015), joint optimization of both caching policies and multicast transmission schemes were developed to reduce serving cost and energy cost efficiently. A joint wireless and backhaul load balancing (JWBLB) optimization framework was developed for distributed and cooperative caches (DCC) enabled SCs to minimize the system content transmission delay in (Li et al., 2015). A joint routing and caching optimization algorithm was developed to maximize the fraction of contents served by local SCs in (Poularakis, Iosifidis, Sourlas, & Tassiulas, 2014). A joint optimization on uplink and caching power consumption was proposed for SCs by employing integer linear programming (ILP) (Kantarci, 2015). The joint optimization of content caching and pushing through broadcasting was shown to effectively alleviate cellular data bottleneck and resolve randomness of content request arrivals (S. Zhou, et al., 2015; Wang, Chen, & Liu, 2014). The trade-off between caching size and energy-harvesting capability was studied for energy-harvesting SCs in (Kumar & Saad, 2015). A joint bandwidth allocation and power control optimization problem was formulated as the infinite horizon dynamic programming problem and solved numerically by employing the value iteration algorithm.

In summary, how to implement caching at the ultra-dense SCs in an energy-efficient way while satisfying various practical system constraints such as transmission power, backhaul capacity, storage capacity, QoS requirement, etc., is still an open problem and requires further research efforts. In the following sections, we will introduce the proposed energy-efficient context-aware resource allocation algorithm in details.

III. SYSTEM MODEL AND PROBLEM FORMULATION

We consider a two-tier heterogeneous cellular network as shown in Fig. 1. The first tier is the macro cell with a conventional high-power MBS, and the second tier is composed of K low-power SBSs. We assume that SBSs are ultra-densely deployed, i.e., the number of SBSs is comparable or even higher than

Figure 1. System model of ultra-dense small cells

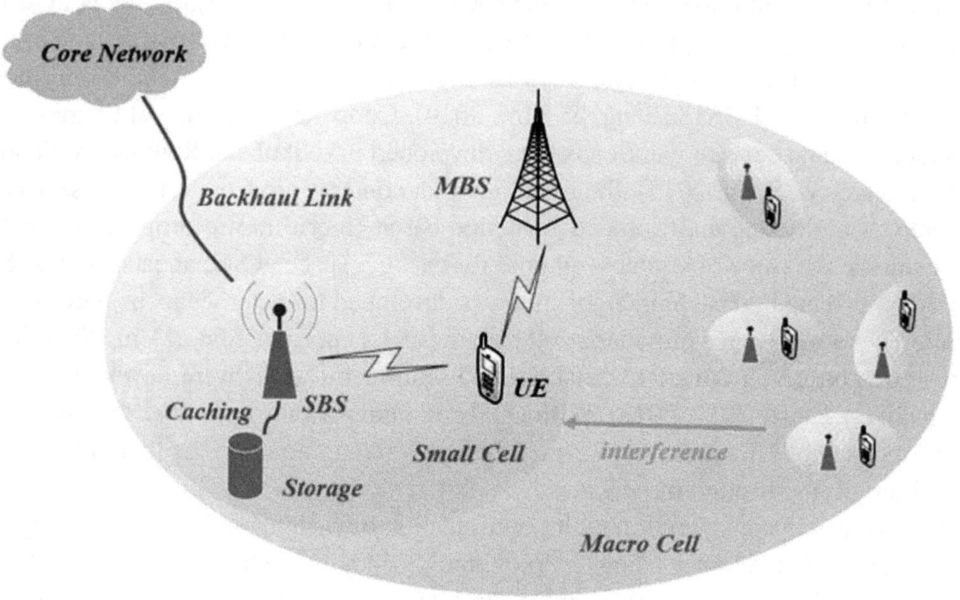

that of UEs, i.e., $K > M$ (Hwang, Song, & Soliman, 2013), and each SBS provides services for at most one UE. The MBS, which is equipped with sophisticated signal processing techniques and high-speed backhaul links, is used to provide converge for the whole cellular network. On the contrary, SBSs, which have limited signal processing, storage, and backhaul capacities, are deployed close to UEs to offload traffic and guarantee QoS requirements.

Throughout the chapter, we use the two different words "file" and "content" interchangeably to represent the same meaning. We denote the sets of SBSs and UEs as $S = \{s_1,\ldots,s_k,\ldots,s_K\}$ and $\mathcal{U} = \{u_1,\ldots,u_m,\ldots,u_M\}$, respectively. The storage capacity of the SBS s_k is denoted as D_k. The total set of files cached in D_k is denoted as $\mathcal{C}_k \subseteq \mathcal{C}$, where $\mathcal{C}$ is the total set of contents in the network. We assume that the size of each file is the same and denoted as s. The file popularity is assumed to follow the Zipf distribution (Breslau et al., 1999; S. Zhou et al., 2015). Each SBS can download popular files via its backhaul link up to its maximum storage capacity D_k if its backhaul bandwidth is available.

The macro-cell spectrum channel is divided into several orthogonal sub-channels (e.g., an orthogonal resource block in LTE), and each SBS $s_k \in S$ is allocated with a bandwidth of w_k. In order to improve SE, SBSs reuse the same spectrum after certain geographical distance by employing dynamic frequency allocation (Ali & Leung, 2009). SBSs and UEs are properly matched with each other to optimize EE performances under the constraint of QoS. We define the partner selection decisions of SBS and UE as:

Definition 1: The SBSs' $K \times M$ partner selection matrix is denoted as X, where the (k, m)-th element $x_{k,m} \in \{0,1\}$ represents the partner selection decision of the SBS-UE partnership (s_k, u_m) for the

SBS s_k, $\forall s_k \in S$, $\forall u_m \in \mathcal{U}$. If $x_{k,m} = 0$, s_k prefers not to form a partnership with u_m, and if $x_{k,m} = 1$, otherwise.

Definition 2: The UEs' $M \times K$ partner selection matrix is denoted as Y, where the (m, k)-th element $y_{m,k} \in \{0,1\}$ represents the partner selection decision of the SBS-UE partnership (u_m, s_k) for the UE u_m, $\forall s_k \in S$, $\forall u_m \in \mathcal{U}$. If $y_{m,k} = 0$, u_m prefers not to form a partnership with s_k, and if $y_{m,k} = 1$, otherwise.

Remark 1: s_k and u_m can form a SBS-UE partnership if and only if $x_{k,m} = y_{m,k} = 1$.

During a period of T, each UE $u_m \in \mathcal{U}$ proposes to its serving SBS s_k to obtain the needed files $\mathcal{F}_m = \{f_1^m, \ldots, f_{N_m}^m\}$, $\mathcal{F}_m \subset \mathcal{C}$. t_c and $\lfloor T / t_c \rfloor$ are defined as the channel coherence time and the gross time instants, respectively. t_c is also the time interval between two successive time instants. The requested files can be transmitted from s_k to u_m without going through the core network if these files are stored in s_k's cache. At time instant t, the maximum achievable transmission rate between s_k and u_m is given by

$$r_{k,m}(t) = w_k \log_2\left(1 + \frac{x_{k,m} y_{m,k} p_k(t) g_{k,m}(t)}{N_0 + \sum_{j \neq k, j \in S_k} p_j(t) g_{j,m}(t)}\right), \tag{1}$$

where $g_{k,m}$ and p_k denote the channel gain from s_k to u_m and the transmission power of s_k, respectively. $\sum_{j \neq k, j \in S_k} p_j(t) g_{j,m}(t)$ is the cumulative interference from multiple SBSs S_k sharing the same channel simultaneously. $g_{j,m}$ denotes the interference channel gain from the SBS s_j to u_m. We consider both the fast fading and slow fading caused by multi-path propagation and shadowing, respectively (Goldsmith, 2005). For example, $g_{k,m}$ is as rewritten as

$$g_{k,m} = \varpi \beta_{k,m} \zeta_{k,m} d_{k,m}^{-\alpha}, \tag{2}$$

where ϖ and α denote the pathloss constant and exponent, respectively. $\beta_{k,m}$ and $\zeta_{k,m}$ represent the fast-fading and slow-fading gain following the exponential and log-normal distribution respectively. $d_{k,m}$ denotes the transmission distance.

Denoting the number of files in s_k's cache as $\|\mathcal{F}_m \cap \mathcal{C}_k\|$, the estimated time duration to transmit $\|\mathcal{F}_m \cap \mathcal{C}_k\|$ files to u_m is calculated as

$$\tilde{T}_{k,m}^{cache} = \frac{\mathcal{F}_m \cap \mathcal{C}_k \quad s}{\bar{r}_k(t)}, \tag{3}$$

where $\bar{r}_k(t)$ is the average transmission rate of s_k up to the time instant *t*. If $\tilde{T}_{k,m}^{cache} > T$, u_m can receive only part of the cached files. Hence, the time duration needed for successfully transmitting the cached files is defined as:

$$T_{k,m}^{\max} = \min\left\{\tilde{T}_{k,m}^{cache}, T\right\}. \tag{4}$$

In the case that the requested files are not available locally, i.e., $\mathcal{F}_m \setminus \mathcal{F}_m \cap \mathcal{C}_k$, these files needed to be downloaded firstly from the core network by s_k through its backhaul link and then transmitted to u_m. The maximum transmission rate between s_k and u_m is given by

$$r_{k,m}^{\max}(t) = \min\left\{r_{k,m}(t), x_{k,m} y_{m,k} B_k\right\}. \tag{5}$$

Due to limited backhaul capacity and short transmission distance, B_k may not be large enough to accommodate $r_{k,m}(t)$, i.e., $B_k \ll r_{k,m}(t)$, which causes significant delay for u_m regardless of the wireless channel state.

The total throughput (bits) between s_k and u_m during *T* is defined as

$$U_{k,m} = \left(\sum_{t=1}^{T_{k,m}^{\max}/t_c} r_{k,m}(t) + \sum_{t=T_{k,m}^{\max}/t_c}^{T/t_c} r_{k,m}^{\max}(t)\right) t_c, \tag{6}$$

where x denotes the largest integer not greater than x.

The total energy consumption (J) of s_k during *T* is given by

$$E_k = \sum_{t=1}^{T/t_c} \left(\frac{1}{\eta} p_k(t) + p_k^{cir}\right) t_c, \tag{7}$$

where η is the power amplifier (PA) efficiency, i.e., $0 < \eta < 1$. p_k^{cir} is the circuit power and is assumed as constant.

We define the *utility* EE (bits/J) of s_k as the ratio of throughput to total energy consumption (Kwon & Birdsall, 1986):

$$U_k^{EE} = \frac{\sum_{u_m \in \mathcal{U}} U_{k,m}}{E_k} = \frac{\sum_{t=1}^{T_{k,m}^{\max}/t_c} r_{k,m}(t) + \sum_{t=T_{k,m}^{\max}/t_c}^{T/t_c} r_{k,m}^{\max}(t)}{\sum_{t=1}^{T/t_c} \frac{1}{\eta} p_k(t) + p_k^{cir}}. \tag{8}$$

As for u_m, due to the equivalent total downlink throughput to s_k during T defined in (6), e.g., $U_{m,k} = U_{k,m}$, the total energy consumption is just only considered and obtained by:

$$E_m = p_m^{cir} T, \tag{9}$$

where p_m^{cir} is the constant circuit power of u_m needed to receive data.

The EE of u_m is given by

$$U_m^{EE} = \frac{\sum_{s_k \in S} U_{k,m}}{E_m} = \frac{\sum_{t=1}^{T_{k,m}^{\max}/t_c} r_{k,m}(t) + \sum_{t=T_{k,m}^{\max}/t_c}^{T/t_c} r_{k,m}^{\max}(t) t_c}{p_m^{cir} T}. \tag{10}$$

The challenges faced by the SBSs during the resource allocation design are summarized as follows.

- How to choose UE as SBS-UE partner to optimize its own EE under various practical limitations such as storage capacity, local content availability, backhaul capacity, QoS, etc?
- How to allocate the transmission power for the potential SBS-UE combination in order to optimize EE while guaranteeing the QoS and transmission power requirements?
- Whether the selected UE has the same intension to form a SBS-UE partnership?
- Will the selected UE has the inclination to be attracted by other SBSs meditating to form a partnership with it?

Thus, the energy-efficient context-aware resource allocation problem for SBSs is actually composed of two subproblems, the first one is the *power allocation subproblem*, and the second one is the *partner selection subproblem.*

Define

$$\boldsymbol{X}_k = \left\{x_{k,1}, \ldots, x_{k,m}, \ldots, x_{k,M}\right\}$$

and

$$\boldsymbol{p}_k = \left\{p_k(1), \ldots, p_k(t), \ldots, p_k(T / t_c)\right\}$$

as the strategy sets of partner selection and power allocation, respectively. For any $s_k \in S$, the EE maximization problem is formulated as

$$\max_{(x_k, p_k)} U_k^{EE}\left(\boldsymbol{X}_k, \boldsymbol{P}_k\right)$$

$$s.t.\ C_1: 0 \le p_k(t) \le p_k^{\max}, t = \left[1, \frac{T_{k,m}^{\max}}{t_c}\right],$$

$$C_2: 0 \le p_k(t) \le p_k^B(t), t = \left[\frac{T_{k,m}^{\max}}{t_c}, \frac{T}{t_c}\right],$$

$$C_3: U_{k,m}\left(x_{k,m}, P_k\right) \ge x_{k,m} y_{m,k} U_{k,m}^{\min}, \forall_{u_m} \in \mathcal{U},$$

$$C_4: x_{k,m} = \{0,1\}, \forall_{u_m} \in \mathcal{U},$$

$$C_5: \sum_{u_m \in \mathcal{U}} x_{k,m} \le 1. \tag{11}$$

C_1 and C_2 represent that the transmission power should be less than $p_k^{\max}$ in the time range $\left[1, T_{k,m}^{\max} / t_c\right]$ and less than $p_k^B(t)$ in the time range $\left[T_{k,m}^{\max} / t_c, T / t_c\right]$. $p_k^B(t)$ is the minimum transmission power required to achieve the backhaul capacity B_k, which is calculated as

$$p_k^B(t) = \frac{\left(2^{\frac{B_k}{w_k}} - 1\right)\left(N_0 + \sum_{j \ne k, j \in S_k} p_j(t) g_{j,m}(t)\right)}{g_{k,m}(t)}. \tag{12}$$

C_3 denotes the minimum QoS restriction in terms of throughput. C_4 and C_5 are the partner selection restrictions for s_k.

Similarly, the challenges for UEs during the resource allocation design are listed as follows:

- How to choose a SBS as a UE-SBS partner to maximize its own EE?
- Whether the selected SBS has the same intension to form the partnership considering various implementation constraints?
- Will the selected SBS has the inclination to be attracted by other UEs meditating to form a partnership with it?

Denoting

$$y_m = \left\{y_{m,1}, \ldots, y_{m,k}, \ldots, y_{m,K}\right\}$$

as the partner selection strategy set of u_m, the EE optimization problem is formulated as:

$$\max_{y_m} U_m^{EE}\left(y_m\right)$$

$$s.t.\ C_6 : U_{k,m}\left(x_{k,m}, P_k\right) \geq x_{k,m} y_{m,k} U_{k,m}^{\min}, \forall_{s_k} \in S$$

$$C_7 : y_{m,k} = \left\{0,1\right\}, \forall_{s_k} \in S$$

$$C_8 : \sum_{s_k \in S} y_{m,k} \leq 1 \tag{13}$$

C_6 denotes the QoS constraint, which is similar to C_3. C_7 and C_8 are the partner selection restrictions for u_m.

Remark 2: For the QoS constraints C_3 and C_6, considering (6), it is obvious that the files $\mathcal{F}_m \cap \mathcal{C}_k$ are transmitted at data rate $r_{k,m}$ during the interval [0, $T_{k,m}^{\max}$], and the files $\mathcal{F}_m \setminus \mathcal{F}_m \cap \mathcal{C}_k$ are transmitted at data rate $r_{k,m}^{max}$ during the interval [$\left[T_{k,m}^{\max}, T\right]$. As a result, $\mathcal{F}_m \cap \mathcal{C}_k$ only depends on the instantaneous channel capacity, but $\mathcal{F}_m \setminus \mathcal{F}_m \cap \mathcal{C}_k$ depends on the limited backhaul capacity B_k.

Remark 3: It is noted that the formulated problem in (11) belongs to MINLP, which is NP-hard and computationally intractable. The partner selection and power allocation subproblems are coupled with each other.

To provide a tractable solution for (11), the partner selection and power allocation subproblems are decoupled and reformulated as a *one-to-one matching problem*, which is denoted as a triple $\left(S, \mathcal{U}, \mathcal{P}\right)$. S and $\mathcal{U}$ represent two finite and independent sets of SBSs and UEs participating the match. $\mathcal{P}$ denotes the set of preferences. The detailed definition of the *one-to-one matching* μ is (Roth & Sotomayor, 1991):

Definition 3: In the matching problem $\left(S, \mathcal{U}, \mathcal{P}\right)$, a *matching* μ is a one-to-one correspondence from the set $S \cup \mathcal{U}$ onto itself under preference $\mathcal{P}$ such that for each $s_k \in S$ and $u_m \in \mathcal{U}$, $\mu\left(s_k\right) \in \mathcal{U} \cup \left\{s_k\right\}$ and $\mu\left(u_m\right) \in S \cup \left\{u_m\right\}$. $\mu\left(s_k\right) = u_m$ if and only if $\mu\left(u_m\right) = s_k$.

If u_m or s_k is matched with itself, i.e., $\left(u_m\right) = u_m$ or $\mu\left(s_k\right) = s_k$, it means that the QoS requirement of u_m cannot be satisfied by any SBS, or none UE preferred to be served by s_k. In the first case, the QoS requirement of u_m should be lowered for u_m to be matched with some SBS. In the second case, s_k can be turned into sleeping mode for energy conservation. We have the property that $\mu\left(s_k\right) = s_k$ and $\mu\left(s_k\right) \in \mathcal{U}$ are not feasible simultaneously under this scenario, which also holds for $\mu\left(u_m\right)$. Both s_k and u_m can send a request to its preferred partner according to its preference. We define the set of total proposals generated jointly by K SBS (or M UEs) as a matching. A common assumption is that any SBS or UE only concerns its own matching partner regardless of other SBSs' and UEs' matching results. Hereinafter, we introduce how to establish the two-sided preferences $\mathcal{P}$ and obtain the matching μ with $\mathcal{P}$.

IV. THE ENERGY-EFFICIENT CONTEXT-AWARE MATCHING APPROACH

In this section, we employ an iterative algorithm to establish preference lists and introduce the proposed matching algorithm in detail with the known two-sided preferences. Then, we analyze some properties of the proposed algorithm such as stability, optimality, implementation issues and algorithmic complexity, etc.

The Establishment of Preference Profiles

In order to employ the energy-efficient matching algorithm, we need to establish the preference lists for all the UEs and SBSs denoted as the set $\mathcal{P}$. Taking the SBS s_k as an example, the achievable maximum EE is used as the criteria to represent s_k's preference over u_m under the matching $\mu(s_k) = u_m$, $\mu(u_m) = s_k$, and $x_{k,m} = y_{m,k} = 1$. To obtain the maximum EE, the following power allocation problem need to be addressed:

$$\max_{p_k} U_k^{EE}\left(\boldsymbol{p_k}\right)_{\mu(s_k)=u_m}, s.t.\ C_1, C_2, C_3. \tag{14}$$

Noting that (14) is a nonconvex problem due to the fractional-from objective function. It can be transformed to an equivalent convex optimization problem based on nonlinear fractional programming (Dinkelbach, 1967). Then an iterative algorithm is proposed to address the transformed problem.

The maximum EE under $\mu(s_k) = u_m$ is defined as:

$$q_{k,m}^* = \max_{\boldsymbol{p_k}} U_k^{EE}\left(\boldsymbol{p_k}\right)_{\mu(s_k)=u_m} = \frac{U_{k,m}\left(\boldsymbol{P_k^*}\right)}{E_k\left(\boldsymbol{P_k^*}\right)}, \tag{15}$$

where $\boldsymbol{P_k^*}$ is the optimum power allocation strategy set for s_k. (15) can be transformed into a more tractable form according to the following theorem (Dinkelbach, 1967).

Theorem 1: $q_{k,m}^*$ is achieved if and only if

$$\max_{p_k} U_{k,m}\left(\boldsymbol{p_k}\right) - q_{k,m}^* E_k\left(\boldsymbol{P_k}\right) = U_{k,m}\left(\boldsymbol{P_k^*}\right) - q_{k,m}^* E_k\left(\boldsymbol{P_k^*}\right) = 0 \tag{16}$$

Therefore, instead of solving (14), the following equivalent problem can be solved to obtain the same EE:

$$\max_{p_k} U_{k,m}\left(\boldsymbol{P_k}\right) - q_{k,m}^{*} E_k\left(\boldsymbol{P_k}\right), s.t.\ C_1, C_2, C_3 \tag{17}$$

(17) is actually a weighted sum of U_{km} and E_k, where $q_{k,m}^{*}$ is the negative weight of the power consumption.

Although the equivalent optimization problem (17) is convex, the value of $q_{k,m}^{*}$ remains to be unknown. Consequently, to obtain $q_{k,m}^{*}$, an iterative algorithm is proposed based on Dinkelbach's method, which is illustrated in Algorithm 1. Initially, we set $q_{k,m}$ as a small positive number, e.g., 10^{-4}. At each iteration, the following transformed problem is solved by employing Karush-Kuhn-Tucker (KKT) conditions and Lagrange dual decomposition.

$$\max_{p_k} U_{k,m}\left(\boldsymbol{P_k}\right) - q_{k,m} E_k\left(\boldsymbol{P_k}\right), s.t.\ C_1, C_2, C_3 \tag{18}$$

In case of $t = \left[1, T_{k,m}^{\max} / t_c\right]$ and $t = \left[T_{k,m}^{\max} / t_c, T / t_c\right]$, the optimal value $\hat{p}_k\left(t\right)$ related to $q_{k,m}\left(t\right)$ can be obtained by the following two formulas, respectively:

$$\hat{p}_k\left(t\right) = \left[\frac{\eta\left(1+\vartheta_k\right) w_k t_c \log_2 e}{q_{k,m}\left(t\right) + \eta\lambda_k\left(t\right)} - \frac{I_{k,m}\left(t\right) + N_0}{g_{k,m}\left(t\right)}\right]^{+}, \tag{19}$$

$$\hat{p}_k\left(t\right) = \left[\frac{\eta\left(1+\vartheta_k\right) w_k t_c \log_2 e}{q_{k,m}\left(t\right) + \eta\theta_k\left(t\right)} - \frac{I_{k,m}\left(t\right) + N_0}{g_{k,m}\left(t\right)}\right]^{+}. \tag{20}$$

Equation (19) and (20) indicate a water-filling algorithm, where λ_k, θ_k and ϑ_k represent the price of guaranteeing the transmission power and QoS constraints respectively. Then update the Lagrange multipliers by employing the gradient method as (Blasco & Gunduz, 2014)

$$\lambda_k\left(t,\tau+1\right) = \left[\lambda_k\left(t,\tau\right) + \varepsilon_{k,\lambda}\left(t,\tau\right)\left(\hat{p}_k\left(t,\tau\right) - p_k^{max}\right)\right]^{+},$$

$$\theta_k\left(t,\tau+1\right) = \left[\theta_k\left(t,\tau\right) + \varepsilon_{k,\theta}\left(t,\tau\right)\left(\hat{p}_k\left(t,\tau\right) - p_k^{B}\left(t\right)\right)\right]^{+},$$

$$\vartheta_k\left(\tau+1\right) = \left[\vartheta_k\left(\tau\right) + \varepsilon_{k,\vartheta}\left(\tau\right)\left(U_{k,m}\left(\tau\right) - U_{k,m}^{\min}\right)\right]^{+}, \tag{21}$$

where τ is the iteration index, and ε is the positive step size.

In the end of the iteration, $q_{k,m}$ of the next iteration is updated as $q_{k,m} = U_{k,m}\left(\hat{\boldsymbol{P}}_{\boldsymbol{k}}\right) / E_k\left(\hat{\boldsymbol{P}}_{\boldsymbol{k}}\right)$. If either the stopping criteria Δ or the maximum iteration number $N_{\max}$ is reached, the iteration process will terminate. The optimum $\boldsymbol{P}_{\boldsymbol{k}}^*$ is equal to $\boldsymbol{P}_{\boldsymbol{k}}^*$ obtained in the final iteration, and $q_{k,m}^*$ is calculated by (15).

In a similar way, the preference of u_m over s_k is also defined as the maximum achievable EE under the matching $\mu\left(u_m\right) = s_k$. During [0, *T*], the preference of u_m only lies on the achievable total throughput of s_k, i.e., $U_{k,m}$ due to two reasons, which are that the contents stored by s_k is not available to u_m and the denominator of (10) is constant.

Then we introduce how to establish the preference profile as shown in Algorithm 2. First of all, the maximum EE of s_k is obtained by using Algorithm 1 under all the possible matching with any $u_m \in \mathcal{U}$. Next, a complete, reflexive, and transitive binary relation is introduced to describe the preference for all the $s_k \in S$ and $u_m \in \mathcal{U}$ (Roth & Sotomayor, 1991). For example, $u_m \succ_{s_k} u_{m'}$ represents that s_k prefers u_m to $u_{m'}$:

$$u_m \succ_{s_k} u'_m \Leftrightarrow q_{k,m}^* > q_{k,m'}^*. \tag{22}$$

Similarly, $s_k \succ_{u_m} s_j$ is denoted that u_m prefers s_k to s_j:

$$s_k \succ_{u_m} s_j \Leftrightarrow U_{k,m}\left(\boldsymbol{P}_{\boldsymbol{k}}^*\right) > U_{j,m}\left(\boldsymbol{P}_{\boldsymbol{j}}^*\right). \tag{23}$$

Then, the preferences on the opposite set $\mathcal{U}$ are ranked in descending order to establish the preference list of s_k, which is defined as $P\left(s_k\right) = \left\{\cdots, u_m, u_{m'}, \cdots\right\}$. Similarly, the preference list of u_m is denoted as $P_{u_m} = \left\{\cdots, s_k, s_j, \cdots\right\}$, which can be obtained in the same way. Thus, we denote the total set of preference lists as

$$\mathcal{P} = \left\{P\left(s_1, \ldots P\left(s_K\right), P\left(u_1\right), \ldots P\left(u_M\right)\right)\right\}.$$

The Proposed Matching Algorithm

After the preference list establishment for all the SBSs and UEs, we propose an energy-efficient context-aware matching algorithm to match SBSs and UEs based on the GS algorithm (Gale & Shapley, 1962) as shown in Algorithm 3. Initially, each SBS sends a request to its most preferred UE, which is at the top of its preference list. After receiving the requests, each UE selects the most preferred SBS from all the received proposals and rejects the remaining SBSs. The selected SBS is temporarily kept as a candidate. In the next step, any SBS rejected in the previous step sends a request to its next most preferred UE which has not issued a rejection yet. Each UE selects the most preferred SBSs from all the proposals received in this step and the SBS kept from the previous step, and rejects the remaining SBSs. If a SBS is rejected by all of the UEs in its preference lists, then the SBS terminates to send requests and turns into sleeping mode for energy saving. The process continues until every SBS is either matched

Algorithm 1. The energy-efficient iterative power allocation algorithm

Input: $\mathcal{F}_m$, $g_{k,m}$, $\sum_{j\neq k, j\in S_k} p_j g_{j,m}$, C_k, $\tilde{T}_{k,m}^{cache}$, p_k^{cir}, $p_k^{\max}$, p_k^{B}, $U_{k,m}^{\min}$.

Output: $q_{k,m}^{*}$.

Initialize: $N_{\max}$, Δ, $q_{k,m}$

While $n < N_{\max}$ **do**

obtain $\hat{\boldsymbol{P}}_{\boldsymbol{k}}$ using (19) or (20)

if $U_{k,m}\left(\hat{\boldsymbol{P}}_{\boldsymbol{k}}\right) - q_{k,m}E_k\left(\hat{\boldsymbol{P}}_{\boldsymbol{k}}\right) > \Delta$ **then**

Update: $q_{k,m} = U_{k,m}\left(\hat{\boldsymbol{P}}_{\boldsymbol{k}}\right) / E_k\left(\hat{\boldsymbol{P}}_{\boldsymbol{k}}\right)$

else

$p_k^{*} = \hat{\boldsymbol{P}}_{\boldsymbol{k}}$, and $q_{k,m}^{*} = U_{k,m}\left(\boldsymbol{P}_{\boldsymbol{k}}^{*}\right) / E_k\left(\boldsymbol{P}_{\boldsymbol{k}}^{*}\right)$

end if

Update: $n = n + 1$

end while

with some UE or has been rejected by every UE on its preference list. When the algorithm terminates, a stable matching between SBSs and UEs is produced.

For the *preference indifference* case, i.e., when some SBSs or UEs have the same preference for two or more possible candidates, the matching problem can be solved by introducing some additional *tie-breaking* rules to Algorithm 3 to obtain strict preference lists. For instance, if the SBS s_k likes u_m and $u_{m'}$ equally, i.e., $q_{k,m}^{*} = q_{k,m'}^{*}$, the tie-breaking rule comes into play by differentiating u_m and $u_{m'}$ in accordance with some other rules, such as signal to noise plus interference ratio (SINR), or occupancy time of the backhaul link, i.e., $\min\left\{T - T_{k,m}^{\max}, T - T_{j,m}^{\max}\right\}$. In general, the additional tie-breaking rules enable SBSs to determine which UEs should be proposed to, and enables UEs to specify which SBSs should be kept as candidates.

Another popular case is matching with *incomplete preference lists*, i.e., both SBSs and UEs only know part of the information of the opposite set. For example, it is extremely difficult for a SBS to acquire the full information of every UE in a large-scale network. If s_k cannot obtain the information of u_m, then u_m will be absent on s_k's preference list $P\left(s_k\right)$. The preference lists of $P\left(s_k\right)$ and $P\left(u_m\right)$ are assumed to be consistent for all the s_k and u_m, which means that removing u_m from s_k's preference list $P\left(s_k\right)$ implies removing s_k from u_m's preference list $P\left(u_m\right)$ simultaneously (O'Malley, 2007). Similarly, such a problem can be solved by introducing *preference deletion* rules to Algorithm 3. Taking an example, SBSs or UEs that are not participating in the matching process should be removed from the two-side preference lists. In other words, deleting the SBS-UE pair $\left(s_k, u_m\right)$ represents deleting s_k and u_m from $P\left(u_m\right)$ and $P\left(s_k\right)$, respectively. With the extended tie-breaking and preference deletion rules, the algorithm can continue to operate as in the strict and complete list scenario to obtain a stable matching in polynomial time.

Algorithm 2. Preference establishment algorithm

Input: $\mathcal{S}$, $\mathcal{U}$.
Output: $\mathcal{P}$
for $s_k \in \mathcal{S}$ **do**
for $u_m \in \mathcal{U}$ **do**
calculate mutual preferences of the SBS-UE pair (s_k, u_m) using Algorithm 1.
end for
end for
for $s_k \in \mathcal{S}$ **do**
establish $P(s_k)$ by sorting each $u_m \in \mathcal{U}$ in descending order based on $q^*_{k,m}$.
end for
for $u_m \in \mathcal{U}$ **do**
establish $P(u_m)$ by sorting each $s_k \in \mathcal{S}$ in descending order based on $U_{k,m}$.
end for

Algorithm 3. Energy-efficient stable matching algorithm

Input: $\mathcal{S}$, $\mathcal{U}$, $\mathcal{P}$.
Output: an energy-efficient context-aware matching μ
Initialize: $\mu = \phi$, $\Phi = \mathcal{S}$.
while $\Phi \neq \phi$ **do**
for $s_k \in \Phi$ **do**
s_k proposes to the most preferred UE among those who have not yet rejected it in $P(s_k)$.
end for
for $u_m \in \mathcal{U}$ **do**
if u_m receives a proposal from s_k, and prefers s_k to its currently engaged s_j, i.e., $s_k \succ_{u_m} s_j$ **then**
s_k is kept engaged, while s_j is rejected, i.e., $\mu(u_m) = s_k$;
add s_j into Φ, and remove s_k from Φ.
else
s_j is continually kept engaged, while s_k is rejected, i.e., $\mu(u_m) = s_j$.
end if
end for
end while

Discussions

In this subsection, we provide the detailed discussions and analysis for the stability, optimality, implementation issues and algorithmic complexity of the proposed algorithm. All of the proofs can be found in (Z. Zhou et al., 2015) and are omitted here due to space constraint.

Stability

The *blocking pair* is introduced to prove the stability of the matching. Taking an example, if both s_k and u_m are not matched with each other, i.e., $\mu(s_k) \neq u_m$ and $\mu(u_m) \neq s_k$, (s_k, u_m) can become a blocking pair for the current matching if both s_k and u_m prefer each other to their current matching partners, i.e., $u_m \succ_{s_k} \mu(s_k)$ and $s_k \succ_{u_m} \mu(u_m)$. Thus the current matching is not stable due to the disruption of s_k and u_m.

The *stability* of a matching is defined as follows (Roth & Sotomayor, 1991):

Definition 4: A matching μ is stable if it is not blocked by any individual SBS-UE pair.
Theorem 2: Algorithm 3 can produce a stable matching μ in finite steps.

Optimality

Theorem 3: Algorithm 1 is guaranteed to converge to the unique maximum EE $q^*_{k,m}$ for any s_k in finite iterations.
Theorem 4: For SBSs, the matching result generated by using Algorithm 3 is weak Pareto optimal.

Implementation

Despite the distributed implementation, the proposed algorithm can also be implemented with the help of the powerful MBS in a centralized manner. The primary role of the MBS is to facilitate the matching between SBSs and UEs based on the two-sided preferences, which is presented as follows. Initially, the MBS accumulates the necessary information of all the SBSs and UEs like $\mathcal{F}_m$, $g_{k,m}$, $I_{k,m}$, C_k, $\tilde{T}^{cache}_{k,m}$, p^{cir}_k, $p^{\max}_k$, p^B_k, and $U^{\min}_{k,m}$. Then the preference profiles $P(s_k)$ and $P(u_m)$ are established by employing Algorithm 2. Finally, a stable matching is obtained by employing Algorithm 3 under the established preference lists.

It is also feasible to integrate the proposed algorithm with future cloud-based architectures of cellular networks such as the cloud radio access network (C-RAN) (Luo, Zhang, & Lim, 2014; Liu, Zhao, Zhou, Cheng, & Niu, 2014). In C-RAN, the centralized base baseband unit (BBU) pool is able to manage the dense deployment of remote radio heads (RRHs) with edge storage capabilities. The BBU pool can provide flexible control for information exchange and RRU-UE matching coordinated through centralized network.

Complexity

Algorithm 3 is guaranteed to terminate within a finite number of iterations due to the finite number of SBSs and each SBS proposes only once to any UE. The complexity of Algorithm 1 to establish preference lists for any SBS-UE pair follows the order of $\mathcal{O}\left(L_{loop}^{\max}, L_{dual}^{\max}\right)$. $L_{loop}^{\max}$ and $L_{dual}^{\max}$ represent the final iteration numbers to achieve convergence and solve dual problems, respectively. With respect to the complexity of Algorithm 2, sorting preferences for K SBSs and M UEs is known to have a complexity of $\mathcal{O}\left(KM\log\left(KM\right)\right)$. The algorithmic complexity of Algorithm 3 is in polynomial time $\mathcal{O}\left(KM\right)$ (O'Malley, 2007).

V. PERFORMANCE EVALUATION

In this section, the performance of the proposed algorithm is compared with the context-unaware maximum SINR based UE association algorithm (Goldsmith, 2005; Miao, Himayat, & Li, 2010). We choose simulation parameter values according to (Shanmugam et al., 2013; Pantiano et al., 2015; Pantiano et al., 2014a; Breslau et al., 1999; Pantisano et al., 2014b), which are given in Table 1. The cell radius is 500 m. The results are obtained by taking average of 500 simulations. In each simulation, K SBSs and M UEs are randomly distributed over the area as shown in Figure 2. Each SBS is allocated with $w_k = 2$ MHz bandwidth from the total 100 MHz bandwidth. Each UE requests $\mathcal{F}_m = 1.5 \times 10^3$

Figure 2. A snapshot of the locations of SBSs and UEs (K = M = 150, the cell radius is 500 m)

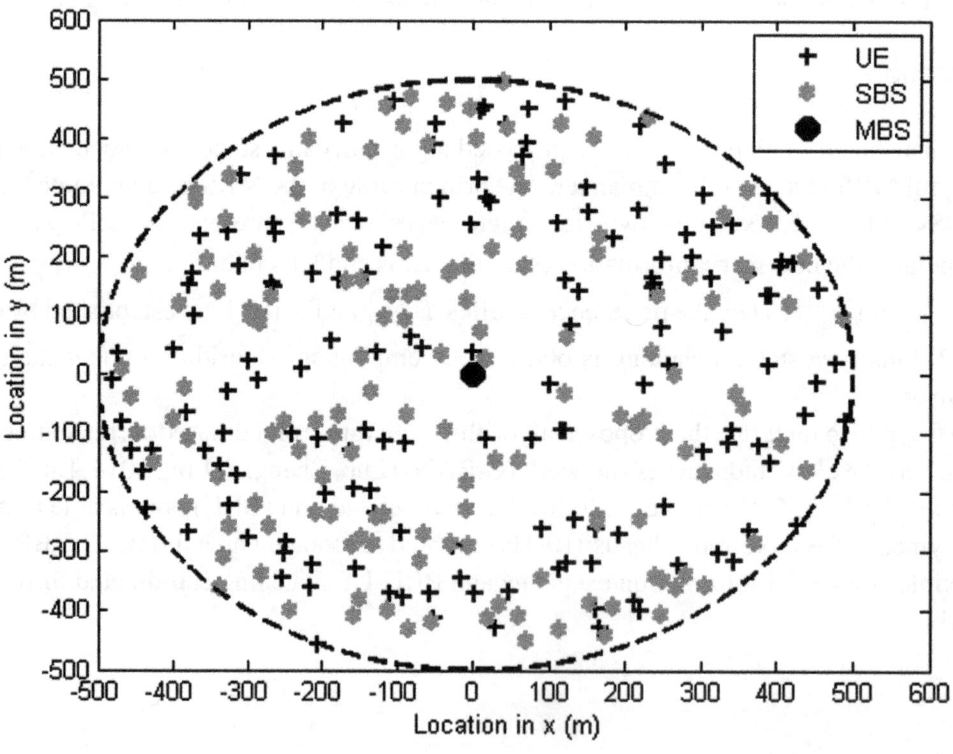

Table 1. Simulation parameter

Parameter	Value
Cell radius	500m
Pathloss exponent ±	4
Pathloss constant ϖ	10^{-2}
Shadowing $\zeta_{k,m}$ (standard deviation of a log-normal distribution)	8 dB
Multi-path fading $\beta_{k,m}$ (the mean of an exponential distribution)	1
Maximum transmission power p_k^{max}	33 dBm
SBS circuit power p_k^{cir}	41.6 dBm
UE constant circuit power p_m^{cir}	20 dBm
Noise power N_0	-114 dBm
Number of SBSs K	100~250
Number of UEs M	100~250
PA efficiency η	35%
Bandwidth per SBS w_k	2 MHz
Total number of files C	1.5×10^9
Number of files requested per UE $\mathcal{F}_m$	1.5×10^3
File size s	2 KB
Storage capacity per SBSs D_k	0~5 TB
Backhaul capacity per SBSs B_k	0.5~10 Mbps
Zipf parameter	0.4
QoS requirement U_k^{min}	$0.5 \times \mathcal{F}_m$

files from the total set of files $C = 1.5 \times 10^9$. The size of each file in *C* is assumed to be equal, i.e., 2 KB. The storage capacity and backhaul capacity for all the SBSs are chosen from the interval $D_k = [0, 5]$ TB and $B_k = [0.5, 10]$ Mbps, respectively. The duration of the training phase is supposed to be 600 seconds before the performance evaluation, where a set of popular files C_k is downloaded by s_k using backhaul link according to file popularity.

The performance of the proposed algorithm is evaluated with respects to EE performance and matching satisfaction. The matching satisfactions of SBSs are based on cumulative distribution functions (CDFs) of their matched UEs. For example, we assume $u_{m'}$ as the s_k's satisfaction threshold, which used as criteria to evaluate s_k's satisfaction. If s_k is matched with some UE that ranks higher than $u_{m'}$ in $P(s_k)$, s_k is satisfied with the current matching μ, i.e., $\mu(s_k) \succeq_{s_k} u_{m'}$. On the contrary, s_k is discontented with μ if it is matched to some

UE which ranks lower than $u_{m'}$ in $P(s_k)$, i.e., $u_{m'} \succ_{s_k} \mu(s_k)$. The CDF of the satisfaction is defined as $\Pr\{\mu(s_k) \succeq_{s_k} u_{m'}\}$, which describes the probability that $\mu(s_k)$ ranks higher or at least equal to $u_{m'}$ in $P(s_k)$.

Figure 3 and Figure 4 show the average EE performance versus the storage capacity D_k of SBS under two different backhaul capacities $B_k = 2, 4$ Mbps. There are $K = 100$ SBSs and $M = 100$ UEs in the network. Both of the two figures show that the proposed algorithm benefits significantly and obtains an exponential increase in the performance gain during a storage-capacity limited regime, i.e.,

Figure 3. The relationship between the average energy efficiency performance and the storage capacity per SBS (K = M = 100, $B_k = 4$ Mbps $D_k = 0 \sim 5$ TB)

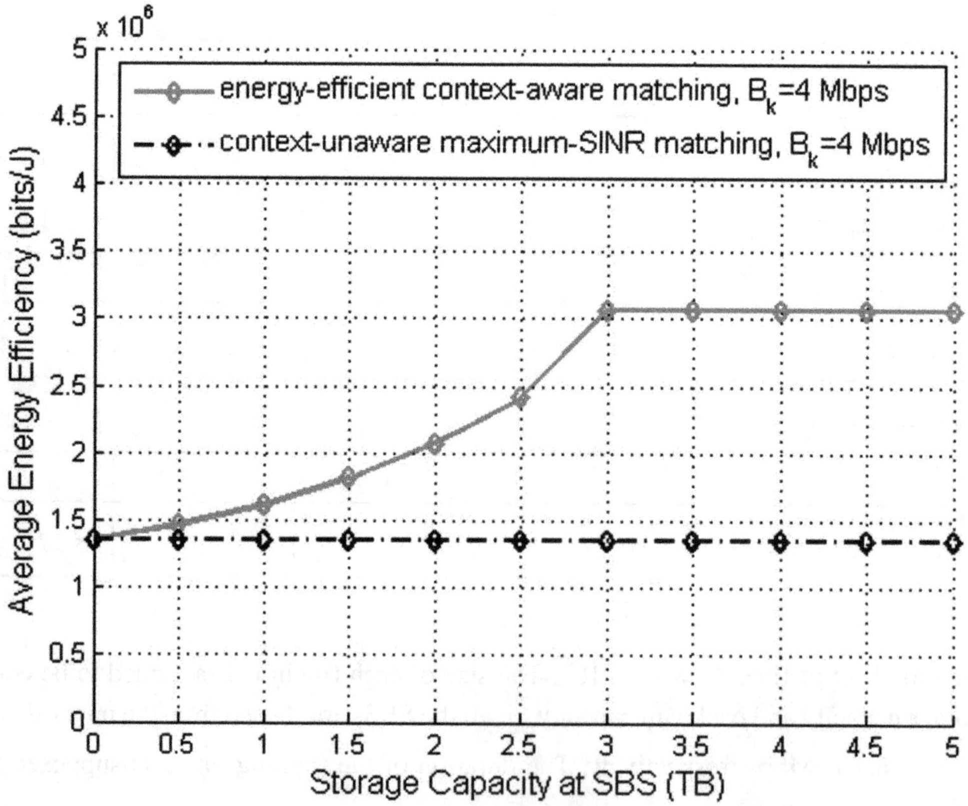

Figure 4. The relationship between the average energy efficiency performance and the storage capacity per SBS (K = M = 100, $B_k = 2$ *Mbps* $D_k = 0 \sim 5$ *TB)*

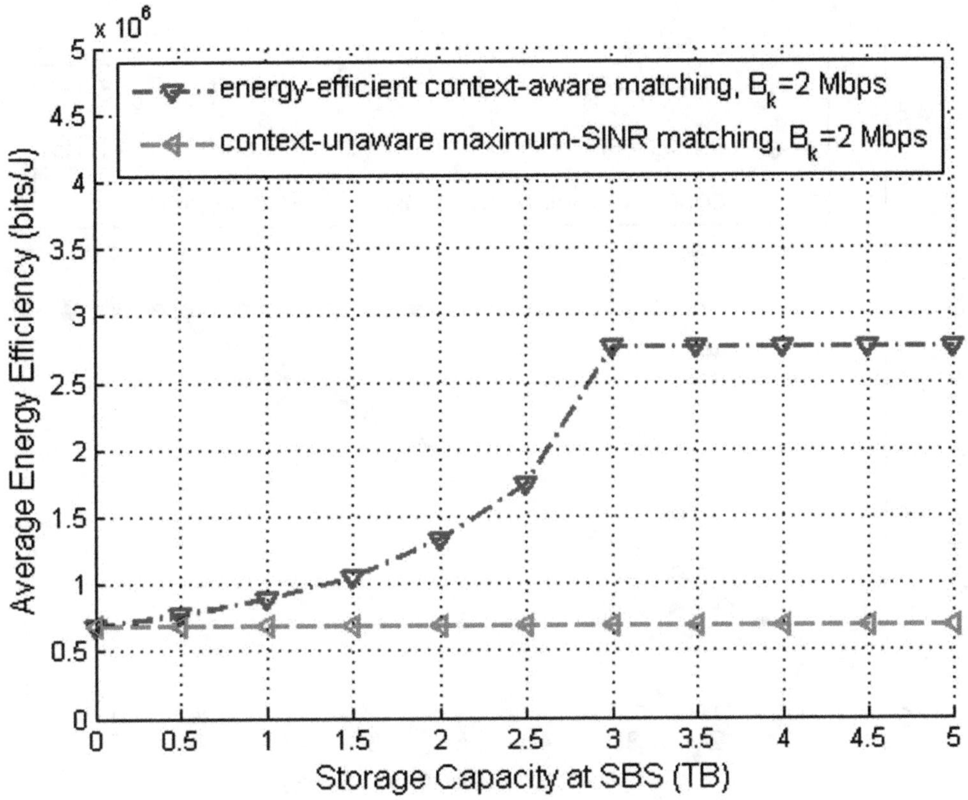

$D_k \leq 3$ TB. This is because that increasing the storage capacity leads to a higher probability of serving the requested files locally. Considering that the transmission rate for local cached files is much higher than the limited backhaul capacity, the required duration for transmitting the request files and the corresponding circuit energy consumption can be reduced dramatically. The maximum performance gains can be obtained by the proposed algorithm are 305% and 126% under $B_k = 2$ Mbps and $B_k = 4$ Mbps, respectively. The average EE performance can be substantially improved by the proposed algorithm which is able to exploit the availability of local content in particularly under the backhaul-capacity limited scenarios. When the storage capacity becomes large enough to cache all of the available files, i.e., $D_k > 3$ TB, the caching gain achieved by the proposed algorithm reaches saturation.

Figure 5 shows the average EE performance versus the backhaul capacity B_k of SBSs. We also assume that there are $K = 100$ SBSs and $M = 100$ UEs in the network. We can see that under D_k =1, 3 TB, the optimum EE performance gains can be obtained are as high as 133% and 928%, respectively. When B_k =0.5 Mbps, the performance gains are the maximum, which decrease monotonically with the increase of backhaul capacity. However, the performance gain due to caching becomes negligible when the backhaul capacity is large enough, because backhaul capacity is not a bottleneck for data transmission any more.

Figure 5. The relationship between the average energy efficiency performance and the backhaul capacity per SBS (K = M = 100, $B_k = 0.5 \sim 10$ *Mbps,* $D_k = 1,3$ *TB)*

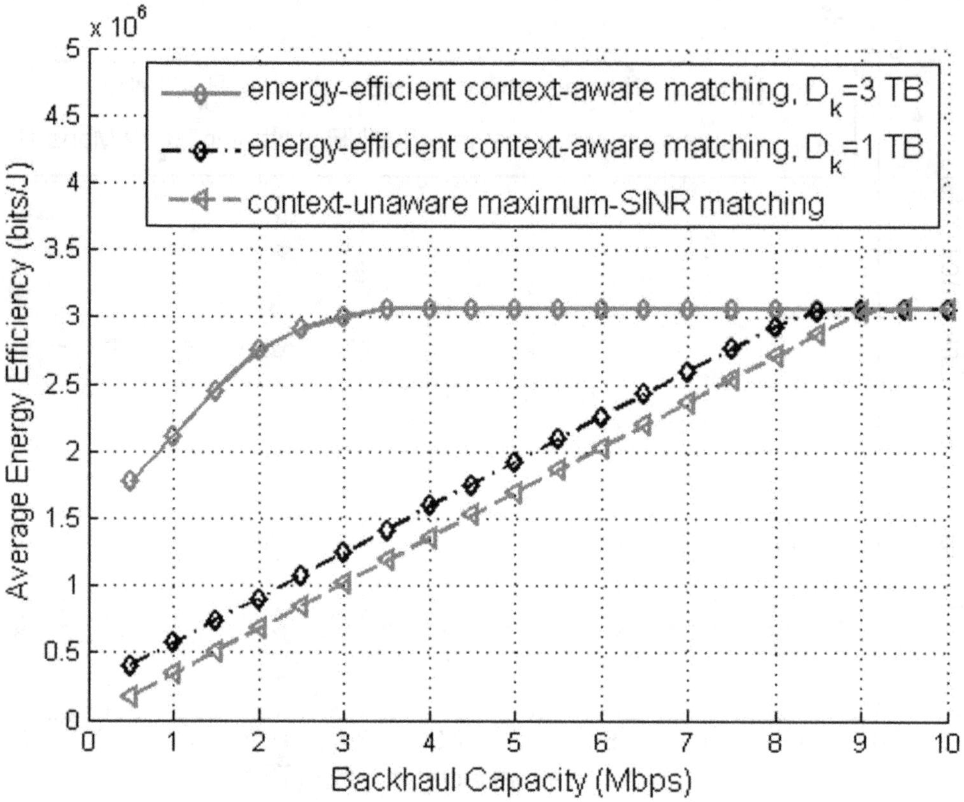

Figure 6 shows the SBSs' satisfactions versus satisfaction threshold. The proposed algorithm is compared with context-unaware random matching under $B_k = 2$ Mbps, $D_k = 2$ TB, $K = M = 150, 250$. The maximum SINR algorithm is not used for compassion because SBSs' preferences are completely ignored. Using a Monte-Carlo approach, we obtain the CDF in a statistical way by conducting simulation for 10^4 times. The probability to be matched with the top three choices for SBSs is as high as 46.12% for the proposed algorithm, while the corresponding probability for the context-unaware random matching algorithm is only 2.01%. As the numbers SBSs and UEs are increased from 150 to 250, the probabilities of being matched with their first three choices of SBSs are 43.5% and 1.17% for the proposed matching and random matching respectively. The satisfaction performance achieved by the proposed algorithm is an order of magnitude higher than the random approach. In conclusion, the proposed algorithm is capable of exploiting the spatial diversity provided by the ultra-dense deployment, and enhancing matching satisfaction for a wide range of thresholds.

VI. CONCLUSION

In this chapter, a thorough introduction and summary of the related works and current research progress in the field of edge-caching based resource allocation were provided. The existing literatures were

Figure 6. The relationship between the CDF of SBSs' satisfactions and the satisfaction threshold (N = K= 150, 250, $B_k = 2$ *Mbps* $D_k = 2$ *TB,* 10^4 *simulations)*

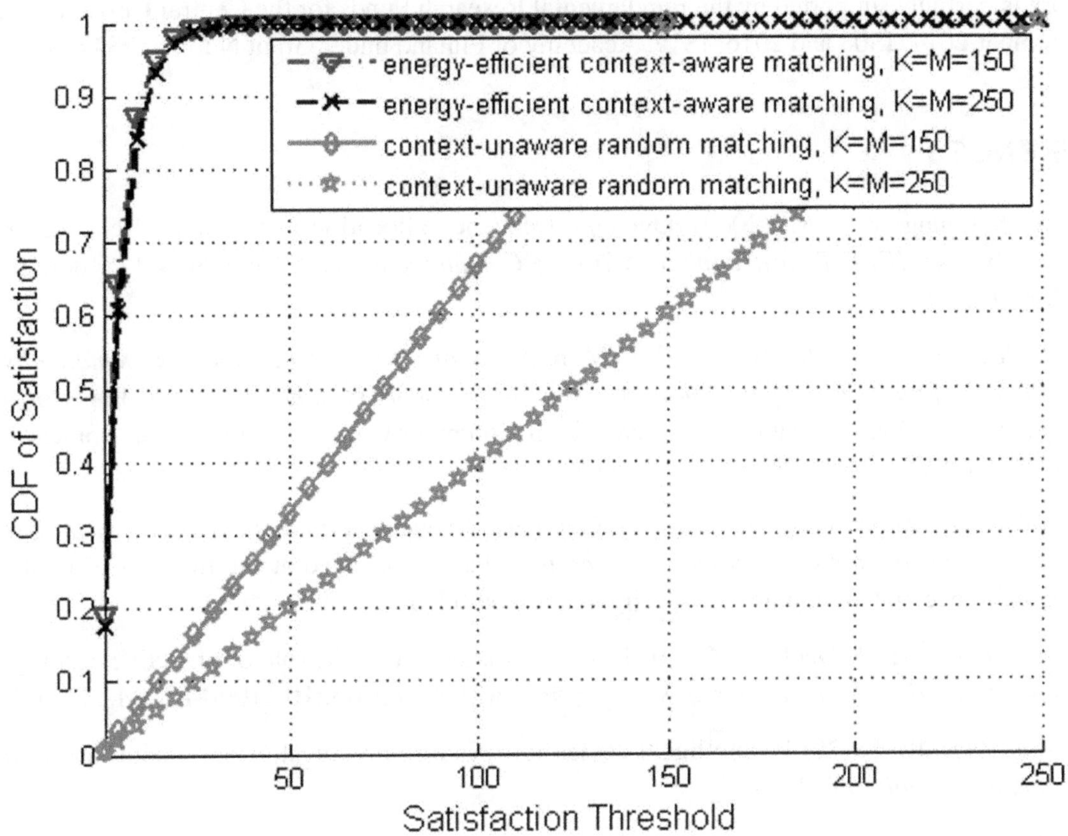

categorized and classified based on research motivations and application scenarios. Then we focused on the under-explored open issues of energy-efficient context-aware resource allocation design in ultra-dense small cells (SCs). We proposed an energy-efficient context-aware resource allocation algorithm for solving the two-sided matching problem. The NP-hard joint partner selection and power allocation problem was formulated as a one-to-one matching problem based on the preferences of SBSs and UEs. An iterative power allocation algorithm was proposed to optimize EE by combining nonlinear fractional programming and Lagrange dual decomposition. Inspired by the Gale-Shapley (GS) algorithm, an energy-efficient algorithm was proposed to match SBSs and UEs, which was proved to be stable for all the two sets and also weak Pareto optimal for SBSs. Additional preference deletion and tie-breaking rules were proposed to solve the matching problem with incomplete preference lists or ties. The properties of the proposed algorithm have been discussed and analyzed in detail. Finally, the proposed algorithm was compared with conventional context-unaware heuristic algorithms and the efficiency and superiority were demonstrated through simulation results. Future works on context-aware resource allocation can go into the directions of integrating service/traffic/user-specific characteristics, and take green collaborative frameworks into consideration and employ some joint performance metrics such as EE and transmission delay, etc.

ACKNOWLEDGMENT

This work is partially supported by the Fundamental Research Funds for the Central Universities under Grant Number 2014MS08 and 2016MS17, Academy of Finland under Grant Number 284748, 288473.

REFERENCES

Ali, S. H., & Leung, V. C. M. (2009). Dynamic frequency allocation in fractional frequency reused OFDMA networks. *IEEE Transactions on Wireless Communications*, *8*(8), 4286–4295. doi:10.1109/TWC.2009.081146

Bai, F., Sadagopan, N., & Helmy, A. (2003, March). *Important: A framework to systematically analyze the impact of mobility on performance of routing protocols in adhoc networks.* Paper presented at the meeting of IEEE International Conference on Computer Communications, San Francisco, CA. doi:10.1109/INFCOM.2003.1208920

Bao, Y., Wang, X., Zhou, S., & Niu, Z. (2012, October). *An energy-efficient client pre-caching scheme with wireless multicast for video-on-demand services.* Paper presented at the meeting of IEEE Asia-Pacific Conference on Communications, Jeju Island, South Korea.

Bastug, E., Bennis, M., & Debbah, M. (2014). Living on the edge: The role of proactive caching in 5G wireless networks. *IEEE Communications Magazine*, *52*(8), 82–89. doi:10.1109/MCOM.2014.6871674

Bello, O., & Zeadally, S. (2014). Intelligent device-to-device communications in the internet of things. *IEEE Systems Journal,* (99), 1–11.

Blasco, P., & Gunduz, D. (2014, June). *Learning-based optimization of cache content in a small cell base station.* Paper presented at the meeting of IEEE International Conference on Communications, Sydney, Australia. doi:10.1109/ICC.2014.6883600

Boccardi, F., & Braun, V. (2014). Scenarios for 5G mobile and wireless communications: The vision of the METIS project. *IEEE Communications Magazine*, *52*(5), 26–35. doi:10.1109/MCOM.2014.6815890

Boyd, S., & Vandenberghe, L. (2004). *Convex Optimization.* Cambridge, UK: Cambridge University Press. doi:10.1017/CBO9780511804441

Braun, L., Klein, A., Carle, G., Reiser, H., & Eisl, J. (2012, April). *Analyzing caching benefits for You Tube traffic in edge networks.* Paper presented at the meeting of IEEE Network Operations and Management Symposium, Maui, HI.

Breslau, L., Cao, P., Fan, L., Phillips, G., & Shenker, S. (1999, March). *Web caching and Zipf-like distributions: evidence and implications.* Paper presented at the meeting of IEEE International Conference on Computer Communications, New York, NY. doi:10.1109/INFCOM.1999.749260

Bu, S., Yu, F. R., Cai, Y., & Liu, X. (2012). When the smart grid meets energy-efficient communications: Green wireless cellular networks powered by the smart grid. *IEEE Transactions on Wireless Communications*, *11*(8), 3014–3024.

Chan, Y., Zhang, S., & Xu, S. (2011). Fundamental trade-offs on green wireless networks. *IEEE Communications Magazine*, *49*(6), 30–37. doi:10.1109/MCOM.2011.5783982

Chen, Z., & Kountouris, M. (2015, June). *Cache-enabled small cell networks with local user interest correlation*. Paper presented at the meeting of IEEE International Workshop on Signal Processing Advances in Wireless Communications, Stockholm, Sweden.

Dinkelbach, W. (1967). On nonlinear fractional programming. *Management Science*, *13*(7), 492–498. doi:10.1287/mnsc.13.7.492

El-Hajj, A. M., Dawy, Z., & Saad, W. (2012, June). *A stable matching game for joint uplink/downlink resource allocation in OFDMA wireless networks*. Paper presented at the meeting of IEEE International Conference on Communications, Ottawa, Canada.

ElBamby, M. S., Bennis, M., Saad, W., & Latvaaho, M. (2014, August). *Content-aware user clustering and caching in wireless small cell networks*. Paper presented at the meeting of IEEE International Symposium on Wireless Communications Systems, Barcelona, Spain. doi:10.1109/ISWCS.2014.6933489

Feng, X., Sun, G., Gan, X., Yang, F., Tian, X., Wang, X., & Guizani, M. (2014). Cooperative spectrum sharing in cognitive radio networks: A distributed matching approach. *IEEE Transactions on Communications*, *62*(8), 2651–2664. doi:10.1109/TCOMM.2014.2322352

Gale, D., & Shapley, L. S. (1962). College admissions and the stability of marriage. *The American Mathematical Monthly*, *69*(1), 9–15. doi:10.2307/2312726

Gao, G., Zhang, W., Wen, Y., Wang, Z., & Zhu, W. (2015). Cost-efficient video transcoding in media cloud by leveraging user viewing pattern. *IEEE Transactions on Multimedia,* (99), 1–10.

Gerami, M., Ming, X., & Skoglund, M. (2014). Partial repair for wireless caching networks with broadcast channels. *IEEE Wireless Communications Letters*, *4*(2), 145–148. doi:10.1109/LWC.2014.2386879

Goldsmith, A. (2005). *Wireless Communications*. Cambridge, UK: Cambridge University Press. doi:10.1017/CBO9780511841224

Golrezaei, N., Molisch, A. F., Dimakis, A. G., & Caire, G. (2013). Femtocaching and device-to-device collaboration: A new architecture for wireless video distribution. *IEEE Communications Magazine*, *51*(4), 142–149. doi:10.1109/MCOM.2013.6495773

Golrezaei, N., Molisch, M., Ji, A. F., Dimakis, A. G., & Caire, G. (2012. November). *Device-to-device communications for wireless video delivery*. Paper presented at the meeting of IEEE Asilomar Conference on Signals, Systems and Computers, Pacific Grove, CA.

Gu, Y., Zhang, Y., Pan, M., & Han, Z. (2014, December). *Cheating in matching of device to device pairs in cellular networks*. Paper presented at the meeting of IEEE Global Communications Conference, Austin, TX. doi:10.1109/GLOCOM.2014.7037583

Gu, Y., Zhang, Y., Pan, M., & Han, Z. (2015, March). *Student admission matching based content-cache allocation*. Paper presented at the meeting of IEEE Wireless Communications and Networking Conference, New Orleans, LA.

Guan, Y., Xiao, Y., Feng, H., Shen, C. C., & Cimini, L. J. (2014, December). *MobiCacher: mobility-aware content caching in small-cell networks*. Paper presented at the meeting of IEEE Global Communiocations Conference, Austin, TX. doi:10.1109/GLOCOM.2014.7037523

Hamidouche, K., Saad, W., & Debbah, M. (2014, May). *Many-to-many matching games for proactive social-caching in wireless small cell networks*. Paper presented at the meeting of IEEE International Symposium on Modeling and Optimization in Mobile, Ad Hoc, and Wireless Networks, Hammamet, Tunisia. doi:10.1109/WIOPT.2014.6850348

Hu, Z., Zheng, Z., Wang, T., & Song, L. (2015, April). *Small-cell caching mechanism for multi-service providers*. Paper presented at the meeting of IEEE Conference on Computer Communications Workshops, Hong Kong, China.

Hwang, I., Song, B., & Soliman, S. S. (2013). A holistic view on hyper-dense heterogeneous and small cell networks. *IEEE Communications Magazine*, *51*(6), 20–27. doi:10.1109/MCOM.2013.6525591

Kantarci, M. E. (2014, April). *Multicast-aware caching for small cell networks*. Paper presented at the meeting of IEEE Wireless Communications and Networking Conference, Istanbul, Turkey.

Kantarci, M. E. (2015, March). *Content caching in small cells with optimized uplink and caching power*. Paper presented at the meeting of IEEE Wireless Communications and Networking Conference, New Orleans, LA.

Kumar, A., & Saad, W. (2015, June). *On the tradeoff between energy harvesting and caching in wireless networks*. Paper presented at the meeting of IEEE Communicaitons Conference, London, UK. doi:10.1109/ICCW.2015.7247470

Kwon, H., & Birdsall, T. (1986). Channel capacity in bits per joule. *IEEE Journal of Oceanic Engineering*, *11*(1), 97–99. doi:10.1109/JOE.1986.1145138

Kwon, Y. M., Oh, S. M., Shin, J., & Chung, M. Y. (2015, November). *Radio resource management for moving small-cells with proactive content cache*. Paper presented at the meeting of IEEE Region 10 Conference, Macao, China.

Li, H., Wang, Z., & Hu, D. (2015, August). *Joint wireless and backhaul load balancing in cooperative caches enabled small-cell networks*. Paper presented at the meeting of IEEE Personal, Indoor and Mobile Radio Communications, Hong Kong, China. doi:10.1109/PIMRC.2015.7343607

Li, J., Chen, W., Xiao, M., Shu, F., & Liu, X. (2015). Efficient video pricing and caching in heterogeneous networks. *IEEE Transactions on Vehicular Technology,* (99), 1-8.

Li, Q., Niu, H., Papathanassiou, A., & Wu, G. (2014). 5G network capacity: Key elements and technologies. *IEEE Vehicular Technology Magazine*, *9*(1), 71–76. doi:10.1109/MVT.2013.2295070

Lim, S. H., Ko, Y. B., Jung, G. H., Kim, J., & Jang, M. W. (2014). Inter-chunk popularity-based edge-first caching in content-centric networking. *IEEE Communications Letters*, *18*(8), 1331–1334. doi:10.1109/LCOMM.2014.2329482

Liu, J., Zhao, T., Zhou, S., Cheng, Y., & Niu, Z. (2014). CONCERT: A cloud-based architecture for next-generation cellular systems. *IEEE Wireless Communications*, *21*(6), 14–22. doi:10.1109/MWC.2014.7000967

Luo, S., Zhang, R., & Lim, T. J. (2014). Downlink and uplink energy minimization through user association and beamforming in C-RAN. *IEEE Transactions on Wireless Communications*, *14*(1), 494–508. doi:10.1109/TWC.2014.2352619

Miao, G., Himayat, N., & Li, Y. (2010). Energy-efficient link adaptation in frequency-selective channels. *IEEE Transactions on Communications*, *58*(2), 545–554. doi:10.1109/TCOMM.2010.02.080587

Nikolaou, S., Renesse, R. V. & Schiper, N. (2015). Proactive cache placement on cooperative client caches for online social networks and applications. *IEEE Transactions on Parallel Distribution Systems,* (99), 1–10.

Niyato, D., Wang, P., Pink, T. H., Saad, W., & Kim, D. I. (2014). Cooperation in delay tolerant networks with wireless energy transfer: Performance analysis and optimization. *IEEE Transactions on Vehicular Technology,* (99), 3740-3754.

O'Malley, G. (2007). *Algorithmic aspects of stable matching problems* (Unpublished doctoral dissertation). University of Glasgow, Glasgow, UK.

Oh, E., Krishnamachari, B., Liu, X., & Niu, Z. (2011). Toward dynamic energy-efficient operation of cellular network infrastructure. *IEEE Communications Magazine*, *49*(6), 56–61. doi:10.1109/MCOM.2011.5783985

Pantiano, F., Bennis, M., Saad, W., & Debbah, M. (2015, June). *Match to cache: joint user association and backhaul allocation in cache-aware small cell networks*. Paper presented at the meeting of IEEE International Conference on Communications, London, UK.

Pantisano, F., Bennis, M., Saad, W., & Debbah, M. (2014a, May). *Cache-aware user association in backhaul-constrained small cell networks*. Paper presented at the meeting of International Symposium on Modeling and Optimization in Mobile, Ad Hoc, and Wireless Networks, Hammamet, Tunisia. doi:10.1109/WIOPT.2014.6850276

Pantisano, F., Bennis, M., Saad, W., & Debbah, M. (2014b, November). *In-network caching and content placement in cooperative small cell networks*. Paper presented at the meeting of International Conference on 5G Ubiquitous Connectivity, Levi, Finland. doi:10.4108/icst.5gu.2014.258230

Perabathini, B., Bastug, E., Kountouris, M., Debbah, M., & Conte, A. (2015, June). *Caching at the edge: a green perspective for 5G networks*. Paper presented at the meeting of IEEE International Conference on Communications, London, UK. doi:10.1109/ICCW.2015.7247608

Pingyod, A., & Somchit, Y. (2015, May). *Content updating method in femtocaching*. Paper presented at the meeting of IEEE International Joint Conference Computer Science and Software Engineering, Chonburi, Thailand.

Poularakis, K., Iosifidis, G., Sourlas, V., & Tassiulas, L. (2014). Approximation algorithms for mobile data caching in small cell networks. *IEEE Transactions on Communications*, *62*(10), 3665–3677. doi:10.1109/TCOMM.2014.2351796

Poularakis, K., Iosifidis, G., Sourlas, V., & Tassiulas, L. (2016). Exploiting caching and multicast for 5G wireless networks. *IEEE Transactions on Wireless Communications,* (99), 1–13.

Roth, A. E., & Sotomayor, M. (1991). *Two sided matching: A study in game-theoretic modeling and analysis*. Cambridge, UK: Cambridge University Press.

Rusek, F., Persson, D., Lau, B. K., & Larsson, E. G. et al.. (2013). Scaling up MIMO: Opportunities and challenges with very large arrays. *IEEE Signal Processing Magazine*, *30*(1), 40–60. doi:10.1109/MSP.2011.2178495

Shanmugam, K., Golrezaei, N., Dimakis, A. G., Molisch, A. F., & Caire, G. (2013). Femtocaching: Wireless content delivery through distributed caching helpers. *IEEE Transactions on Information Theory*, *59*(12), 8402–8413. doi:10.1109/TIT.2013.2281606

Shim, J. N., Min, B. Y., Kim, K., Jang, J., & Kim, D. K. (2014, May). *Advanced femto-caching file placement technique for overlapped helper coverage*. Paper presented at the meeting of IEEE Vehicular Technology Conference, Seoul, South Korea.

Song, J., Song, H., & Choi, W. (2015, June). *Optimal caching placement of caching system with helpers*. Paper presented at the meeting of IEEE International Conference on Communications, London, UK. doi:10.1109/ICC.2015.7248590

Su, Z., & Xu, Q. (2015). Content distribution over content centric mobile social networks in 5G. *IEEE Communications Magazine*, *53*(6), 66–72. doi:10.1109/MCOM.2015.7120047

Wang, K., Chen, Z., & Liu, H. (2014). Push-based wireless converged networks for massive multimedia content delivery. *IEEE Transactions on Wireless Communications*, *13*(5), 2894–2905. doi:10.1109/TWC.2014.040914.131422

Wang, T., Song, L., & Han, Z. (2015, March). *Dynamic femtocaching for mobile users*. Paper presented at the meeting of IEEE Wireless Communications and Networking Conference, New Orleans, LA.

Xylomenos, G., Vasilakos, X., Tsilopoulos, C., Siris, V. A., & Polyzos, G. C. (2012). Caching and mobility support in a publish-subscribe internet architecture. *IEEE Communications Magazine*, *50*(7), 52–58. doi:10.1109/MCOM.2012.6231279

Zhang, F., Xu, C., & Zhang, Y. (2015, June). *EdgeBuffer: caching and prefetching content at the edge in the mobility first future Internet architecture*. Paper presented at the meeting of IEEE Symposium on a World of Wireless, Mobile and Multimedia Networks, Boston, MA.

Zhou, S., Gong, J., Zhou, Z., Chen, W., & Niu, Z. (2015). GreenDelivery: Proactive content caching and push with energy harvesting-based small cells. *IEEE Communications Magazine*, *53*(4), 142–149. doi:10.1109/MCOM.2015.7081087

Zhou, Z., Dong, M., Ota, K., & Chang, Z. (2015). Energy-efficient context-aware matching for resource allocation in ultra-dense small cells. *IEEE Access*, *3*, 1849–1860. doi:10.1109/ACCESS.2015.2478863

Chapter 3
Challenges in Energy-Efficient Communications as Enablers for Green Solutions on the 5G Heterogeneous Networks

Irma Uriarte-Ramírez
Universidad Autónoma de Baja California, Mexico

Norma A. Barboza-Tello
Universidad Autónoma de Baja California, Mexico

Paul Medina-Castro
Universidad Autónoma de Baja California, Mexico

ABSTRACT

One of the most representative challenges of multi-tier heterogeneous network architectures is the interference management among different Radio Access Technologies. In this sense, in order to achieve the potential increase on network capacity forecasted for 5G, the issue of interference must be addressed. Intelligent energy management, as proposed in the growing area of Green Communications could serve as an appealing solution to the interference issue. However developing an optimal energy management plan requires the understanding of the characteristics of 5G cellular networks that could be exploited to improve energy and resources efficiency. In this work we present a study about challenges in energy-efficient communications and technologies as enablers for green solutions and how this challenges can be extended to meet those of the 5G heterogeneous networks, in order to identify possible solutions to address the energy efficiency and interference mitigation issues.

INTRODUCTION

Nowadays, there is an interest on technology development in the growing area of green communications for reducing energy consumption in wireless communications. Some works reported in literature claim that there will be an increase of CO_2 equivalent emissions by a factor of three until 2020 compared to 2007 (Uriarte-Ramírez & Barboza-Tello, 2014). Societal development will lead to changes in the way

DOI: 10.4018/978-1-5225-1712-2.ch003

mobile and wireless communication systems are used. This has accelerated the implementation of 5G networks that can provide exponentially more capacity, lower latency, ubiquitous connectivity, as well as increased reliability and availability. Those 5G future services such as e-banking, e-learning, and e-health will continue to proliferate. On-demand information and entertainment (e.g. in the form of augmented reality) will progressively be delivered over mobile and wireless communication systems. These developments will lead to an avalanche of mobile and wireless traffic volume predicted to increase a thousand-fold over the next decade (Osseiran, et al., 2014). Thus, those multimedia application services must cope with high data rate demands. However, higher data rates necessitate more energy per bit for a given bit error rate (BER), which in turn increases the overall energy consumption of the system and the production of CO_2 emissions that are threat to global warming. Thus, researching energy-efficient designs for high data rate wireless communication systems is a burning issue. Due to escalating expansion of wireless network infrastructures and exponential growth in traffic rate, a considerable amount of worldwide energy is consumed by Information and Communications Technology (ICT) of which more than 70% is being used by radio access part/radio frequency (Mohammad S. & Alagan Anpalagan, 2013) . Climate change and the need to reduce global energy usage are now accepted. Today 4 billion of people on the earth have access to a mobile phone; effective access to information and communications is a proven stimulus to economic development. Recent studies have shown the potential for IT (Information Technology) to reduce energy consumption in other industries. As a key enabler of green solutions on environmental an economic grounds, more energy efficient radio access networks are required. When it is about the power, the consumption of a mobile handset per subscriber is lower than the base station. However manufacturing energy is a larger component in the mobile handset that in the base station. It is mainly due to the fact that lifetime of a base station is longer (typically 10-15 years) compared to a handset (approximately 2 years) (Pollin S., et al., 2008).

Green Communication technologies currently receive a lot of attention. Wireless systems have been optimized in terms of spectrum efficiency and transmission reliability. However, most of the recent efforts have ignored the importance of wireless networks´ environmental responsibility, e.g. energy efficiency and environmental impact. Information and Communications Technology (ICT) already represents around 2% of total of carbon emissions, and this is expected to increase from 0.53 billion tonnes (Gt) carbon dioxide equivalent (CO_2e) in 2002 to 1.43 Gt CO_2e in 2020. Optimizing the energy efficiency of wireless communication not only reduces environmental impact, but also cuts overall network costs and help make communication more practical and affordable in a pervasive setting. In cellular networking, base station consumes a maximum portion of the total energy used in the system, more than 50% of total energy is consumed by the radio access part, where 50-80% is used for the power amplifier (PA). (Feng, Jiang, Lim, Cimini Jr., & Feng, 2013). There are currently more than 4 millions base stations serving mobile users, each consuming an average of 25 MWh per year. To address this problem, Green Networking has focused bringing architectural changes in base station and network planning. On this last point (network planning) Heterogeneous Networks (HetNets) have emerged as a possible solution for energy efficiency at the base station. In this sense in work we describe 5G Future Network, Green Comm Projects and stand out HetNets as green solutions. The chapter begins with a brief description on 5G Challenges and potential green solutions. Later the Green Comm Projects are described. The following section gives an overview in HetNets and describes how it could be an energy efficient deployment strategy. Next, we introduce challenges in Het Nets and possible solution. Future work and conclusions are given at the end.

5G CHALLENGES AND POTENCIAL GREEN SOLLUTIONS

A hot topic in wireless communications is the Fifth Generation (5G) system, this concept is on track since 2013. In 5G concept, networks became more sophisticated, deployment became denser, and the anytime anywhere goal had almost been achieved (Wang & Rangapillai, 2012). Then, 5G cellular network must address the following challenges (Osseiran, et al., 2014):

- Higher capacity (1000 times increased data volume per area),
- Higher data rate (10 to 100 times increased typical user data rate),
- Lower end to end latency (5 times reduced Ent-to-Ent (E2E) latency),
- Massive device connectivity (10 to 100 times increased number of connected devices),
- Reduced cost (Sustainable),
- Quality of Experience provisioning (Consistent).

Figure 1 shows the pyramidal relationship among these challenges.

In 5G it is expected that mobile and wireless traffic volume increase a thousand-fold over the next decade which will be driven by the expected 50 billion devices connected to the cloud by 2020 an all of them will need to access and share data, anywhere and anytime. With the rapid increase in the numbers of connected devices, some challenges will appear which should be responded by increasing capacity and by improving energy efficiency, cost and spectrum utilization as well providing.

The 5G cellular architecture is heterogeneous, it includes macrocells, microcells, small cells, and relays. This architecture proposes the interconnectivity among the different emerging technologies like Massive MIMO network, Cognitive Radio network, mobile and static small-cell networks. This proposal also explains the role of network function virtualization (NFV) cloud in the 5G cellular network architecture. The concept of Device to Device (D2D) communication, small cell access points and Internet of things (IoT) has also been incorporated in this architecture. The mobile small cell concept is an integral

Figure 1. Challenges in 5G networks
Mastrangelo, 2015.

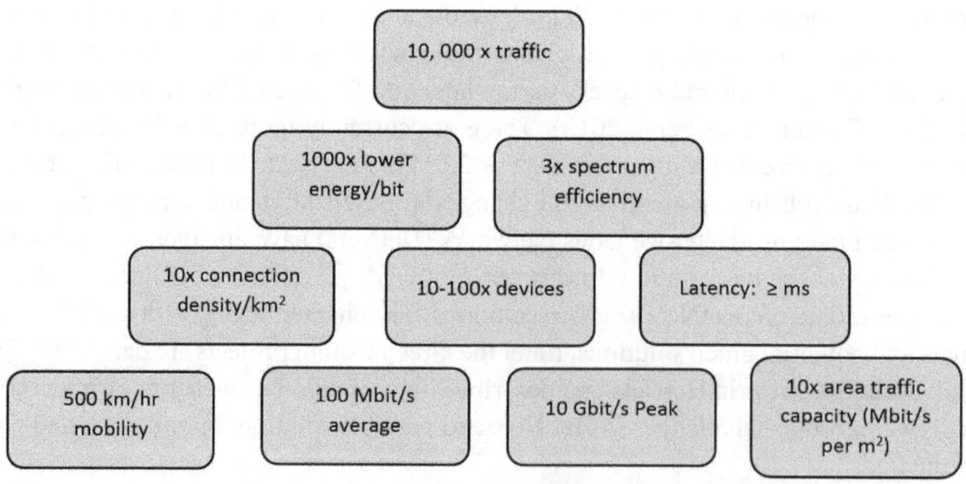

part of 5G wireless cellular network and partially comprises of mobile relay and small cell concepts. The 5G architecture consist in only two logical layers: a radio network and a network cloud.

In addition to these challenges, there is another one: to meet the future requirements and expectation in an affordable and sustainable way. Low energy consumption is the key to achieve this goal. It is an important design target for mobile communication system in the future (Olson, et al., 2013).

Some of the most important challenges in 5G energy efficiency from the green design perspective will be discussed in the following subsections.

Challenges in Energy Efficiency

Some of the major challenges and requirements that 5G mobile networks need to be able to handle in the future with respect to energy efficiency are:

1. **Data Traffic Volume:** Future predictions say that data traffic volumes will experience and exponential growth in the coming years (around 10 times between 2012 and 2018). Also, these predict that per-user data rates are expected to grow by a factor up to 50-100 and the density of mobile Internet users is expected to increase by a factor of up to 10, implying a factor 1000x capacity demand in the 2020 time frame. Therefore, mobile systems in the future need to be capable of delivering significantly more capacity than today. This trend is important also from a green design perspective, since the mobile network evolution should be performed by avoiding the risk of an unreasonable over-provisioning of the network that in this scenario will become more unsustainable in term of cost (Olson, et al., 2013).
2. **Number of Connected Devices:** In the future there are going to be connected to different kinds of machines. In addition to mobile phones, laptop computers or tablets, machines as smart grid devices, sensors, and surveillance cameras will be connected to the networks. This is usually referred to as Internet-of-things or machine-to-machine (M2M) communication. This kind of evolution will also introduce new characteristics to the traffic in the networks. That is why the need of an efficient signaling management will be crucial (Olson, et al., 2013).
3. **Diverse Requirements:** In the evolutionary scenario, 5G will include a lot of new applications. Some of them may require low latency, very high reliability, etc. This challenge need to be taken care of in the system design, as this set up new and varying quality-of-service (QoS) requirements. This may have a significant impact on green design, as the minimization of network power consumption should not have impacts on the correct and efficient management of the QoS in the system (Olson, et al., 2013).
4. **Energy Consumption:** Low energy consumption is very important, it should at least be kept at the same level as today, despite the traffic growth, the massive amount of devices, and new requirements, but that is possible.

Potential Green Solutions

System Architecture (Energy Savings in Base Station)

The system architecture constitutes a fundamental limit on how low energy consumption is achieved. An energy efficient system needs to be efficient both when transmitting data as well as when it is not transmitting data. The system architecture particularly affects the latter because the devices (explicitly or implicitly) exchanges a certain amount of mandatory idle mode messages with network functionalities. In current cellular systems these by far dominate energy consumption, and hence this is an important area to address.

Nowadays, the green researchers are focusing in base station, because there are millions of BSs around the world that increase greenhouse gases and pollution, in addition the higher energy cost to operate them. Therefore, BS equipment manufacturers have begun to offer a number of eco and cost friendly solution to reduce power demands of BS. There are various methods dealing with improved transmitter efficiency, system features, fresh air-cooling, renewable energy sources and energy saving during low traffic (Hasan, Boostanimehr, & Bhargava, 2011).

System Design

The research on technologies such as cognitive radio and cooperative relaying has received a significant attention. While cognitive radio is an intelligent and adaptive wireless communication system that enables us to utilize the radio spectrum in more efficient manner, cooperative relays can provide a lot of improvement in throughput and coverage for futuristic networks. Therefore developments in both these technologies also enable us to solve the problem of energy efficiency via smart radio transmission and distributed signal processing (Olson, et al., 2013).

Network Deployment

The exponential growth in demand for higher data rates and other services in wireless networks requires a more dense deployment of base stations within cellular networks. Energy efficiency improvements through network deployment strategies have been touched upon in several projects for state-of art technologies. Examples of these deployment strategies are different topologies of cells, distributed antenna system and base station cooperation. Especially heterogeneous network deployment.

Radio Transmission

In order to be energy efficient when transmitting MIMO and OFDMA are proposed like schemes to achieve that. MIMO can provide diversity gain and multiplexing gain, it employs adaptive modulation and coding, and also precoded multiple-input multiple-output transmission which allow both multi-layer transmission for increased data rates, and beamforming to certain extend. Meanwhile, OFDMA will be the dominant multiple access scheme for the next generation wireless networks, since utility-based energy-efficient design can significantly improve energy efficiency (Feng, Jiang, Lim, Cimini Jr., & Feng, 2013).

The summary of potential green solutions in 5G networks are shown in Table 1.

GREEN COMMUNICATIONS PROJECTS

Telecommunication equipment typically contains a considerable amount of scares materials and heavy metals. Both the extraction of these materials, typically through mining, and the treatment of the waste represent a large environmental challenge. Next to the concern about material use in ICT there is closely

Table 1. Opportunity green areas in 5G networks

Opportunity Areas	Potential Green Solution
System architecture	• Energy saving in base stations • Minimizing BS energy consumption • Using renewables energy resources • Energy-aware cooperative BS
System design	• Enabling technologies • Green comm Cognitive Radio • Green comm Cooperative Relay
Network deployment	Heterogeneous networks
Radio transmission	• MIMO • OFDMA

Hasan, Boostanimehr, & Bhargava, 2011; Olson, et al., 2013.

related to carbon emissions. (Williams, et al., 2008). In addition to, with the rapid evolution of ICT the corresponding energy consumption is also growing at a staggering rate. Mobile networks represent about 0.2% of the total of carbon emission. In addition to the environmental problem, energy cost also represent a significant portion of networks operators overall expenditures (OPEX). (Hasan, Boostanimehr, & Bhargava, 2011) . Therefore, from the operators perspective, energy efficiency (EE) not only has great ecological benefits and represents social responsibility in fighting climate change, but also has significant economic benefits. (Feng, Jiang, Lim, Cimini Jr., & Feng, 2013).

Reducing energy cost and environment effects has become a trend that prompted researchers to explore a new area of research "Energy Efficient or Green Cellular Networking". The purpose of this new area is to reduce energy consumption, lower electromagnetic radiation, improve resource utilization and make resource consumption and environmental impact to a minimum (Wie, Song, Liu, Wang, & Li, 2011). Therefore Green Communication can be defined as striving to reduce energy cost while still maintaining Quality of Service (QoS) in terms of coverage needs, capacity and user needs (He, et al., 2011) . Topics of green wireless networks include green cellular base station, energy-efficient mobile terminals, hierarchical and distributed techniques for energy distribution and management, green ad hoc and sensor network, green cognitive radios, electromagnetic pollution mitigation, energy-efficient signal processing techniques for wireless links, resource efficient cross-layer optimization methods for wireless networking, opportunistic spectrum sharing without causing harmful interference pollution (Wu, 2012). This trend has stimulated the interest of researchers in an innovative new research. In this regard, "Green Radio" is a vast research discipline that needs to cover all the layer of the protocol stack and various system architectures and it is important to identify the fundamental trade-offs linked to energy efficiency and the overall performance.

First efforts to reduce power consumption in wireless networks were called Green Radio (GR) project. This trend has simulated the interest of researchers in an innovative new research. In this regard, "Green Radio" is a vast research discipline that needs to cover all layers of the protocol stack and various system architectures, then it is important to identify the fundamental trade-offs related to energy efficiency and the overall performance (Hasan, Boostanimehr, & Bhargava, 2011).

The term "green" come from the efforts to reducing greenhouse gases (CO_2) emission from industries. Another motivation, especially for operators, is to gain commercial benefits, mainly by reducing operating expense related to energy cost. Efforts in previous researches, from different groups can be broadly classified into the following categories (Wu, Zhang, Zukerman, & Yung, 2015):

1. Improving energy efficiency of hardware components.
2. Turning off components selectively.
3. Optimizing energy efficiency of the radio transmission process.
4. Planning and deploying heterogeneous cells.
5. Adopting renewable energy resource.

Different efforts coming from international research groups dedicated to energy-efficient wireless communications, are being carried out like EARTH (Energy Aware Radio and NeTwork Technologies), TREND (Towards Real Energy-Efficient Network Design), C2POWER (Cognitive Radio and Cooperative Strategies for POWER saving in multi-standard wireless devices), OPERA-Net (Optimizing Power Efficiency in mobile Radio Networks), and eWIN (Energy –Efficient wireless Networking). See Table 2.

Due to the increase of projects related to green communications and hence the increase in the number of diverse energy efficiency metrics, standards developing organizations, such as European Technical

Table 2. Solutions for energy efficient networks

Project	Solutions
Green Radio	• Maximize equipment and base station-reuse. • Large vs. Small cell deployment. • Overlay source • SISO vs. MIMO with packet scheduling • Applying dynamic spectrum Access.
EARTH	• Optimization of cell size • Heterogeneous network deployment • Relay and cooperative communication • Cooperative scheduling, interference coordination. • Multi-RAT • MIMO, OFDM
Green Touch	• Green Meter Assessment for mobile Access networks. • Large Scale Antenna Systems. • Enhanced Wireless receiver to mitigate poor channel quality and interference. • Interference alignment in cellular networks.
TREND	• Assessment of power consumption in ICT. • Energy-efficiency in Access and home networks. • Instrumenting the network for power saving.
C2POWER	Cognitive radio and cooperative strategies.
OPERA NET	• Energy saving in base station. • Scalable MIMO-detection. • Energy aware device. • Mobile radio Access network's end-to-end efficiency.
eWIN	• Architectural designs for low-energy Wireless Access. • Dynamic and flexible spectrum management techniques. • Radio resource management for cooperative and competitive heterogeneous environments.
5GREEN	• Green deployment strategies (HetNets). • Massive MIMO as a green solution.

Feng, Jiang, Lim, Cimini Jr., & Feng, 2013.

Standard Institute (ETSI) and Alliance for Telecommunications Industry Solutions (ATIS), are currently making efforts to define energy efficiency metrics for wireless networks. These metrics can be classified into three categories: facility-level, equipment-level and network-level metrics (Hasan, Boostanimehr, & Bhargava, 2011).

The next-generation wireless networks are expected to provide high-speed Internet access anywhere and anytime. The exponentially growing data traffic and the requirement for ubiquitous access have triggered a dramatic expansion of network infrastructures and fast escalation of energy demands. The escalation of energy consumption in wireless networks directly results in increased greenhouse gas emission, which has been recognized as a major threat to environmental protection and sustainable development.

The deployment of cellular networks is typically optimized to provide ubiquitous access and service to mobile user. Thus, significant number of base stations is primarily used to provide coverage, meaning that they usually do not operate at full load, even at peak traffic hours. During off-peak hours all base stations operate at lower load, or might not even serve any user at all. That leads to conclude that the energy efficiency of base stations is particularly poor in this condition, which can be attributed to several components.

The specific objective of the Green Projects is to find out innovative methods for the reduction of the total energy needed to operate the radio access networks and to identify appropriate radio architectures that enable such power reduction.

Green Radio Project

The first efforts to reduce power consumption in wireless networks were called Green Radio (GR) project. This project defines two different perspectives: to examine alternative to the existing cellular network and to study novel techniques that can be used in base station (Pollin S., et al., 2008). Since the term Green Radio was first coined by Mobile VCE's Visions Group in early 2007, it has become widely used in the industry. The goal of this research is to secure 100x reduction in energy requirements for delivery of high data rate services, thereby reducing CO_2 emissions and OPEX (operating expenditure) costs while enabling new services. It is investigating both architectural and individual technological approaches.

GR research is a large and comprehensive area that covers all layers and protocol stack of wireless access networks as well as the architectures and techniques. Four fundamentals tradeoffs in green wireless network design and are identified in order to improve energy efficiency while guaranteeing satisfactory quality-of-service/cost (Blume, Zeller, & Barth, 2010):

- **Deployment Efficiency (DE)-Energy Efficiency (EE) Trade-Off:** To balance the deployment cost, throughput and energy consumption in a network as a whole.
- **Spectrum Efficiency (SE)-EE Trade-Off:** Given bandwidth available, to balance the achievable rate and energy consumption of the system.
- **Bandwidth (BW)-Power (PW) Trade-Off:** Given a target transmission rate, to balance the bandwidth utilized and the power needed for transmission.
- **Delay (DL)-PW Trade-Off:** To balance the average end-to-end service delay and average power consumed in transmission.

Finally, a typical mobile phone network in the UK may consume approximately 40MW, excluding the power consumed by the users' handsets. From an operator perspective, reduced energy consumption translates directly to lower operating expenditure, OPEX. The need for reduced carbon emissions and OPEX costs for wireless networks are the two key drivers of Mobile VCE's Green Radio program (Green Radio, 2016)

EARTH Project

In January 2010 a consortium started EARTH project (Energy Aware Radio and Network Technologies). This is a highly ambitious and unique FP7 IP project, investigating the energy efficiency of mobile communication systems. It is applying an integrated approach and is committed to the development of a new generation of energy efficient equipment, comprising components, deployment strategies and network management solutions (EARTH, 2014)

The EARTH is focusing on mobile broadband networks with future key technologies LTE (3GPP Long Term Evolution) and LTE-Advanced. Its objective is to provide tangible results such as enhanced network architectures and deployment strategies (Pejanovic-Djurisic, Stovrag, & Ilic-Delibasic, Fundamental optimization criteria for Green wireless communications, 2012).

EARTH targeted for improvements beyond current development trends. The technologies developed by EARTH enable wireless communications systems with unprecedented energy efficiency, significant reduction in environment pollution and operating cost. Thereby the economical barrier to offer mobile broadband coverage will be substantially reduced including countryside areas which are not yet reached by mobile broadband services. The tangible results of the EARTH project are:

1. Energy efficient deployment strategies,
2. Energy efficient network architectures,
3. New network management mechanisms, adaptive to load variations with time,
4. Innovative component designs with energy efficient adaptive operating points,
5. New radio and network resource management protocols for multi-cell cooperative networking.

The new techniques and their integration into Integrated Solutions have been validated using sophisticated simulation tools and in a mobile network test plant. EARTH created valuable IPR and provided timely contributions to standardization and regulations processes (EARTH, 2014).

Green Touch

In response to research needs and opportunities for sustainable network growth, Bell Labs, in collaboration with several other partners, founded Green Touch a global consortium whose goal is to demonstrate by 2015 key components, technologies, and architectures that will make the network a thousand-fold (100x) more energy efficient from current levels. To achieve this ambitious goal, Green Touch brings together leading experts from the ICT industry to engage in collaborative, pre-competitive research that is focused on increasing the energy efficiency of communications network and significantly reducing the carbon footprint of the components on which they are based.

Starting from principles of information theory and computer sciences Green Touch has analyzed the fundamental limit of global physical limits would allow designing a system that is several orders of magnitude more efficient than today´s systems. The aim is to completely rethink communication systems, to analyze theoretical potentials, to tackle fundamental research in information theory and electronic devices. Within five years the consortium wants to demonstrate proof-of-concept to build confidence in ICT industry in such completely new system designs (GreenTouch, 2016).

TREND

TREND is a project coordinated by *Politecnico di Torino*. TREND aims at integrating the activities of major European actors in networking, including manufacturers, operator, research centers, to quantitatively asses the energy demand of current and future telecom infrastructures, as well as to design energy efficient, scalable, and sustainable networks. This project wants to laying dawn the bases for a new holistic approach to energy-efficient networking, investigating effective strategies and mechanism to reduce energy consumption in current and future networks in general, and the future internet in particular. The principal goal of TREND is to establish the integration in the European Union research community in green networking with a long term perspective to consolidate the European leadership in the field (Trend, 2016).

C2POWER

The main objective of this project is to research, develop and demonstrate energy saving technologies for multi-standard wireless mobile devices, exploiting the combination of cognitive radio and cooperative strategies while still enabling the required performance in terms of data rate and QoS to support active applications.

The promise of a truly mobile experience is to have the freedom to roam around anywhere and not be bound to a single location. However, the energy required to keep mobile devices connected to the network over extended periods of time quickly dissipates. In fact, the operational time, has been identified as the number one criteria by the majority of the consumers purchasing a mobile device. Moreover, concern about exhausting battery lifetime is also one of the main reasons why users do not opt to use advanced multimedia services on their mobiles more frequently.

To avoid the foreseen 4G "energy trap" C2POWER project will investigate, develop and demonstrate how cognition and cooperative strategies can be extended to decrease the overall energy consumption of mobile devices while still enabling the required performance in terms of QoS (C2Power, 2016).

In particular, C2POWER will investigate two complementary techniques to increase power efficiency at the wireless interface of handsets:

- Cooperative power saving strategies between neighboring nodes using low power short range communications.
- Cognitive handover mechanisms to select the Radio Access Technology which has the lowest energy demand in heterogeneous environments.

OPERA Net

Heterogeneous Broadband Wireless Network deployment is expected to create an explosive growth of new and existing services, providing the user with a rich broadband experience through true ubiquitous, real-time, multi-media communications on the move. It is a reality however that new high-bandwidth, high-spectral efficiency modulation schemes being developed to support these broad-band services are fundamentally 'RF-unfriendly', and one consequence is the power demanded by future networks is predicted to dramatically increase beyond available and sustainable power envelopes. This fact coupled with increasing environmental pressures presents a significant barrier to mobile network deployment that, if unaddressed, will severely restrict technical functionality, delay network roll-out and consequently limit the economic and social benefit to the European Community. It is the consortium's view that if the efficiency problem remains unaddressed, there is a significant risk that Europe's world-leading position and competitive edge in this market segment will significantly impacted.

Despite ongoing research into energy and power efficiency within academia and the wireless network industry, this is highly fragmented and tends to focus on improving and optimizing individual subsystems in isolation. In contrast, the OPERA-Net project aims to address this issue through a holistic approach considering a complete end-to-end system, identifying all relevant network elements and their interdependencies. The fundamental premise is that the efficiency of these individual system elements is highly inter-related, and in order to achieve the radical efficiency improvements envisaged, these must be considered in unison. The project will address the following criteria and objectives (OPERA-Net 2, 2016):

- Radical improvement in energy efficiency at system, infrastructure and terminal level
- Develop metrics and Key Performance Indicators for mobile network efficiency to direct the development and stimulate competition for efficiency through easy system comparability
- Enable EU industry to take leadership in environmentally sustainable mobile networks

eWIN

With the ICT industry consuming up to 10% of the world's energy and doubling every five years, there is a mounting pressure to engineer "Green ICT systems" where energy consumption is seen as a first-class system property.

Although such green ICT systems certainly require efficient components, much of the wasteful energy expenditure is due to poor resource management, unfortunate protocol interactions, and a mindset of offering high throughput "at any price". This project aims at developing energy-efficient wireless networking principles, both in terms of minimizing energy usage in current and upcoming wireless standards, and in terms of clean-slate architectures for energy-efficient wireless communication (wireless@kth, 2016).

5GrEEN

5GrEEN is a joint effort of partners tightly connected to the METIS project representing the telecom vendor perspective, the mobile operator view, and leading academic institution. 5GrEEN will specifically focus on energy efficiency aspects of 5G mobile networks, and will hence contribute significantly to the important design target of low energy consumption (Olson, et al., 2013) .

The obvious of this project challenge is to provide this thousand-fold capacity increase to billions of devices at today's energy consumption and operating cost levels (5GrEEn Toward Green 5G Mobile Networks, 2016).

- The development of this new generation systems that will be put in service during the next decade has now started. Low energy performance needs to be taken into account when the standardization process takes off. The present approach suggests a new clean slate system architecture with a logical separation between the ability to establish availability of the network and the ability to provide functionality or service.
- Such architecture, referred to as 5G, will be studied in 5GrEEn. The results originating from this project (e.g., standardization in 3GPP and ETSI, development of energy efficient backhauling solutions, energy efficiency assessment of current networks, energy-QoE tradeoff evaluation) will be applicable in 4G solutions, paving the way for the next generation networks.

HETEROGENOUS NETWORKS AS AN ENERGY-EFFICCIENT NETWORK DEPLOYMENT STRATEGIE

As we mentioned before, environmental concerns and operational cost efficiency recently diverted the attention of wireless broadband research activities toward energy efficient deployment. Network deployment strategies are always among the hot topics in cellular communications. However, early work in this area is mainly focused on network performance, such as coverage, spectral efficiency, and capacity. Recently, as energy consumption has become a primary concern, energy efficiency network deployment strategies have attracted increasing interest. The optimal cell size in terms of energy consumption where the tradeoff between EE and deployment cost is also discussed. In addition to energy efficient base station and core network design, emerging deployment strategies are to be considered that incorporate heterogeneous networks with cells of various size (mix of macrocells, microcells, picocells and femtocells) (Auer, et al., 2010). Heterogeneous networks have emerged as one of the key network architecture for addressing the aggressive capacity and coverage demands of future 5G networks. Significant network capacity gain can be achieved at low cost, not only through aggressive reuse of spectrum across the multiple tiers in the network, but also through harnessing additional spectrum in un-licensed bands by integrating WiFi and network. (Himayat, et al., 2014). The fifth generation (5G) cellular wireless network is treated as a promising solution for high traffic volume demands. 5G wireless networks are expected to be a mixture of network tiers of different sizes, transmissions powers, backhaul connections, and radio access technologies (Liu, Zhang, & Xie, 2015).

In contrast to conventional homogeneous macro cell deployment, HetNets reduces energy consumption in the network by shortening the propagation distance between nodes in the network and utilizing higher frequency band to support higher data rates. For fulfilling the requirements of green communication and maintaining the quality of service (QoS), heterogenous network (HetNet) consisting of generous low-power and small-coverage nodes has been seen as an efficient approach for improving the spectrum and energy efficiency (EE). The major constraint of these approaches is the extra small cells bring additional radio interferences as compared to conventional homogeneous macro cell networks, which might negatively affect user experience. Meanwhile, if too many micro, pico or femto cells are deployed, the trend of savings may even be reversed because of extra embodied energy consumed by newly deployed cells as well as overhead introduced in transmission. Therefore, the number of extra smaller cells, as well as their location, needs to be carefully planned in order to reduce total energy consumption (Wu, Zhang, Zukerman, & Yung, 2015).

Heterogenous Networks (HetNets)

A HetNet is a network consisting of infrastructure points with various wireless access technologies, each of them having different capabilities, constraints, and operating functionalities. HetNets elements are (López-Pérez, et al., 2011):

- **Macrocells:** Consist of conventional operator-installed BSs, providing open public access and a wide area coverage typically on the order of few kilometers.
- **Picocells:** These are low power operator installed cell towers with the same backhaul and access features as macrocells. They usually deployed in centralized way, serving of a few tens of users within a radio range of 300 m or less, and have a typical transmit power range from 23 to 30 dBm.
- **Femtocells:** They are also known as home BSs or home eNBs, they are low cost, low power user-deployed access point, offloading data traffic using consumer´s broadband connection (digital subscriber line [DSL], cable, of fiber), and serving a dozen active user in homes or enterprises.
- **Relays:** They are usually operator-deployed access points that route data from the macro BS to end users and vice versa.
- **RRHs:** They are compact-size, high-power, and low-weight units, which are mounted outside the conventional macro BS and connected to it generally through a fiber optic cable, thus creating a distributed BS.

Current cellular networks are typically deployed as homogeneous networks using a macro-centric planned process. A homogeneous cellular system is a network of base-stations in a planned layout have similar transmit power levels, antenna patterns, receiver nose floors, and similar backhaul connectivity on the (packet) data network (Khandekar, Bhushan, Tingfang, & Vanghi, 2010).

Picocells and femtocells are also promising deployment strategies to provide cost-effective services. Picocells and femtocells are usually installed within buildings for better indoor coverage. Indoor picocells and femtocells bring receivers closer transmitters and femtocells bring receivers closer to the transmitters and effectively reduce the penetration loss and path loss; thus, energy consumption can be significantly reduced.

The total network energy consumption in urban areas for high-data-rate demand can be reduced by up to 60% with user deployed residential picocells. The achievable gain in system EE with the deployment of femtocells in existing macrocells is significant due to smaller path loss, lower transmit power and hence lower energy requirement (Feng, Jiang, Lim, Cimini Jr., & Feng, 2013).

On the other hand, one of the major challenges in small cell deployment is the incursion of inter-tier, intra-tier, and cross-tier interference. Therefore, interference mitigation between different layers is one of the key issues that need to be resolved to reach the successful deployment of HetNets such a green comm solution.

The interference problem in HetNets is especially challenging due to the following reasons:

1. **Unplanned Deployment:** Low-power nodes such as femtocells are typically deployed in an ad hoc manner by users. They can even be moved or switched on/off at any time. Hence, traditional network planning and optimization becomes inefficient because operators do not control neither the number nor the location of these cells. This motivates the need for new decentralized interfer-

ence avoidance schemes that operate independently in each cell, utilizing only local information, whereas achieving an efficient solution for the entire network.

2. **CSG (Closes Subscriber Group) Access:** The fact that some cell may operate in CSG mode, in which cell Access is restricted and nonsubscribers are thus not always connected to the nearest BS (base station), originates significant cross-tier interference components.
3. **Power Difference Between Nodes:** Picocells and relays usually operate in open Access mode, meaning that all users of given operator can access them. Open Access helps minimize downlink (DL) interference as end users always connect to the strongest cell, thus avoiding the CSG interference issue. However, in HetNets, being attached to the cell that provides the strongest DL received signal strength (RSS) may not be the best strategy since users tend to connect to macrocells, not to the cells at the shortest path loss distance. This is due to the large difference in transmissions power between macrocells and low-power nodes.
4. **Range Expended Users:** To address the problem arising due to the power difference between the nodes in HetNets, new cells selection methods that allow user association with cell that provide weaker DL pilot signal. (López-Pérez D., et al., 2011)

To deal with the mentioned interference, a lot of interference mitigation schemes have been proposed. Some of these schemes are described in the following section.

Inter-Cell Interference Coordination (ICIC)

The interference problems may significantly degrade the overall HetNet performance, which requires the use of ICIC schemes to guarantee its proper operation.

As a result, Inter-cell Interference Coordination (ICIC) is critical to heterogenous network deployment. A basic ICIC technique involves resource coordination amongst interfering base-stations, where an interfering base-station gives up use of some resources in order to enable control and data transmissions to the victim user terminal. More generally, interfering base-station can coordinate on transmission powers and/or spatial beams with each other in order to enable control and data transmissions to their corresponding user terminals (Bodanese, 2011).

Basic ICIC techniques involve resource partitioning in the frequency domain. This means that only partial use of resources in frequency direction and/or adaptation of power levels. ICIC has different schemes, e.g. full frequency reuse, hard frequency reuse, fractional reuse and soft reuse.

- **Full Frequency Reuse:** It means that no frequency partitioning is performed between involved NodeBs (eNBs) of the same network. eNBs is this configuration transmit with uniform power over the entire system bandwidth. The main difficult to this configuration is that cell-edge users experience heavy interference from neighboring cells in the DL and create heavy interference in the uplink, greatly degrading their communication performance (Pauli, Naranjo, & Seidel, 2010).
- **Hard Frequency Reuse:** This method is typically seen in GSM networks, when it comes to distribution of frequencies among the cells. The subcarriers are divided into 3, 4 o 7 disjoints sets. These sets of subcarriers are assigned to the individual eNBs, this reduces the interference at the edge of any pair of cells. While user interference at cell edge is maximally reduced, the spectrum efficiency drops by a factor equal to the reuse factor (Pauli, Naranjo, & Seidel, 2010).

- **Fractional Frequency Reuse:** This is a hybrid frequency partitioning combining the concepts of the Full and Hard schemes. It consists of dividing the spectrum into two parts which will have different frequency reuse. One section of the system spectrum is used in all cells, while the other part of the spectrum is divided among different eNBs. The idea is that the eNB would assign the fully-reused frequency chunks to center cell user equipment (UEs) and the other chunks to the cell-edge UEs. This scheme is useful for ICIC in the uplink, where severe interference situations can occur where the user is located close to a strong interferer in the neighbor cell (Pauli, Naranjo, & Seidel, 2010)
- **Soft Frequency Reuse:** In this frequency partitioning method an eNB transmit in the whole system bandwidth, but using a non-uniform power spectrum. This scheme leads to improved SINR (signal-to-interference-plus-noise ratio) levels for cell-edge in the high-transmit power regions, and the other regions still have suitable SINR levels for cell-centers UEs (Pauli, Naranjo, & Seidel, 2010).

Enhanced Inter-Cell Interference Coordination (eICIC)

In order to address such dominant interference scenarios, enhanced inter-cell interference coordination (eICIC) techniques have recently been developed for Release 10, which can be grouped under three major categories:

- Time-domain techniques.
- Frequency-domain techniques.
- Power control techniques.

Time-domain techniques: eICIC methods transmissions of the victim users are scheduled in time-domain resources where the interference from other nodes is mitigated. They can be classified into four categories (López-Pérez D., et al., 2011):

1. **Subframe Aligament:** When the subframes of macro eNB and home eNB are aligned, their control and data channel overlap with each other. Thus, in order not to interfere to the control channel, control chanel eICIC needs to be implemented at the femtocells.
2. **OFDM Symbol Shift:** The subframe boundary of a home eNB is shifted by a number of orthogonal frequency-division multiplexing (OFDM) symbols with respect to the subframe boundary of a macro eNB in order to prevent overlap between the control channels of femtocell and macrocell signals.
3. **Frequency-Domain Techniques:** In these schemes, control channels and physical signals of different cells are scheduled in reduced bandwidths in order to have totally orthogonal transmission of these signals at different cells. While frequency-domain orthogonalization may be achieved in a static manner, it may also implemented dynamically through victim UE detection (López-Pérez D., et al., 2011).
4. **Power Control Techniques:** One last approach that has been heavily discussed in HetNets for handling dominant interferences scenarios is to apply different power control techniques at femtocell. While reducing the radiated power at femtocell users, it may significantly improve the performance of interference (López-Pérez D., et al., 2011).

Cross-Tier Cooperative Sensing Model

This technique wants to address spectral efficiency and interference issues among different network tiers. It can help devices or small cells, and also the system information of other cells in tiers.

The main components of an energy and spectrum harvesting driven 5G system include energy harvesting cognitive radio (CR) devices, picocells networks, femtocells networks and macro cells.

Cooperative sensing in one cell: in the one-cell case, CR devices sense channels at the same time. Synchronization among cooperative CR devices usually achieved in an infrastructure-based architecture. The cooperative CR devices could cooperatively sense the licensed spectrum in the picocell. After cooperative sensing, the participating devices send the sensing results to the CR picocell BS for a final decision (Liu, Zhang, & Xie, 2015).

Cooperative sensing in multiple cells: For the multi-cell case, diverse Cognitive Radio devices from different networks may be involved in cooperative sensing. The cooperative devices sensing moments and location are different as well as their sensing abilities. When the cooperative sensing procedure is evaluates the spectrum and energy using priority queuing analysis that considers the wireless environment, traffic characteristics and the competing users' behaviors (Liu, Zhang, & Xie, 2015).

FUTURE RESEARCH DIRECTIONS

The schemes briefly described in the previous section are early efforts to mitigate interference. Nowadays, the different consortiums that form the projects described in the previous section are seeking for enhanced schemes. For example, the EARTH project is developing a holistic approach for the design of energy efficient communications networks. One of its goals are heterogeneous network deployment strategies. On the other hand Mobile Virtual Centre of Excelence (VCE) with the Green radio project has chosen the Third Generation Partnership Project (3GPP) Long Term Evolution (LTE) as the baseline technology for the research program in where HetNets are considered. As a mentioned before, HetNets are energy efficient network because the short distance between transmitter and receiver offered by small cell sizes that suggest an enormous potential for energy savings (Auer, et al., 2010). That's because they are receiving a lot of attention from the international consortium, especially in mitigate interference that it the principal obstacle to use this new deployment strategy.

CONCLUSION

This chapter is intended to highlight challenges in energy-efficient communications as enablers for green solutions for 5G heterogeneous networks. For some green projects HetNets are an excellent solution for these problem related with energy efficient. There is global concern in reduce CO_2 emissions. The Information and Communication Technologies (ICT including data networks, fixed and wireless networks and end-user equipment) are responsible for about 2% of global CO_2, that is why new research in this trend had emerged, in order to intensified efforts to reduce climate change.

REFERENCES

C2Power. (2016, March 16). Retrieved from Cognitive Radio and Cooperative Strategies for POWER saving in multi-standard wireless devices: http://ict-power.eu

Auer, G., Gódor, I., Hévizi, L., Imran, M. A., Malmodin, J., Fazekas, P., & Tafazolli, R. (2010). *Enablers for Energy Efficient Wireless Networks. In Vehicular Technology Conference Fall (VTC 2010-Fall)* (pp. 1–5). Ottawa, ON: IEEE.

Blume, O., Zeller, D., & Barth, U. (2010). *Approaches to energy efficient wireless access networks. In Communications, Control and Signal Processing (ISCCSP)* (pp. 1–5). Limassol: IEEE.

Bodanese, E. (2011). *A Brief Introduction to Heterogeneous Networks (HetNets) and its Challenges. In Conmmunication Technology and Application, ICCTA* (pp. 605–609). Beijing: IEEE.

Congzheng, H., Harrold, T., Armour, S., Krikidis, I., Vide, S., Grant, P. M., & Hanzo, L. et al. (2011). Green Radio: Radio Techniques to Enable Energy-Efficient Wireless Networks. *IEEE Communications Magazine*, 46–54.

EARTH. (2014, April 1). Retrieved from EARTH: http://ict-earth.eu

Feng, D., Jiang, C., Lim, G., Cimini, L. J. Jr, & Feng, G. (2013). A Survey of Energy-Efficient Wireless Communications. *IEEE Communications Surveys and Tutorials*, *15*(1), 167–177. doi:10.1109/SURV.2012.020212.00049

Green Radio. (2016, March 21). Retrieved from Mobile vce Green Radio: www.mobilevce.com

GreenTouch. (2016, March 21). Retrieved from GreenTouch Web site: http://greentouch.org

Hasan, Z., Boostanimehr, H., & Bhargava, V. K. (2011). Green Cellular Networks: A Survey, Some Research Issues and Challenges. *IEEE Communications Surveys and Tutorials*, *13*(4), 524–540. doi:10.1109/SURV.2011.092311.00031

He, A., Amanna, A., Tsou, T., Chen, X., Dinesh, D., Gaeddert, J., & Bose, T. et al. (2011). Green Communications: A Call for Power Efficient Wireless Systems. *Journal of Communication*, *6*(4), 340–351.

Himayat, N., Yeh, S.-p., Panah, A. Y., Talwar, S., Gerasimenko, M., Andreev, S., & Koucheryavy, Y. (2014). Multi-Radio Heterogeneous Networks: Architectures and Performance. *2014 Internacional Conference on Computing Networking and Communicatio*. doi:10.1109/ICCNC.2014.6785341

Khandekar, A., Blushan, N., Tingfang, J., & Vanghi, V. (2010). LTE-Advanced: Heterogeneous Networks. *2010 European Wireless Conference* (pp. 978-982). Lucca, Italy: IEEE. doi:10.1109/EW.2010.5483516

Liu, Y., Zhang, Y., & Xie, S. (2015). Integrated Energy and Spectrum Harvesting for 5G Wireless Communications. *IEEE Network*, *29*(3), 75–81. doi:10.1109/MNET.2015.7113229

López-Pérez, D., Guvenc, I., De la Roche, G., Kountouris, M., & Zhang, J. (2011). Enhanced Intecell Interference Coordination Challenges in Heterogeneous Networks. *IEEE Wireless Communications*, 22-30.

López-Pérez, D., Guvenc, I., de la Roche, G. d., Kountouris, M., Quek, T. Q., & Zhang, J. (2011). *Enhanced Intercell Interference Coordination Challenges in Heterogeneous Networks*. IEEE Wireless Communication.

Mohammad, S. O., & Alagan Anpalagan, I. W. (2013). Handbook of Green Information and Communications Systems. Academic Press.

OPERA-Net 2. (2016, March 16). Retrieved from Celtic-Plus Smart Connected World: http://projects.celticplus.eu

Osseiran, A., Boccardi, F., Braun, V., Kasume, K., Marsh, P., Maternia, M., & Fallgren, M. et al. (2014). Scenarios for 5G Mobile and Wireless Communications: The Vision of the METIS Project. *IEEE Communications Magazine*, *52*(5), 26–35. doi:10.1109/MCOM.2014.6815890

Pauli, V., Naranjo, J. D., & Seidel, E. (2010). Heterogeneous LTE Networks and Inter-Cell Interference Coordination. *Nobel Mobile Radio*, 1-9.

Pejanovic-Djurisic, M., Stovrag, E., & Ilic-Delibasic, M. (2012). *Fundamental optimization criteria for Green wireless communications. In MIPRO* (pp. 733–736). Opatija: IEEE.

Pejanovic-Djurisic, M. E. S., & Ilic-Delibasic, M. (2012). Fundamental Optimization Criteria for Green Wireless Communications. *MIPRO*, *2012*, 734–736.

Pollin, S., Mangharam, R., Bougard, B., Van den Perre, L., Moerman, I., & Rajkumar, R. (2007). MEERA: Cross-Layer Methodology for Energy Efficient Resource Allocation in Wirelles Networks. *IEEE Transactions on Wireless Communications*, *6*(2), 617–628. doi:10.1109/TWC.2007.05356

Pollin, S., Mangharam, R., Bourgard, B., Moerman, I., Rajkumar, R., & Catthoor, F. (2008). MEERA: Cross-Layer Methodology for Energy Efficient Resource Allocation in Wireless Network. *IEEE Transactions on Wireless Communications*, *7*(1), 98–109. doi:10.1109/TWC.2008.05356

Rittenhouse, G., Goya, S. N., & Samuel, S. (2011). Sustainable Telecommunications. *Tecnical Symposium at ITU* (pp. 19-23). Geneva: IEEE.

Trend. (2016, March 19). Retrieved from Trend Toward Real Energy-efficient Network Design: www.fp7-trend.eu

Wie, Y., Song, M., Liu, B., Wang, X., & Li, Y. (2011). Energy Efficient Cooperative Relaying and Cognitive Radio Technologies to Deliver Green Communication. In *Pervasive Computing and Applications (ICPCA), 2011 6th International Conference* (pp. 105-109). Port Elizabeth: IEEE.

Williams, E., Kahhat, R., Allenby, B., Kavazanjian, E., Kim, J., & Xu, M. (2008). Environmental, Social, and Economic Implications of Global Reuse and Recycling of Personal Computers. *Environmental Science & Technology*, *42*(17), 6446–6454. doi:10.1021/es702255z PMID:18800513

wireless@kth. (2016, March 16). Retrieved from https://wireless.kth

Wu, J. (2012). Green Wireless Communications: From concept to reality (Industry Perspectives). *IEEE Wireless Communications*, *19*(4), 4–5. doi:10.1109/MWC.2012.6272415

Wu, J., Zhang, Y., Zukerman, M., & Yung, E. K.-N. (2015). Energy-Efficient Base-Station Sleep-Mode Techniques in Green Cellular Networks: A Survey. *IEEE Communications Surveys and Tutorials*, *17*(2), 803–826. doi:10.1109/COMST.2015.2403395

Section 2

Enhanced Interference Management Technology with Featured Characteristics

Chapter 4
Interference Management for Full-Duplex Massive MIMO Relaying System with Hardware Impairments

Kui Xu
PLA University of Science and Technology, China

Youyun Xu
PLA University of Science and Technology, China

Xiaochen Xia
PLA University of Science and Technology, China

Dongmei Zhang
PLA University of Science and Technology, China

ABSTRACT

Massive MIMO full-duplex relaying (MM-FDR), where multiple source-destination pairs communicate simultaneously with the help of a common full-duplex relay equipped with very large antenna arrays, is studied in this chapter. Different from the traditional MM-FDR protocol, a general model where sources/destinations are allowed to equip with multiple antennas is considered. The effect of hardware impairments is taken into consideration, and is modeled using transmit/receive distortion noises. We propose a low complexity hardware impairments aware transceiver scheme (named as HIA scheme) to mitigate the distortion noises by exploiting the statistical knowledge of channels and antenna arrays at sources and destinations. A joint degree of freedom and power optimization algorithm is presented to further optimize the spectral efficiency of HIA based MM-FDR. The results show that the HIA scheme can mitigate the ``ceiling effect" appears in traditional MM-FDR protocol, if the numbers of antennas at sources and destinations can scale with that at the relay.

INTRODUCTION

In multi-user MIMO systems, one main challenge is the increased complexity and energy consumption of the signal processing to mitigate the interferences between multiple co-channel users. To achieve energy efficient transmission, the multi-user MIMO system with very large antenna arrays at each base

DOI: 10.4018/978-1-5225-1712-2.ch004

station (known as "massive MIMO" system) has been advocated by the authors of T. L. Marzetta. (2010) recently. The key result shown by T. L. Marzetta. (2010) and H. Q. Ngo, E. G. Larsson. (2013) is that, with very large antenna arrays at each base station, both the intracell and intercell interferences can be substantially reduced with simple linear beamforming (BF) processing.

On the other hand, full-duplex relaying (FDR) is a promising approach to improve the spectral efficiency (SE) of relaying network while retains the merits of half-duplex relaying (HDR). In FDR, it is shown in D. Bharadia, E. McMilin. (2013) that the relay transmits and receives simultaneously at the same frequency and time, but at the cost of a strong echo interference (EI) due to signal leakage between the relay output and input. To mitigate EI, three approaches have been investigated, i.e.:

1. Passive cancellation,
2. Time-domain cancellation by D. Bharadia, E. McMilin. (2013), and
3. Spatial suppression by H. A. Suraweera, I. Krikidis. (2014) and C. Y. A. Shang, P. J. Smith. (2014).

The passive cancellation relies on a combination of path loss, cross-polarization and antenna directionality. The time-domain cancellation is based on the fact that EI signal is known at full-duplex node. Thus, the authors in D. Bharadia, E. McMilin. (2013) shown that it can be regenerated and removed in time-domain. In spatial suppression, H. A. Suraweera, I. Krikidis. (2014) showed in their previous work that EI is mitigated with the multiple transmit/receive antennas by approaches such as null-space projection. Inspired by these works, a number of works have dedicated to the study of FDR protocol on both theory and testbed (See the works of D. Nguyen, L.-N. Tran. (2014) and A. Sabharwal, et al. (2014) and the references therein). To achieve spectral and energy efficient transmissions of multiple source-destination pairs, recent works of H. Cui, L. Song. (2014), H. A. Suraweera, N. Hien Quoc. (2013), H. Q. Ngo, H. A. Suraweerat. (2013), H. Q. Ngo, H. A. Suraweera. (2014) and X. Xia, W. Xie. (2015) sought to incorporate both HDR and FDR with massive MIMO.

However, the aforementioned works on massive MIMO are actually based on the assumption that the base stations or relays are equipped with a large number of high-quality transmit/receive radio frequency (RF) chains (which are expensive and power-hungry). In contrast to conventional MIMO system (e.g., at most 8 antennas in LTE system), the authors in E. G. Larsson, O. Edfors. (2014) showed in their work that massive MIMO must be built with low-cost components since the deploy cost and energy consumption of circuits will increase dramatically as the number of antennas grows very large. Such low-cost components are prone to hardware imperfections (e.g., phase noise, nonlinear power amplifier, I/Q imbalance, nonlinear low-noise amplifier and ADC impairments), which must be considered in the design of practical massive MIMO system.

This chapter focuses on the transceiver design for massive MIMO full-duplex relaying (MM-FDR) with hardware impairments. The effect of hardware impairments is modeled using transmit/receive distortion noises by the authors in B. P. Day, A. R. Margetts. (2012) and G. Zheng, I. Krikidis. (2013). There are several challenges in the design of practical transceiver scheme in the considered system. The first is: *How to deal with the EI cancellation without instantaneous EI channel?* EI cancellation is a critical problem in MM-FDR transceiver design which is not only important to reduce EI power, but also useful to reduce distortion noises caused by hardware impairments at the relay (as will be shown in section IV). Different from FDR with small-scale relay antenna arrays, that is discussed in D. Bharadia, E. McMilin. (2013), E. Everett, A. Sahai. (2014), H. A. Suraweera, I. Krikidis. (2014), C. Y. A. Shang, P. J. Smith. (2014), D. Nguyen, L.-N. Tran. (2014) and A. Sabharwal, et al. (2014), the instantaneous EI channel is usually not easy to obtain in MM-FDR. This is because the learning of EI channel requires training sequence with length not less than the number of relay antennas, which is prohibitive in MM-FDR as the channel coherent time is limited. The lack of EI channel makes traditional EI cancellation techniques (e.g., time-domain cancellation and spatial suppression) difficult to apply. Although passive cancellation does not relies on EI channel estimation, it usually cannot provide satisfactory performance when used alone. Another problem is: *How to suppress distortion noises caused by hardware impairments at sources and destinations.* Different from multi-user interference (MUI), the distortion noise caused by transmit imperfection of source can be viewed as an interference signal with the same channel as transmit data. Thus, it cannot be suppressed by relay antenna arrays during coherent combining (Similar problem appears in reception at destinations). This causes performance ceiling on achievable rate as the number of relay antennas grows large, which degrades the gain of massive MIMO significantly.

Notations: $\mathbb{E}(\cdot)$ and $\mathrm{var}(\cdot)$ denote the expectation and variance. I_n is $n \times n$ identity matrix. $(\cdot)^*$, $(\cdot)^T$ and $(\cdot)^H$ denote conjugate, transpose and conjugate-transpose, respectively. $\rho(\boldsymbol{A})$, $Tr(\boldsymbol{A})$, $[\boldsymbol{A}]_{ij}$, $\lambda_l(\boldsymbol{A})$ and $u_l(\boldsymbol{A})$ denote the spectral radius, trace, (i,j) th element of matrix, l th largest eigenvalue and eigenvector with respect to l th largest eigenvalue of matrix $\boldsymbol{A}$, respectively. a scales with b means $0 < \lim_{b\to\infty} \frac{a}{b} < \infty$.

RELATED WORK

The design and performance of using unlimited number of antennas at the base station in cellular system were first considered by T. L. Marzetta. (2010) and H. Q. Ngo, E. G. Larsson. (2013) for independent antennas, and soon extended by the authors in J. Hoydis, S. Brink. (2013), J. Zhang, C. K. Wen. (2013) and A. Adhikary, J. Nam. (2013), to the scenarios with spatial correlated antennas. The asymptotic SINR for single cell cellular system was analyzed in H. Q. Ngo, E. G. Larsson. (2013). It has been shown that, with very large antenna arrays at the base station, a deterministic SINR which depends only on the large-scale fading of channels can be achieved, if the transmit power is scaled by N with perfect CSI and $\sqrt{N}$ with imperfect CSI (N denotes the number of base station antennas). The authors in J. Hoydis, S. Brink. (2013) have done considerable work to derive the "deterministic equivalent" of SINR for massive MIMO system with spatial correlated antennas. The asymptotic analysis and performance optimization for massive MIMO system with distributed sets of correlated antennas were considered in J. Zhang, C. K. Wen. (2013), A. Adhikary, J. Nam. (2013) and H. Yin, D. Gesbert. (2014).

In the field of massive MIMO relaying, the SE and EE of one-way and two-way HDR with very large relay antenna arrays were investigated in H. Cui, L. Song. (2014) and H. A. Suraweera, N. Hien Quoc. (2013). The MM-FDR with decode-and-forward (DF) relay was first introduced in H. Q. Ngo, H. A. Suraweerat. (2013) and H. Q. Ngo, H. A. Suraweera. (2014), and analyzed in end-to-end achievable rate as linear processing is employed. In our previous work, the asymptotic performance of amplify-and-forward based MM-FDR was considered in X. Xia, W. Xie. (2015) and the scaling behavior of the infinitely repeating echo interference was determined.

A constant envelope signal design has been proposed in S. Mohammed & E. Larsson. (2013) to facilitate the use of power-efficient RF power amplifiers. The authors in C. Studer, & E. Larsson. (2013)

presented a low peak to average power ratio (PAR) precoding solution to enable efficient implementation using non-linear RF components in massive MIMO system. In FDR with small-scale antennas, the optimal precoding design under limited ADC dynamic range was considered in B. P. Day, A. R. Margetts. (2012) and G. Zheng, I. Krikidis. (2013). However, these works cannot provide much insight for the effect of hardware impairments on MM-FDR.

SYSTEM MODEL

Consider the network with K source-destination (S-D) pairs and a single full-duplex relay R, where source S_k wishes to communicate with destination D_k ($k \in \{1,\cdots,K\}$) with the help of R. The relay adopts the DF policy. It is assumed that the sources and destinations are equipped with N_S and N_D antennas respectively, while the relay is equipped with $N_R + N_T$ antennas (N_R for reception and N_T for transmission). We are interested in the large-(N_R, N_T) regime, i.e., $\min\{N_R, N_T\} \to \infty$. N_S, N_D and K can be either fixed or scale with $\min\{N_R, N_T\}$.

Let $\boldsymbol{H}_{SR,k} \in \mathbb{C}^{N_R \times N_S}$ be the channel matrix from S_k to receive antenna array of relay and let $\boldsymbol{H}_{RD,k} \in \mathbb{C}^{N_T \times N_D}$ be the channel matrix from D_k to transmit array of relay. Let $\boldsymbol{H}_{EI} \in \mathbb{C}^{N_R \times N_T}$ denote the EI channel matrix between transmit and receive arrays of relay. The spatial correlation of each MIMO channel is characterized by the Kronecker model shown in J. H. Kotecha, & A. M. Sayeed. (2004). Thus, $\boldsymbol{H}_{SR,k}$ can be expressed as $\boldsymbol{H}_{SR,k} = \sqrt{\beta_{SR,k}}\boldsymbol{C}_{SR,k}^{1/2}\boldsymbol{X}_{SR,k}\tilde{\boldsymbol{C}}_{SR,k}^{1/2}$, where $\beta_{SR,k}$ represents the large-scale fading. $\boldsymbol{C}_{SR,k} \in \mathbb{C}^{N_R \times N_R}$ and $\tilde{\boldsymbol{C}}_{SR,k} \in \mathbb{C}^{N_S \times N_S}$ characterize the spatial correlation of received signals across receive array of relay and that of transmitted signals across transmit array of S_k. $\boldsymbol{X}_{SR,k} \in \mathbb{C}^{N_R \times N_S}$ consists of the random components of channels whose elements are i.i.d with distribution $CN(0,1)$. Based on the same model, the channel matrices $\boldsymbol{H}_{RD,k}$ and $\boldsymbol{H}_{EI}$ can be expressed as $\boldsymbol{H}_{RD,k} = \sqrt{\beta_{RD,k}}\boldsymbol{C}_{RD,k}^{1/2}\boldsymbol{X}_{RD,k}\tilde{\boldsymbol{C}}_{RD,k}^{1/2}$ and $\boldsymbol{H}_{EI} = \sqrt{\beta_{EI}}\boldsymbol{C}_{EI}^{1/2}\boldsymbol{X}_{EI}\tilde{\boldsymbol{C}}_{EI}^{1/2}$. To facilitate the analysis, we assume the following conditions on correlation matrices, i.e., $\forall \boldsymbol{C} \in \left\{\boldsymbol{C}_{SR,k}, \tilde{\boldsymbol{C}}_{SR,k}, \boldsymbol{C}_{RD,k}, \tilde{\boldsymbol{C}}_{RD,k}, \boldsymbol{C}_{EI}, \tilde{\boldsymbol{C}}_{EI}\right\}$, where $k \in \{1,\cdots,K\}$,

A1: The spectral radius of $\boldsymbol{C}$ is bounded by a constant, i.e., $\rho(\boldsymbol{C}) \le C$.
A2: $\boldsymbol{C}$ is a Hermitian and Teoplitz matrix and has unit diagonal elements.

The former is a common assumption in the previous studies of massive MIMO, J. Hoydis, S. Brink. (2013) and J. Zhang, C. K. Wen. (2013), which follows from energy conservation and **A2** corresponds to the case of uniform linear (ULA) in J. H. Kotecha, & A. M. Sayeed. (2004).

To characterize the effect of hardware imperfections, we adopt the new signal model from B. P. Day, A. R. Margetts. (2012) and G. Zheng, I. Krikidis. (2013).

Imperfect Transmit RF Chain

We model the effect of imperfect transmit RF chain by adding, per transmit antenna, an independent zero-mean Gaussian "transmit distortion noise". A key property is that the power of the distortion noise at each antenna is proportional to the signal power transmitted at this antenna. The experimental results in G. Santella, & F. Mazzenga. (1998) and H. Suzuki, T. Tran. (2008) have shown that the independent Gaussian distortion noise model closely captures the joint effect of additive power amplifier noise, non-linearities in the DAC and power amplifier, and oscillator phase noise. Let $\boldsymbol{x}_{S,k}\left[u\right] \in \mathbb{C}^{N_S \times 1}$ and $\boldsymbol{x}_R\left[u\right] \in \mathbb{C}^{N_T \times 1}$ denote the transmit vectors of source S_k and relay at time instant u. Based on the above model, the distorted transmit signals can be expressed as

$$\tilde{\boldsymbol{x}}_{S,k}\left[u\right] = \boldsymbol{x}_{S,k}\left[u\right] + \boldsymbol{t}_{S,k}\left[u\right] and \ _{R}\left[u\right] = \boldsymbol{x}_R\left[u\right] + \boldsymbol{t}_R\left[u\right] \tag{1}$$

where the distortion noises

$$\boldsymbol{t}_{S,k}\left[u\right] \sim CN\left(0, \nu_{S,k} diag\left(\mathbb{E}\left[\boldsymbol{x}_{S,k}\left[u\right]\boldsymbol{x}_{S,k}^H\left[u\right]\right]\right)\right)$$

and

$$\boldsymbol{t}_R\left[u\right] \sim CN\left(0, \nu_R \times diag\left(\mathbb{E}\left[\boldsymbol{x}_R\left[u\right]\boldsymbol{x}_R^H\left[u\right]\right]\right)\right).$$

Note that $\nu_{S,k} > 0$ ($\nu_R > 0$) characterizes the level of transmit imperfection. For example, $\nu_{S,k} = 0$ ($\nu_R = 0$) corresponds to the conventional assumption of perfect transmit RF chains. The quality of transmit RF chains degrades as $\nu_{S,k}$ (ν_R) increases.

Imperfect Receive RF Chain

We model the effect of imperfect receive RF chain by adding, per receive antenna, an independent zero-mean Gaussian "receive distortion noise", whose variance is proportional to the signal power received at that antenna. More precisely, let $\boldsymbol{y}_R\left[u\right] \in \mathbb{C}^{N_R \times 1}$ and $\boldsymbol{y}_{D,k}\left[u\right] \in \mathbb{C}^{N_D \times 1}$ be the undistorted received signals of the relay and destination D_k at time instant u, the distorted received signals can be expressed as

$$\tilde{\boldsymbol{y}}_R\left[u\right] = \boldsymbol{y}_R\left[u\right] + \boldsymbol{r}_R\left[u\right] and \ \tilde{\boldsymbol{y}}_{D,k}\left[u\right] = \boldsymbol{y}_{D,k}\left[u\right] + \boldsymbol{r}_{D,k}\left[u\right] \tag{2}$$

where the distortion noises

$$\boldsymbol{r}_R\left[u\right] \sim CN\left(0, \mu_R diag\left(\mathbb{E}\left[\boldsymbol{y}_R\left[u\right]\boldsymbol{y}_R^H\left[u\right]\right]\right)\right)$$

and

$$r_{D,k}[u] \sim CN\left(0, \mu_{D,k} \times diag\left(\mathbb{E}\left[y_{D,k}[u] y_{D,k}^H[u]\right]\right)\right).$$

The expectations are over the transmit signal vectors, additive white Gaussian noises (AWGNs) and fading channels. $\mu_R > 0$ ($\mu_{D,k} > 0$) characterizes the level of receive imperfection. The experimental studies in W. Namgoong. (2005) have shown that the independent Gaussian noise model is a good approximation to the joint effect of additive automatic gain control (AGC) noise, quantization noise in the ADC, and oscillator phase noise in practical hardware.

At time instant u, all sources transmit signals $\mathbf{x}_{S,k}[u]$ (k=1,$\cdots$,K) to the relay simultaneously. Meanwhile, the relay broadcasts the decoded signals to the destinations. Based on the models in (1) and (2), the received signals at the relay and D_k can be expressed as

$$\begin{aligned} \tilde{y}_R[u] &= \sum_{k=1}^{K} H_{SR,k}\left(x_{S,k}[u] + t_{S,k}[u]\right) + H_{EI}\left(x_R[u] + t_R[u]\right) + r_R[u] + n_R[u] \\ \tilde{y}_{D,k}[u] &= H_{RD,k}^H\left(x_R[u] + t_R[u]\right) + r_{D,k}[u] + n_{D,k}[u] \end{aligned} \tag{3}$$

To keep the complexity low, it is assumed that a single data steam is transmitted at each source and each node employs only linear processing. In particular, S_k transmits the unit-power symbol $s_k[u]$ using the unitary BF vector $p_{S,k}$. Therefore, the transmit vector of S_k can be expressed as

$$x_{S,k}[u] = \sqrt{E_{S,k}} p_{S,k} s_k[u],$$

where $E_{S,k}$ denotes the transmit power. Relay combines the received signal by multiplying the receive BF matrix

$$W_R = \left[w_{R,1}, \cdots, w_{R,K}\right],$$

i.e., $\breve{y}_R[u] = W_R^H \tilde{y}_R[u]$. The kth element of $\breve{y}_R[u]$

$$\begin{aligned} \breve{y}_{R,k}[u] &= \sqrt{E_{S,k}} w_{R,k}^H H_{SR,k} p_{S,k} s_k[u] + \sqrt{E_{S,j}} \sum_{j=1, j\neq k}^{K} w_{R,k}^H H_{SR,j} p_{S,j} s_j[u] + w_{R,k}^H \sum_{j=1}^{K} H_{SR,j} t_{S,j}[u] \\ &+ w_{R,k}^H H_{EI}\left(x_R[u] + t_R[u]\right) + w_{R,k}^H r_R[u] + w_{R,k}^H n_R[u] \end{aligned} \tag{4}$$

is used to decode the symbol of S_k. Relay retransmits the decoded symbol using transmit BF matrix $W_T = \left[w_{T,1}, \cdots, w_{T,K}\right]$. Thus, the transmit vector of the relay can be expressed as $x_R[u] = W_T \Lambda_R^{1/2} s[u-d]$, where

$$\boldsymbol{s}\left[u-d\right]=\left[s_1\left[u-d\right],\cdots,s_K\left[u-d\right]\right]^T$$

and

$$\boldsymbol{\Lambda}_R = diag\left(E_{R,1},\cdots,E_{R,K}\right)$$

denotes power allocation matrix at the relay. d denotes the processing delay. Here we have assumed the delay is large enough, so that the receive and transmit signals of the relay at each time instant are uncorrelated. This is a common assumption in the context of full-duplex system as shown in C. Y. A. Shang, P. J. Smith. (2014), D. Nguyen, L.-N. Tran. (2014) and H. Q. Ngo, H. A. Suraweera. (2014). To meet the power constraint at the relay, $\boldsymbol{x}_R\left[u\right]$ must satisfy

$$Tr\left(\mathbb{E}\left[\boldsymbol{x}_R\left[u\right]\boldsymbol{x}_R^H\left[u\right]\right]\right)=\sum_{l=1}^{K}E_{R,l}\leq E_R^{\max}.$$

D_k uses the unitary BF vector $\boldsymbol{p}_{D,k}$ to combine the received signal $\tilde{\boldsymbol{y}}_{D,k}\left[u\right]$. Using the expression of $\boldsymbol{x}_R\left[u\right]$, the combined signal is expressed as

$$\begin{aligned}\tilde{y}_{D,k}\left[u\right] &= \sqrt{E_{R,k}}\boldsymbol{p}_{D,k}^H\boldsymbol{H}_{RD,k}^H\boldsymbol{w}_{T,k}s_k\left[u-d\right]+\boldsymbol{p}_{D,k}^H\sum_{j=1,j\neq k}^{K}\sqrt{E_{R,j}}\boldsymbol{H}_{RD,k}^H\boldsymbol{w}_{T,j}s_j\left[u-d\right]\\ &+\boldsymbol{p}_{D,k}^H\boldsymbol{H}_{RD,k}^H\boldsymbol{t}_R\left[u\right]+\boldsymbol{p}_{D,k}^H\boldsymbol{r}_{D,k}\left[u\right]+\boldsymbol{p}_{D,k}^H\boldsymbol{n}_{D,k}\left[u\right]\end{aligned} \tag{5}$$

LIMITATION OF MM-FDR WITH SINGLE ANTENNA SOURCES AND DESTINATIONS

Upper Bound on Achievable Rate

As $N_S = N_D = 1$, we set $p_{S,k} = p_{D,k} = 1$, replace $\boldsymbol{H}_{i,k}$ ($i \in \{SR, RD\}$) with $\boldsymbol{h}_{i,k}$, and let $\boldsymbol{H}_i = \left[\boldsymbol{h}_{i,1},\cdots,\boldsymbol{h}_{i,K}\right]$. Moreover, we let

$$\boldsymbol{t}_S\left[u\right]=\left[\boldsymbol{t}_{S,1}^T\left[u\right],\cdots,\boldsymbol{t}_{S,K}^T\left[u\right]\right]^T.$$

We consider a block-fading channel with coherence time T (symbol times), where τ are used for uplink training of each source/destination, and the remaining $T-2K\tau$ are used for data transmission. The upper bound on achievable rate is obtained by assuming perfect knowledge of $\boldsymbol{h}_{SR,k}$, $\boldsymbol{h}_{RD,k}$ and $\boldsymbol{H}_{EI}$ can be provided with pilot signals, and meanwhile, the MUIs (i.e., the second terms of right-hand side of (4) and (5)) and EI term $\boldsymbol{w}_{R,k}^{H}\boldsymbol{H}_{EI}\boldsymbol{x}_{R}\left[u\right]$ can be somehow cancelled. This gives us the following upper bound on end-to-end achievable rate

$$R_{k}^{Upper} = \max_{\left\|w_{R,k}\right\|=\left\|w_{D,k}\right\|=1} \min\left\{R_{SR,k}, R_{RD,k}\right\} \tag{6}$$

where $\mathrm{R}_{SR,k}$ and $\mathrm{R}_{RD,k}$ denote the achievable rates of $S_k \rightarrow R$ and $R \rightarrow D_k$ channels, respectively. According to (4), (5) and the assumptions in the above, we have

$$\begin{aligned} R_{SR,k} &= \frac{T-2K\tau}{T}\log_2\left(1+\frac{E_{S,k}\boldsymbol{w}_{R,k}^{H}\boldsymbol{h}_{SR,k}\boldsymbol{h}_{SR,k}^{H}\boldsymbol{w}_{R,k}}{\boldsymbol{w}_{R,k}^{H}\boldsymbol{Q}\boldsymbol{w}_{R,k}}\right) \\ R_{RD,k} &= \frac{T-2K\tau}{T}\log_2\left(1+\frac{E_{R,k}\boldsymbol{w}_{T,k}^{H}\boldsymbol{h}_{RD,k}\boldsymbol{h}_{RD,k}^{H}\boldsymbol{w}_{T,k}}{\boldsymbol{h}_{RD,k}^{H}\boldsymbol{\Theta}_{R}^{T}\boldsymbol{h}_{RD,k}+\mathbb{E}\left[\left\|\boldsymbol{r}_{D,k}\left[u\right]\right\|^{2}\right]+1}\right) \end{aligned} \tag{7}$$

where $\boldsymbol{Q}$ is given as

$$\boldsymbol{Q} = \boldsymbol{H}_{SR}\boldsymbol{\Theta}_{S}^{T}\boldsymbol{H}_{SR}^{H} + \boldsymbol{H}_{EI}\boldsymbol{\Theta}_{R}^{T}\boldsymbol{H}_{EI}^{H} + \boldsymbol{\Theta}_{R}^{R} + \boldsymbol{I}_{N_R}$$

with $\boldsymbol{\Theta}_{S}^{T} = \mathbb{E}\left[\boldsymbol{t}_{S}[u]\boldsymbol{t}_{S}^{H}[u]\right]$, $\boldsymbol{\Theta}_{R}^{T} = \mathbb{E}\left[\boldsymbol{t}_{R}[u]\boldsymbol{t}_{R}^{H}[u]\right]$ and $\boldsymbol{\Theta}_{R}^{R} = \mathbb{E}\left[\boldsymbol{r}_{R}[u]\boldsymbol{r}_{S}^{H}[u]\right]$ denoting the covariance matrices of distortion noises $\boldsymbol{t}_{S}[u]$, $\boldsymbol{t}_{R}[u]$ and $\boldsymbol{r}_{R}[u]$, respectively.

1. **Achievable Rate of $S_k \rightarrow R$ Channel:** Since $w_{R,k}$ is only related with $R_{SR,k}$, it can be optimized separately to maximize $R_{SR,k}$. From (7), the optimization problem is equivalent to the generalized Rayleigh quotient problem, which can be solved as $w_{R,k} = \frac{\boldsymbol{Q}^{-1}\boldsymbol{h}_{SR,k}}{\left\|\boldsymbol{Q}^{-1}\boldsymbol{h}_{SR,k}\right\|}$. The resultant upper bound on $R_{SR,k}$ (denoted by $R_{SR,k}^{Upper}$) can be expressed as

$$R_{SR,k}^{Upper} = \frac{T-2K\tau}{T}\log_2\left(1+E_{S,k}\boldsymbol{h}_{SR,k}^{H}\boldsymbol{Q}^{-1}\boldsymbol{h}_{SR,k}\right) = \frac{T-2K\tau}{T}\log_2\left(1+\frac{E_{S,k}\boldsymbol{h}_{SR}^{H}\boldsymbol{Q}_{k}^{-1}\boldsymbol{h}_{SR,k}}{1+\nu_{S,k}E_{S,k}\boldsymbol{h}_{SR,k}\boldsymbol{Q}_{k}^{-1}\boldsymbol{h}_{SR,k}^{H}}\right) \tag{8}$$

where $\boldsymbol{Q}_k$ is defined as

$$\boldsymbol{Q}_{k} = \boldsymbol{Q} - \nu_{S,k}p_{S,k}\boldsymbol{h}_{SR,k}\boldsymbol{h}_{SR,k}^{H}$$

and is independent of $\boldsymbol{h}_{SR,k}$. The second step follows from the matrix inversion lemma, which is shown in J. W. Silverstein, & Z. D. Bai. (1995). The bound in (8) is not amenable to simple analysis. However, it enables us to derive further bounds on achievable rate which helps us understand the effect of hardware impairments.

Theorem 1: Assuming that the assumptions **A1** and **A2** hold. As $N_S = N_D = 1$, the achievable rate of $S_k \to R$ channel in the large-(N_R, N_T) regime is bounded as

$$R_{SR,k} \leq R_{SR,k}^{Upper} = \frac{T-2K\tau}{T}\log_2\left(1+\frac{E_{S,k}\beta_{SR,k}Tr\left(\boldsymbol{C}_{SR,k}\boldsymbol{\Psi}_k\right)}{1+\nu_{S,k}E_{S,k}\beta_{SR,k}Tr\left(\boldsymbol{C}_{SR,k}\boldsymbol{\Psi}_k\right)}\right) \tag{9}$$

where $\rho = \beta_{EI}Tr\left(\tilde{\boldsymbol{C}}_{EI}\boldsymbol{\Theta}_R^T\right)+1$, $\boldsymbol{\Psi}_k\left(\rho\right)$ is determined by the following fixed-point algorithm with $\delta_{k,l}^{(0)}\left(\rho\right) = \frac{1}{\rho}$

$$\begin{aligned}
\boldsymbol{\Psi}_k\left(\rho\right) &= \left(\frac{1}{N_R}\sum_{l=1,l\neq k}^{K}\frac{\nu_{S,l}p_{S,l}\beta_{SR,l}\boldsymbol{C}_{SR,l}}{1+\delta_{k,l}\left(\rho\right)}+\boldsymbol{\Theta}_R^R+\rho\boldsymbol{I}_{N_R}\right)^{-1}\\
\boldsymbol{\Theta}_R^R &= \mu_R\sum_{l=1}^{K}\nu_{S,l}p_{S,l}\beta_{SR,l}\boldsymbol{I}_{N_R}+\beta_{EI}Tr\left(\tilde{\boldsymbol{C}}_{EI}\boldsymbol{\Theta}_R^T\right)\boldsymbol{I}_{N_R}+\mu_R\boldsymbol{I}_{N_R}\\
\delta_{k,l}\left(\rho\right) &= \lim_{n\to\infty}\delta_{k,l}^{(n)}\left(\rho\right)\\
\delta_{k,l}^{(n)}\left(\rho\right) &= \frac{\nu_{S,l}p_{S,l}\beta_{SR,l}}{N_R}Tr\left(\boldsymbol{C}_{SR,l}\left(\frac{1}{N_R}\sum_{j=1,j\neq k}^{K}\frac{\nu_{S,j}p_{S,j}\beta_{SR,j}\boldsymbol{C}_{SR,j}}{1+\delta_{k,l}^{(n-1)}\left(\rho\right)}+\boldsymbol{\Theta}_R^R+\rho\boldsymbol{I}_{N_R}\right)^{-1}\right)
\end{aligned} \tag{10}$$

Proof: See the Appendix.

The convergence of the fixed-point algorithm in *Theorem 1* has been proved in S. Wagner, R. Couillet. (2012).

2. **Achievable Rate of $R \to D_k$ Channel:** From (7), the achievable rates of $S_k \to R$ and $R \to D_k$ channels are coupled through

$$\boldsymbol{\Theta}_R^T = \mathbb{E}\left[\boldsymbol{t}_R[u]\boldsymbol{t}_R^H[u]\right] = diag\left(\mathbb{E}\left[\boldsymbol{W}_T\boldsymbol{\Lambda}_R\boldsymbol{W}_T^H\right]\right),$$

which makes the design of $\boldsymbol{w}_{T,k}$ very challenging. However, with the following theorem we show that this coupling disappears in the large-(N_R, N_T) regime, which allows a simple upper bound for achievable rate of $R \to D_k$ channel.

Theorem 2: Assuming that the assumptions **A1** and **A2** hold. As $N_S = N_D = 1$, the achievable rate of $S_k \rightarrow R$ channel is independent of $\boldsymbol{w}_{T,k}$. The achievable rate of $R \rightarrow D_k$ channel in the large-(N_R, N_T) regime is bounded as

$$\mathrm{R}_{RD,k} \le \mathrm{R}_{RD,k}^{\mathrm{Upper}}$$
$$= \frac{T-2K\tau}{T}\log_2\left(1+\frac{N_T\beta_{RD,k}E_{R,k}}{\nu_R\beta_{RD,k}\sum_{l=1}^{K}E_{R,l}+\mu_{D,k}\left(N_T\beta_{RD,k}E_{R,k}+\nu_R\beta_{RD,k}\sum_{l=1}^{K}E_{R,l}+1\right)+1}\right) \quad (11)$$

which is achieved by the eigen BF scheme $\boldsymbol{w}_{T,k} = \boldsymbol{h}_{RD,k}/\left\|\boldsymbol{h}_{RD,k}\right\|$.

Proof: See the Appendix.

With Theorems 1 and 2, the upper bound of end-to-end achievable at D_k in the large-(N_R, N_T) regime can be expressed as

$$R_k^{Upper} = \min\left\{R_{SR,k}^{Upper}, R_{RD,k}^{Upper}\right\} \quad (12)$$

Remark 1: The bound in (12) is derived by assuming linear processing and fixed transmit powers at sources and relay, which is in general not capacity achieving. One exception is when $K \ll \min\left\{N_R, N_T\right\}$ and $\min\left\{N_R, N_T\right\} \rightarrow \infty$. As shown in Theorem 2, the design of receive and transmit BF matrices is decoupled as $\min\left\{N_R, N_T\right\} \rightarrow \infty$. Moreover, the distortion noises are assumed to be circularly symmetric complex Gaussian distributed and independent of the desired signal. Thus, based on the results in E. Telatar. (1999) and E. Bjornson, P. Zetterberg. (2013), linear processing along with power control is sufficient to achieve the capacity. When $K \ll \min\left\{N_R, N_T\right\}$, the terms

$$\frac{1}{N_R}\sum_{l=1,l\neq k}^{K}\frac{\nu_{S,l}E_{S,l}\beta_{SR,l}\boldsymbol{C}_{SR,l}}{1+\delta_{k,l}(\rho)}$$

and

$$\frac{1}{N_R}\sum_{j=1,j\neq k}^{K}\frac{\nu_{S,j}E_{S,j}\beta_{SR,j}\boldsymbol{C}_{SR,j}}{1+\delta_{k,l}^{(n-1)}(\rho)}$$

in (10) vanish. By neglecting the low order term (in $\min\left\{N_R, N_T\right\}$), the bound in (12) reduces to

$$\frac{T-2K\tau}{T}\log_2\left(1+\min\left\{v_{S,k}^{-1},\mu_{D,k}^{-1}\right\}\right),$$

which is independent of $E_{S,k}$ and $E_{R,k}$. Thus, (12) is the upper bound for capacity of $S_k \to R \to D_k$ channel as $K \ll \min\left\{N_R, N_T\right\}$ and $\min\left\{N_R, N_T\right\} \to \infty$.

Limitation with Hardware Impairments

Different from Remark 1, we consider the general case in which K can be either fixed or scales with $\min\left\{N_R, N_T\right\}$. With Theorem 1 and Theorem 2, the effect of hardware impairments is still hard to analysis since $\boldsymbol{\Psi}_k$ is not in closed-form. To make the analysis tractable, we assume $\boldsymbol{C}_{SR,l} = \boldsymbol{C}_{SR}$, $\forall l \in \{1,\cdots,K\}$. Using eigenvalue decomposition $\boldsymbol{C}_{SR} = \boldsymbol{U}\boldsymbol{\Sigma}_{SR}\boldsymbol{U}^H$ on $\delta_{k,l}^{(n)}(\rho)$ in (10), we have

$$\delta_{k,l}^{(n)}\left(\rho\right)=\frac{\nu_{S,l}p_{S,l}\beta_{SR,l}}{N_R}Tr\left[\Sigma_{SR}\left(\frac{1}{N_R}\sum_{j=1,j\neq k}^{K}\frac{\nu_{S,j}p_{S,j}\beta_{SR,j}}{1+\delta_{k,l}^{(n-1)}\left(\rho\right)}\Sigma_{SR}+\boldsymbol{\Theta}_R^R+\rho\boldsymbol{I}_{N_R}\right)^{-1}\right] \tag{13}$$

By examining (13) and the expression of $\boldsymbol{\Theta}_R^R$, $\delta_{k,l}^{(n)}(\rho)$ is an $O(1)$ term. Note that this is valid for arbitrary $n > 0$. Thus, we can conclude that $\delta_{k,l}(\rho) = O(1)$. Replacing $\delta_{k,l}\left(\rho\right)$ in $\boldsymbol{\Psi}_k$ with the symbol $O(1)$ and substituting the result into (9), $R_{SR,k}$ can be written as

$$R_{SR,k}^{Upper}$$

$$=\frac{T-2K\tau}{T}\log_2\left(1+\frac{E_{S,k}\beta_{SR,k}\sum_{l=1}^{N_R}\lambda_l\left(\boldsymbol{C}_{SR}\right)\left(\frac{\lambda_l\left(\boldsymbol{C}_{SR}\right)}{N_R}\sum_{l=1,l\neq k}^{K}\frac{\nu_{S,l}E_{S,l}\beta_{SR,l}}{1+O\left(1\right)}+\frac{Tr\left(\boldsymbol{\Theta}_R^R\right)}{N_T}+\rho\right)^{-1}}{1+\nu_{S,k}p_{S,k}\beta_{SR,k}\sum_{l=1}^{N_R}\lambda_l\left(\boldsymbol{C}_{SR}\right)\left(\frac{\lambda_l\left(\boldsymbol{C}_{SR}\right)}{N_R}\sum_{l=1,l\neq k}^{K}\frac{\nu_{S,l}E_{S,l}\beta_{SR,l}}{1+O\left(1\right)}+\frac{Tr\left(\boldsymbol{\Theta}_R^R\right)}{N_T}+\rho\right)^{-1}}\right) \tag{14}$$

With this, we are now ready to analyze effect of hardware impairments. For convenience, we further assume

$$\left\{\beta_{SR,l},\beta_{RD,l},E_{S,l},E_{R,l},\nu_{S,l},\mu_{D,l}\right\}=\left\{\beta_{SR},\beta_{RD},E_S,E_R,\nu_S,\mu_D\right\}.$$

Note that this assumption does not effect the basis conclusions of the analysis. By neglecting the low order terms (in $\min\{N_R, N_T\}$) in (14) and (11), the upper bound on end-to-end achievable rate reduces to

$$R_k^{Upper} = \frac{T-2K\tau}{T}\log_2\left(1+\min\left\{\frac{1}{\nu_S+\frac{K}{N_R}\mu_R\nu_S+\frac{K}{N_R}\nu_R\frac{E_R\beta_{EI}}{E_S\beta_{SR}}},\frac{1}{\mu_{D,k}+\frac{K}{N_T}\nu_R\left(1+\mu_{D,k}\right)}\right\}\right) \tag{15}$$

One can make several observations from (15):

1. As $K \ll \min\{N_R, N_T\}$, the effect of hardware impairments at the relay and EI disappears in the large-(N_R, N_T) regime and the upper bound converges to that in Remark 1. The end-to-end achievable rate is limited by distortion noises caused by hardware impairments at sources and destinations. The explanation is that the transmit distortion can be viewed as an interference signal with the same channel as data of S_k from (3). Thus, it cannot be suppressed after combining by $\boldsymbol{w}_{R,k}$. The power of distortion due to receive imperfection of D_k is proportional to that of desired receive signal (after eigen BF) and also cannot be reduced by the transmit BF scheme of relay. Thus, there is a finite ceiling on end-to-end achievable rate as $\min\{N_R, N_T\} \to \infty$. In fact, the above factor poses a major limitation to the MM-FDR with single antenna sources and destinations, which degrades the gain of massive array of relay significantly.
2. As K scales with $\min\{N_R, N_T\}$, the effect of hardware impairments at the relay and EI is not negligible. Since the EI channel is typically stronger than desired channels, the $S_k \to R$ channel becomes the bottleneck of end-to-end achievable rate. This indicates that it is of great importance to suppress EI to avoid the bottleneck effect.

HARDWARE IMPAIRMENTS AWARE TRANSCEIVER DESIGN

Transceiver Scheme Description

The optimal transceiver scheme for MM-FDR to optimize achievable rate is hard to find since the transmissions of $S_k \to R$ and $R \to D_k$ channels are deeply coupled due to the presence of EI and distortion noises. Even without EI and distortion noises, the problem has been proved NP-hard by J. W. Silverstein, & Z. D. Bai. (1995). As $\boldsymbol{H}_{EI}$ is an $N_R \times N_T$ matrix, the learning of $\boldsymbol{H}_{EI}$ requires training sequence with length no less than $\min\{N_R, N_T\}$. This task becomes prohibitive as the duration of channel coherent time is limited. In this work, we assume that passive EI cancellation, e.g., a combination of path loss, cross-polarization and antenna directionality, has been used and instantaneous EI channel is not available. A low complexity HIA scheme is proposed to mitigate EI (after passive cancellation) and distortion noises. The scheme exploits the statistical channel knowledge and antenna array of sources/destinations, and allows closed-form solution for BF matrix at each node. Moreover, it reduces the feedback of system and enables simple channel estimation scheme. As will be shown in simulations, the HIA scheme along with passive EI cancellation provides satisfactory gain compared with HDR.

1. **Transceiver Design at the Relay:** As the variance of EI channel is typically stronger than desired channels, it is of great importance to control the EI power. As seen in section IV-B, EI suppression is also important in reducing distortion noise due to hardware impairments of relay, especially when K is large. To achieve this, we consider a two-stage BF scheme at the relay. The receive and transmit BF matrices are expressed as $\boldsymbol{W}_R = \boldsymbol{P}_R\underline{\boldsymbol{W}}_R$ and $\boldsymbol{W}_T = \boldsymbol{P}_T\underline{\boldsymbol{W}}_T$.
2. **Outer BF Matrices:** The outer BF matrices $\boldsymbol{P}_R \in \mathbb{C}^{N_R \times A_R}$ ($K \le A_R \le N_R$) and $\boldsymbol{P}_T \in \mathbb{C}^{N_T \times A_T}$ ($K \le A_T \le N_T$) are designed to suppress EI, where A_R and A_T are design parameters. Due to the lack of $\boldsymbol{H}_{EI}$, $\boldsymbol{P}_R$ and $\boldsymbol{P}_T$ are allowed to depend only on the statistical knowledge of EI channels. With (4), we design an average EI power minimization problem, i.e.,

$$\left(\dot{\boldsymbol{P}}_R, \dot{\boldsymbol{P}}_T\right) = \arg\min_{\boldsymbol{P}_R, \boldsymbol{P}_T} \underbrace{\mathbb{E}\left[\underline{\boldsymbol{w}}_{R,k}^H \boldsymbol{P}_R^H \boldsymbol{H}_{EI}\left(\boldsymbol{P}_T \underline{\boldsymbol{W}}_T \Lambda_R \underline{\boldsymbol{W}}_T^H \boldsymbol{P}_T^H + \boldsymbol{\Theta}_R^T\right) \boldsymbol{H}_{EI}^H \boldsymbol{P}_R \underline{\boldsymbol{w}}_{R,k}\right]}_{\mathrm{EI}_k} \tag{16}$$

In (16), the design of $\boldsymbol{P}_R$ and $\boldsymbol{P}_T$ is coupled with the inner BF matrix, which makes closed-form solution inaccessible. Thus, it is of great interest to find new target function to decouple the problem. For such, using Cauchy-Schwarz inequality twice, an upper bound on average EI power can be obtained as

$$\begin{aligned} EI_k &\le EI_k^{Upper} \\ &= \left\|\underline{\boldsymbol{w}}_{R,k}\right\|^2 Tr\left(\underline{\boldsymbol{W}}_T \boldsymbol{\Lambda}_R \underline{\boldsymbol{W}}_T^H\right) Tr\left(\boldsymbol{P}_R^H \boldsymbol{H}_{EI} \boldsymbol{P}_T \boldsymbol{P}_T^H \boldsymbol{H}_{EI}^H \boldsymbol{P}_R\right) + \nu_R \sum_{l=1}^{K} E_{R,k} \left\|\underline{\boldsymbol{w}}_{R,k}\right\|^2 Tr\left(\boldsymbol{P}_R^H \boldsymbol{H}_{EI} \boldsymbol{H}_{EI}^H \boldsymbol{P}_R\right) \end{aligned} \tag{17}$$

Then the problem (16) can be rewritten approximately as

$$\begin{aligned} &\left(\dot{\boldsymbol{P}}_R, \dot{\boldsymbol{P}}_T\right) \\ &= \arg\min_{\boldsymbol{P}_R, \boldsymbol{P}_T} \left\{ \begin{aligned} &Tr\left(\underline{\boldsymbol{W}}_T \Lambda_R \underline{\boldsymbol{W}}_T^H\right) \mathbb{E}\left[Tr\left(\boldsymbol{P}_R^H \boldsymbol{H}_{EI} \boldsymbol{P}_T \boldsymbol{P}_T^H \boldsymbol{H}_{EI}^H \boldsymbol{P}_R\right)\right] \\ &+\nu_R \sum_{l=1}^{K} E_{R,k} \mathbb{E}\left[Tr\left(\boldsymbol{P}_R^H \boldsymbol{H}_{EI} \boldsymbol{H}_{EI}^H \boldsymbol{P}_R\right)\right] \end{aligned} \right\} \\ &= \arg\min_{\boldsymbol{P}_R, \boldsymbol{P}_T} \left\{ \begin{aligned} &Tr\left(\underline{\boldsymbol{W}}_T \Lambda_R \underline{\boldsymbol{W}}_T^H\right) Tr\left(\boldsymbol{P}_T^H \tilde{\boldsymbol{C}}_{EI} \boldsymbol{P}_T\right) Tr\left(\boldsymbol{P}_R^H \boldsymbol{C}_{EI} \boldsymbol{P}_R\right) \\ &+\nu_R N_T \sum_{l=1}^{K} E_{R,k} Tr\left(\boldsymbol{P}_R^H \boldsymbol{C}_{EI} \boldsymbol{P}_R\right) \end{aligned} \right\} \end{aligned} \tag{18}$$

where the multiplicative term $\left\|\boldsymbol{w}_{R,k}\right\|^2$ has been removed since it has no effect on the optimal solution. The second step follows from a similar derivation of (59). From (18), the columns of $\dot{\boldsymbol{P}}_R$ ($\dot{\boldsymbol{P}}_T$) are composed of the eigenvectors corresponding to the A_R (A_T) smallest eigenvalues of $\boldsymbol{C}_{EI}$ ($\tilde{\boldsymbol{C}}_{EI}$), i.e.,

$$\dot{\boldsymbol{P}}_R = \left[\boldsymbol{u}_{N_R-A_R+1}\left(\boldsymbol{C}_{EI}\right), \boldsymbol{u}_{N_R-A_R+2}\left(\boldsymbol{C}_{EI}\right), \cdots, \boldsymbol{u}_{N_R}\left(\boldsymbol{C}_{EI}\right)\right]$$
$$\dot{\boldsymbol{P}}_T = \left[\boldsymbol{u}_{N_T-A_T+1}\left(\tilde{\boldsymbol{C}}_{EI}\right), \boldsymbol{u}_{N_T-A_T+2}\left(\tilde{\boldsymbol{C}}_{EI}\right), \cdots, \boldsymbol{u}_{N_T}\left(\tilde{\boldsymbol{C}}_{EI}\right)\right] \quad (19)$$

In (19), A_R and A_T can be interpreted as parameters to balance the allowable EI power and available degree of freedom (DOF) for data transmission, which should be optimized with respect to specific performance metric. This will be considered in the next section.

3. **Inner BF Matrix:** The inner BF matrices $\underline{\boldsymbol{W}}_R \in \mathbb{C}^{A_R \times K}$ and $\underline{\boldsymbol{W}}_T \in \mathbb{C}^{A_T \times K}$ are designed to realize the multi-user communication. $\underline{\boldsymbol{W}}_R$ and $\underline{\boldsymbol{W}}_T$ can be designed with different criteria, e.g., maximizing the desired signal power which corresponding to the eigen BF, or minimizing the MUI which corresponding to the zero-forcing (ZF) scheme, as shown in H. Q. Ngo, H. A. Suraweera. (2014). We adopt the latter one since the ZF scheme is known to approach the asymptotic limit (in $\min\{N_R, N_T\}$) of achievable rate faster as the number of relay antennas increase, as shown in H. Cui, L. Song, & B. Jiao. (2014). For given BF vectors at sources and destinations, define effective channel vectors of $S_k \to R$ and $R \to D_k$ channels as $\underline{\boldsymbol{h}}_{SR,k} = \boldsymbol{P}_R^H \boldsymbol{H}_{SR,k} \boldsymbol{p}_{S,k}$ and $\underline{\boldsymbol{h}}_{RD,k} = \boldsymbol{P}_T^H \boldsymbol{H}_{RD,k} \boldsymbol{p}_{D,k}$, and let $\underline{\boldsymbol{H}}_{SR} = \left[\underline{\boldsymbol{h}}_{SR,1}, \cdots, \underline{\boldsymbol{h}}_{SR,K}\right]$ and $\underline{\boldsymbol{H}}_{RD} = \left[\underline{\boldsymbol{h}}_{RD,1}, \cdots, \underline{\boldsymbol{h}}_{RD,K}\right]$, the inner BF matrices can be written as

$$\underline{\boldsymbol{W}}_R = \underline{\boldsymbol{H}}_{SR}\left(\underline{\boldsymbol{H}}_{SR}^H \underline{\boldsymbol{H}}_{SR}\right)^{-1} \textit{and } \underline{\boldsymbol{W}}_T = \underline{\boldsymbol{H}}_{RD}\left(\underline{\boldsymbol{H}}_{RD}^H \underline{\boldsymbol{H}}_{RD}\right)^{-1} \boldsymbol{Y}^{-1/2} \quad (20)$$

where $\boldsymbol{Y}$ is a diagonal normalized matrix with $[\boldsymbol{Y}]_{l,l} = \boldsymbol{e}_l^H \left(\underline{\boldsymbol{H}}_{RD}^H \underline{\boldsymbol{H}}_{RD}\right)^{-1} \boldsymbol{e}_l$. In practice, $\boldsymbol{H}_{SR,k}$ and $\boldsymbol{H}_{RD,k}$ should be estimated in order to compute (20). This will be considered in section V-B. Note that (20) requires $K \le \min\{N_R, N_T\}$ so that A_R and A_T can be selected to ensure the invertibility of $\underline{\boldsymbol{H}}_{SR}^H \underline{\boldsymbol{H}}_{SR}$ and $\underline{\boldsymbol{H}}_{RD}^H \underline{\boldsymbol{H}}_{RD}$.

4. **Transceiver Design at the Sources and Destinations:** Similar to J. Hoydis, S. Brink. (2013), we assume no instantaneous knowledge of channels at sources and destinations. This is reasonable since the amount of feedback could be very huge and unaffordable in MM-FDR. However, it is assumed that the local statistical knowledge of channels (i.e., $\left(\boldsymbol{C}_{SR,k}, \tilde{\boldsymbol{C}}_{SR,k}\right)$ for S_k and $\left(\boldsymbol{C}_{RD,k}, \tilde{\boldsymbol{C}}_{RD,k}\right)$ for D_k) can be obtained. As observed from (15), when $N_S = N_D = 1$, the achievable rates of $S_k \to R$ channel and $R \to D_k$ channel are limited by the transmit distortion noise at S_k and receive distortion noise at D_k. This motivates us to design $\boldsymbol{p}_{S,k}$ and $\boldsymbol{p}_{D,k}$ to suppress these negative factors with the antenna arrays of S_k and D_k.
5. **Design of $\boldsymbol{p}_{S,k}$:** Intuitively, according to (3) we can design the following problem

$$\begin{aligned}\dot{\boldsymbol{p}}_{S,k} &= \arg\max_{\|p_{S,k}\|=1} \frac{E_{S,k}\mathbb{E}\left[Tr\left(\boldsymbol{H}_{SR,k}\boldsymbol{p}_{S,k}\boldsymbol{p}_{S,k}^{H}\boldsymbol{H}_{SR,k}^{H}\right)\right]}{\mathbb{E}\left[Tr\left(\boldsymbol{H}_{SR,k}\boldsymbol{\Theta}_{S,k}^{T}\boldsymbol{H}_{SR,k}^{H}\right)\right]} \\ &= \arg\max_{\|p_{S,k}\|=1} \frac{\boldsymbol{p}_{S,k}^{H}\tilde{\boldsymbol{C}}_{SR,k}\boldsymbol{p}_{S,k}}{\nu_{S,k}Tr\left(\tilde{\boldsymbol{C}}_{SR,k}diag\left(\boldsymbol{p}_{S,k}^{H}\boldsymbol{p}_{S,k}\right)\right)} \\ &= \arg\max_{\|p_{S,k}\|=1} \frac{1}{\nu_{S,k}}\boldsymbol{p}_{S,k}^{H}\tilde{\boldsymbol{C}}_{SR,k}\boldsymbol{p}_{S,k}\end{aligned} \quad (21)$$

where $\boldsymbol{\Theta}_{S,k}^{T}$ is the covariance matrix of $\boldsymbol{t}_{S,k}$, i.e.,

$$\boldsymbol{\Theta}_{S,k}^{T} = \mathbb{E}\left[\boldsymbol{t}_{S,k}[u]\boldsymbol{t}_{S,k}^{H}[u]\right] = \nu_{S,k}diag\left(\boldsymbol{p}_{S,k}\boldsymbol{p}_{S,k}^{H}\right).$$

The second step follows from a similar derivation with that in (59) and the last step is based on assumption **A2**. From (3), the right-hand side of (21) can be interpreted as the average signal to transmit distortion noise ratio at S_k. Obviously, the solution of problem (21) is

$$\dot{\boldsymbol{p}}_{S,k} = \boldsymbol{u}_1\left(\tilde{\boldsymbol{C}}_{SR,k}\right) \quad (22)$$

6. **Design of $\boldsymbol{p}_{D,k}$:** Similarly, based on (3), the received BF vector at D_k can be derived by solving the problem

$$\dot{\boldsymbol{p}}_{D,k} = \arg\max_{\|p_{R,k}\|=1} \frac{E_{R,k}\mathbb{E}\left[Tr\left(\boldsymbol{H}_{RD,k}\boldsymbol{p}_{D,k}\boldsymbol{p}_{D,k}^{H}\boldsymbol{H}_{RD,k}^{H}\right)\right]}{\mu_{D,k}\boldsymbol{p}_{D,k}^{H}diag\left(\mathbb{E}\left[\boldsymbol{H}_{RD,k}^{H}\boldsymbol{H}_{RD,k}\right]\right)\boldsymbol{p}_{D,k}},$$

which results in

$$\dot{\boldsymbol{p}}_{D,k} = \boldsymbol{u}_1\left(\tilde{\boldsymbol{C}}_{RD,k}\right) \quad (23)$$

Reduced Dimension Channel Estimation

With HIA scheme, it is sufficient to estimate the effective channel vectors $\underline{\boldsymbol{h}}_{SR,k}$ and $\underline{\boldsymbol{h}}_{RD,k}$. This allows a reduced dimension estimation. For brevity, we consider the estimation of $\underline{\boldsymbol{h}}_{SR,k}$. Let $\phi \in \mathbb{C}^{\tau\times 1}$ and E_T denote the pilot sequence and power of each pilot symbol. ϕ is multiplied by $\dot{\boldsymbol{p}}_{S,k}$ and transmitted by S_k. The received pilot matrix at the relay is expressed as

$$\boldsymbol{Z}_{SR,k} = \boldsymbol{H}_{SR,k}\left(\dot{\boldsymbol{p}}_{S,k}\phi^{T} + \boldsymbol{T}_{SR,k}\right) + \boldsymbol{R}_{SR,k} + \boldsymbol{N}_{SR,k} \quad (24)$$

where $\boldsymbol{T}_{SR,k} \in \mathbb{C}^{N_S \times \tau}$ denotes the transmit distortion noise of S_k, whose l th column has distribution (based on model (1))

$$CN\left(0, \nu_{S,k} E_T diag\left(\dot{\boldsymbol{p}}_{S,k} \dot{\boldsymbol{p}}_{S,k}^H\right)\right).$$

$\boldsymbol{R}_{SR,k} \in \mathbb{C}^{N_R \times \tau}$ denotes receive distortion noise at the relay. With model (2), the l th column of $\boldsymbol{R}_{SR,k}$ has distribution

$$CN\left(0, \mu_R diag\left(\mathbb{E}\left[\boldsymbol{Z}_{SR,k} \boldsymbol{e}_l \left(\boldsymbol{Z}_{SR,k} \boldsymbol{e}_l\right)^H\right]\right)\right).$$

$\boldsymbol{N}_{SR,k}$ is the AWGN matrix, whose elements are i.i.d and distributed as $CN(0,1)$. By multiplying both side of (24) with $\dot{\boldsymbol{P}}_R^H$, we have

$$\tilde{\boldsymbol{Z}}_{SR,k} = \underline{\boldsymbol{h}}_{SR,k} \phi^T + \dot{\boldsymbol{P}}_R^H \boldsymbol{H}_{SR,k} \boldsymbol{T}_{SR,k} + \dot{\boldsymbol{P}}_R^H \boldsymbol{R}_{SR,k} + \dot{\boldsymbol{P}}_R^H \boldsymbol{N}_{SR,k} \tag{25}$$

From (25), as a merit of HIA scheme, the total length of pilot sequences to obtain the estimates of all effective channels can be as less as $2K$ (when τ is set to 1). Moreover, only $A_R \times A_R$ matrix inversion is needed in order to compute the estimates of $\underline{\boldsymbol{h}}_{SR,k}$, when the LMMSE channel estimator in the follow theorem is adopted.

Theorem 3: The LMMES estimator of effective channel $\underline{\boldsymbol{h}}_{SR,k}$ can be expressed as

$$\hat{\underline{\boldsymbol{h}}}_{SR,k} = \underline{\boldsymbol{C}}_{SR,k} \boldsymbol{\Gamma}_{SR,k} \tilde{\boldsymbol{z}}_{SR,k} \tag{26}$$

where $\underline{\boldsymbol{C}}_{SR,k}$ is the covariance matrix of $\underline{\boldsymbol{h}}_{SR,k}$, which can be expressed as

$$\underline{\boldsymbol{C}}_{SR,k} = \beta_{SR,k} \lambda_1\left(\tilde{\boldsymbol{C}}_{SR,k}\right) \dot{\boldsymbol{P}}_R^H \boldsymbol{C}_{SR,k} \dot{\boldsymbol{P}}_R.$$

$\tilde{\boldsymbol{z}}_{R,k}$ and $\boldsymbol{\Gamma}_{SR,k}$ are given by

$$\begin{aligned} \tilde{\boldsymbol{z}}_{SR,k} &= \underline{\boldsymbol{h}}_{SR,k} + \frac{1}{\tau E_T} \dot{\boldsymbol{P}}_R^H \boldsymbol{H}_{SR,k} \boldsymbol{T}_{SR,k} \phi^* + \frac{1}{\tau E_T} \dot{\boldsymbol{P}}_R^H \boldsymbol{R}_{SR,k} \phi^* + \frac{1}{\tau E_T} \dot{\boldsymbol{P}}_R^H \boldsymbol{N}_{SR,k} \phi^* \\ \boldsymbol{\Gamma}_{SR,k} &= \left(\underline{\boldsymbol{C}}_{SR,k} + \frac{\nu_{S,k}}{\lambda_1\left(\tilde{\boldsymbol{C}}_{SR,k}\right)\tau} \underline{\boldsymbol{C}}_{SR,k} + \frac{1}{\tau E_T} \boldsymbol{I}_{A_R} + \frac{\mu_R}{\tau}\left(\frac{1}{E_T} + \beta_{SR,k}\left(\lambda_1\left(\tilde{\boldsymbol{C}}_{SR,k}\right) + \nu_{S,k}\right)\right) \boldsymbol{I}_{A_R} \right)^{-1} \end{aligned} \tag{27}$$

The real effective channel $\underline{\boldsymbol{h}}_{SR}$ can be decomposed as $\underline{\boldsymbol{h}}_{SR,k} = \hat{\underline{\boldsymbol{h}}}_{SR,k} + \Delta\underline{\boldsymbol{h}}_{SR,k}$ with $\Delta\underline{\boldsymbol{h}}_{SR,k}$ denoting the estimation error. $\hat{\underline{\boldsymbol{h}}}_{SR,k}$ and $\Delta\underline{\boldsymbol{h}}_{SR,k}$ are uncorrelated whose covariance matrices can be expressed as $\hat{\underline{\boldsymbol{C}}}_{SR,k} = \underline{\boldsymbol{C}}_{SR,k}\boldsymbol{\Gamma}_{SR,k}\underline{\boldsymbol{C}}_{SR,k}$ and $\underline{\boldsymbol{C}}_{SR,k} - \hat{\underline{\boldsymbol{C}}}_{SR,k}$, respectively.

Proof: A sketch of the proof is presented in the Appendix.

Different from that with ideal hardware, the LMMSE estimator in Theorem 3 is not equivalent to MMSE estimator, since the received pilot signal is corrupted by the term $\dot{\boldsymbol{P}}_R^H \boldsymbol{H}_{SR,k}\boldsymbol{T}_{SR,k}$ (due to transmit imperfection of S_k), which is not independent with the channel to be estimated. There might exist non-linear estimator that results in smaller MSE. However, the difference should be small since the distortion noises are relatively weak.

The LMMSE estimator of $\underline{\boldsymbol{h}}_{RD,k}$ can be obtained using the same approach. For further analysis, we define $\underline{\boldsymbol{C}}_{RD,k}$ and $\hat{\underline{\boldsymbol{h}}}_{RD,k}$ as the covariance matrix and LMMSE estimates of $\underline{\boldsymbol{h}}_{RD,k}$, respectively. Moreover, we define $\hat{\underline{\boldsymbol{C}}}_{RD,k} = \underline{\boldsymbol{C}}_{RD,k}\boldsymbol{\Gamma}_{RD,k}\underline{\boldsymbol{C}}_{RD,k}$ as the covariance matrix of $\hat{\underline{\boldsymbol{h}}}_{RD,k}$, where $\boldsymbol{\Gamma}_{RD,k}$ is given by

$$\boldsymbol{\Gamma}_{RD,k} = \left(\underline{\boldsymbol{C}}_{RD,k} + \frac{\nu_{D,k}}{\lambda_1\left(\tilde{\boldsymbol{C}}_{RD,k}\right)\tau}\underline{\boldsymbol{C}}_{RD,k} + \frac{1}{\tau E_T}\boldsymbol{I}_{A_T} + \frac{\mu_R}{\tau}\left(\frac{1}{E_T} + \beta_{RD,k}\left(\lambda_1\left(\tilde{\boldsymbol{C}}_{RD,k}\right) + \nu_{D,k}\right)\right)\boldsymbol{I}_{A_T}\right) \tag{28}$$

Achievable Rate Analysis

This subsection analyzes the achievable rate of HIA-MM-FDR. The achievable rate expressions are derived based on the bounding technique in B. Hassibi, & B. M. Hochwald. (2003).

1. **Achievable Rate of $S_k \to R$ Channel:** By treating

$$\sqrt{E_{S,k}}\mathbb{E}\left[\boldsymbol{w}_{R,k}^H \boldsymbol{H}_{SR,k}\dot{\boldsymbol{p}}_{S,k}\right]s_k[u]$$

as the desired signal at $S_k \to R$ channel, and approximating the remaining terms in (4), i.e.,

$$\breve{y}_{R,k}[u] - \sqrt{E_{S,k}}\mathbb{E}\left[\boldsymbol{w}_{R,k}^H \boldsymbol{H}_{SR,k}\dot{\boldsymbol{p}}_{S,k}\right]s_k[u]$$

using the worst-case uncorrelated additive Gaussian noise with the same variance, the following rate is achievable for $S_k \to R$ channel

$$R_{SR,k}^{HIA} = \frac{T-2K\tau}{T}\log_2\left(\frac{E_{S,k}\left|\mathbb{E}\left[\boldsymbol{w}_{R,k}^H \boldsymbol{H}_{SR,k}\dot{\boldsymbol{p}}_{S,k}\right]\right|^2}{\begin{array}{l}E_{S,k}\,\mathrm{var}\left(\boldsymbol{w}_{R,k}^H \boldsymbol{H}_{SR,k}\dot{\boldsymbol{p}}_{S,k}\right)+\mathbb{E}\left[EI_k\right]+\mathbb{E}\left[MUI_{R,k}\right]\\ +\mathbb{E}\left[D_{R,k}^T\right]+\mathbb{E}\left[D_{R,k}^R\right]+\mathbb{E}\left[\left\|\boldsymbol{w}_{R,k}\right\|^2\right]\end{array}}\right) \tag{29}$$

where EI_k, $MUI_{R,k}$, $D_{R,k}^T$ and $D_{R,k}^R$ denote the EI, MUI, effective distortion due to transmit imperfection of sources and effective distortion due to receive imperfection of relay respectively, which are given by

$$\begin{aligned}
EI_k &= \boldsymbol{w}_{R,k}^H \boldsymbol{H}_{EI}\left(\boldsymbol{W}_T\Lambda_R\boldsymbol{W}_T^H+\mathbb{E}\left[\boldsymbol{t}_R[u]\boldsymbol{t}_R^H[u]\right]\right)\boldsymbol{H}_{EI}^H\boldsymbol{w}_{R,k}\\
MUI_{R,k} &= \sum_{j=1,j\neq k}^{K} E_{S,j}\left|\boldsymbol{w}_{R,k}^H\boldsymbol{H}_{SR,j}\dot{\boldsymbol{p}}_{S,j}\right|^2\\
D_{R,k}^T &= \sum_{j=1}^{K}\boldsymbol{w}_{R,k}^H\boldsymbol{H}_{SR,j}\mathbb{E}\left[\boldsymbol{t}_{S,j}[u]\boldsymbol{t}_{S,j}^H[u]\right]\boldsymbol{H}_{SR,j}^H\boldsymbol{w}_{R,k}\\
D_{R,k}^R &= \boldsymbol{w}_{R,k}^H\mathbb{E}\left[\boldsymbol{r}_R[u]\boldsymbol{r}_R^H[u]\right]\boldsymbol{w}_{R,k}
\end{aligned} \tag{30}$$

wherein, the BF matrices at the relay are given by

$$\boldsymbol{W}_R = \dot{\boldsymbol{P}}_R\hat{\underline{\boldsymbol{H}}}_{SR}\left(\hat{\underline{\boldsymbol{H}}}_{SR}^H\hat{\underline{\boldsymbol{H}}}_{SR}\right)^{-1} \text{ and } \boldsymbol{W}_T = \dot{\boldsymbol{P}}_T\hat{\underline{\boldsymbol{H}}}_{RD}\left(\hat{\underline{\boldsymbol{H}}}_{RD}^H\hat{\underline{\boldsymbol{H}}}_{RD}\right)^{-1}\hat{\boldsymbol{\Upsilon}}^{-1/2} \tag{31}$$

where $\hat{\underline{\boldsymbol{H}}}_i = \left[\hat{\underline{\boldsymbol{h}}}_{i,1},\cdots,\hat{\underline{\boldsymbol{h}}}_{i,K}\right]$ ($i\in\{SR,RD\}$) and $\hat{\boldsymbol{\Upsilon}}$ is a diagonal normalized matrix with $[\hat{\boldsymbol{\Upsilon}}]_{l,l} = \boldsymbol{e}_l^H\left(\hat{\underline{\boldsymbol{H}}}_{RD}^H\hat{\underline{\boldsymbol{H}}}_{RD}\right)^{-1}\boldsymbol{e}_l$. Note that we have replaced the effective channels in inner BF matrices with their estimates obtained in Theorem 3.

Theorem 4: Assume that the assumptions **A1** and **A2** hold, and A_R is selected so that $Tr\left(\underline{\boldsymbol{C}}_{SR,k}\right)=O(N_R)$.

With HIA scheme, the achievable rate of $S_k \to R$ channel in the large-(N_R, N_T) regime is given by (29), where the powers of desired signal and AWGN are

$$E_{S,k}\left|\mathbb{E}\left[\boldsymbol{w}_{R,k}^H\boldsymbol{H}_{SR,k}\dot{\boldsymbol{p}}_{S,k}\right]\right|^2 = E_{S,k}$$

and

$$\mathbb{E}\left[\left\|\boldsymbol{w}_{R,k}\right\|^2\right] = \left(Tr\left(\hat{\underline{\boldsymbol{C}}}_{SR,k}\right)\right)^{-1},$$

and

$$\begin{cases} E_{S,k}\,\mathrm{var}\left(\boldsymbol{w}_{R,k}^{H}\boldsymbol{H}_{SR,k}\dot{\boldsymbol{p}}_{S,k}\right)=E_{S,k}\left(Tr\left(\hat{\underline{\boldsymbol{C}}}_{SR,k}\right)\right)^{-2}\delta_{k}^{SR} \\ \mathbb{E}\left[EI_{k}\right]=\beta_{EI}\left(Tr\left(\hat{\underline{\boldsymbol{C}}}_{SR,k}\right)\right)^{-2}Tr\left(\tilde{\boldsymbol{C}}_{EI}\boldsymbol{\Omega}_{R}\right)Tr\left(\hat{\underline{\boldsymbol{C}}}_{SR,k}\dot{\boldsymbol{P}}_{R}^{H}\boldsymbol{C}_{EI}\dot{\boldsymbol{P}}_{R}\right) \\ \mathbb{E}[MUI_{R,k}]=\left(Tr\left(\hat{\underline{\boldsymbol{C}}}_{SR,k}\right)\right)^{-2}\sum_{j=1,j\neq k}^{K}E_{S,j}Tr\left(\left(\underline{\boldsymbol{C}}_{SR,j}-\hat{\underline{\boldsymbol{C}}}_{SR,j}\right)\hat{\underline{\boldsymbol{C}}}_{SR,k}\right) \\ \mathbb{E}[D_{R,k}^{T}]=\left(Tr\left(\hat{\underline{\boldsymbol{C}}}_{SR,k}\right)\right)^{-2}\nu_{S,k}E_{S,k}g_{kk}+\left(Tr\left(\hat{\underline{\boldsymbol{C}}}_{SR,k}\right)\right)^{-2}\sum_{j=1,j\neq k}^{K}\nu_{S,j}E_{S,j}g_{kj} \\ \mathbb{E}\left[D_{R,k}^{R}\right]=\mu_{R}\left(Tr\left(\hat{\underline{\boldsymbol{C}}}_{SR,k}\right)\right)^{-1}\left(\sum_{j=1}^{K}E_{S,j}\beta_{SR,j}Tr\left(\tilde{\boldsymbol{C}}_{SR,j}\boldsymbol{\Omega}_{S,j}\right)+\beta_{EI}Tr\left(\tilde{\boldsymbol{C}}_{EI}\boldsymbol{\Omega}_{R}\right)+1\right) \end{cases} \tag{32}$$

where δ_{k}^{SR} is defined as

$$\delta_{k}^{SR}=\mathbb{E}[|\,\Delta\underline{\mathbf{h}}_{SR,k}^{H}\hat{\underline{\mathbf{h}}}_{SR,k}\,|^{2}]$$

and is given by Lemma 1 in the Appendix. g_{kk} and g_{kj} are given by (73) and (74) in the Appendix. $\boldsymbol{\Omega}_{S,k}$ and $\boldsymbol{\Omega}_{R}$ are expressed as

$$\begin{aligned} \boldsymbol{\Omega}_{S,k}&=\dot{\boldsymbol{p}}_{S,k}\dot{\boldsymbol{p}}_{S,k}^{H}+\nu_{S,k}diag\left(\dot{\boldsymbol{p}}_{S,k}\dot{\boldsymbol{p}}_{S,k}^{H}\right) \\ \boldsymbol{\Omega}_{R}&=\sum_{l=1}^{K}E_{R,l}\left(Tr\left(\hat{\underline{\boldsymbol{C}}}_{RD,l}\right)\right)^{-1}\left(\dot{\boldsymbol{P}}_{T}\hat{\underline{\boldsymbol{C}}}_{RD,l}\dot{\boldsymbol{P}}_{T}^{H}+\nu_{R}diag\left(\dot{\boldsymbol{P}}_{T}\hat{\underline{\boldsymbol{C}}}_{RD,l}\dot{\boldsymbol{P}}_{T}^{H}\right)\right) \end{aligned} \tag{33}$$

The scaling behaviors of above factors are shown in Table 1.

Proof: See the Appendix.

Several important results on the effect of EI, MUI and hardware impairments on achievable rate of $S_k \rightarrow R$ channel can be obtained from Theorem 4 and Table 1.

2. **Effect of EI:** With the proposed HIA scheme, the scaling behavior of EI is high correlated with $\lambda_{N_R-A_R+1}\left(\boldsymbol{C}_{EI}\right)$ and $\lambda_{N_T-A_T+1}(\tilde{\boldsymbol{C}}_{EI})$. With the assumption **A1**, we have $\lambda_{N_R-A_R+1}(\boldsymbol{C}_{EI})\leq O(1)$ and $\lambda_{N_T-A_T+1}(\tilde{\boldsymbol{C}}_{EI})\leq O(1)$. In fact, under specific spatial correlation model (e.g., the physical channel model with a fixed number of angular bins), it is possible to make $\lambda_{N_R-A_R+1}\left(\boldsymbol{C}_{EI}\right)< O(1)$ and $\lambda_{N_T-A_T+1}(\tilde{\boldsymbol{C}}_{EI})< O(1)$ by selecting proper A_R and A_T. In this case, the power of EI can decrease faster than $O(KN_R^{-1})$ in the large-(N_R, N_T) regime. Note that this is in contrast with the result in H. Q. Ngo, H. A. Suraweera, M. Matthaiou, & E. G. Larsson. (2014), which shown that the power of EI decreases exactly with $O(KN_R^{-1})$.

Table 1. Scaling behaviors of different factors in Theorem 4

	Desired Signal	AWGN	$E_{S,k}\,\mathrm{var}\left(\boldsymbol{w}_{R,k}^{H}\boldsymbol{H}_{SR,k}\dot{\boldsymbol{p}}_{S,k}\right)$	$\mathbb{E}\left[EI_{k}\right]$	$\mathbb{E}\left[MUI_{D,k}\right]$	$\mathbb{E}\left[D_{D,k}^{T}\right]$	$\mathbb{E}\left[D_{D,k}^{R}\right]$
Perfect Hardware	$O(1)$	$O\left(N_R^{-1}\right)$	$O\left(N_R^{-1}\right)$	$O\left(KN_R^{-1}\lambda_{N_R-A_R+1}\left(\boldsymbol{C}_{EI}\right)\times\lambda_{N_T-A_T+1}\left(\tilde{\boldsymbol{C}}_{EI}\right)\right)$	$O\left(N_R^{-1}\right)$	0	0
Perfect Hardware	$O(1)$	$O\left(N_R^{-1}\right)$	$\begin{cases}O(1), & N_S \text{ is fixed}\\ O\left(N_S^{-1}\right), & N_S \text{ scales with } N_R\end{cases}$	$O\left(KN_R^{-1}\lambda_{N_R-A_R+1}\left(\boldsymbol{C}_{EI}\right)\times\lambda_{N_T-A_T+1}\left(\tilde{\boldsymbol{C}}_{EI}\right)\right)$	$O\left(N_R^{-1}\right)$	$\begin{cases}O(1), & N_S \text{ is fixed}\\ O\left(\max\left\{N_S^{-1}, KN_R^{-1}\right\}\right), & N_S \text{ scales with } N_R\end{cases}$	$O\left(KN_R^{-1}\right)$

3. **Effect of MUI:** The scaling behavior of MUI confirms well with the classic result in MM-FDR with perfect hardware, which is given in H. Q. Ngo, H. A. Suraweera, M. Matthaiou, & E. G. Larsson. (2014), i.e., the MUI diminishes as $N_R \to \infty$ if $K \ll N_R$.
4. **Effect of Transmit Imperfection of Sources:** (i) When N_S is fixed, an interesting observation is that the term $E_{S,k}\,\mathrm{var}\left(\boldsymbol{w}_{R,k}^H \boldsymbol{H}_{SR,k} \dot{\boldsymbol{p}}_{S,k}\right)$, which is widely convinced $\mathcal{O}(N_R^{-1})$ terms with perfect hardware, scales as $\mathcal{O}(1)$. As a result, the achievable rate of $S_k \to R$ channel is limited by the joint effect of $E_{S,k}\,\mathrm{var}\left(\boldsymbol{w}_{R,k}^H \boldsymbol{H}_{SR,k} \dot{\boldsymbol{p}}_{S,k}\right)$ and the effective distortion due to the transmit imperfection of sources, i.e., $D_{R,k}^{\mathrm{T}}$. In fact, the variation of scaling behavior of $E_{S,k}\,\mathrm{var}\left(\boldsymbol{w}_{R,k}^H \boldsymbol{H}_{SR,k} \dot{\boldsymbol{p}}_{S,k}\right)$ can also be viewed as a "negative effect" of transmit imperfection of sources. To see this, inserting the expression of δ_k^{SR} (given by Lemma 1 in the Appendix) into (32) and letting $N_R \to \infty$, we can obtain

$$E_{S,k}\,\mathrm{var}\left(\boldsymbol{w}_{R,k}^H \boldsymbol{H}_{SR,k} \dot{\boldsymbol{p}}_{S,k}\right) = \frac{\nu_{S,k}}{\tau} E_{R,k} Tr\left(\hat{\boldsymbol{C}}_{SR,k}\right) \dot{\boldsymbol{p}}_{S,k}^H diag\left(\dot{\boldsymbol{p}}_{S,k}^H \dot{\boldsymbol{p}}_{S,k}\right) \dot{\boldsymbol{p}}_{S,k} + O\left(N_R^{-1}\right),$$

which demonstrates that $E_{S,k}\,\mathrm{var}\left(\boldsymbol{w}_{R,k}^H \boldsymbol{H}_{SR,k} \dot{\boldsymbol{p}}_{S,k}\right)$ scales as $O(1)$ as long as $\nu_{S,k} > 0$. (ii) When N_S scales with N_R and $K \ll N_R$, it is seen that the proposed HIA scheme reduces the effect of transmit imperfection at sources by a factor of $\frac{1}{N_S}$. The reason is that the transmit distortion is pre-suppressed by the transmit BF vector $\dot{\boldsymbol{p}}_{S,k}$.

5. **Effect of Receive Imperfection of the Relay:** By comparing the result in Table 1 and (15), it is seen that the HIA scheme does not change the scaling behavior of distortion noise caused by receive imperfection at the relay. However, we point out that the receive distortion noise power is still be reduced, since the power of EI (which is a main contributor of receive distortion noise) is suppressed.
6. **Achievable Rate of $R \to D_k$ Channel:** According to (5) and the bounding technique in B. Hassibi, & B. M. Hochwald. (2003), the achievable rate of $R \to D_k$ channel can be expressed as

$$R_{RD,k}^{HIA} = \frac{T - 2K\tau}{T} \log_2\left(1 + \frac{E_{R,k}\left|\mathbb{E}\left[\dot{\boldsymbol{p}}_{D,k}^H \boldsymbol{H}_{RD,k}^H \boldsymbol{w}_{T,k}\right]\right|^2}{E_{R,k}\,\mathrm{var}\left(\dot{\boldsymbol{p}}_{D,k}^H \boldsymbol{H}_{RD,k}^H \boldsymbol{w}_{T,k}\right) + \mathbb{E}\left[MUI_{D,k}\right] + \mathbb{E}\left[D_{D,k}^T\right] + D_{D,k}^R + 1}\right) \tag{34}$$

where $MUI_{D,k}$, $D_{D,k}^T$ and $D_{D,k}^R$ denote the MUI, effective distortion due to transmit imperfection of relay and effective distortion due to receive imperfection of destinations, respectively, which are given by

$$
\begin{aligned}
MUI_{D,k} &= \sum_{j=1,j\neq k}^{K} E_{R,j}\left|\dot{\boldsymbol{p}}_{D,k}^{H}\boldsymbol{H}_{RD,k}^{H}\boldsymbol{w}_{T,j}\right|^{2} \\
D_{D,k}^{T} &= \dot{\boldsymbol{p}}_{D,k}^{H}\boldsymbol{H}_{RD,k}^{H}\boldsymbol{\Theta}_{R}^{T}\boldsymbol{H}_{RD,k}\dot{\boldsymbol{p}}_{D,k} \\
D_{D,k}^{R} &= \dot{\boldsymbol{p}}_{D,k}^{H}\boldsymbol{\Theta}_{D,k}^{R}\dot{\boldsymbol{p}}_{D,k}
\end{aligned} \tag{35}
$$

Theorem 5: Assume that assumptions **A1** and **A2** hold, and A_T is selected so that $Tr\left(\underline{\boldsymbol{C}}_{RD,k}\right) = O(N_T)$.

With HIA scheme, the achievable rate of $R \to D_k$ channel in the large-(N_R, N_T) regime is given by (34), where the power of desired signal is

$$
E_{R,k}\left|\mathbb{E}\left[\dot{\boldsymbol{p}}_{D,k}^{H}\boldsymbol{H}_{RD,k}^{H}\boldsymbol{w}_{T,k}\right]\right|^{2} = E_{R,k}Tr\left(\hat{\underline{\boldsymbol{C}}}_{RD,k}\right),
$$

and

$$
\begin{cases}
E_{R,k}\,\mathrm{var}\left(\dot{\boldsymbol{p}}_{D,k}^{H}\boldsymbol{H}_{RD,k}^{H}\boldsymbol{w}_{T,k}\right) = \left(Tr\left(\hat{\underline{\boldsymbol{C}}}_{RD,k}\right)\right)^{-1}E_{R,k}\delta_{k}^{RD} \\
\mathbb{E}\left[MUI_{D,k}\right] = \sum_{j=1,j\neq k}^{K}\left(Tr\left(\hat{\underline{\boldsymbol{C}}}_{RD,j}\right)\right)^{-1}E_{R,j}Tr\left(\left(\underline{\boldsymbol{C}}_{RD,k} - \hat{\underline{\boldsymbol{C}}}_{RD,k}\right)\hat{\underline{\boldsymbol{C}}}_{RD,j}\right) \\
\mathbb{E}\left[D_{D,k}^{T}\right] = \nu_{R}\sum_{l=1}^{K}E_{R,l}\left(Tr\left(\hat{\underline{\boldsymbol{C}}}_{RD,l}\right)\right)^{-1}\lambda_{1}\left(\tilde{\boldsymbol{C}}_{RD,k}\right)Tr\left(\boldsymbol{C}_{RD,k}diag\left(\dot{\boldsymbol{P}}_{T}\hat{\underline{\boldsymbol{C}}}_{RD,l}\dot{\boldsymbol{P}}_{T}^{H}\right)\right) \\
D_{D,k}^{R} = \dfrac{\mu_{D,k}E_{R,k}}{Tr\left(\hat{\underline{\boldsymbol{C}}}_{RD,k}\right)}\left(\dfrac{\dot{\boldsymbol{p}}_{D,k}^{H}diag\left(\dot{\boldsymbol{p}}_{D,k}\dot{\boldsymbol{p}}_{D,k}^{H}\right)\dot{\boldsymbol{p}}_{D,k}}{\left(Tr\left(\hat{\underline{\boldsymbol{C}}}_{RD,k}\right)\right)^{-2}}\right. \\
+\dfrac{\mu_{R}}{\tau}\left(\dfrac{1}{E_{T}} + \beta_{RD,k}\left(\lambda_{1}\left(\tilde{\boldsymbol{C}}_{RD,k}\right) + \nu_{D,k}\right)\right)\dfrac{Tr\left(\underline{\boldsymbol{C}}_{RD,k}^{3}\boldsymbol{\Gamma}_{RD,k}^{2}\right)}{\lambda_{1}\left(\tilde{\boldsymbol{C}}_{RD,k}\right)} + \dfrac{1}{\tau E_{T}}\dfrac{Tr\left(\underline{\boldsymbol{C}}_{RD,k}^{3}\boldsymbol{\Gamma}_{RD,k}^{2}\right)}{\lambda_{1}\left(\tilde{\boldsymbol{C}}_{RD,k}\right)} \\
\left.+\sum_{j=1,j\neq k}^{K}\dfrac{E_{R,j}}{E_{R,k}}\beta_{RD,j}Tr\left(\boldsymbol{C}_{RD,k}\dot{\boldsymbol{P}}_{T}\hat{\underline{\boldsymbol{C}}}_{RD,j}\dot{\boldsymbol{P}}_{T}^{H}\right) + \dfrac{Tr\left(\hat{\underline{\boldsymbol{C}}}_{RD,k}\right)}{E_{R,k}}\right)
\end{cases} \tag{36}
$$

δ_{k}^{RD} is defined as $\delta_{k}^{RD} = \mathbb{E}\left[\left|\Delta\underline{\boldsymbol{h}}_{RD,k}^{H}\hat{\underline{\boldsymbol{h}}}_{RD,k}\right|^{2}\right]$ and is given by Lemma 1 in the Appendix. The scaling behaviors of above factors are shown in Table 2.

Proof: The proof is similar with that for Theorem 4, and thus it is neglected.

From Theorem 5 and Table 2, one can observe a strong similarity between the effects of hardware impairments on $R \to D_k$ channel and that on $S_k \to R$ channel. In particular, the effective distortion due to the transmit imperfection of relay $D_{D,k}^{\mathrm{T}}$ is suppressed by the large transmit array of the relay, and it diminishes as $N_T \to \infty$ and $K \ll N_T$. Moreover, when N_D is fixed, the achievable rate of $R \to D_k$

Table 2. Scaling behaviors of different factors in Theorem 5

	Desired Signal	AWGN	$E_{S,k}\ \mathrm{var}\left(\boldsymbol{p}_{D,k}^{H}\boldsymbol{H}_{RD,k}^{H}\boldsymbol{w}_{T,k}\right)$	$\mathbb{E}\left[MUI_{D,k}\right]$	$\mathbb{E}\left[D_{D,k}^{T}\right]$	$\mathbb{E}\left[D_{D,k}^{R}\right]$
Perfect Hardware	$O(N_T)$	$O(1)$	$O(1)$	$O\left(N_T^{-1}\right)$	0	0
Perfect Hardware	$O(N_T)$	$O(1)$	$\begin{cases} O(N_T), & N_D \text{ is fixed} \\ O\left(\frac{N_T}{N_D}\right), & N_D \text{ scales with } N_T \end{cases}$	$O\left(N_T^{-1}\right)$	$O(K)$	$\begin{cases} O(N_T), & N_D \text{ is fixed} \\ O\left(\max\left\{\frac{N_T}{N_D}, K\right\}\right), & N_D \text{ scales with } N_T \end{cases}$

channel is limited by the term $E_{R,k}\ \mathrm{var}\left(\dot{\boldsymbol{p}}_{D,k}^{H}\boldsymbol{H}_{RD,k}^{H}\boldsymbol{w}_{T,k}\right)$ and distortion noise caused by receive imperfection of D_k. However, as N_D scales with N_T and $K \ll N_T$, these factors are suppressed by $\frac{1}{N_D}$. This is because that applying $\dot{\mathbf{p}}_{D,k}$ can be viewed as a post-suppression on receive distortion noise.

With HIA scheme, the following end-to-end rate for $S_k \to R \to D_k$ channel is achievable in the large-(N_R, N_T) regime

$$R_k^{HIA} = \min\left(R_{SR,k}^{HIA}, R_{RD,k}^{HIA}\right) \tag{37}$$

Different from that for upper bound in the last section, R_k^{HIA} can not be expressed in a simple form like (15). However, based on the analysis under Theorem 4 and Theorem 5, one can simply deduce that R_k^{Lower} will converge to a finite ceiling if N_S and N_D are fixed. However, with HIA scheme, the end-to-end achievable rate grows without bound as $\min(N_R, N_T) \to \infty$ if N_S and N_D scale with $\min(N_R, N_T)$ and $K \ll \min(N_R, N_T)$. This is the same to the situation with ideal hardware, which is discussed in H. Q. Ngo, H. A. Suraweera, M. Matthaiou, & E. G. Larsson. (2014).

Discussion on Hardware Design

Although the achievable rate expressions hold for arbitrary $K \le \min\{N_R, N_T\}$, we will restrict our analysis to $K \ll \min\{N_R, N_T\}$. This is a fundamental condition to ensure the benefit of massive MIMO, i.e., the MUI is suppressed by a large surplus of degrees of freedom, which is discussed in E. G. Larsson, O. Edfors. (2014). In MM-FDR with hardware impairments, it is also essential to control the number of S-D pairs to suppress the EI and distortions, as shown in Theorem 4 and Theorem 5.

1. **Hardware Design of Sources and Destinations:** As N_S and N_D are fixed and small, from Theorem 4 and 5, it is of great importance to reduce $\nu_{S,k}$, $\nu_{D,k}$ and $\mu_{D,k}$ in order to alleviate the "ceiling effect" on achievable rate. This indicates that it is beneficial to use high-quality hardware at sources and destinations in this case. Note that, different from the relay side, the use of high-quality hardware at sources and destinations is affordable, since the increased cost scales with N_S and N_D, instead of $\min\{N_R, N_T\}$.

When N_S and N_D scale with $\min\{N_R, N_T\}$, a key observation from Theorem 4 and 5 is that the distortion noises due to hardware impairments at sources and destinations are lower order terms in $\min\{N_R, N_T\}$ when compared with the desired signal powers. This means that the sources and destinations can actually decrease hardware quality to some extent without degrading the performance.

2. **Hardware Design of Relay:** Similar to that for sources and destinations (when N_S and N_D scale with $\min\{N_R, N_T\}$), the distortion noises caused by hardware impairments at relay are lower order

terms in $\min\{N_R, N_T\}$ when compares with the desired signal powers, which makes it possible to decrease the hardware quality without hurting the performance greatly.

Example: To get a clear insight, we consider the achievable rate of $S \rightarrow R$ channel given by Theorem 4. We assume that there is only one source-destination pair and let $\nu_R = \mu_R$. When N_S scales with N_R, the achievable rate of $S_k \rightarrow R$ channel reduces to

$$R_{SR,k}^{HIA} = \frac{T-2K\tau}{T}\log_2\left(1+\frac{1}{c_1\nu_{S,1}N_R^{-1}+c_2\mu_R N_R^{-1}+O\left(N_R^{-1}\right)}\right) \tag{38}$$

where c_1 and c_2 are positive constants independent of $\nu_{S,1}$, μ_R and $\{N_S, N_R\}$. In (38), we have replaced N_S with the product of N_R and some constant. Assume we wish to achieve rate R_T. According to (38), we have (as $N_R \rightarrow \infty$)

$$c_1\nu_{S,1}+c_2\mu_R = \left(\left(2^{\frac{T}{T-2K\tau}R_T}-1\right)^{-1}-O\left(N_R^{-1}\right)\right)N_R \approx \left(2^{\frac{T}{T-2K\tau}R_T}-1\right)^{-1}N_R \tag{39}$$

The result in (39) is encouraging since it implies that we can increase $\nu_{S,1}$ and μ_R linearly by increasing N_R (with N_S scales with N_R) without degrading the achievable rate of $S_k \rightarrow R$ channel. As a result, inexpensive MM-FDR is possible. For example, let us assume that the receive distortion noise at the relay is dominated by the ADC quantization error. Let P denote the signal power, if the ADC quantizes the received signal to a B bit resolution and automatic gain control is used, the output of ADC can be modeled as a combination of desired signal with reduced power $(1-2^{-2B})P$ and an additive uncorrelated quantization error with power $2^{-2B}P$, which is shown in A. Mezghani, N. Damak. (2010). Normalizing the desired signal power with factor $\frac{1}{1-2^{-2B}}$, the quantization process can be characterized using the imperfect receive RF chain model (2), where $\mu_R = \frac{2^{-2B}}{1-2^{-2B}}$ or

$$B = \frac{1}{2}\log_2\frac{1+\mu_R}{\mu_R} \approx \frac{1}{2}\log_2\frac{1}{\mu_R} \tag{40}$$

where the approximation is due to $\mu_R \ll 1$. Assuming $c_1\nu_{S,1}$ in (39) is comparable with $c_2\mu_R$, (39) can be rewritten as

$$c_2\mu_R \approx c_3\left(2^{\frac{T}{T-2K\tau}R_T}-1\right)^{-1}N_R$$

with $0 < c_3 < 1$. With this and (40), if relay uses 64 antennas instead of one, we can reduce the ADC resolution per antenna by 3 without degrading the performance, which is a considerable reduction in hardware cost.

JOINT DEGREE OF FREEDOM AND POWER OPTIMIZATION

Let $E_{S,k}^{\max}$ and $E_R^{\max}$ be the maximum transmit power constraints at S_k and relay, respectively, the SE optimization problem can be formulated as follows

$$\max_{E_{S,1},\cdots,E_{S,K};E_{R,1},\cdots,E_{R,K};A_R,A_T} SE = \sum_{k=1}^{K} \min\left\{R_{SR,k}^{HIA}, R_{RD,k}^{HIA}\right\} s.t. \begin{cases} 0 \le E_{S,k} \le E_{S,k}^{\max}, k = 1,\cdots,K \\ \sum_{k=1}^{K} E_{R,k} \le E_R^{\max} \\ A_R \in \mathcal{A}_R, A_T \in \mathcal{A}_T \end{cases} \tag{41}$$

where $\mathcal{A}_R = \{K, K+1,\cdots,N_R\}$ and $\mathcal{A}_T = \{K, K+1,\cdots,N_T\}$.

To facilitate the analysis, $R_{SR,k}^{HIA}$ and $R_{RD,k}^{HIA}$ should be expressed as more mathematically tractable form. For such, according to Theorem 4 and Theorem 5, we rewrite $R_{SR,k}^{HIA}$ and $R_{RD,k}^{HIA}$ as

$$R_{SR,k}^{HIA} = \frac{T - 2K\tau}{T}\log_2\left(1 + \gamma_{SR,k}\right)$$

and

$$R_{RD,k}^{HIA} = \frac{T - 2K\tau}{T}\log_2\left(1 + \gamma_{RD,k}\right),$$

where $\gamma_{SR,k}$ and $\gamma_{RD,k}$ denote the effective received SINRs of $S_k \to R$ and $R \to D_k$ channels respectively, which can be expressed as

$$\begin{aligned} \gamma_{SR,k} &= \frac{E_{S,k}}{\sum_{j=1}^{K} a_{k,j}^R E_{S,j} + \sum_{j=1}^{K} b_{k,j}^R E_{R,j} + \left(\mu_R + 1\right)\left(Tr\left(\hat{\underline{C}}_{SR,k}\right)\right)^{-1}} \\ \gamma_{RD,k} &= \frac{Tr\left(\hat{\underline{C}}_{RD,k}\right) E_{R,k}}{\sum_{j=1}^{K} b_{k,j}^D E_{R,j} + \mu_{D,k} + 1} \end{aligned} \tag{42}$$

where

$$a_{kj}^{R} = \left(Tr\left(\hat{\underline{\boldsymbol{C}}}_{SR,k}\right)\right)^{-2} \delta_{kj}^{SR} + \nu_{S,k}\left(Tr\left(\hat{\underline{\boldsymbol{C}}}_{SR,k}\right)\right)^{-2} g_{kj} + \mu_R \beta_{SR,j} \frac{Tr\left(\tilde{\boldsymbol{C}}_{SR,j}\boldsymbol{\Omega}_{S,j}\right)}{Tr\left(\hat{\underline{\boldsymbol{C}}}_{SR,k}\right)}$$

$$b_{kj}^{R} = \frac{\beta_{EI} Tr\left(\tilde{\boldsymbol{C}}_{EI}\left(\dot{\boldsymbol{P}}_T \hat{\underline{\boldsymbol{C}}}_{RD,j} \dot{\boldsymbol{P}}_T^H + \nu_R diag\left(\dot{\boldsymbol{P}}_T \hat{\underline{\boldsymbol{C}}}_{RD,j} \dot{\boldsymbol{P}}_T^H\right)\right)\right)}{Tr\left(\hat{\underline{\boldsymbol{C}}}_{SR,k}\right) Tr\left(\hat{\underline{\boldsymbol{C}}}_{RD,j}\right)} \left(\frac{Tr\left(\hat{\underline{\boldsymbol{C}}}_{SR,k} \dot{\boldsymbol{P}}_R^H \boldsymbol{C}_{EI} \dot{\boldsymbol{P}}_R\right)}{Tr\left(\hat{\underline{\boldsymbol{C}}}_{SR,k}\right)} + \mu_R\right)$$

$$b_{kj}^{D} = \begin{cases} \dfrac{\delta_{kk}^{RD}}{Tr\left(\hat{\underline{\boldsymbol{C}}}_{RD,k}\right)} + \dfrac{\nu_R Tr\left(\boldsymbol{C}_{RD,k} diag\left(\dot{\boldsymbol{P}}_T \hat{\underline{\boldsymbol{C}}}_{RD,k} \dot{\boldsymbol{P}}_T^H\right)\right)}{\left(\lambda_1\left(\tilde{\boldsymbol{C}}_{RD,k}\right)\right)^{-1} Tr\left(\hat{\underline{\boldsymbol{C}}}_{RD,k}\right)} + \dfrac{\mu_{D,k} \dot{\boldsymbol{p}}_{D,k}^H diag\left(\dot{\boldsymbol{p}}_{D,k} \dot{\boldsymbol{p}}_{D,k}^H\right) \dot{\boldsymbol{p}}_{D,k}}{\left(Tr\left(\hat{\underline{\boldsymbol{C}}}_{RD,k}\right)\right)^{-1}} \\ + \left(\dfrac{1}{E_T} + \beta_{RD,k}\left(\lambda_1\left(\tilde{\boldsymbol{C}}_{RD,k}\right) + \nu_{D,k}\right) + \dfrac{1}{E_T \mu_R}\right) \dfrac{\mu_{D,k} \mu_R Tr\left(\underline{\boldsymbol{C}}_{RD,k}^3 \boldsymbol{\Gamma}_{RD,k}^2\right)}{\tau \lambda_1\left(\tilde{\boldsymbol{C}}_{RD,k}\right) Tr\left(\hat{\underline{\boldsymbol{C}}}_{RD,k}\right)}, j = k \\ \dfrac{\delta_{kj}^{RD}}{Tr\left(\hat{\underline{\boldsymbol{C}}}_{RD,j}\right)} + \nu_R \dfrac{Tr\left(\boldsymbol{C}_{RD,k} diag\left(\dot{\boldsymbol{P}}_T \hat{\underline{\boldsymbol{C}}}_{RD,j} \dot{\boldsymbol{P}}_T^H\right)\right)}{\left(\lambda_1\left(\tilde{\boldsymbol{C}}_{RD,k}\right)\right)^{-1} Tr\left(\hat{\underline{\boldsymbol{C}}}_{RD,j}\right)} + \dfrac{\mu_{D,k} \beta_{RD,j} Tr\left(\boldsymbol{C}_{RD,k} \dot{\boldsymbol{P}}_T \hat{\underline{\boldsymbol{C}}}_{RD,j} \dot{\boldsymbol{P}}_T^H\right)}{Tr\left(\hat{\underline{\boldsymbol{C}}}_{RD,k}\right)}, j \neq k \end{cases}$$

According to the results in the above and the formula

$$\frac{T - 2K\tau}{T} \sum_{k=1}^{K} \log_2\left(1 + \gamma_k\right) = \frac{T - 2K\tau}{T} \log_2\left(\prod_{k=1}^{K}\left(1 + \gamma_k\right)\right)$$

(where γ_k is defined as $\gamma_k = \min\{\gamma_{SR,k}, \gamma_{RD,k}\}$), problem (41) can be rewritten as

$$\min_{E_{S,1},\cdots,E_{S,K};E_{R,1},\cdots,E_{R,K};A_R,A_T} \prod_{k=1}^{K}\left(1 + \gamma_k\right)^{-1}$$
$$s.t. \begin{cases} C_1 : \dfrac{E_{S,k}}{\sum_{j=1}^{K} a_{k,j}^R E_{S,j} + \sum_{j=1}^{K} b_{k,j}^R E_{R,j} + \left(\mu_R + 1\right)\left(Tr\left(\hat{\underline{\boldsymbol{C}}}_{SR,k}\right)\right)^{-1}} \geq \gamma_k, k = 1,2,\cdots,K \\ C_2 : \dfrac{Tr\left(\hat{\underline{\boldsymbol{C}}}_{RD,k}\right) E_{R,k}}{\sum_{j=1}^{K} b_{k,j}^D E_{R,j} + \mu_{D,k} + 1} \geq \gamma_k, k = 1,2,\cdots,K \\ C_3 : E_{S,k} \leq E_{S,k}^{\max}, k = 1,2,\cdots,K \\ C_4 : \sum_{j=1}^{K} E_{R,j} \leq E_R \\ C_5 : A_R \in \mathcal{A}_R, A_T \in A_T \end{cases} \tag{43}$$

By using some algebra manipulations on the inequality constraints, C_1, C_2 can be rewritten as

Algorithm 1. Solve the power control problem with fixed A_R and A_T by GP

1: Initialization: Let $\hat{\gamma}_k^{(i)}$ denote the solution of γ_k after the ith iteration. Compute the initial value $\hat{\gamma}_k^{(0)}$ using (42). Set a tolerance ε and the maximum iteration times L_{GP}.

2: For the (i+1)th iteration:

a. Computer $\omega_k = \hat{\gamma}_k^{(i)} / \left(1+\hat{\gamma}_k^{(i)}\right)$ and $\theta_k = \left(\hat{\gamma}_k^{(i)}\right)^{-b_k}\left(1+\hat{\gamma}_k^{(i)}\right)$.

b. Solve the following GP problem $\min\limits_{E_{S,1},\cdots,E_{S,K};E_{R,1},\cdots,E_{R,K}} \prod_{k=1}^{K} \theta_k^{-1}\gamma_k^{-\omega_k}, s.t.\ C_1 \sim C_5$.

3: If $\max\limits_{k=1,2,\cdots,K}\left|\hat{\gamma}_k^{(i+1)} - \hat{\gamma}_k^{(i)}\right| < \varepsilon$ or $i+1 = L_{GP}$, stop. Otherwise, set $i=i+1$ and go back to step 2.

$$
\begin{aligned}
&C_1 : \sum_{j=1}^{K} a_{k,j}^R E_{S,k}^{-1} E_{S,j}\gamma_k + \sum_{j=1}^{K} b_{k,j}^R E_{S,k}^{-1} E_{R,j}\gamma_k + \left(\mu_R + 1\right)\left(Tr\left(\underline{\hat{C}}_{SR,k}\right)\right)^{-1} E_{S,k}^{-1}\gamma_k \le 1 \\
&C_2 : \sum_{j=1}^{K} b_{k,j}^D E_{R,k}^{-1} E_{R,j}\gamma_k + \left(\mu_{D,k} + 1\right) E_{R,k}^{-1}\gamma_k \le Tr\left(\underline{\hat{C}}_{RD,k}\right)
\end{aligned} \tag{44}
$$

Problem (43) is a combinatorial optimization problem which is in general NP hard. Moreover, A_R and A_T do not explicitly appear in the target function, which makes the problem more challenging. To solve (43), we propose a JDPO algorithm to find suboptimal solution. Our strategy is as follows. First, using the similar approach as that in P. C. Weeraddana, M. Codreanu. (2011), we show that the power control problem with fixed A_R and A_T can be approximated as a geometric programming (GP) and solved efficiently using the convex optimization tools, as shown in S. Boyd, S.-J. Kim. (2007). Then we present a heuristic approach based on sequential optimization to solve A_R and A_T.

3. **Power Control with Fixed A_R and A_T** : From (43) and (44), the inequality constraints are posynomial functions, which is proven in S. Boyd, S.-J. Kim. (2007). As shown in P. C. Weeraddana, M. Codreanu. (2011), for any $\gamma_k > 0$, $1+\gamma_k$ can be approximated using a monomial function $\theta_k\gamma_k^{\omega_k}$ near a point $\hat{\gamma}_k$, where $\omega_k = \hat{\gamma}_k / \left(1+\hat{\gamma}_k\right)$ and $\theta_k = \hat{\gamma}_k^{-\omega_k}\left(1+\hat{\gamma}_k\right)$. By using this, the target function of (43) can be approximated as a monomial function, i.e., $\prod_{k=1}^{K}\theta_k^{-1}\gamma_k^{-\omega_k}$. As a result, the solution for power control problem (with fixed A_R and A_T) can be obtained by solving several GPs. The algorithm is summarized in Algorithm 1.
4. **Optimization of A_R and A_T** : An obvious approach to obtain A_R and A_T is via 2-D search. However, the complexity is very high since the power control problem described in the above should be computed for each (A_R, A_T). To reduce the complexity, we present a heuristic approach to find suboptimal A_R and A_T. In this approach, A_R (A_T) is searched over a subset $\tilde{A}_R$ ($\tilde{A}_T$), whose

Algorithm 2. JDPO algorithm for SE optimization

```
Initialization: Set an initial A_T, i.e., A_T^(0), and a maximum repeat times L.
Select the subsets Ã_R and Ã_T .
for all l=1,…,L do
Set A_T = A_T^(l−1), and compute SE using Algorithm 1 for all A_R ∈ Ã_R.
Update A_R^(l) = arg max_{A_R} SE.
Set A_R = A_R^(l) and compute SE using Algorithm 1 for all A_T ∈ Ã_T .
Update A_T^(l) = arg max_{A_T} SE.
end for
```

elements are sampled from A_R (A_T). For example, $\tilde{A}_R$ can be selected as $\left\{K, 2K, \cdots, \left\lfloor \frac{N_R}{K} \right\rfloor K\right\}$.

The motivation of this idea is that the target function does not change dramatically as A_R and A_T increase/decrease by a small step as observed from the numerical results. Moreover, instead of 2-D search, A_R and A_T are optimized sequentially over the sampled subsets $\tilde{A}_R$ and $\tilde{A}_T$ via 1-D search. The process repeats several times (much less than $\min\left\{\left|\tilde{A}_R\right|, \left|\tilde{A}_T\right|\right\}$) to improve the solution.

A summary for the JDPO algorithm is presented in Algorithm 2. Note that the algorithm converges to a local optimum since the target function is improved in each iteration. The complexity of Algorithm 2 is upper bounded as $L\left(\left|\tilde{A}_R\right|, \left|\tilde{A}_T\right|\right) L_{GP} C_{GP}$, where C_{GP} is the complexity to solve GP when $A_R = N_R$ and $A_T = N_T$. Usually, GP is solved using inner point method with polynomial time. The exact expression of C_{GP} is quite difficult and related with the structure of the problem. Some insights on C_{GP} can be found in S. Boyd, & L. Vandenberghe. (2004), Sec. 11.5.

Remark 2: (JDPO for EE Optimization) Instead of the SE optimization, we can also formulate an EE optimization problem subject to a target SE R_T

$$\max_{E_{S,1},\cdots,E_{S,K};E_{R,1},\cdots,E_{R,K};A_R,A_T} \frac{1}{\sum_{k=1}^{K} E_{S,k} + \sum_{k=1}^{K} E_{R,k}} \sum_{k=1}^{K} \min\left\{R_{SR,k}^{HIA}, R_{RD,k}^{HIA}\right\}$$

$$s.t. \begin{cases} \sum_{k=1}^{K} \min\left\{R_{SR,k}^{HIA}, R_{RD,k}^{HIA}\right\} = R_T \\ 0 \le E_{S,k} \le E_{S,k}^{\max}, k = 1, \cdots, K \\ \sum_{k=1}^{K} E_{R,k} \le E_R^{\max} \end{cases} \tag{45}$$

Using (42), the above problem is equivalent to

$$\max_{E_{S,1},\cdots,E_{S,K};E_{R,1},\cdots,E_{R,K};A_R,A_T} \sum_{k=1}^{K} E_{S,k} + \sum_{k=1}^{K} E_{R,k}, s.t.\ C_1 \sim C_5$$

and

$$\prod_{k=1}^{K}\left(1+\gamma_k\right) = 2^{\frac{T}{T-2K\tau}R_T}.$$

With the technique in P. C. Weeraddana, M. Codreanu. (2011), the equality constraint can be converted to a monomial, and the power control problem becomes a GP.

SIMULATION RESULTS AND DISCUSSION

Throughout this section, we set $\nu_{S,k} = \nu_S$, $\nu_{D,k} = \nu_D$, $\mu_{D,k} = \mu_D$ for convenience. The correlation matrices of desired channels are generated with the exponential correlation model, as given in S. Loyka. (2001)

$$\left[\boldsymbol{C}_{i,k}\ or\ \tilde{\boldsymbol{C}}_{i,k}\right]_{l,j} = \begin{cases} r_{i,k}^{j-l}, l \le j \\ \left(r_{i,k}^{l-j}\right)^{*}, l > j \end{cases}, i \in \left\{SR, RD\right\} \tag{46}$$

The model approximates the property ULA, where the correlation between adjacent antennas is $\left|r_{i,k}\right| \in [0,1]$ and the phase of $r_{i,k}$ describes the angle of arrival/departure as seen from the array. E. Bjornson, D. Hammarwall. (2009) shows how to map some of the practical parameters of ULA to the above model. The correlation matrices of EI channel $\boldsymbol{C}_{EI}$ and $\tilde{\boldsymbol{C}}_{EI}$ are generated similarly with a parameter r_{EI}. For convenience, we let $\left|r_{i,k}\right| = r_0$, $\forall k \in \{1,\cdots,K\}$. The phases of $r_{i,k}$ and r_{EI} are uniformly selected from $[0,\pi]$.

We assume that 25~35dB EI cancellation can be provided by passive EI suppression techniques. In all simulations, the variances of EI channel and desired channels are selected as $\beta_{EI} / \beta_{i,k} \in [0,25]$dB. Based on simplified path loss model, which is given in A. Goldsmith. (2005), p. 75 (for desired channels) and free-space path loss model, which is given in A. Goldsmith. (2005), p. 32 (for EI channel), the above range corresponds to the setup with the distances from sources/destinations to relay varying from 250m to 500m and 10m segregation between relay transmit and receive arrays.

Impact of Hardware Impairments

This subsection considers the effect of hardware impairments on SE of MM-FDR. The channel coherent time is set to $T = 300$ and the length of pilot sequence is $\tau = 2$.

Figure 1 shows the SE of MM-FDR with single antenna at sources/destinations with different levels of hardware impairments. The SEs based on transceiver scheme in section IV and HIA scheme are simulated. From Fig. 1, the SE is more sensitive to hardware impairments at sources and destinations. When $\nu_S = \nu_D = \mu_D = 0.2^2$, the SE approaches to a finite ceiling quickly as the number of relay antennas increases. Similar results can be observed when sources and destinations are equipped with multiple but fixed number of antennas in Fig. 2. This matches well with the conclusion in sections IV and V. However, the result changes when N_S and N_D scale with $\min\{N_R, N_T\}$ and HIA scheme is used. From Figure 2, as $N_S = \left\lfloor \frac{N_R}{K} \right\rfloor$ and $N_D = \left\lfloor \frac{N_T}{K} \right\rfloor$, similar SEs are achieved as sources/destinations or relay

Figure 1. SE of traditional MM-FDR with single antenna at sources/destinations, where $N_S = N_D = 1$, $K = 10$, $E_{S,k} = E_{R,k} = 8\,dB$, $E_T = 10\,dB$, $\beta_{SR,k} = \beta_{RD,k} = \beta_{EI} = 1$, $r_0 = 0.2$, $|r_{EI}| = 0.8$, $A_R = \max\left\{K, \left\lfloor \frac{2}{3} N_R \right\rfloor\right\}$, $A_T = \max\left\{K, \left\lfloor \frac{2}{3} N_T \right\rfloor\right\}$.

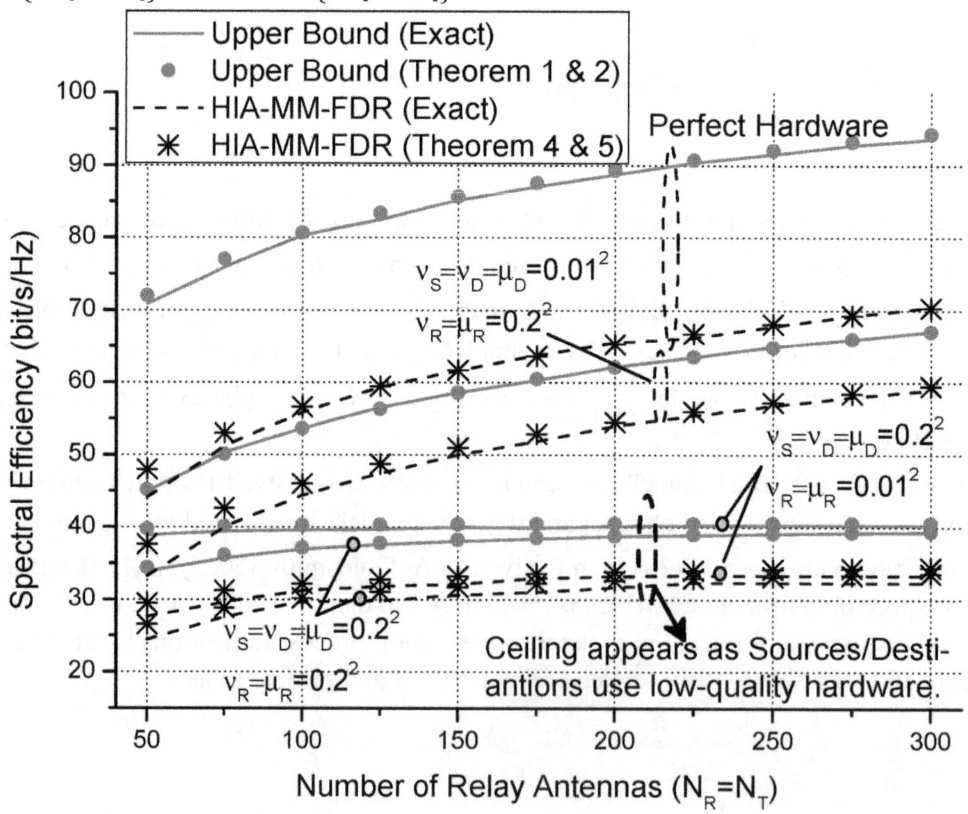

Figure 2. SE of MM-FDR with multiple antennas at sources/destinations. The setup is the same with Figure 1.

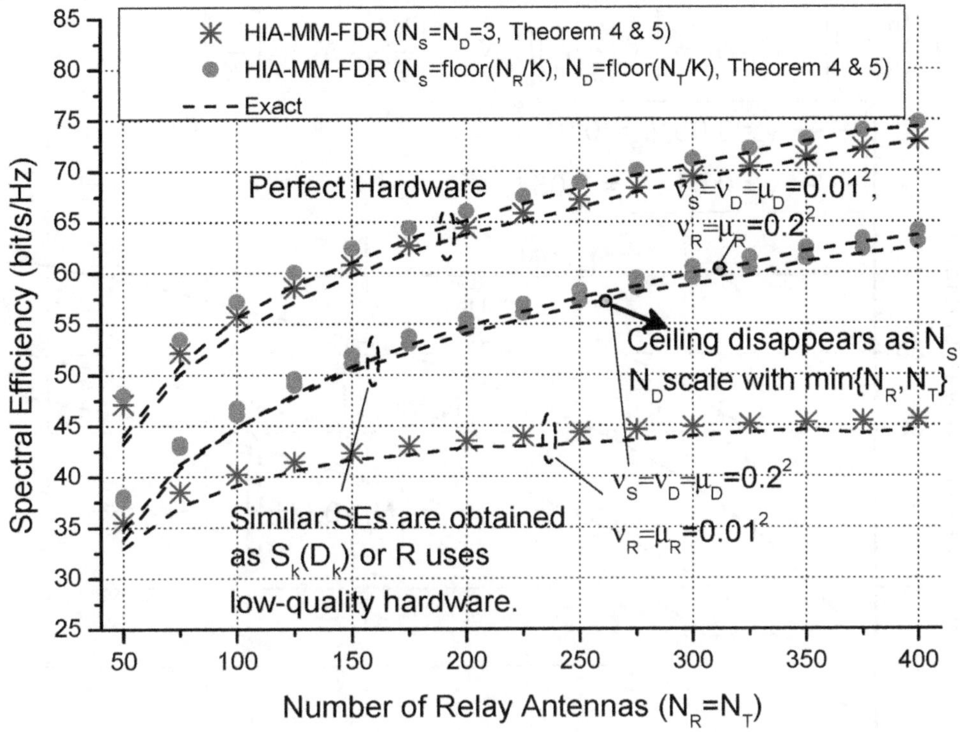

employ low-quality hardware, and no performance ceiling appears. This demonstrates the validness of HIA scheme. At last, it is seen that the asymptotic results in Theorem 1 & 2 and Theorem 4 & 5 match well with exact results.

As the performance is affected by impairments of both transmit and receive RF chains at relay, we compare the effect of transmit and receive imperfections on SE of HIA-MM-FDR in Figure 3. We assume that the sources and destinations use high-quality hardware ($\nu_S = \nu_R = \mu_R = 0.01^2$). The number of receive antennas at the relay is set to $N_R = 0.5N_T$, $N_R = N_T$ and $N_R = 2N_T$. From the figure, the effect of receive imperfection is more detrimental. The reason is that, with EI at the relay, the power of distortion noise caused by receive imperfection is much stronger than that due to transmit imperfection. The performance difference decreases as N_R increases from $0.5N_T$ to $N_R = 2N_T$, since the power of effective receive distortion scales as $O\left(KN_R^{-1}\right)$ as shown in Theorem 4. This indicates that use relatively higher-quality hardware or more antennas at the receive side of relay is beneficial to improve SE. Using a similar setup, one can obtain a parallel conclusion for destination, i.e., the receive imperfection is more harmful than transmit imperfection. This is because that the transmit imperfection of destination only induces larger channel estimator errors (see Theorem 3), which is a part of the receive signal at the destination. Thus, according to model (2), the distortion due to receive imperfection will be stronger than that due to transmit imperfection.

Figure 3. Comparison between transmit/receive imperfections at the relay, where $\nu_S = \nu_D = \mu_D = 0.01^2$, $K = 10$, $E_{S,k} = E_{R,k} = 5\,dB$, $E_T = 10\,dB$, $\beta_{SR,k} = \beta_{RD,k} = 1$, $\beta_{EI} = 5\,dB$, $r_0 = 0.2$, $|r_{EI}| = 0.8$, $A_R = \max\left\{K, \left\lfloor \frac{2}{3} N_R \right\rfloor\right\}$, $A_T = \max\left\{K, \left\lfloor \frac{2}{3} N_T \right\rfloor\right\}$, $N_S = \left\lfloor \frac{N_R}{K} \right\rfloor$, $N_D = \left\lfloor \frac{N_T}{K} \right\rfloor$.

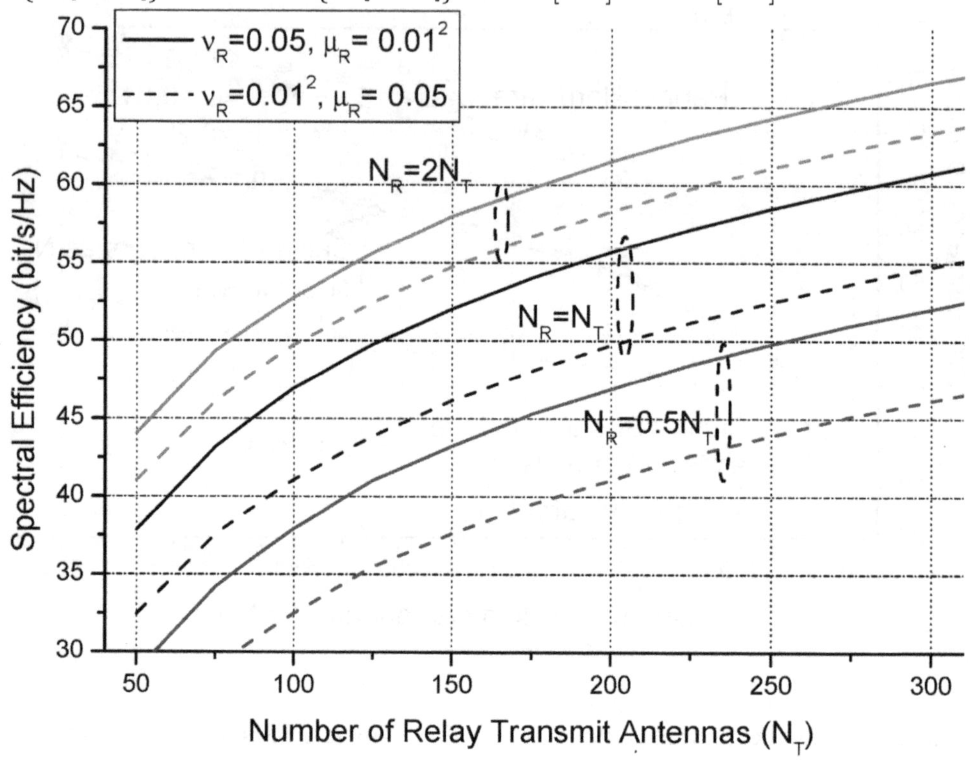

Impact of Number of S-D Pairs and Length of Channel Coherent Time

Figure 4 shows the SE as a function of the number of S-D pairs K, where the length of pilot sequence of each source or destination is set to 1 (minimum length required). When $T = 300$, the figure reveals that the SE will not increase without bound as the number of S-D pairs increases since the SE is ultimately limited by the channel coherent time. In fact, as the number of S-D pair approaches to $T/2$, the SE converges to zero since all time is allocated to pilot phase. On the other hand, the SE per S-D pair is a strictly decreasing function of K, since less time is used for data transmission. Similar result can be observed as the channel coherent time is very long ($T \to \infty$). This is because that the MUI, EI and distortions caused by impairments of relay become limiting factors when K is large as seen from the scaling behaviors in Table 1 and Table 2. The result indicates that the number of S-D pairs should be limited in order to ensure a SE guarantee for each S-D pair, and this number decreases as the quality of hardware degrades.

Figure 4. SE v.s. number of S-D pair K, where $\nu_S = \nu_D = \nu_R = \nu$, $\mu_D = \mu_R = \mu$, $E_{S,k} = E_{R,k} = 5\,dB$, $E_T = 10\,dB$, $\beta_{SR,k} = \beta_{RD,k} = 1$, $\beta_{EI} = 5\,dB$, $r_0 = 0.4$, $|r_{EI}| = 0.7$, $N_R = N_T = 200$, $N_S = N_D = 10$, $A_R = \max\left\{K, \left\lfloor \frac{2}{3}N_R \right\rfloor\right\}$, $A_T = \max\left\{K, \left\lfloor \frac{2}{3}N_T \right\rfloor\right\}$.

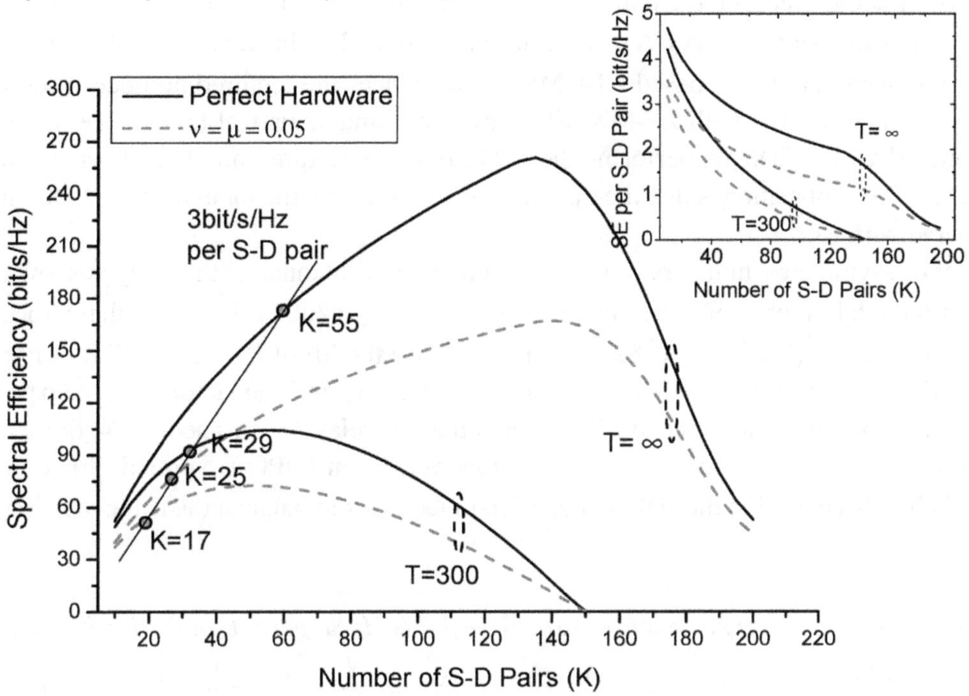

Comparison with Relevant Schemes

In this subsection, the SE of HIA-MM-FDR is compared with the massive MIMO HDR (MM-HDR) in H. A. Suraweera, N. Hien Quoc. (2013) and MM-FDR with ZF-based transceiver (ZF-MM-FDR) in H. Q. Ngo, H. A. Suraweera. (2014). The channel coherent time is set to $T = 300$ and the length of pilot sequence is $\tau = 2$. The power of pilot symbol is $E_T = 10$ dB. It is assumed that the BF vectors of sources and destinations for MM-HDR and ZF-MM-FDR are also computed using (22) and (23), since the BF design at sources and destinations is not considered in H. Q. Ngo, H. A. Suraweera. (2014). For HIA-MM-FDR, the maximum power constraints at sources and relay are set to $E_{S_k}^{\max} = 5$ dB and $E_R^{\max} = KE_{S_k}^{\max}$. Without JDPO, we set $E_{S,k} = E_{S,k}^{\max}$, $E_{R,k} = E_R^{\max} / K$ dB. Moreover, A_R and A_T are set to $A_R = \max\left\{K, \left\lfloor \frac{2}{3}N_R \right\rfloor\right\}$ and $A_T = \max\left\{K, \left\lfloor \frac{2}{3}N_R \right\rfloor\right\}$. When JDPO is applied, $E_{S,k}$, $E_{R,k}$, A_R and A_T are determined by Algorithm 2. The maximum repeat time L of JDPO is set to 3 and the elements of subset $\tilde{A}_R$ are picked uniformly from A_R with step $\max\left\{10, \left\lfloor \frac{N_R}{K} \right\rfloor\right\}$. $\tilde{A}_T$ is obtained with the similar approach. The above parameters are chosen so that the performance loss due to the suboptimal search approach to obtain A_R and A_T is negligible.

Figure 5 simulates the number of relay antennas required to achieve 3bit/s/Hz SE per S-D pair with different levels of hardware quality, where we set $\nu_S = \nu_D = \nu_R = \nu$ and $\mu_D = \mu_R = \mu$. The figure clearly shows a tradeoff between the number of antennas and hardware quality, that is (as discussed in section V-D), by increasing N_R and N_T, we can reduce ν and μ linearly without degrading the SE. Meanwhile, it is seen that the proposed HIA-MM-FDR reduces the required number of relay antennas significantly when compared with ZF-MM-FDR, by exploiting statistical knowledge of EI channels. Moreover, the HIA-MM-FDR outperforms the MM-HDR (in required number of relay antennas) for large ν and μ. The explanation is that the distortion noises become the main limiting factor in this case when compared with EI.

The effect of asymmetric numbers of transmit and receive antennas at the relay is shown in Fig. 6, where we set the total number of relay antennas as $N_R + N_T = 200$. It is shown that allocating more antennas to the receive side of relay is beneficial to improve the SE of HIA-MM-FDR (without JDPO) and ZF-MM-FDR. The reason is that, in the considered setup, the limiting factors of system performance are EI and distortion noise due to receive imperfection at the relay. From Theorem 4, these factors can be suppressed by receive antenna array of relay. However, when JDPO is applied, it is optimal to set $N_R = N_T$. This is because that the JDPO algorithm in fact tries to balance the achievable rates of each

Figure 5. Number of relay antennas required to achieve 3bit/s/Hz SE per S-D pair with different hardware qualities, where K=10, $\beta_{SR,k} = \beta_{RD,k} = 1$, $\beta_{EI} = 5\,dB$, $N_S = \left\lfloor \frac{N_R}{K} \right\rfloor$, $N_D = \left\lfloor \frac{N_T}{K} \right\rfloor$, $r_0 = 0.4$, $|r_{EI}| = 0.7$.

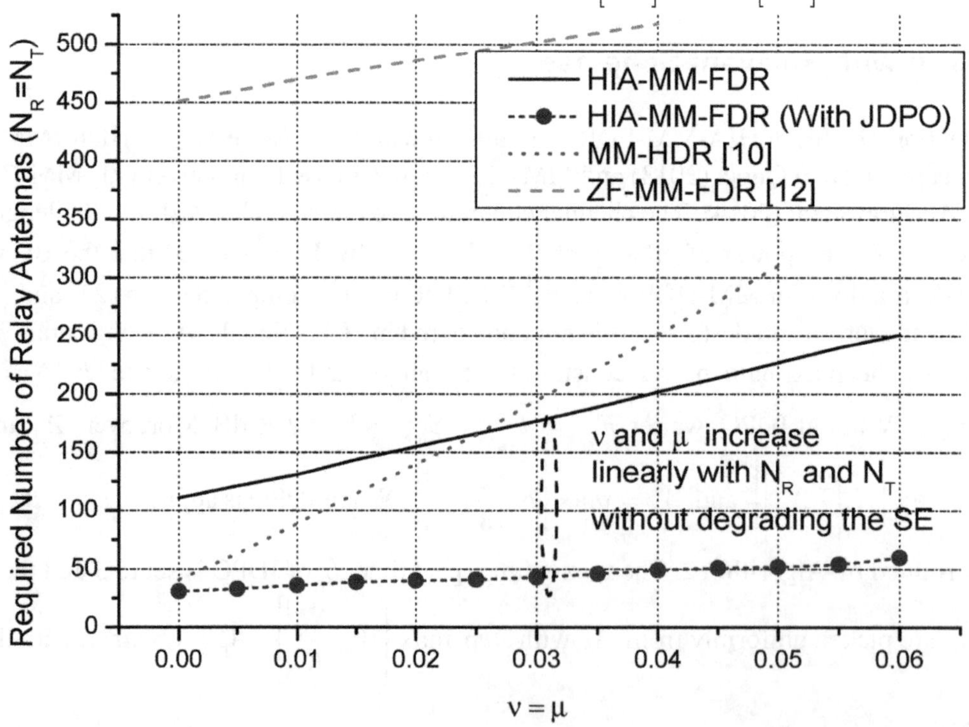

Figure 6. SE comparison with asymmetric numbers of transmit and receive relay antennas, where K=10, $N_S = \left\lfloor \frac{N_R}{K} \right\rfloor$, $N_D = \left\lfloor \frac{N_T}{K} \right\rfloor$, $\beta_{SR,k} = \beta_{RD,k} = 1$, $\beta_{EI} = 5\,dB$, $r_0 = 0.4$, $|r_{EI}| = 0.7$, $\nu_S = \nu_D = \nu_R = 0.05$, $\mu_D = \mu_R = 0.05$.

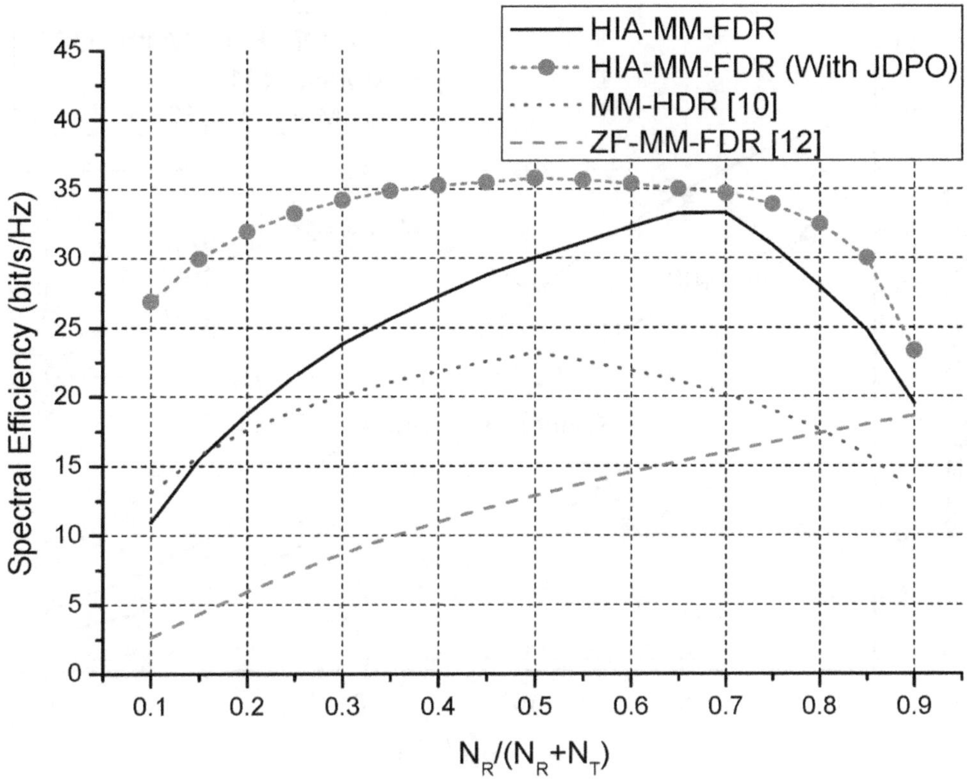

hop by power control and adjusting A_R and A_T. This makes it unnecessary to allocate more antennas to receive side since the EI and receive distortion have been well suppressed.

Figure 7 shows the SE of HIA-MM-FDR as a function of variance for EI channel β_{EI}. As expected, there exists a switching point between HIA-MM-FDR and MM-HDR as β_{EI} increases. By increasing N_R and N_T, the constraint on β_{EI} for HIA-MM-FDR to achieve a performance gain relaxes, which indicates that HIA-MM-FDR becomes more attractive when the number of relay antennas is large. Moreover, Figure 7 demonstrates that the proposed JDPO algorithm can reduce the constraint on β_{EI} significantly. In particular, as $N = 200$, a SE gain of 7.5bit/s/Hz can be achieved by HIA-MM-FDR when compared to MM-HDR as β_{EI} is 20dB. In 2.4GHz frequency band, assuming a 30dB passive EI cancellation and using the path loss models discussed in the above, this indicates that HIA-MM-FDR with JDPO can support the transmission with 7.5bit/s/Hz gain when the distances from sources/destinations to relay are about 400m (which resulting in $\beta_{EI} / \beta_{SR,k}$ *and* $\beta_{EI} / \beta_{RD,k} \approx 20$ dB).

Figure 7. SE v.s. variance of echo interference channel β_{EI}*, where K=10,* $N_S = \left\lfloor \frac{N_R}{K} \right\rfloor$, $N_D = \left\lfloor \frac{N_T}{K} \right\rfloor$, $\beta_{SR,k} = \beta_{RD,k} = 1$, $r_0 = 0.4$, $|r_{EI}| = 0.7$, $\nu_S = \nu_D = \nu_R = 0.05$, $\mu_D = \mu_R = 0.05$.

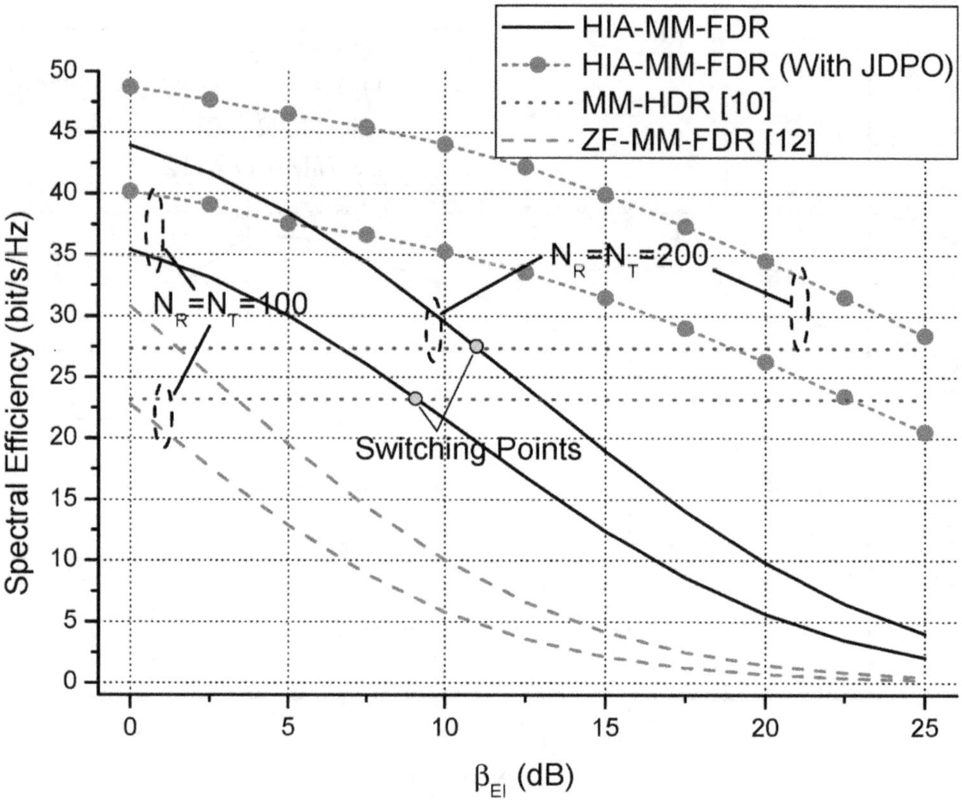

CONCLUSION

This chapter considers the transceiver design of MM-FDR with hardware impairments. A low complexity HIA scheme is proposed to mitigate the distortion noises by exploiting the statistical knowledge of channels and antenna arrays at sources and destinations. A joint degree of freedom and power optimization algorithm is presented to further optimize the SE of HIA-MM-FDR. The analytic results demonstrate that the proposed scheme can mitigate the "ceiling effect" appears in traditional MM-FDR protocol, if the numbers of antennas at sources and destinations can scale with that at the relay. Moreover, simulation results show that the HIA-MM-FDR outperforms MM-FDR with traditional transceiver scheme.

REFERENCES

Marzetta, T. L. (2010). Noncooperative cellular wireless with unlimited numbers of base station antennas. *IEEE Transactions on Wireless Communications*, *9*(11), 3590–3600. doi:10.1109/TWC.2010.092810.091092

Ngo, H. Q., Larsson, E. G., & Marzetta, T. L. (2013). Energy and spectral efficiency of very large multiuser MIMO systems. *IEEE Transactions on Communications*, *61*(4), 1436–1449. doi:10.1109/TCOMM.2013.020413.110848

D. Bharadia, E. McMilin, & S. Katti. (2013). Full duplex radios. *ACM Sigcomm*, 1-5.

Everett, E., Sahai, A., & Sabharwal, A. (2014). Passive self-interference suppression for full-duplex infrastructure nodes. *IEEE Transactions on Wireless Communications*, *24*(2), 680–694. doi:10.1109/TWC.2013.010214.130226

Suraweera, H. A., Krikidis, I., Zheng, G., Yuen, C., & Smith, P. J. (2014). Low-complexity end-to-end performance optimization in MIMO full-duplex relay systems. *IEEE Transactions on Wireless Communications*, *13*(2), 913–927. doi:10.1109/TWC.2013.122313.130608

Shang, C. Y. A., Smith, P. J., Woodward, G. K., & Suraweera, H. A. (2014). Linear transceivers for full duplex MIMO relays. *in Proc.Australian Communications Theory Workshop (AusCTW 2014)*. doi:10.1109/AusCTW.2014.6766420

Nguyen, D., Tran, L.-N., Pirinen, P., & Latva-aho, M. (2014). Precoding for full duplex multiuser MIMO systems: Spectral and energy efficiency maximization. *IEEE Transactions on Signal Processing*, *61*(16), 4038–4050. doi:10.1109/TSP.2013.2267738

Sabharwal, A., Schniter, P., Guo, D., Bliss, D. W., Rangarajan, S., & Wichman, R. (2014). In-band full-duplex wireless: Challenges and opportunities. *IEEE Journal on Selected Areas in Communications*, *32*(9), 1637–1652. doi:10.1109/JSAC.2014.2330193

Cui, H., Song, L., & Jiao, B. (2014). Multi-pair two-way amplify-and-forward relaying with very large number of relay antennas. *IEEE Transactions on Wireless Communications*, *13*(5), 2636–2645. doi:10.1109/TWC.2014.032514.130885

Suraweera, H. A., Hien Quoc, N., & Duong, T. Q. et al. (2013). Multi-pair amplify-and-forward relaying with very large antenna arrays. *Proceedings of the IEEE, ICC*, 4635–4640.

Ngo, H. Q., Suraweerat, H. A., & Larsson, E. G. (2013). Multipair massive MIMO full-duplex relaying with MRC/MRT processing. *Proceedings of the IEEE, ICC*, 4807–4813.

Ngo, H. Q., Suraweera, H. A., Matthaiou, M., & Larsson, E. G. (2014). Multipair full-duplex relaying with massive arrays and linear processing. *IEEE Journal on Selected Areas in Communications*, *32*(9), 1721–1737. doi:10.1109/JSAC.2014.2330091

Xia, X., Xie, W., & Zhang, D. et al. (2015). Multi-pair full-duplex amplify-and-forward relaying with very large antenna arrays. In *Proc. IEEE WCNC*.

Larsson, E. G., Edfors, O., Tufvesson, F., & Marzetta, T. (2014). Massive MIMO for next generation wireless systems. *IEEE Communications Magazine*, *52*(2), 186–195. doi:10.1109/MCOM.2014.6736761

Day, B. P., Margetts, A. R., Bliss, D. W., & Schniter, P. (2012). Full-duplex MIMO relaying: Achievable rates under limited dynamic range. *IEEE Journal on Selected Areas in Communications*, *30*(9), 1541–1553. doi:10.1109/JSAC.2012.120921

Zheng, G., Krikidis, I., & Ottersten, B. (2013). Full-duplex cooperative cognitive radio with transmit imperfections. *IEEE Transactions on Wireless Communications*, *12*(8), 2498–2511. doi:10.1109/TWC.2013.032513.121464

Hoydis, J., Brink, S., & Debbah, M. (2013). Massive MIMO in UL/DL of cellular networks: How many antennas do we need? *IEEE Journal on Selected Areas in Communications*, *31*(2), 160–171. doi:10.1109/JSAC.2013.130205

Zhang, J., Wen, C. K., Jin, S., Gao, X., & Wong, K. (2013). On capacity of large-scale MIMO multiple access channels with distributed sets of correlated antennas. *IEEE Journal on Selected Areas in Communications*, *31*(2), 133–148. doi:10.1109/JSAC.2013.130203

Adhikary, A., Nam, J., Ahn, J.-Y., & Caire, G. (2013). Joint spatial division and multiplexing: The large-scale array regime. *IEEE Transactions on Information Theory*, *59*(10), 6441–6463. doi:10.1109/TIT.2013.2269476

Yin, H., Gesbert, D., & Cottatellucci, L. (2014). Dealing with interference in distributed large-scale MIMO systems: A statistical approach. *IEEE Journal of Selected Topics in Signal Processing*, *8*(5), 942–953. doi:10.1109/JSTSP.2014.2322583

Mohammed, S., & Larsson, E. (2013). Per-antenna constant envelope precoding for large multi-user MIMO systems. *IEEE Transactions on Communications*, *61*(3), 1059–1071. doi:10.1109/TCOMM.2013.012913.110827

Studer, C., & Larsson, E. (2013). PAR-aware large-scale multi-user MIMO OFDM downlink. *IEEE Journal on Selected Areas in Communications*, *31*(2), 303–313. doi:10.1109/JSAC.2013.130217

Kotecha, J. H., & Sayeed, A. M. (2004). Transmit signal design for optimal estimation of correlated MIMO channels. *IEEE Transactions on Signal Processing*, *52*(2), 546–557. doi:10.1109/TSP.2003.821104

Santella, G., & Mazzenga, F. (1998). A hybrid analytical-simulation procedure for performance evaluation in M-QAM-OFDM schemes in presence of nonlinear distortions. *IEEE Trans. Veh. Tech.*, *47*(2), 142–151. doi:10.1109/25.661041

Suzuki, H., Tran, T., & Collings, I. B. et al. (2008). Transmitter noise effect on the performance of a MIMO-OFDM hardware implementation achieving improved coverage. *IEEE Journal on Selected Areas in Communications*, *26*(8), 867–876. doi:10.1109/JSAC.2008.080804

Namgoong, W. (2005). Modeling and analysis of nonlinearities and mismatches in AC-coupled direct-conversion receiver. *IEEE Transactions on Wireless Communications*, *4*(1), 163–173. doi:10.1109/TWC.2004.840239

Silverstein, J. W., & Bai, Z. D. (1995). On the empirical distribution of eigenvalues of a class of large dimensional random matrices. *Journal of Multivariate Analysis*, *54*(2), 175–192. doi:10.1006/jmva.1995.1051

Wagner, S., Couillet, R., Debbah, M., & Slock, D. T. M. (2012). Large system analysis of linear precoding in correlated MISO broadcast channels under limited feedback. *IEEE Transactions on Information Theory*, *58*(7), 4509–4537. doi:10.1109/TIT.2012.2191700

Telatar, E. (1999). Capacity of multi-antenna Gaussian channels. *European Trans. Telecom., 10*(6), 585–595. doi:10.1002/ett.4460100604

Bjornson, E., Zetterberg, P., Bengtsson, M., & Ottersten, B. (2013). Capacity limits and multiplexing gains of MIMO channels with transceiver impairments. *IEEE Communications Letters, 17*(1), 91–94. doi:10.1109/LCOMM.2012.112012.122003

Shomorony, I., & Avestimehr, A. S. (2014). Degrees of Freedom of Two-Hop Wireless Networks: Everyone Gets the Entire Cake. *IEEE Transactions on Information Theory, 60*(5), 2417–2431. doi:10.1109/TIT.2014.2308172

Cheng, Z., & Devroye, N. (2014). The degrees of freedom of the K-pair-user full-duplex two-way interference channel with a MIMO relay. In *Proc. IEEE Int. Symp. Inf. Theory (ISIT)*, (pp. 2714-2718).

Qian, L., Li, K. H., & The, K. C. (2014). Achieving Optimal Diversity-Multiplexing Tradeoff for Full-Duplex MIMO Multihop Relay Networks. *IEEE Transactions on Information Theory, 57*(1), 303–316.

Hassibi, B., & Hochwald, B. M. (2003). How much training is needed in multiple-antenna wireless links? *IEEE Transactions on Information Theory, 49*(4), 951–963. doi:10.1109/TIT.2003.809594

Mezghani, A., Damak, N., & Nossek, J. A. (2010). Circuit aware design of power-efficient short range communication systems.*In Proc. IEEE Int. Symp. Wireless Commun. Syst. (ISWCS)*, (pp. 869-873). doi:10.1109/ISWCS.2010.5624403

Weeraddana, P. C., Codreanu, M., Latva-aho, M., & Ephremides, A. (2011). Resource allocation for cross-layer utility maximization in wireless networks. *IEEE Trans. Veh. Tech., 60*(10), 2790–2809. doi:10.1109/TVT.2011.2157544

Boyd, S., Kim, S.-J., Vandenberghe, L., & Hassibi, A. (2007). A tutorial on geometric programming. *Optimization and Engineering, 8*(1), 67–127. doi:10.1007/s11081-007-9001-7

Boyd, S., & Vandenberghe, L. (2004). *Convex Optimization*. Cambridge, UK: Cambridge Univ. Press. doi:10.1017/CBO9780511804441

Loyka, S. (2001). Channel capacity of MIMO architecture using the exponential correlation matrix. *IEEE Communications Letters, 5*(9), 369–371. doi:10.1109/4234.951380

Bjornson, E., Hammarwall, D., & Ottersten, B. (2009). Exploiting quantized channel norm feedback through conditional statistics in arbitrarily correlated MIMO systems. *IEEE Transactions on Signal Processing, 57*(10), 4027–4041. doi:10.1109/TSP.2009.2024266

Goldsmith, A. (2005). *Wireless communications*. Cambridge, UK: Cambridge Univ. Press. doi:10.1017/CBO9780511841224

APPENDIX

Useful Lemmas Related to the Channel Estimates

Lemma 1: Let $\hat{\underline{h}}_{SR,k}$ and $\hat{\underline{h}}_{RD,k}$ be the LMMSE estimates of $\underline{h}_{SR,k}$ and $\underline{h}_{RD,k}$, respectively. Define

$$\delta_k^{SR} = \mathbb{E}\left[\left|\Delta \underline{h}_{SR,k}^H \hat{\underline{h}}_{SR,k}\right|^2\right]$$

and

$$\delta_k^{RD} = \mathbb{E}\left[\left|\Delta \underline{h}_{RD,k}^H \hat{\underline{h}}_{RD,k}\right|^2\right].$$

In the large-(N_R, N_T) regime, we have

$$\delta_k^{SR} = \frac{\nu_{S,k}}{\tau}\left(Tr\left(\hat{\underline{C}}_{SR,k}\right)\right)^2 \dot{\boldsymbol{p}}_{S,k}^H diag\left(\dot{\boldsymbol{p}}_{S,k}^H \dot{\boldsymbol{p}}_{S,k}\right)\dot{\boldsymbol{p}}_{S,k} + \frac{\mu_R}{\tau}\left(\frac{1}{E_T} + \beta_{SR,k} Tr\left(\tilde{\boldsymbol{C}}_{SR,k}\boldsymbol{\Omega}_{S,k}\right)\right) Tr\left(\underline{\boldsymbol{C}}_{SR,k}^3 \boldsymbol{\Gamma}_{SR,k}^2\right) + \frac{1}{\tau E_T} Tr\left(\underline{\boldsymbol{C}}_{SR,k}^3 \boldsymbol{\Gamma}_{SR,k}^2\right) \quad (47)$$

The expression of δ_{kj}^{RD} can be obtained by replacing $\underline{C}_{SR,k}$, $\hat{\underline{C}}_{SR,k}$, $\boldsymbol{\Gamma}_{SR,k}$, $\nu_{S,k}$ and $\dot{\boldsymbol{p}}_{S,k}$ in (47) with $\underline{C}_{RD,k}$, $\hat{\underline{C}}_{RD,k}$, $\boldsymbol{\Gamma}_{RD,k}$, $\nu_{D,k}$ and $\dot{\boldsymbol{p}}_{D,k}$, respectively.

Proof: We show the proof for δ_k^{SR} and the derivation for δ_k^{RD} is similar. According to the uncorrelation between $\Delta \underline{h}_{SR,k}^H$ and $\hat{\underline{h}}_{SR,k}$, we have

$$\delta_k^{SR} = \mathbb{E}\left[\left|\Delta \underline{h}_{SR,k}^H \hat{\underline{h}}_{SR,k}\right|^2\right] = \mathbb{E}\left[\left|\underline{h}_{SR,k}^H \hat{\underline{h}}_{SR,k}\right|^2\right] - \left(Tr\left(\hat{\underline{C}}_{SR,k}\right)\right)^2 \quad (48)$$

With (26) and (27), $\mathbb{E}\left[\left|\underline{h}_{SR,k}^H \hat{\underline{h}}_{SR,k}\right|^2\right]$ can be written as

$$\mathbb{E}\left[\left|\underline{h}_{SR,k}^H \hat{\underline{h}}_{SR,k}\right|^2\right] = \mathbb{E}\left[\left|\underline{h}_{SR,k}^H \underline{\boldsymbol{C}}_{SR,k}\boldsymbol{\Gamma}_{SR,k}\underline{h}_{SR,k}\right|^2\right] + \left(\frac{1}{\tau E_T}\right)^2 \mathbb{E}\left[\left|\underline{h}_{SR,k}^H \underline{\boldsymbol{C}}_{SR,k}\boldsymbol{\Gamma}_{SR,k}\dot{\boldsymbol{P}}_R^H \boldsymbol{H}_{SR,k}\boldsymbol{T}_{SR,k}\phi^*\right|^2\right] + \left(\frac{1}{\tau E_T}\right)^2 \mathbb{E}\left[\left|\underline{h}_{SR,k}^H \underline{\boldsymbol{C}}_{SR,k}\boldsymbol{\Gamma}_{SR,k}\dot{\boldsymbol{P}}_R^H \boldsymbol{R}_{SR,k}\phi^*\right|^2\right] + \left(\frac{1}{\tau E_T}\right)^2 \mathbb{E}\left[\left|\underline{h}_{SR,k}^H \underline{\boldsymbol{C}}_{SR,k}\boldsymbol{\Gamma}_{SR,k}\dot{\boldsymbol{P}}_R^H \boldsymbol{N}_{SR,k}\phi^*\right|^2\right] \quad (49)$$

In the large-(N_R, N_T) regime, the result in J. Hoydis, S. Brink, & M. Debbah. (2013) (Lemma 4 (iv)) shows that

$$\mathbb{E}\left[\left|\underline{\boldsymbol{h}}_{SR,k}^H \underline{\boldsymbol{C}}_{SR,k} \boldsymbol{\Gamma}_{SR,k} \underline{\boldsymbol{h}}_{SR,k}\right|^2\right] = \left(Tr\left(\hat{\underline{\boldsymbol{C}}}_{SR,k}\right)\right)^2.$$

Replacing $\underline{\boldsymbol{h}}_{SR,k}$ with $\dot{\boldsymbol{P}}_R^H \boldsymbol{H}_{SR,k} \dot{\boldsymbol{p}}_{S,k}$ and using the following large N_R approximation (obtained based on J. Hoydis, S. Brink, & M. Debbah. (2013) (Lemma 4 (ii))

$$\begin{aligned}&\frac{1}{N_R}\boldsymbol{H}_{SR,k}\dot{\boldsymbol{P}}_R\underline{\boldsymbol{C}}_{SR,k}\boldsymbol{\Gamma}_{SR,k}\dot{\boldsymbol{P}}_R^H\boldsymbol{H}_{SR,k}=\frac{1}{N_R}\beta_{SR,k}\tilde{\boldsymbol{C}}_{SR,k}^{1/2}\boldsymbol{X}_{SR,k}^H\boldsymbol{C}_{SR,k}^{1/2}\dot{\boldsymbol{P}}_R\underline{\boldsymbol{C}}_{SR,k}\boldsymbol{\Gamma}_{SR,k}\dot{\boldsymbol{P}}_R^H\boldsymbol{C}_{SR,k}^{1/2}\boldsymbol{X}_{SR,k}\tilde{\boldsymbol{C}}_{SR,k}^{1/2}\\&=\frac{1}{N_R}\beta_{SR,k}Tr\left(\underline{\boldsymbol{C}}_{SR,k}\boldsymbol{\Gamma}_{SR,k}\dot{\boldsymbol{P}}_R^H\boldsymbol{C}_{SR,k}\dot{\boldsymbol{P}}_R\right)\tilde{\boldsymbol{C}}_{SR,k}\end{aligned} \tag{50}$$

It can be shown that

$$\mathbb{E}\left[\left|\underline{\boldsymbol{h}}_{SR,k}^H\underline{\boldsymbol{C}}_{SR,k}\boldsymbol{\Gamma}_{SR,k}\dot{\boldsymbol{P}}_R^H\boldsymbol{H}_{SR,k}\boldsymbol{T}_{SR,k}\phi^*\right|^2\right]=\nu_{S,k}\tau E_T^2\left(Tr\left(\hat{\underline{\boldsymbol{C}}}_{SR,k}\right)\right)^2\frac{\dot{\boldsymbol{p}}_{S,k}^H\tilde{\boldsymbol{C}}_{SR,k}diag\left(\dot{\boldsymbol{p}}_{S,k}^H\dot{\boldsymbol{p}}_{S,k}\right)\tilde{\boldsymbol{C}}_{SR,k}\dot{\boldsymbol{p}}_{S,k}}{\left(\lambda_1\left(\tilde{\boldsymbol{C}}_{SR,k}\right)\right)^2} \tag{51}$$

Using the similar approach on the remaining terms in (49), Lemma 1 is obtained.

Lemma 2: Let $\underline{\boldsymbol{C}}_{i,k}$ and $\hat{\underline{\boldsymbol{C}}}_{i,k}$ ($i \in \{SR, RD\}$) be the covariance matrices of effective channel $\underline{\boldsymbol{h}}_{i,k}$ and its estimates $\hat{\underline{\boldsymbol{h}}}_{i,k}$ given by Theorem 3. If $Tr(\underline{\boldsymbol{C}}_{SR,k}) = O(N_R)$ and $Tr(\underline{\boldsymbol{C}}_{RD,k}) = O(N_T)$, we have $Tr(\hat{\underline{\boldsymbol{C}}}_{SR,k}) = O(N_R)$ and $Tr(\hat{\underline{\boldsymbol{C}}}_{RD,k}) = O(N_T)$. Moreover, $\lim_{N_R\to\infty}\frac{A_R}{N_R} > 0$ and $\lim_{N_T\to\infty}\frac{A_T}{N_T} > 0$ (i.e., A_R and A_T scale with $O(N_R)$ and $O(N_T)$, respectively).

Proof: We present the proof for scaling behaviors of $Tr(\hat{\underline{\boldsymbol{C}}}_{SR,k})$ and A_R. Using definition of $\hat{\underline{\boldsymbol{C}}}_{SR,k}$ in Theorem 3 and eigenvalue decomposition $\underline{\boldsymbol{C}}_{SR,k} = \underline{\boldsymbol{U}}\underline{\Sigma}_{SR,k}\underline{\boldsymbol{U}}^H$, it can be shown that

$$\begin{aligned}&Tr\left(\hat{\underline{\boldsymbol{C}}}_{SR,k}\right)=Tr\left(\underline{\boldsymbol{C}}_{SR,k}\boldsymbol{\Gamma}_{SR,k}\underline{\boldsymbol{C}}_{SR,k}\right)\\&=Tr\left(\Sigma_{SR,k}\left(\omega_1\underline{\Sigma}_{SR,k}+\omega_2\boldsymbol{I}_{A_R}\right)^{-1}\Sigma_{SR,k}\right)\\&=\sum_{l=1}^{A_R}\left(\lambda_l\left(\underline{\boldsymbol{C}}_{SR,k}\right)\right)^2\left(\omega_1\lambda_l\left(\underline{\boldsymbol{C}}_{SR,k}\right)+\omega_2\right)^{-1}\end{aligned} \tag{52}$$

where ω_1 and ω_2 are positive constants independent of A_R. From (52), $Tr(\hat{\underline{\boldsymbol{C}}}_{SR,k})$ is bounded as

$$\sum_{l=1}^{A_R} \frac{\left(\lambda_l\left(\underline{\mathbf{C}}_{SR,k}\right)\right)^2}{\omega_1\lambda_1\left(\underline{\mathbf{C}}_{SR,k}\right)+\omega_2} \le \mathrm{Tr}\left(\hat{\underline{\mathbf{C}}}_{SR,k}\right) \le \frac{1}{\omega_1}\mathrm{Tr}\left(\underline{\mathbf{C}}_{SR,k}\right) \tag{53}$$

In (53), the upper bound is an $O(N_R)$ term. With the definition of $\underline{\boldsymbol{C}}_{SR,k}$ in the Theorem 3, we have

$$\begin{aligned} &\lambda_1\left(\underline{\boldsymbol{C}}_{SR,k}\right) = \beta_{SR,k}\lambda_1\left(\tilde{\boldsymbol{C}}_{SR,k}\right)\boldsymbol{u}_1^H\left(\underline{\boldsymbol{C}}_{SR,k}\right)\dot{\boldsymbol{P}}_R^H\boldsymbol{C}_{SR,k}\dot{\boldsymbol{P}}_R\boldsymbol{u}_1\left(\underline{\boldsymbol{C}}_{SR,k}\right) \\ &\le \beta_{SR,k}\lambda_1\left(\tilde{\boldsymbol{C}}_{SR,k}\right)\lambda_1\left(\boldsymbol{C}_{SR,k}\right) = O(1) \end{aligned} \tag{54}$$

where the last step is based on assumption **A1**. Since $Tr\left(\underline{\boldsymbol{C}}_{SR,k}\right) = O(N_R)$, we can conclude that $\lambda_1\left(\underline{\boldsymbol{C}}_{SR,k}\right) = O(1)$ (otherwise $Tr\left(\underline{\boldsymbol{C}}_{SR,k}\right) \le A_R\lambda_1\left(\underline{\boldsymbol{C}}_{SR,k}\right) < O(N_R)$). Thus, the lower bound in (53) also scales with $O(N_R)$. Moreover, based on the definition of $\underline{\boldsymbol{C}}_{SR,k}$ in the Theorem 3, we have

$$Tr\left(\underline{\boldsymbol{C}}_{SR,k}\right) = \beta_{SR,k}\lambda_1\left(\tilde{\boldsymbol{C}}_{SR,k}\right)\sum_{l=1}^{A_R}\dot{\boldsymbol{p}}_{R,l}^H\boldsymbol{C}_{SR,k}\dot{\boldsymbol{p}}_{R,l} \le A_R\beta_{SR,k}\lambda_1\left(\tilde{\boldsymbol{C}}_{SR,k}\right)\lambda_1\left(\boldsymbol{C}_{SR,k}\right) \tag{55}$$

where $\dot{\boldsymbol{p}}_{R,l}$ is the l th column of $\dot{\boldsymbol{P}}_R$\$. From (55), A_R must scale $O(N_R)$. Otherwise, $Tr\left(\underline{\boldsymbol{C}}_{SR,k}\right) < O(N_R)$ based on assumption **A1**.

Proof of Theorem 1

To facilitate analysis, we first derive the covariance matrices of distortion noises $\boldsymbol{t}_S[u]$ and $\boldsymbol{r}_R[u]$. Based on the model (1), we have

$$\boldsymbol{\Theta}_S^T = diag\left(\nu_{S,1}E_{S,1},\cdots,\nu_{S,K}E_{S,K}\right).$$

Moreover, define

$$\boldsymbol{x}_S[u] = \left[\boldsymbol{x}_{S,1}^T[u],\cdots,\boldsymbol{x}_{S,K}^T[u]\right]^T,$$

$\boldsymbol{\Theta}_R^R$ can be expressed based on model (2) as

$$\boldsymbol{\Theta}_R^R = \mu_R diag\left(\mathbb{E}\left[\boldsymbol{H}_{SR}\boldsymbol{\Omega}_S\boldsymbol{H}_{SR}^H\right]\right) + \mu_R diag\left(\mathbb{E}\left[\boldsymbol{H}_{EI}\boldsymbol{\Omega}_R\boldsymbol{H}_{EI}^H\right]\right) + \mu_R\boldsymbol{I}_{N_R} \tag{56}$$

where

$$\begin{aligned}\boldsymbol{\Omega}_S &= \mathbb{E}\left[\left(\boldsymbol{x}_S[u]+\boldsymbol{t}_S[u]\right)\left(\boldsymbol{s}[u]+\boldsymbol{t}_S[u]\right)^H\right] = diag\left(\left(1+\nu_{S,1}\right)E_{S,1},\cdots,\left(1+\nu_{S,K}\right)E_{S,K}\right)\\ \boldsymbol{\Omega}_R &= \mathbb{E}\left[\left(\boldsymbol{x}_R[u]+\boldsymbol{t}_R[u]\right)\left(\boldsymbol{x}_R[u]+\boldsymbol{t}_R[u]\right)^H\right] = \mathbb{E}\left[\boldsymbol{x}_R[u]\boldsymbol{x}_R^H[u]\right]+\boldsymbol{\Theta}_R^T\end{aligned} \tag{57}$$

For further analysis, we approximate $\mathbb{E}\left[\boldsymbol{x}_R[u]\boldsymbol{x}_R^H[u]\right]$ in (57) with

$$diag\left(\mathbb{E}\left[\boldsymbol{x}_R[u]\boldsymbol{x}_R^H[u]\right]\right) = \frac{1}{\nu_R}\boldsymbol{\Theta}_R^T.$$

Note that the approximation results in a new upper bound on $R_{SR,k}^{Upper}$ (The new bound will also be referred to as $R_{SR,k}^{Upper}$ for convenience). Straight-forward computations yield to

$$\mathbb{E}\left[\boldsymbol{H}_{SR}\boldsymbol{\Omega}_S\boldsymbol{H}_{SR}^H\right] = \sum_{l=1}^{K}\nu_{S,l}E_{S,l}\mathbb{E}\left[\boldsymbol{h}_{SR,l}\boldsymbol{h}_{SR,l}^H\right] = \sum_{l=1}^{K}(1+\nu_{S,l})E_{S,l}\beta_{SR,l}\boldsymbol{C}_{SR,l} \tag{58}$$

Following the spatial correlation model of $\boldsymbol{H}_{EI}$ in section III and the approximation under (57), the expression of $\mathbb{E}\left[\boldsymbol{H}_{EI}\boldsymbol{\Omega}_R\boldsymbol{H}_{EI}^H\right]$ can be obtained as

$$\begin{aligned}\mathbb{E}\left[\boldsymbol{H}_{EI}\boldsymbol{\Omega}_R\boldsymbol{H}_{EI}^H\right] &\approx \left(1+\frac{1}{\nu_R}\right)\beta_{EI}\mathbb{E}\left[\boldsymbol{C}_{EI}^{1/2}\boldsymbol{X}_{EI}\tilde{\boldsymbol{C}}_{EI}^{1/2}\boldsymbol{\Theta}_R^T\tilde{\boldsymbol{C}}_{EI}^{1/2}\boldsymbol{X}_{EI}^H\boldsymbol{C}_{EI}^{1/2}\right]\\ &= \left(1+\frac{1}{\nu_R}\right)\beta_{EI}\sum_{l=1}^{N_T}\mathbb{E}\left[\boldsymbol{C}_{EI}^{1/2}\boldsymbol{X}_{EI}\tilde{\boldsymbol{C}}_{EI}^{1/2}\left(\boldsymbol{\Theta}_R^T\right)^{1/2}\boldsymbol{e}_l\boldsymbol{e}_l^T\left(\boldsymbol{\Theta}_R^T\right)^{1/2}\tilde{\boldsymbol{C}}_{EI}^{1/2}\boldsymbol{X}_{EI}^H\boldsymbol{C}_{EI}^{1/2}\right]\\ &= \left(1+\frac{1}{\nu_R}\right)\beta_{EI}\sum_{l=1}^{N_T}\mathbb{E}\left[\left\|\left(\tilde{\boldsymbol{C}}_{EI}^{1/2}\left(\boldsymbol{\Theta}_R^T\right)^{1/2}\boldsymbol{e}_l\right)^T\otimes\boldsymbol{C}_{EI}^{1/2}vec\left(\boldsymbol{X}_{EI}\right)\right\|^2\right]\\ &= \left(1+\frac{1}{\nu_R}\right)\beta_{EI}\sum_{l=1}^{N_T}\left\|\left(\tilde{\boldsymbol{C}}_{EI}^{1/2}\left(\boldsymbol{\Theta}_R^T\right)^{1/2}\boldsymbol{e}_l\right)^T\otimes\boldsymbol{C}_{EI}^{1/2}\right\|^2 = \left(1+\frac{1}{\nu_R}\right)\beta_{EI}Tr\left(\tilde{\boldsymbol{C}}_{EI}\boldsymbol{\Theta}_R^T\right)\boldsymbol{C}_{EI}\end{aligned} \tag{59}$$

where the third step follows from the formula

$$vec\left(\boldsymbol{A}\boldsymbol{B}\boldsymbol{c}\right) = \left(\boldsymbol{c}^T\otimes\boldsymbol{A}\right)vec\left(\boldsymbol{B}\right)$$

and the fourth step is due to the fact that the elements of $\boldsymbol{X}_{EI}$ are i.i.d with distribution $CN(0,1)$. Substituting (58) and (59) into (56), the expression of $\boldsymbol{\Theta}_R^R$ in (10) is obtained.

Then we derive the asymptotic bound in Theorem 1. Note that (8) is in a similar structure with the achievable rate of the regularized zero-forcing based downlink multi-user MIMO system, which is discussed in J. Hoydis, S. Brink. (2013). This motivates us to use the asymptotic method in J. Hoydis, S.

Brink. (2013) (Theorem 1) to derive $R_{SR,k}^{Upper}$ in the large-(N_R, N_T) regime. However, the major difficulty here is that (8) does not match perfectly with the condition of J. Hoydis, S. Brink. (2013) (Theorem 1) due to the presence of residual EI term $\boldsymbol{H}_{EI}\boldsymbol{\Theta}_R^T\boldsymbol{H}_{EI}^H$ (introduced by $\boldsymbol{Q}_k$). To address this problem, by replacing $\boldsymbol{H}_{EI}\boldsymbol{\Theta}_R^T\boldsymbol{H}_{EI}^H$ in $\boldsymbol{Q}_k$ with $diag\left(\boldsymbol{H}_{EI}\boldsymbol{\Theta}_R^T\boldsymbol{H}_{EI}^H\right)$, we approximate $\boldsymbol{Q}_k$ as

$$\boldsymbol{Q}_k = \boldsymbol{H}_{SR}\boldsymbol{\Theta}_S^T\boldsymbol{H}_{SR}^H - \nu_{S,k}E_{S,k}\boldsymbol{h}_{SR,k}\boldsymbol{h}_{SR,k}^H + diag\left(\boldsymbol{H}_{EI}\boldsymbol{\Theta}_R^T\boldsymbol{H}_{EI}^H\right) + \boldsymbol{\Theta}_R^R + \boldsymbol{I}_{N_R} \tag{60}$$

The approximation results in a new upper bound on $R_{SR,k}^{Upper}$, since it reduces the power of residual EI after combining by $\boldsymbol{w}_{R,k}$ (The new bound will also be referred to as $R_{SR,k}^{Upper}$ for convenience). Based on J. Hoydis, S. Brink. (2013) (Lemma 4 (ii)), we can replace $diag\left(\boldsymbol{H}_{EI}\boldsymbol{\Theta}_R^T\boldsymbol{H}_{EI}^H\right)$ with its deterministic equivalence in the large-N_T regime, i.e.,

$$\frac{1}{N_T} diag\left(\boldsymbol{H}_{EI}\boldsymbol{\Theta}_R^T\boldsymbol{H}_{EI}^H\right) = \frac{1}{N_T}\beta_{EI} Tr\left(\tilde{\boldsymbol{C}}_{EI}\boldsymbol{\Theta}_R^T\right)\boldsymbol{I}_{N_R} \tag{61}$$

Finally, with (60) and (61), the asymptotic expression of $R_{SR,k}^{Upper}$ in Theorem 1 can be obtained by first applying J. Hoydis, S. Brink. (2013) (Lemma 4 (ii)) on term $\boldsymbol{h}_{SR}^H\boldsymbol{Q}_k^{-1}\boldsymbol{h}_{SR,k}$, and then using the result in J. Hoydis, S. Brink. (2013) (Theorem 1).

Chapter 5
Interference Mitigation for Satellite:
Terrestrial Heterogeneous Coexistence Cognitive MIMO System Based on Digital Beamforming

Yong Liao
Chongqing University, China

Yufeng Li
Chongqing University, China

Shumin Zhang
Chongqing University, China

Ming Zhao
Chongqing University, China

Xin Zhou
Chongqing University, China

Ling Chen
Chongqing University, China

Xuanfan Shen
Chongqing University, China

Yi Hu
Chongqing University, China

ABSTRACT

For the coexistence and increasing interference of satellite-terrestrial network and terrestrial wireless network, we analyze a typical scenario where the GEO satellite-terrestrial network and the 4G mobile communication network coexist heterogeneously. Besides, a multi-user cognitive system model that secondary satellite terminals interfere the primary MIMO 4G base stations is also proposed, with whose general signal processing is deduced. Meanwhile, DBF technology in 4G base station system is adopted to minimize the cognitive interference caused by multi-antennas and multi-users. And we propose an OBW-FAI. Weight vector is only related to the azimuth of the interferences, thus the proposed algorithm does not need real-time and repeat calculations, and has small complexity. Finally, the numerical simulation results verify that the proposed system and algorithm can effectively reduce interference between satellite-terrestrial network and terrestrial wireless network to a certain extent.

DOI: 10.4018/978-1-5225-1712-2.ch005

INTRODUCTION

Cognitive Radio (CR) has been widely studied and applied, which is the most effective technology to solve the serious shortage of spectrum resources and utilization in wireless communications recently (Haykin, 2005). There are typically two main approaches for CR to access the licensed spectrum. One is that secondary users will access the spectrum after sensing free band of primary users. Once the primary users are back to their spectrum, and secondary users need to release the spectrum. The other one is that secondary users share the spectrum with primary users without influence on communications of primary users when secondary users sense that primary users are using the spectrum. Currently, the second method has been widely studied, so how to ensure the quality of service (QoS) of primary users, as well as the interference of secondary users have no impact on minimum signal to interference plus noise ratio (SINR) of primary users has become research hotspots and issues.

BACKGROUND

Currently, the existing anti-jamming technologies can be implemented in frequency, time, spatial domain and anti-jamming technology integrated frequency domain, time domain and spatial domain. Anti-jamming technology in spatial domain is mainly implemented by smart antennas whose key technology is digital beamforming (DBF) technology. DBF can dynamically produce a variety of desired beams in the digital domain. When the antenna is in a receiving state, it is ensured that the gain in the direction of expected signal is not affected. Moreover, adaptive zero of antenna pattern is aligning to interference signal in order to suppress interference (Wang, Pan, & Song, 2013). This technique has been widely employed in wireless communication because of properties above.

In the field of cognitive radio, some researchers have applied DBF technology to various fields, such as interference control (Yiu, Vu, & Tarokh, 2009), maximize the channel capacity (Luan, Gao, Zhang, Li, James, & Lei, 2012), SINR balancing (Cumanan, Musavian, Lambotharab, & Gershman, 2010) and so on. However, it is still a challenge to employ the DBF under the condition of the secondary user and the primary user sharing the spectrum. In addition, the channel capacity of cognitive Multiple Input Multiple Output (MIMO) system has also been investigated widely (Akin & Gursoy, 2013; Z. Li, J. D. Li, Liu, & Shen, 2014), but few studies focus on cognitive systems based on DBF-MIMO.

With the recent people's increasing communication needs of 5W (Whoever, Wherever, Whenever, Whomever and Whatever), particularly in the field of disaster prevention and mitigation, satellite communication has become more important than ever. And the utilization of satellite communication has also been gradually improved. Satellite network is a typical resource-constrained network. How to improve satellite network intelligently and efficiently through cognitive technology? The key point is a currently hot research direction. University of Surrey, Luxembourg University and other six universities in European Union (EU) carried out the "Cognitive Radio for Satellite communications (CoRaSat)" project (CoRaSat, n.d.), aiming to use satellite spectrum flexibly and intelligently through cognitive radio, and some research results have been achieved (Liolos, et al, 2013). China Aerospace Science and Industry Corporation (CASIC) (Li, 2007), Maritime Satellite TT&C Department of China (Liu & Zhu, 2008), Shanghai Engineering Center for Micro-satellites (Ma, Chen, Liu, & Liang, 2010), China Academy of Space Technology (CAST) (Liu, Bi, Sun, & Liu, 2012), China Academy of Space Technology (Xi'an) (Chen, Qiu, & Wang, 2012) and the China Electronics Technology Group Corporation (CETC) carried out the research in this field at different degrees (Zhang, Jiang, & Zheng, 2013). However, their works mainly focus on some system framework and concept levels. Overall, many technical challenges of the cognitive technology applications in satellite communication have not been widely studied (Zhang, et al, 2013). Among them, the heterogeneous cognitive fusion of satellite network and terrestrial network would be a trend of development.

In the existing researches of coexistence between heterogeneous networks, what a majority of researchers considered is only the terrestrial wireless communication network heterogeneous coexistence scenarios, such as the coexistence between Wireless Personal Area Network (WPAN) and Wireless Local Area Network (WLAN) (Zhang & Shin, 2013), between WLAN and Worldwide Interoperability for Microwave Access (WiMAX) and so on (Sun & Zhang, Chen, Qin & Teng 2013). In terms of the application of coexistence between satellite network and terrestrial wireless communication, in 2005, C. Tawil et al. (2005) proposed a satellite-terrestrial scenario, and they also provided a relative method of sharing Ku-band spectrum, this method is that terrestrial transmissions simultaneously along with direct broadcast satellite transmissions on a common frequency. And it is suggested to adopt different directional antennas at user location to achieve spectrum multiplexing. However, the proposed scenario only considered the interference caused by direct broadcast downlink in Geosynchronous Earth Orbit (GEO) and terrestrial broadcast signals to the same terrestrial receivers. In order to avoid interference, independent directional antennas were aligned to receive different signals with a terrestrial azimuth range. But it neither investigated how to reduce co-channel interference impacted on terrestrial receivers by satellite earth station uplink, nor employed the spectrum sharing policy in CR technology. K. Sithamparanathan et al. (2010) presented the concept of Cognitive Satellite Terrestrial Radios (SCTR) for Hybrid Satellite Terrestrial System (HSTS) in 2010. Then two particular HSTS applications were used to illustrate the SCTR concept. The hybrid satellite Ultra Wide Band (UWB) communication system for WPAN with short range on terrestrial communications, and the hybrid satellite-Wireless Regional Area Networks (WRAN) for long range on terrestrial communications were presented. In both the applications the satellite uplink and the terrestrial radios adopt the CR functionalities. The research results indicated that the future satellite terrestrial terminals need to integrate and co-exist with the extremely crowded terrestrial wireless systems. Adopting CR is an effective way to solve the problem of sharing finite wireless spectrum. But the authors didn't consider mobile communication system constructed by satellites and terrestrial operators, so it was a lack of signal processing analysis of CR-based heterogeneous network communication system. Furthermore, no methods were provided to reduce co-channel interference. It is obvious that some European countries and America have done some research and applications in the coexistence of satellite terrestrial networks, but overall, the studies on applications are very few. In China, the research in this field has just started, many basic theories and key technologies are remaining to be overcome.

After years of requirements demonstration (Zhang, Gan & Li, 2010) (Zhang & Gan, 2015), China is about to construct the satellite mobile communication system which is based on GEO and working in S-band (Ming, 2012). With the construction and operation of this system, it can greatly make up for lack of terrestrial cellular mobile communication. Meanwhile, a problem should be paid attention to is that, with the increasing satellite terrestrial-terminal users and frequently used in the future, when GEO satellite mobile communication systems use cognitive style and share the same frequency band with terrestrial wireless networks, it will inevitably interfere with terrestrial wireless communications systems because of the scarcity of spectrum, typically 4G mobile communication systems. Due to 4G Base Station (BS) equipped with multiple antennas that has MIMO functionalities generally locates in a certain height off the ground, if Tianlian[1] satellite terrestrial-terminal users in a certain area nearly have uplink communication with Tianlian satellite at the same time, it will cause interference with reception for mobile BS when terrestrial mobile users communicate with 4G BS uplink.

To solve the above co-existing interference problems, on the base of existing similar satellite-terrestrial heterogeneous coexistence scenarios and advanced research of cognitive satellite communications in Europe and America, as well as in China, in terms of a typical coexistence scenarios between GEO satellite-terrestrial network and terrestrial 4G mobile communication network, this chapter proposes a heterogeneous multi-user cognitive system model, which describes the interference caused by terrestrial secondary user (satellite terrestrial station) to terrestrial primary user (MIMO 4G BS), and deduces its general signal processing. At the meantime, the 4G BS system adopts DBF to minimize the cognitive interference of multi-user multi-antenna, so that the authors design an optimal beam weight based on fixed azimuth interference (OBW-FAI) which can resist fixed azimuth interference source. Finally, Monte-Carlo method is adopted for numerical analysis of the system's channel capacity and SINR performance induces.

SYSTEM AND SIGNAL MODEL

System Model

The authors consider the coexistence scenarios of satellite network and terrestrial network, both of them are working on the uplink, satellite link is as secondary user link, and terrestrial link works as primary user link. In this section, they consider the GEO satellite mobile communication systems which are portable (vehicle-mounted, backpack and so on) Very Small Aperture Terminal (VSAT), work in S-band, and assume that the position of VSAT can move temporarily according to the user needs. However, its position is fixed relatively when communicating with GEO satellite. Owing to small antenna aperture and wide beam, VSAT can easily interfere with other terrestrial communication systems. Meanwhile, terrestrial 4G network can provide mobile users access to S-band. The interference caused by 4G mobile users to GEO satellite can be ignored, but the interference caused by satellite terrestrial station to 4G BS will be focused on in this section.

For example, China is in the northern hemisphere, when the satellite terrestrial stations which are set in China communicate with GEO satellite, they are all oriented to south. Figure 1 is the heterogeneous coexistence of GEO satellite-terrestrial network and 4G network. In this scenario, the VSAT is deployed in the north of the 4G BS, and the authors adopt cognitive mode to share S frequency band with the 4G mobile network. Then when the uplink signal passes the 4G BS, the co-frequency interference occurs for 4G BS to receive the uplink signals of mobile users.

The authors use the signal processing system based on DBF-MIMO to reduce the receiving interference from the secondary MIMO satellite terrestrial station to the primary 4G BS.

Signal Model

The signal processing model of the proposed system is shown in Figure 2. The authors consider the 4G fixed access user with multiple antennas (such as 4G MIMO wireless router) of a convenient to analysis

Figure 1. Heterogeneous coexistence topology of GEO satellite-terrestrial network and 4G network

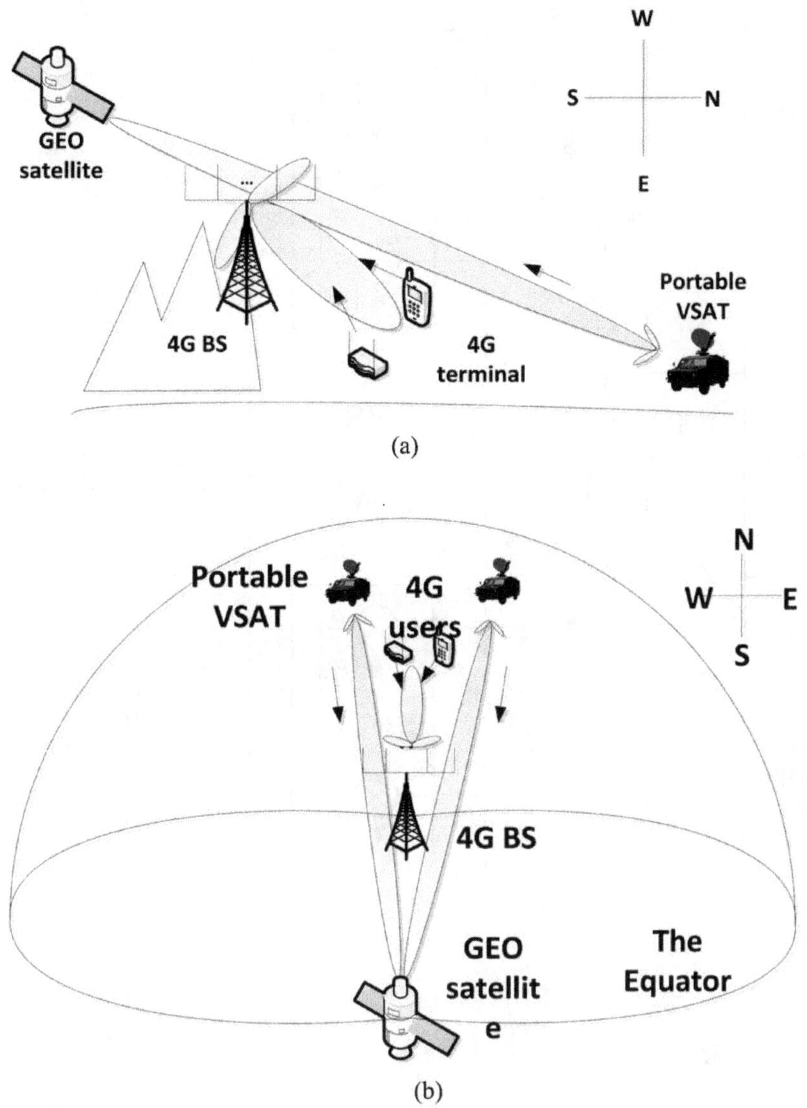

of MIMO multiple user. As the primary user, the general 4G mobile user is a single antenna system, which can be considered as a special MIMO primary user.

Assume 4G BS has M antenna arrays, each antenna array has L array elements, there are K primary users and secondary users totally.

On the uplink, each user has N antennas, BS uses DBF to distinguish different user data. The signal vector $\boldsymbol{y}$ BS received can be written as

$$\boldsymbol{y} = \sum_{k=1}^{K} \boldsymbol{h}^{k} \boldsymbol{a}\left(\boldsymbol{\theta}^{k}\right) \boldsymbol{x}^{k} + \boldsymbol{z} \tag{1}$$

Figure 2. The uplink model of MIMO multiple user to 4G DBF-MIMO BS

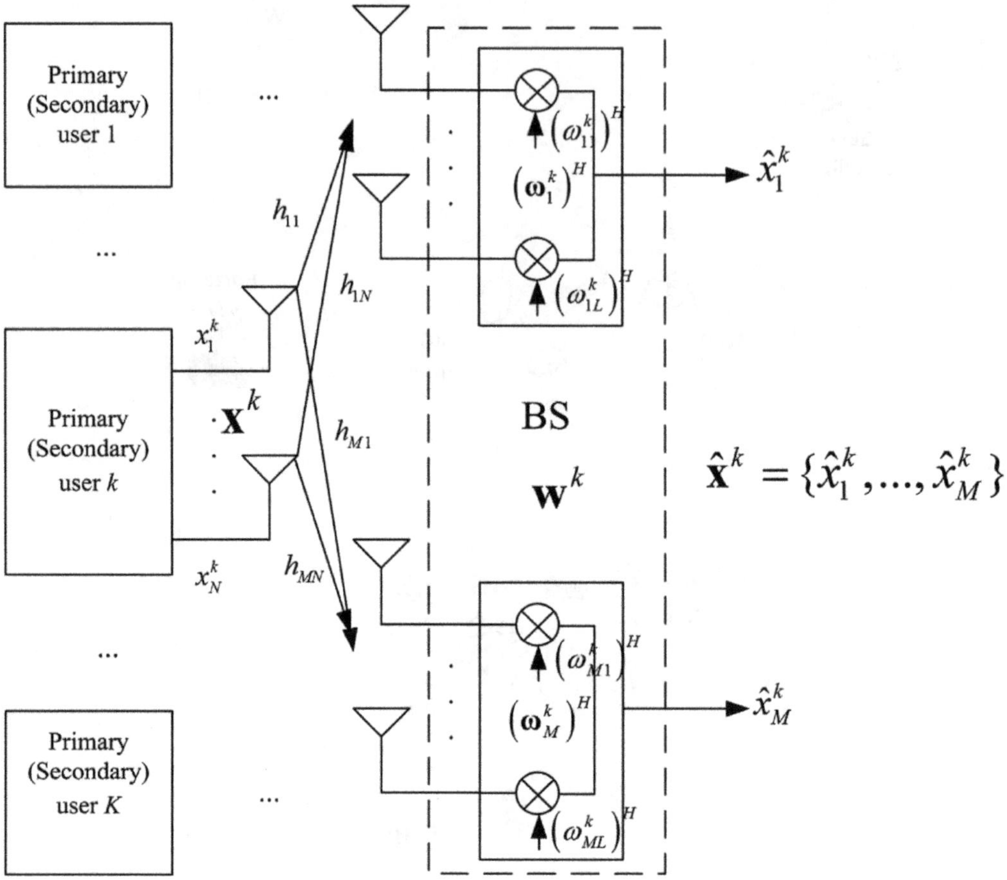

where

$$\boldsymbol{h}^k = \left[h_{mn}^k\right], n \in N, m \in M$$

is the $M \times N$ channel gain matrix of the kth user to BS. h_{mn}^k stands for the channel fading coefficients between the nth transmitting antenna and the mth receiving antenna array of the kth user. And it is assumed that the antennas of the user and all the antenna arrays are the line of sight transmission, then h_{mn}^k is constant. $\boldsymbol{x}^k = \left[x_n^k\right], n \in N$, is the transmission signal vector of the kth user to BS, and it represents the transmission signal of the kth user to the nth antenna. $\boldsymbol{a}\left(\boldsymbol{\theta}^k\right) = \left[\boldsymbol{a}\left(\boldsymbol{\theta}_{mn}^k\right)\right]$ is the $M \times N$ array response matrix of the kth user. θ_{mn}^k is the azimuth angle from the nth antenna of the kth user to the mth antenna array. z is the $M \times N$ independent and identically distributed Gauss noise matrix. $\boldsymbol{a}\left(\boldsymbol{\theta}_{mn}^k\right)$ is the element in the array response matrix of the kth users, and it can be expressed as (Katkovnik, Lee, & Kim, 2006)

$$a\left(\theta_{mn}^{k}\right) = \left[1, e^{\frac{-j2\pi d \sin\left(\theta_{mn}^{k}\right)}{\lambda}}, \ldots, e^{\frac{-j2\pi (M-1) d \sin\left(\theta_{mn}^{k}\right)}{\lambda}}\right]^{\mathrm{T}} \tag{2}$$

where d is the interval between BS antenna arrays, λ represents the carrier wave length of the transmit antenna, and $(.)^{\mathrm{T}}$ is transpose.

For the convenience of analysis, the authors consider the situation of a primary user (number is t) and K-1 interference secondary users

$$\boldsymbol{y} = \boldsymbol{h}^{t}\boldsymbol{a}\left(\boldsymbol{\theta}^{t}\right)\boldsymbol{x}^{t} + \boldsymbol{q} \tag{3}$$

where h^{t} is the channel gain matrix of the target user t to BS antenna array. $\boldsymbol{a}\left(\boldsymbol{\theta}^{t}\right)$ is the array response matrix of the target user t. $\boldsymbol{x}^{t} = \left\{x_{1}^{t}, \ldots, x_{N}^{t}\right\}$ stands for transmission signal of the primary user, and $\boldsymbol{q} = \sum_{k=2}^{K} \boldsymbol{h}^{k}\boldsymbol{a}\left(\boldsymbol{\theta}^{k}\right)\boldsymbol{x}^{k} + \boldsymbol{z}$. In order to obtain signal of the target user t among the received signals, received signal matrix $\boldsymbol{y}$ generates array output $\hat{\boldsymbol{x}}^{t}$ through the weight matrix $\boldsymbol{W}$ of BS

$$\hat{\boldsymbol{x}}^{t} = \boldsymbol{W}^{\mathrm{H}}\boldsymbol{y} \tag{4}$$

where weight matrix $W = \left\{\boldsymbol{w}^{k}\right\}, \boldsymbol{k} \in \boldsymbol{K}$, $\boldsymbol{w}^{k} = \left[\omega_{ml}^{k}\right]$, $l \in L$, $m \in M$. H is Hamilton conjugate transpose. The selection principle of $\boldsymbol{W}$ is to make the useful signal of the formula (3) be the largest and the interference be the least as far as possible.

THE DESIGN OF OPTIMAL BEAM WEIGHT BASED ON FIXED POSITION

In the design scenario, the interference of 4G BS mainly comes from the satellite terrestrial terminal, the terminal is near to BS with a fixed position. Thus, the authors design an optimal beam weight OBW-FAI algorithm based on fixed azimuth to against the fixed position interference.

The ideal beam weight coefficient should meet the following conditions.

1. It can maximize the output power of beam in target azimuth.
2. It forces the output of interference azimuth beam to 0.
3. It does not affect the output SINR. Set the target azimuth of 4G users to θ, and other fixed azimuth of the satellite terrestrial station is θ_i (i=1,2,…,N), and beam weight coefficient vector denotes $\boldsymbol{W}$.

The beamforming optimization problem can be described as

$$\begin{aligned}\max\left(P_\theta\right) &= \max\left(E\left[\left|\hat{x}^d(t)\right|^2\right]\right)\\ &= \max\left(\boldsymbol{W}^{\mathrm{H}}\left[\boldsymbol{y}(t)\boldsymbol{y}^{\mathrm{H}}(t)\right]\boldsymbol{W}\right)\\ &= \max\left(E\left[x(t)\right]^2\left[\boldsymbol{W}\boldsymbol{a}(\theta)\boldsymbol{a}^{\mathrm{H}}(\theta)\boldsymbol{W}\right]+\sigma^2\left\|\boldsymbol{W}\right\|\right)\end{aligned} \tag{5}$$

where P_θ is the power of array antenna whose receive azimuth is θ, $E\left[x(t)\right]^2$ is signal energy, σ^2 is the white noise variance, and ||.|| is norm. Obviously, the maximum of (5) is equal to the follows

$$\max\left(E\left[x(t)\right]^2\left[\boldsymbol{W}\boldsymbol{a}(\theta)\boldsymbol{a}^{\mathrm{H}}(\theta)\boldsymbol{W}\right]\right) \tag{6}$$

The constraints are as follows

$$\begin{cases}\boldsymbol{W}^{\mathrm{H}}\boldsymbol{a}(\theta_1)=0,\\ \boldsymbol{W}^{\mathrm{H}}\boldsymbol{a}(\theta_2)=0,\\ \dots\\ \boldsymbol{W}^{\mathrm{H}}\boldsymbol{a}(\theta_{N-1})=0,\\ \boldsymbol{W}^{\mathrm{H}}\boldsymbol{a}(\theta_N)=0,\\ \left\|\boldsymbol{W}\right\|=\boldsymbol{W}^{\mathrm{H}}\boldsymbol{W}=1.\end{cases} \tag{7}$$

The solution of the object function (6) with the constraints (7) can be solved by the Lagrange multiplier method. Finally, the optimal weight vector can be expressed as

$$\boldsymbol{W}_{opt}=\frac{\boldsymbol{C}\boldsymbol{a}(\theta)}{\sqrt{\boldsymbol{a}^{H}(\theta)\boldsymbol{C}\boldsymbol{a}(\theta)}} \tag{8}$$

where

$$\boldsymbol{C}=\boldsymbol{I}-\boldsymbol{a}(\boldsymbol{\theta})\left[\left(\boldsymbol{a}(\boldsymbol{\theta})\right)^{\mathrm{H}}\left(\boldsymbol{a}(\boldsymbol{\theta})\right)\right]^{-1}\left(\boldsymbol{a}(\boldsymbol{\theta})\right)^{\mathrm{H}},$$

$\boldsymbol{I}$ is a unit vector matrix, $\boldsymbol{a}(\boldsymbol{\theta})=\left[\boldsymbol{a}\left(\theta_{mn}^k\right)\right]$ is the array response matrix of user *k*, and θ_{mn}^k is the antenna array azimuth angle of the user *k* from the antenna *n* to antenna *m*.

The solution of the optimal weight does not depend on the received signals. It is only affected by azimuth of interference sources. Real-time and repeated calculation are not needed, so that the calculation is simplified.

NUMERICAL ANALYSIS

To verify the performance of the proposed beamforming algorithm in this chapter, the OBW-FAI algorithm is compared to classic beamforming Direct Matrix Inversion (DMI) algorithm (Lorenz & Boyd, 2005) in simulation analysis.

Performance Indicators

The authors measure the performance of DBF algorithm mainly from the following three indices.

1. **Throughput:** From (Dai, Chen, & Chuang, 2013), the expression of channel capacity for DBF-MIMO system can be obtained as follow

$$C = \sum_{i=1}^{M} \log_2\left(1 + SINR_{opt}\right) \tag{9}$$

$$SINR_{opt} = f\left(\boldsymbol{W}\right) = \frac{\gamma \left|\boldsymbol{W}^{\mathrm{H}} \boldsymbol{a}(\boldsymbol{\theta}^{d})\right|^{2}}{\boldsymbol{W}^{\mathrm{H}} \boldsymbol{R}_{i+n} \boldsymbol{W}} \propto \frac{L}{N}, \gamma \tag{10}$$

where M is the number of antennas of BS; $SINR_{\text{opt}}$ is the optimal SINR when the weight $\boldsymbol{W}$ gets the optimal value, it is proportional to the array element number L and SNR γ, and inversely proportional to the target users' antenna number N,$\boldsymbol{R}_{i+n}$ is covariance matrix of interference and noise.

2. **SINR:** Based on received signals, the OBW-FAI beam weight can be obtained by (8), and get the SINR of DBF algorithm by (10). If the SINR of the target users received by BS is greater than expected, the communication services of target users can be ensured, as follow

$$SINR_{d} \geq SINR_{th} \tag{11}$$

where $SINR_{\text{th}}$ is the SINR threshold of target users. However, if $SINR_{d} < SINR_{\text{th}}$, the communication services of target users fail.

3. **The Beam Pattern of the Beamforming:** The beam pattern of the beamforming is defined as follow

$$G = 20\lg\left(\left|\boldsymbol{W}^{\mathrm{H}} \boldsymbol{a}(\boldsymbol{\theta}^{d})\right|\right) \tag{12}$$

To ensure the reliability, the authors analysis the data by the Monte Carlo method.

Parameter Settings

The authors consider the parameters as follows: the target users are at the direction of horizontal angle -30 ° of 4G BS. The interference satellite terrestrial station is located in Chongqing (107 ° E, 30 ° N), whose azimuth is distributed in 5° to 80° of BS horizontal angle, and randomly distributed in the circular 85 meters away from BS uniformly. The azimuth of the interference satellite terrestrial station is known, but specific location and the number of interference is unknown. BS is located in the suburbs, and it is built on hills 100 meters high in order to realize a wider coverage in suburbs. The antenna array is the ideal linear array which has $\lambda/2$ spacing of array elements, antenna array spacing is 10λ. Other simulation parameters are shown in Table 1 to Table 3.

Table 1. Parameters of GEO satellite

Parameter	Value
Satellite longitude	115.5°E
Cover range of latitude	3°N~53°N
Cover range of longitude	73°E~135°E
Carrier frequency	2.3 GHz

IGI, 2016.

Table 2. Parameters from VSAT to 4G BS

Parameter	Value
VSAT transmit power	30 dBm
VSAT antenna gain	42.7 dB
VSAT EIRP	72.7 dBm
VSAT antenna aperture	1.2 m
Distance between VSAT and BS	130 m
Path loss	36.42 dB
BS antenna gain	10 dB
Noise power (8MHz)	-104.956 dBm

IGI, 2016.

Table 3. Parameters from target users to 4G BS

Parameter	Value
Target users' transmit power	20 dBm
Target users' antenna gain	10 dB
Distance between target users and BS	0.05~5 km
Range of path loss	48.08-68.08 dB
BS antenna gain	10 dB
Noise power (8MHz)	-104.956 dBm

IGI, 2016.

Numerical Result

1. **Channel Capacity:** From the above definitions it can be known, *N* represents the number of antennas, *M* means the number of antenna arrays in BS, *L* stands for the number of array elements in the antenna array of BS, and the authors use $N\times (M, L)$ represents the system.

Figure 3 shows that when BS is not equipped with antenna arrays, there is no DBF algorithm existed, and the channel capacity is only affected by the number of MIMO antennas. Compared to SISO systems, the capacity of 2x2 MIMO systems doubles almost. When BS is equipped with antenna arrays, it has the ability of digital beamforming. However, due to the difference of performance capacity in generated optimal beam weight between OBW-FAI and DMI, the ability to reduce interference of other secondary users is different under the same SNR, leading to different SINR, which can be reflected in channel capacity that OBW-FAI-MIMO has better channel capacity than DMI-MIMO when both adopting $2\times (2, 4)$ systems.

2. **Beam Pattern:** When the number of users K=10, the number of interference secondary users is 9, the transmit power of each interference user is 30 dBm, the distance and the arrival angle to BS are different. To analyze in aspect of horizontal angle, as shown in Figure 4, in the range of 5°

Figure 3. Relationship between SNR and channel capacity under different antennas and digital beamforming algorithms

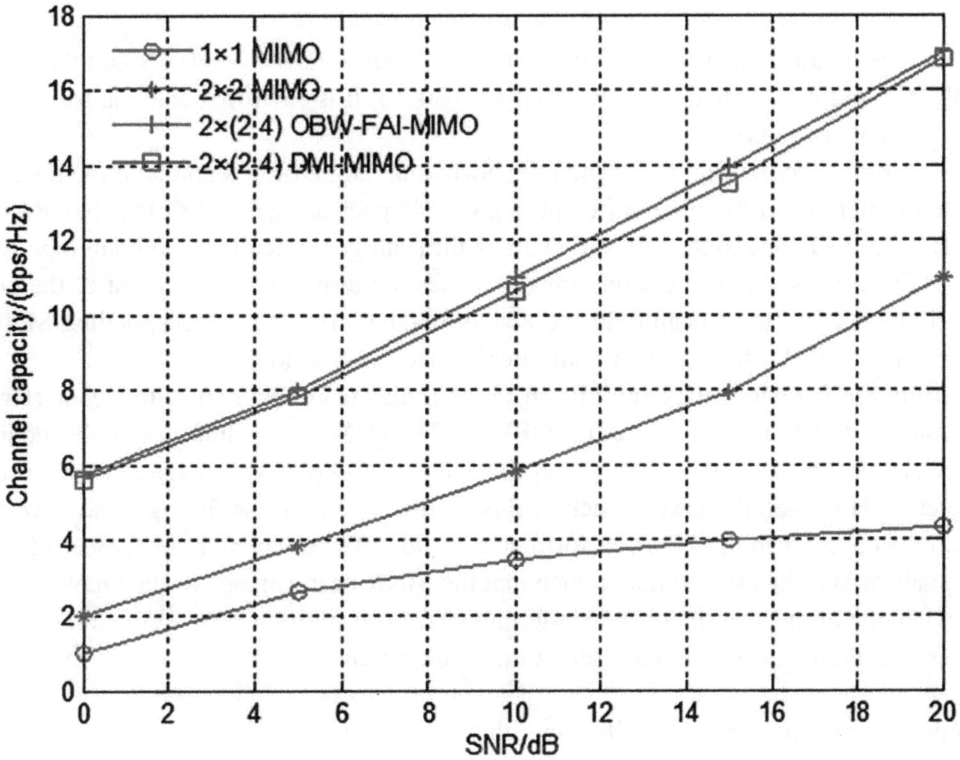

Figure 4. Beam pattern of different beamforming algorithm in 2×(2, 4) system when K=10

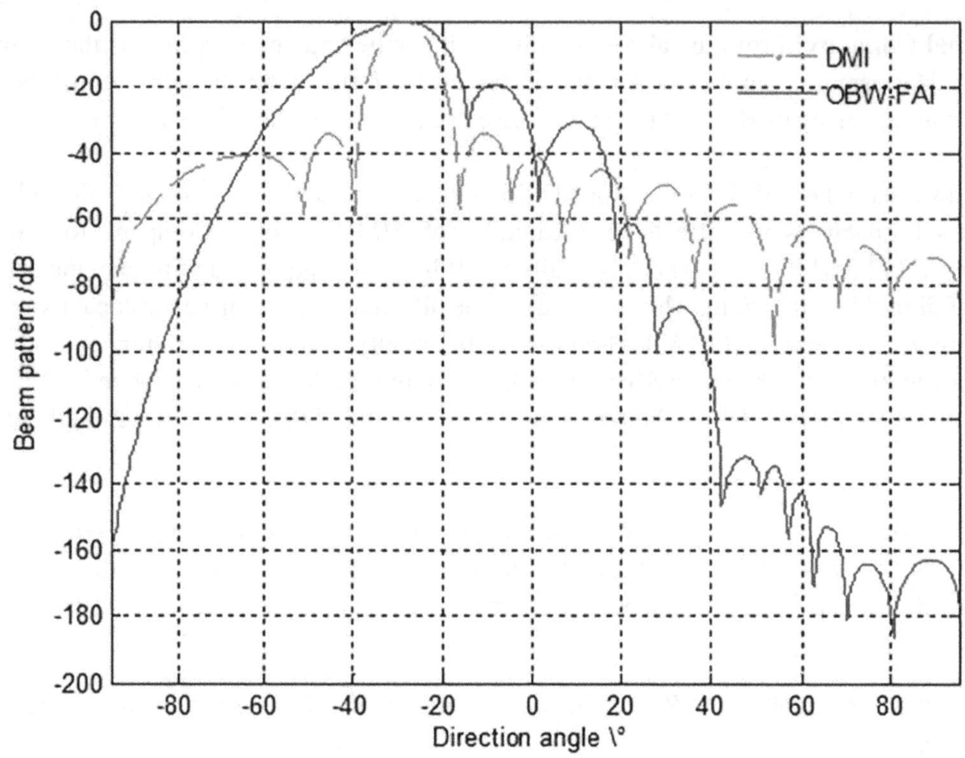

~80°, the beam pattern of DMI algorithm changes from -60 dB to -180 dB and the beam pattern of OBW-FAI algorithm changes from -50 dB to -80 dB. It is clear that antenna gain of OBW-FAI is better than that of DMI.

3. **SINR:** According to the Figure 5, when the horizontal angle of antenna is in the range of 5° and 80°, the interference users have different degrees' impact on the SINR. This SINR is belong to target BS received. Due to the OBW-FAI algorithm can guarantee that the main lobe is not influenced as far as possible, at the same time, it causes the directional zero point of the beam at the given direction. And it also can reduce the side-lobe beam. So the corresponding SINR is better compared to the DMI algorithm without interference suppression.
4. **The Number of Interference Users:** It can be concluded from Figure 6, with the interference users increasing, the SINR performance of the OBW-FAI and DMI algorithms gradually decline. But the falling trend of the SINR is relatively slow, because the OBW-FAI can against fixed user azimuth interference. However, the DMI is lack of this ability. So the SINR falling trend is very obvious. The authors compare to the SINR performance of the OBW-FAI when the number of interference is less than 7. And then the authors notice that the SINR performance of the OBW-FAI decreases faster when the number of interference satellite terrestrial station users is more than 7. The reason is that it is limited to against fixed user azimuth interference.

The aforementioned numerical results show that the authors should fully consider the impact of throughput, SINR and the number of possible interference users in the actual design of BS of DBF-

Figure 5. When k=10, 2×(2, 4) system corresponding to different beamforming algorithm SINR

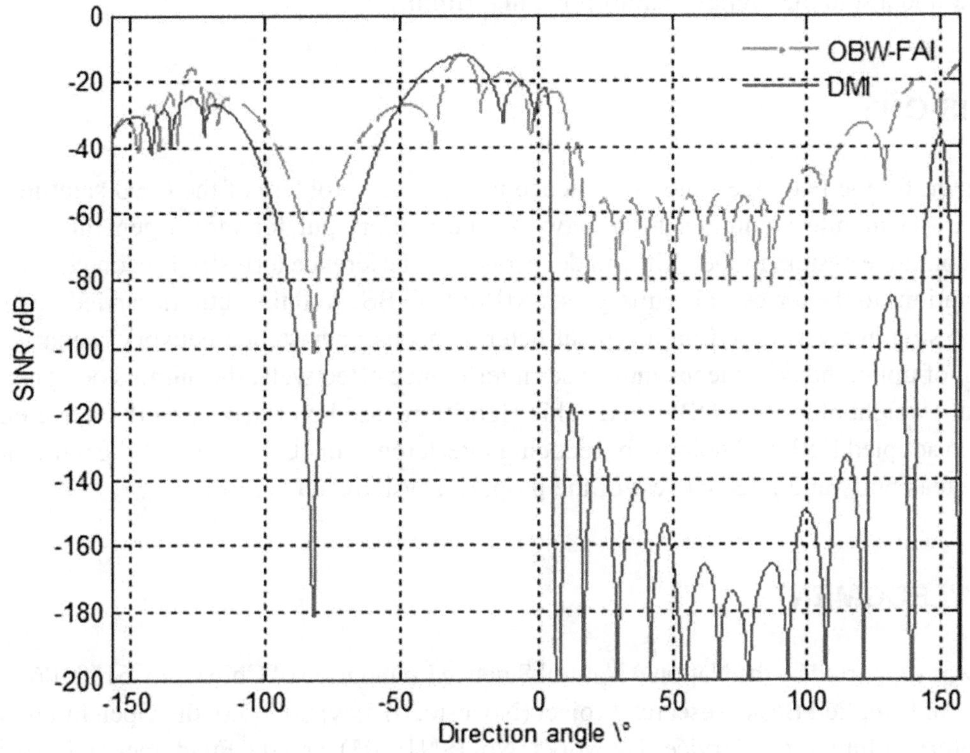

Figure 6. 2×(2,4) system the relationship between SINR and the number of interference users under different beamforming algorithm

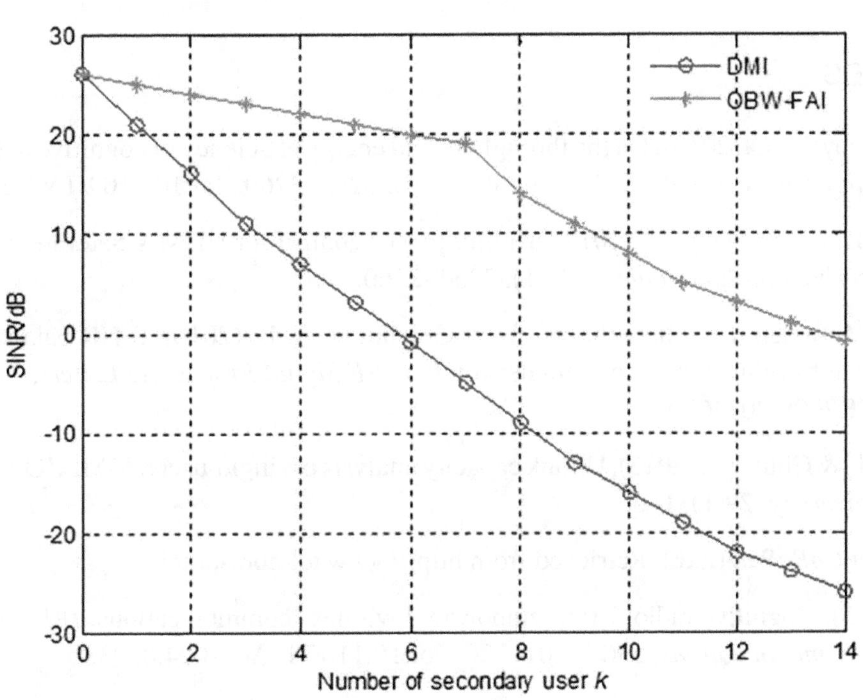

MIMO. It should also be cautious to choose a suitable number of transmit antennas of users, antenna arrays of BS and a suitable digital beamforming algorithm.

CONCLUSION

In this chapter, for the heterogeneous coexistence interference problem of the GEO satellite-terrestrial network and 4G mobile communication network, the authors put forward a general heterogeneous multi-user cognitive system model. This model is one of interference terrestrial secondary user satellite terrestrial stations to the terrestrial primary user MIMO 4G BS. And the authors carried out a complete signal processing analysis based on a typical heterogeneous coexistence network scenario. To solve the problem of uplink heterogeneous multi-user interference effectively, the authors designed a kind of optimal beam weight algorithm OBW-FAI. This algorithm based on fixed azimuth interference. In addition, it also adopted DBF technology for reducing interference in 4G BS. At last, the numerical results verified the feasibility and effectiveness of the proposed system and algorithm.

ACKNOWLEDGMENT

This work was supported by the National Natural Science Foundation of China (No. 61501066), Chongqing Frontier and Applied Basic Research Project (No. cstc2015jcyjA40003), the Open Fund of the State Key Laboratory of Integrated Services Networks (No. ISN16-03), and the Fundamental Research Funds for the Central Universities (CDJZR165505). The authors would like to thank the Editor Dr. Chungang Yang (Xidian University, China) for his generous comments and helpful advice to improve the quality of this chapter.

REFERENCES

Akin, S., & Gursoy, M. C. (2013). On the throughput and energy efficiency of cognitive MIMO transmissions. *IEEE Transactions on Vehicle Technology*, *62*(7), 3245–3260. doi:10.1109/TVT.2013.2254142

Chen, P., Qiu, L. D., & Wang, Y. (2012). Uplink power control for CDMA Satellite cognitive radio. *Systems Engineering and Electronics*, *34*(11), 2354–2360.

Cumanan, K., Musavian, L., Lambotharab, S., & Gershman, A. B. (2010). SINR balancing technique for downlink beamforming cognitive radio networks. *IEEE Signal Processing Letters*, *17*(2), 133–136. doi:10.1109/LSP.2009.2033650

Dai, J., Chen, M., & Chung, P. (2013). Uplink capacity analysis of single-user SA-MIMO system. *Journal of Southeast University*, *29*(1), 1–6.

EU FP7 Project CoRaSat. (n.d.). Retrieved from http://www.ict-corasat.eu

Haykin, S. (2005). Cognitive radio: Brain-empowered wireless communications. *IEEE Journal on Selected Areas in Communications*, *23*(2), 201–220. doi:10.1109/JSAC.2004.839380

Katkovnik, V., Lee, M., & Kim, Y. (2006). Performance study of the minimax robust phased array for wireless communications. *IEEE Transactions on Communications*, *54*(4), 608–613. doi:10.1109/TCOMM.2006.873068

Li, X. H. (2007). Space information acquisition based on software-defined radio and cognitive radio. *Aerospace Electronic Warfare*, *23*(6), 33–35.

Li, Z., Li, J. D., Liu, Q., & Shen, B. (2014). Interference alignment based spectrum sharing and user scheduling for cognitive radio MIMO system. *Journal of Communication*, *35*(1), 167–172.

Liolos, K., Schlueter, G., Krause, J., Zimmer, F., Combelles, L., Grotz, J., … Vanelli, C. A. (2013). Cognitive radio scenarios for satellite communications: The CoRaSat approach. In Proceedings of Future Network and Mobile Summit. Lisbon: IEEE.

Liu, H. X., & Zhu, J. K. (2008). Dynamic spectrum allocation of little satellite group communication based on cognitive radio. *Communications Technology*, *41*(7), 1–3.

Liu, Q. J., Bi, S. J., Sun, J., & Liu, T. X. (2012). Potential applications of cognitive radio technology in satellite navigation system. In *Proceedings of 3rd Annual Conference of China's Satellite Navigation*. Guangzhou: CSNC.

Lorenz, R. G., & Boyd, S. P. (2005). Robust minimum variance beamforming. *IEEE Transactions on Signal Processing*, *53*(5), 1684–1696. doi:10.1109/TSP.2005.845436

Luan, T., Gao, F. F., Zhang, X. D., Li, C. F., & Lei, M. (2012). Rate maximization and beamforming design for relay-aided multiuser cognitive networks. *IEEE Transactions on Vehicle Technology, 61*(4), 1940-1945.

Ma, L., Chen, X. T., Liu, H. J., & Liang, X. W. (2010). The application of cognitive radio technology in LEO satellite system. In *Proceedings of 6th Annual Conference of Satellite Communications New Business New Technology*. Beijing: ChinaInstitute of Communications.

Ming, C. N., (2012). Some suggestions on demand for research and construction of China satellite mobile communication system. *Space International*, 1-9.

Sithamparanathan, K., Nardis, L. D., Benedetto, M. G., Guidotti, A., Corazza, G., & Corazza, G. E. (2010). Cognitive satellite terrestrial radio. In *2010 IEEE Global Telecommunications Conference* (pp. 1-6).

Sun, W., Zhang, P., Chen, Y., Qin, Z., & Teng, D. (2013). A tight coupling cooperation scheme in WiFi/WiMAX heterogeneous mesh networks. *International Journal of Computers, Communications & Control*, *8*(3), 460–468. doi:10.15837/ijccc.2013.3.10

Tawil, C., & Tawil, S. (2009). *Apparatus and method for transmitting terrestrial signals on a common frequency with satellite transmissions*. USA: 6892050 B1.

Wang, H. X., Pan, C. S., & Song, J. H. (2013). *Spaceborne smart antenna beamforming technology*. Beijing, China: Defense Industry Press.

Yiu, S., Vu, M., & Tarokh, V. (2009). Interference and noise reduction by beamforming cognitive networks. *IEEE Transactions on Communications*, *57*(10), 3144–3153. doi:10.1109/TCOMM.2009.10.080501

Zhang, G. X., & Gan, Z. M. (2005). Study on the development ideas and strategies of China satellite mobile communication system. *Digital Communication World, 95*(7), 24–27.

Zhang, G. X., Gan, Z. M., & Li, G. X. (2010). Study on the development of China satellite mobile communication. *Satellite & Networks, 94*(3), 32-43.

Zhang, J., Jiang, B. Q., & Zheng, L. (2013). The application of cognitive radio network in the satellite communication. *Journal of Guilin University of Electronic Technology, 33*(4), 284–287.

Zhang, X., & Shin, K. G. (2013). Cooperative carrier signaling: Harmonizing coexisting WPAN and WLAN devices. *IEEE/ACM Transactions on Networking, 21*(2), 426–439. doi:10.1109/TNET.2012.2200499

KEY TERMS AND DEFINITIONS

Cognitive System: An intelligent network which can sense the external environment, as well as adjust the internal configuration of the communication network in real time, and adapt to the changes of the external environment by understanding and studying of the external environment.

Digital Beamforming: A combination of digital signal process and beamforming to control the direction of waves and signals.

Geosynchronous Earth Orbit: Satellites are located in an orbit with a height of 36 thousand kilometers, in that orbit, the satellites have a speed same with the speed of earth rotation.

Interference Mitigation: A technology to reduce the interference.

Multi-Users Multi-Input Multi-Output: Base station serves for multi-users, the base station and users are equipped with multi-antennas.

Optimal Beam Weight: Coefficients of beams to maximize the output power in the target direction, and to make the interference azimuth beam equals to 0 without impact on Signal to Interference plus Noise Ratio.

Satellite-Terrestrial Heterogeneous Coexistence: A scenario where the Geostationary Earth Orbit (GEO) satellite-terrestrial network and the 4G mobile communication network coexist heterogeneously.

ENDNOTE

1 "Tianlian" series of satellites are the Chinese Data Relay Satellites in GEO. They are mainly developed by the China Academy of Space Technology affiliated to China Aerospace Science and Technology Corporation, and used to provide data relay and TT&C (Tracking Telemetering & Command) services for Chinese Shenzhou manned spacecraft and subsequent manned spacecrafts. Moreover, they also provide data relay service for Chinese medium and low orbit resource satellites, and provide TT&C support for spacecraft launching. At present, the networking system of three satellites, Tianlian I,II and III, has been established and run successfully, which makes China become the second country who possess quasi global coverage range Data Relay Satellite System after the United States.

Chapter 6
The Combination of Resource Allocation and Interference Alignment for Ultra-Dense Heterogeneous Cellular Networks

Yun Meng
Chang'an University, China

Yuan Dong
Chang'an University, China

Song Shi
Chang'an University, China

ABSTRACT

Different with the traditional resource allocation problems, the consistency of resource occupation and the diversity of interference of participants in interference alignment (IA) clusters bring about the complexity in the combination of resource allocation and IA. Therefore, this chapter gives a transformed conflict graph-based solution framework which considers the low complexity of chordal graph, where the selection criteria of IA clusters are determined by the influence of IA on the resource occupation and the interference. The simulation results show that the proposed schemes can improve the network performance.

INTRODUCTION

The scarcity of spectrum resources and the exponential growth of user data traffic drive the future wireless network increase the spectrum efficiency, therefore, the ultra-dense deployment of heterogeneous cellular networks has been regarded as a promising way by spectrum reuse (Zahir, Arshad, Nakata,

DOI: 10.4018/978-1-5225-1712-2.ch006

& Moessner, 2013; Andrews, Claussen, Dohler, Rangan, & Reed, 2012). However, without proper configuration, the interference will be severe and then hinder the improvement of spectrum efficiency. Hence, the interference management is a critical issue for ultra-dense heterogeneous cellular networks.

Resource allocation and interference alignment (IA) are the two key techniques of interference management. The former allocates the orthogonal resources to the interfering users while decreases the spectrum efficiency. Moreover, in the ultra-dense heterogeneous cellular networks, resource allocation cannot absolutely mitigate the serious interference without rejecting a portion of users. The latter aligns the interferences into special signal dimensions and then the left signal dimensions are interference free where the useful signals can be transmitted (Cadambe & Jafar, 2008; Jafar, 2011). In the multi-input multi-output (MIMO) system, the IA via precoding can increase the degrees of freedom in the interference channel (Cadambe et al., 2008; Jafar, 2011). However, the number of users is restricted by the feasible constraint of IA. Therefore, the intensive interference in the ultra-dense heterogeneous cellular networks cannot be completely eliminated just by IA, and it is needed to select the users to form IA cluster and mitigate the interferences of users out of the IA cluster. In addition, the additional signal dimensions of each participant users are occupied for compressing interference form the other users of the IA cluster. Therefore, compared with the non-participant user, the participant user in IA cluster occupies more resources to achieve the equivalent degree of freedoms. In summary, each of resource allocation and IA has its own advantages and limitations. To improve the performance of ultra-dense heterogeneous cellular network, the joint interference management combing resource allocation and IA become an emerging area of research (Meng, Li, Li, & Pan, 2015; Li, Meng, Li, & Suo, 2015).

The rest of this chapter is organized as follows. In Section 2, we introduce the background by surveying the related works. In Section 3, we analyze the influence of IA on resource allocation. In Section 4, we formulate the graph-based solution framework of the resource allocation combing with IA. In Section 5, we show the simulation results. Finally, we conclude this chapter and give the possible research directions in the future.

THE BACKGROUND

Due to the significance of the joint interference management combing resource allocation and IA, there are many prior works have been studied. This section will introduce the related works from three parts, including the IA feasible constraint, the classification of interference network and the method of IA cluster formation.

The number of users is restricted by the IA feasible constraint. In MIMO network, let $\left(M \times N, d\right)^K$ denote the K-user interference channel where the transmitter equipped with M antennas, intend to send d independent data streams, to the receiver equipped with N antennas. According to the work of Jafar (2011), the interference channel setting above needs to satisfy the following condition:

$$d \leq \frac{M+N}{K+1} \tag{1}$$

Therefore, for the given number of antennas M and N, the maximum number of users in a feasible IA cluster can be expressed as follows:

$$K_{\max} \leq \left\lfloor \frac{M+N}{d} - 1 \right\rfloor \quad (2)$$

In the ultra-dense heterogeneous cellular networks, the interference is serious where the number of interfering links is likely beyond the constraint (2), which means executing IA among all the users to eliminating the interference in network is unachievable.

According to the strength of interference, we classify the networks into two categories as follows:

- **K-User MIMO Interference Channel:** All users are proximity to each other in K-user MIMO interference channel, where all the receivers can receive strong signals from all the transmitters.
- **Asymmetric Attenuation Interference Network:** Different with K-user MIMO interference channel, the geographic distances of interfering links in the asymmetric attenuation interference network have big gaps, hence the interference signal strengths have significant differences resulted by the distance-dependent path loss.

In practical communication network, the achievable data rate is decreased mainly by the strong interference and is not sensitive to the weak interference (Chen & Cheng, 2014). In addition, most of IA methods need extensive channel state information (CSI) to design the precoder/decoder of users (Cadambe et al., 2008), where the size of channel matrix is exponentially growing with the number of links. Hence, the estimation of CSI brings about extensive signaling overhead. Even though the iterative algorithms reduce the required CSI for the design the precoder/decoder (Gomadam, Cadambe, & Jafar, 2011; Peters, & Heath, R. W., 2009), the signaling overhead has not been reduced significantly as the iterative algorithms need to exchange CSI feedback during the iterations. Therefore, the interfering links can achieve markedly rate improvement by IA in strong interference, while using IA to eliminate the weak interference cannot improve the data rates, and eliminating all the interference in the second category of network is not necessary, sometimes even a bad bargain.

Due to the issues represented above, scheduling a part of users to execute IA has been proposed to overcome the issues of IA feasible constraint and the huge signaling overhead.

The idea of IA cluster was first proposed for cellular network in the works of Tresch et al. (2009a) and ad hoc network in the work of Tresch et al. (2010), while only single IA cluster was considered in the works of Tresch et al. (2009a) and no clustering method was mentioned in the work of Tresch et al. (2010). However, the multiple IA clusters must be taken into account as the number of interfering links in ultra-dense heterogeneous cellular networks is large. According to the classification of networks in system model, the works of IA clustering can be classified into the following two cases:

- **IA Clustering in K-User MIMO Interference Channel:** Including the works of Ronasi et al. (2014), Park et al. (2014), and El-Absi et al. (2014) and connected interference network scenario of the work of Peters et al. (2012).

To maximize the network throughput is the uniform optimization object of these studies, while there are different constraint conditions. The work of Ronasi et al. (2014) uses the theory of Lyapunov optimization to guarantee the stability of all queues in interference channel, where the imperfect IA is employed instead of the perfect IA restricted by feasibility condition. The IA algorithm using spatial and frequency domain is considered in the work of Park et al. (2014), where the achievable DOF is restricted by the feasibility conditions. Studies of El-Absi et al. (2014) and the connected interference network scenario of the work of Peters et al. (2012) eliminates interference among IA clusters by the orthogonal resource scheduling on frequency and slot respectively, where the number of user meet the IA feasibility condition.

- **IA Clustering in Asymmetric Attenuation Interference Network:** Including the works of Meng et al. (2015), Li et al. (2015), Guler et al. (2014), Ao et al. (2014), Chen et al. (2014), Pantisano et al. (2013), Nguyen et al. (2013), Lertwiram et al. (2012), and Da et al. (2011) and connected interference network scenario of the work of Peters et al. (2012).

The work of Guler et al. (2014) aligns the uplink interference of femtocell from macrocell as much as possible on the premise of the signal-to-interference ratio (SIR) threshold of macrocell users, where the imperfect IA is also under consideration. Different with the work of Guler et al. (2014), the feasibility conditions are satisfied in the rest of references mentioned in this case. The set of IA clusters in the study of Ao et al. (2014) is given as a known condition and each element in the set executes IA. The formation of IA cluster is optimized by the maximization of the aggregated intra-cluster interference in the work of Chen et al. (2014). Users in the works of Pantisano et al. (2013), Nguyen et al. (2013), Lertwiram et al. (2012), and Da et al. (2011) form the IA clusters according to the interference still existed after resource allocation in the network, where the change brought by IA in resource allocation are not considered. Based on the analysis on the optimization and the constraint resulted by IA, studies of Meng et al. (2015) and Li et al. (2015) select users to optimize the DOFs of users. In the connected interference network scenario of Peters et al. (2012), the partition of users is uniform based on the maximization of interference in cluster.

THE COMPLEXITY OF RESOURCE ALLOCATION INDUCED BY IA

We analyze the complexity of resource allocation induced by IA in two aspects including the resource occupation and the interference expression.

- **The Consistency of Resource Occupation in the IA Cluster in Scheduling:** The theorem in Cadambe et al. (2008) has been shown that users with single antenna in K-user interference channel can almost obtain $K/2$ DOFs per orthogonal time and frequency dimension, where needs the exponential symbol extension of time and frequency. Therefore, IA in MIMO interference channel by beamforming solutions has gained significant attention. In MIMO system, the IA via precoding has been an important implication that all the users in IA cluster use the identical time and frequency resource. Therefore, all the users of an IA cluster need to be scheduled as a whole in resource allocation. In addition, even though the performance of IA cluster can be improved by executing IA, IA cluster needs occupy more resources compared with the non-IA user for the same performance of a single user.
- **The Diversity of Interference of Participants in IA Cluster:** We can divide the interference suffered by the participants in IA cluster into the inner-cluster interference and the outer-cluster in-

terference. If the interferer belongs to the IA cluster, it is the inner-cluster interference. Otherwise, it is outer-cluster interference. On the basis of meeting the feasible constraints, the interference among users of a scheduled IA cluster can be eliminated by perfect IA. As for the IA cluster fail to meet the feasibility conditions, the imperfect IA mitigates a portion of the interference in cluster while there is leakage interference existing among users. In general, the inner-cluster interference can be mitigated by IA. However, it is uncertain for the outer-cluster interferences, which is caused by the diversity of geographical location of interferers for each participant in the IA cluster. Therefore, the conflict relationships in resource allocation for each participant in the IA cluster are different.

An illustrated example of network including 6 femtocells is given in Figure 1. It can be seen that the interferences among UE2, UE3 and UE4 can be eliminated by perfect IA. As shown in Figure 1, UE2 is interfered by FAP1, UE3 is interfered by FAP5, and UE4 is interfered by both FAP 5 and FAP 6. Hence, we can find that the interference signals received by the UE2, UE3 and UE4 are different.

Based on the analysis above, the consistency of resource occupation and the diversity of interference of participants in IA cluster are in coexistence, which bring about the uncertain of improving or deteriorating the performance by IA. Figure 2 shows three illustrate examples of interference network. The maximum DOFs in Figure 2(a), Figure 2(b) and Figure 2(c) without IA can be derived as 4, 2 and 4, respectively. If the IA clusters $\{A, B, D\}$, $\{E, F, H\}$ and $\{I, J, L\}$ perform IA, the maximum DOFs in

Figure 1. An illustrate example of interference network. Each femtocell has one user (UE) and the UE is associated with the corresponding femto access point (FAP). The feasible IA cluster in this network is composed of femtocell 2, 3 and 4.

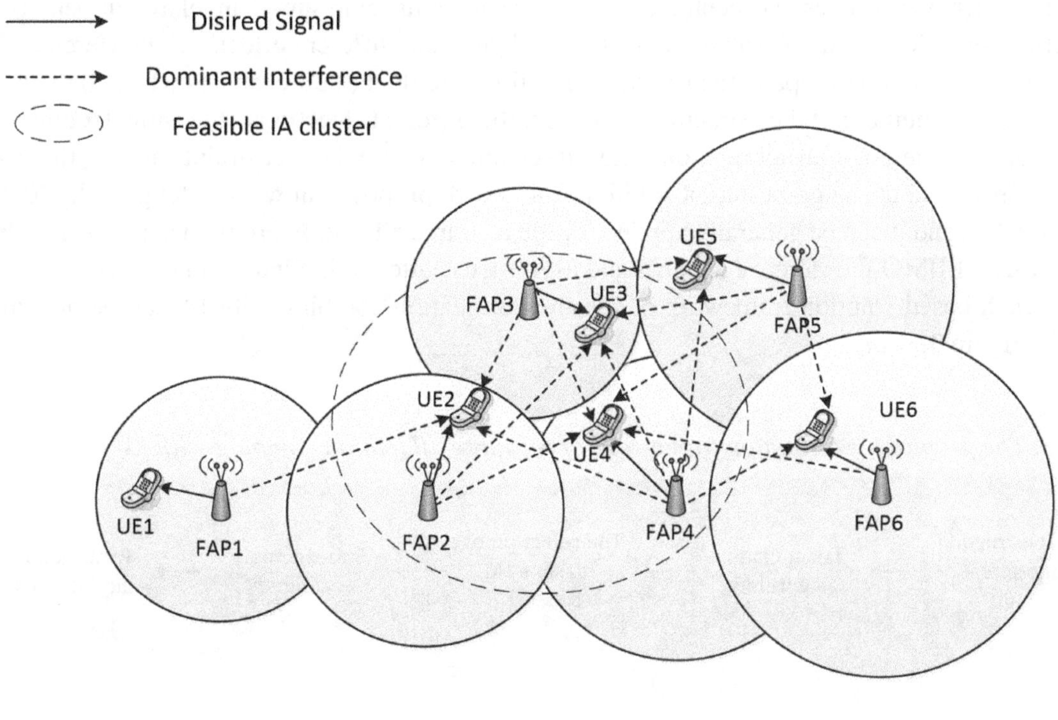

Figure 2. Three examples of interference network. Each vertex corresponds to the downlink form FAP to its associate UE. Each UE has two antennas. The feasible IA clusters are $\{A, B, D\}$ in (a), $\{E, F, H\}$ in (b) and $\{I, J, L\}$ in (c), respectively.

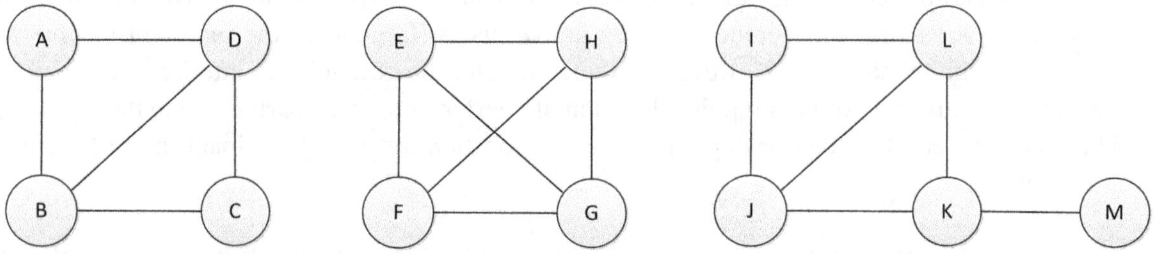

Figure 2(a), Figure 2(b) and Figure 2(c) are 3, 3 and 5, respectively. Therefore, it can be seen that the improving or deteriorating the performance induced by executing IA in an interference network is relevant with the network topology.

THE GRAPH-BASED SOLUTION FRAMEWORK OF RESOURCE ALLOCATION COMBING WITH IA

The resource allocation problem in the interference network has been extensively researched in coloring problem and its variants in graph theory. However, compared with the existing various coloring problem, our resource allocation problem combining with IA is fundamentally different, which can be summarized in the following three aspects. First, IA can eliminate the mutual interferences in the IA cluster, i.e., each two vertices in executed IA cluster are not conflict in sub-channel assignment. Second, the multiple selections of the feasible IA clusters will generate different effects on interference. Third, compared with the nonparticipant, the IA cluster requires double the sub-channels when all the UEs can transmit the same number of data streams. Therefore, the different selections of feasible IA clusters will bring about the different conflict relations and sub-channel assignment constraints among the vertices, which is considered in the graph-based solution framework proposed in works Meng et al. (2015) and Li et al. (2015), and it can be generally applied for the resource allocation problem combined with IA in both of K-user MIMO interference channel and asymmetric attenuation interference network.

The graph-based solution framework of resource allocation combing with IA can be described as shown in Figure 3.

Figure 3. The graph-based solution framework of resource allocation combing with IA

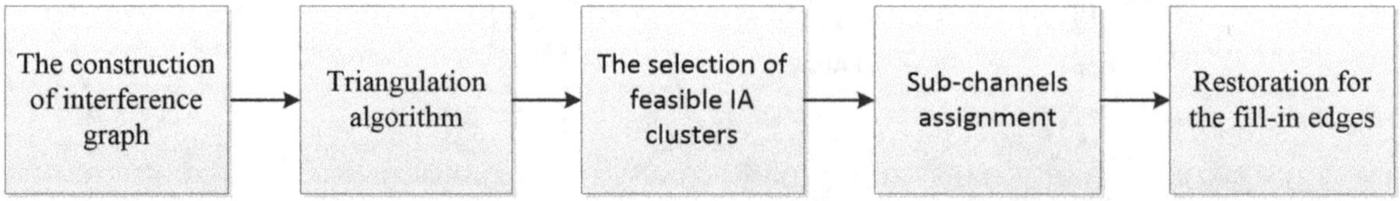

System Model

We consider the downlink in the ultra-dense heterogeneous cellular networks. The multiple UEs associated with the same FAP are allocated with the orthogonal resources, and then there is no intra-cell interference. For the UEs associated with different FAPs, they can be allocated with the same resource. All the FAPs and UEs are equipped with two antennas. Therefore, all the downlinks in the network are 2x2 point to point MIMO channels, where each downlink has two DOFs (Foschini and Gans, 1998; Telatar, 1999). Only the feasible IA clusters are considered in this part, hence the number of UEs in each IA cluster is 3 restricted by the feasible constraints. Therefore, each IA cluster can obtain 3 DOFs with IA via precoding on one sub-channel, where each UE can obtain one DOF. We use g to denote the set of vertices in feasible IA cluster that can be formed by the methods described in 2.3, and its set can be denoted as $S = \{g_1, g_2, \ldots, g_i, \ldots, g_I\}$.We assume that each user has two data streams to transmit. Hence, the user without executing IA requires one sub-channel to satisfy the requirement, while the executed IA cluster needs to occupy two sub-channels to satisfy all the requirements of the participant users.

The dominant interference is defined as it can individually contribute enough interference to cause that the signal strength of victim is lower than the minimum SIR threshold (Weber, Andrews, & Jindal, 2007).

The Construction of Interference Graph and the Transformed Conflict Graph Using IA

The interference network is modeled as the interference graph denoted as $G = (V, E, W)$, where V, E, and W represents the set of vertices, the set of edges and the set of weights of vertices, respectively. Each vertex $v \in V$ denotes a downlink form FAP to its associated UE, and each edge $e \in E$ denotes that there is dominant interference between the two adjacent links. Each weight $w \in W$ denotes the resource requirement of the corresponding UE that it can transmit the expected data streams. According to the assumption in system model, the resource requirement of each user without IA is one. A clique is a set of vertices that each two vertices are adjacent in graph and a maximal clique denoted as C which is a clique and it cannot be a subset of other cliques. A maximal clique can be seen as a maximal resource conflict region as each vertex needs to be allocated with orthogonal resource with each other (Jaiswal, Ganz, & Mettu, 2009). We use CL_n to denote the aggregate resource requirements of users in C_n, hence CL_n is equal to the size of C_n in G. And we use $V(C_n)$ to denote the set of vertices in C_n.

Based on the inclusion relation with the IA cluster g_i, all the maximal cliques in G can be classified as the following three categories (Meng et al., 2015):

- The maximal cliques including g_i, whose set is denoted as

$$\mathcal{C}_1(g_i) = \{C_n | V(C_n) \cap g_i = g_i\}.$$

- The maximal cliques including the nonempty and proper subset of g_i, whose set is denoted as

$$\mathcal{C}_2(g_i) = \{C_n | V(C_n) \cap g_i \neq \varnothing \text{ and } V(C_n) \cap g_i \subset g_i\}.$$

- The maximal cliques having no repetitive element of g_i, whose set is denoted as

$$\mathcal{C}_3(g_i) = \{C_n | V(C_n) \cap g_i = \varnothing\}.$$

The color number in graph theory is the minimum number of the required colors that can color all the vertices in graph without allocating the same color for each two adjacent vertices [28]. According to the assumption in network model, we can use the color number of interference graph to evaluate the resource requirement of all the users in network. Based on the theorem in graph theory, the color number of a clique is equal to its size. Hence, the color number of a feasible IA cluster is 3 when it is not scheduled with IA; otherwise, it reduces to 2.

All the users in *K*-user MIMO interference channel are proximity to each other, where all the interferences in the network are dominant interferences, and each user needs to be allocated with orthogonal resource with each other if they are not scheduled with IA. Therefore, the interference graph of *K*-user MIMO interference channel is a clique with size of *K*. Then the color number of the *K*-user MIMO interference channel is *K* without scheduling IA. Since the color number of IA cluster can be reduced by IA, the color number of the *K*-user MIMO interference channel can be decreased by IA. Therefore, the scheduling of IA cluster in *K*-user MIMO interference channel can improve the system performance, the illustrate example is shown as in Figure 2(b). And then we can allocate the orthogonal resources to each user without IA and the IA clusters.

In the asymmetric attenuation interference network, since the distances between the senders and the receivers are different, there are nonadjacent vertices in graph due to the existence of weak interference. Therefore, the effect of IA on the network performance is uncertain. For examples, the number of maximum DOFs in Figure 2(a) is decreased by IA while it is contrary in Figure 2(c).

To analyze the effect of IA on interference and the resource occupation in the network performance, we construct the transformed conflict graph for the interference graph (Meng et al., 2015; Li et al., 2015). The transformed graph of IA cluster g_i is denoted as $G_{g_i}^T$. And we use $\mathcal{S}_S$ to represent the set of the selected IA clusters. Then the transformed graph of the IA clusters in $\mathcal{S}_S$ can be denoted as $G_{\mathcal{S}_S}^T$. The transform method is described as the following three sides. For the vertices in $G_{g_i}^T$, each clique corresponding to the selected IA cluster g_i is replaced by the IA vertex denoted as v_{g_i}. For the edges in $G_{g_i}^T$, each vertex that is adjacent with one of vertices of g_i in G is adjacent with v_{g_i}. For the weights in $G_{g_i}^T$, the weight of IA vertex v_{g_i} is two since a IA cluster needs two sub-channels to transmit all the expected data streams of its participants.

The effect of IA on the interference and the resource occupation of the network can be analyzed by the comparison in vertices, edges and weights between G and $G_{g_i}^T$. In G, each vertex that is adjacent with one vertex of g_i must belong to the maximal cliques in $\mathcal{C}_1(g_i)$ and $\mathcal{C}_2(g_i)$. In $G_{g_i}^T$, each vertex that is adjacent with v_{g_i} must belong to the maximal clique including v_{g_i}. In addition, the change of weights from the clique g_i to v_{g_i} will result in the variation of resource contention in the maximal cliques including v_{g_i}. Therefore, all the maximal cliques including IA vertex v_{g_i} in $G_{g_i}^T$ need to be enumerated to make comparisons with the associated maximal cliques in G.

According to the classification of the associated maximal clique in G, all the maximal cliques including v_{g_i} in $G_{g_i}^T$ can be classified as the following three categories (Meng et al., 2015):

- The maximal clique transformed for $C_{n'} \in \mathcal{C}_1(g_i)$, whose set is denoted as

$$\mathcal{C}_1^T(g_i) = \left\{ C_n^T(g_i) \middle| V(C_{n'}) \cap V\left(C_n^T(g_i)\right) = V\left(C_n^T(g_i)\right) \setminus v_{g_i}, C_{n'} \in \mathcal{C}_1(g_i) \right\}.$$

- The maximal clique transformed for $C_{n'} \in \mathcal{C}_2(g_i)$, whose set is denoted as

$$\mathcal{C}_2^T(g_i) = \left\{ C_n^T(g_i) \middle| V(C_{n'}) \cap V\left(C_n^T(g_i)\right) = V\left(C_n^T(g_i)\right) \setminus v_{g_i}, C_{n'} \in \mathcal{C}_2(g_i) \right\}.$$

- The maximal clique is newly generated, whose set is denoted as

$$\mathcal{C}_3^T(g_i) = \left\{ C_n^T(g_i) \middle| V\left(C_n^T(g_i)\right) = V(C_{n'}) \cup v_{g_i}, C_{n'} \in \mathcal{C}_3(g_i) \right\}.$$

where we use $C_n^T(g_i)$ to denote the *n*th maximal clique including v_{g_i} in $G_{g_i}^T$.

The resource requirement of $C_n^T(g_i)$ is denoted as $CL_n^T(g_i)$, which changes in different situations compared with the associated maximal clique in G. In $\mathcal{C}_1^T(g_i)$, the resource requirement of the associated maximal clique is decreased by IA of g_i, which can be derived as

$$CL_n^T(g_i) = CL_n - 1.$$

In $\mathcal{C}_2^T(g_i)$, the resource requirement of the associated maximal clique can an increase, decrease, or leave unchanged by IA of g_i, which is determined by the cardinality of the intersection of $V(C_n)$ and g_i, and we can derive that

$$CL_n^T(g_i) = CL_n - \left| V(C_n) \cap g_i \right| + 2.$$

In $\mathcal{C}_3^T(g_i)$, the resource requirement of the associated maximal clique is increased by IA of g_i, which can be derived as

$$CL_n^T(g_i) = CL_n + 2.$$

From the analysis above, we can find that the effects of IA executed in g_i on resource requirements are different in the multiple maximal cliques. To analyze the effect of IA executed in g_i on the resource requirement of network, it is needed to enumerate all the maximal cliques in G and $G_{g_i}^T$, respectively. However, the number of maximal cliques in general graph is growing at an exponential rate with the number of vertices, which cannot be enumerated in polynomial-time. Since the scale of the ultra-dense heterogeneous cellular network is large, the exhaustive search of maximal cliques in G leads to a prohibitive computation time. Chordal graph is the graph that each cycle longer than 3 has chord (West, 2001). Chordal graph has some particular properties. The number of maximal cliques in a chordal graph grows linearly as the number of vertices increases, which can be listed by maximum cardinality search (MCS) algorithm in a polynomial-time (Blair & Peyton, 1993). Therefore, we add the fill-in edges in the interference graph to form a chordal graph, where the maximum cardinality search minimal triangulation algorithm is used (Blair & Peyton, 1993). Then we can derive the following properties.

Property 1: If interference graph G is chordal, the transformed conflict graph $G_{S'}^T$ including one or more feasible IA clusters is also a chordal graph (Meng et al., 2015).

Therefore, the lists of maximal cliques both in G and in $G_{S'}^T$ can be enumerated in polynomial-time, which is beneficial for the selection of IA clusters with a low complexity.

Property 2: The maximal clique $C_n^T\left(g_i\right) \in \mathcal{C}_3^T\left(g_i\right)$ in the transformed conflict graph $G_{g_i}^T$ will only be generated from a non-chordal graph G (Meng et al., 2015).

We use $G\left(g \cup adj_G\left(g\right)\right)$ to denote the subgraph that is consisted of g_i and the adjacent vertices of its participants. Here we can derive that all the maximal cliques in chordal graph G changed by the IA are included in $G\left(g \cup adj_G\left(g\right)\right)$. Therefore, the changes of interference and resource occupation of IA can be analyzed in $G\left(g \cup adj_G\left(g\right)\right)$.

The Selection Criteria of IA Clusters and the Resource Allocation Algorithm for the Maximization Satisfactory Ratio of Users

Here we defined that a user is satisfactory if the two data streams can be transmitted simultaneously. Therefore, one sub-channel is needed for a user without IA to be satisfactory while the executed IA cluster need two sub-channels to guarantee that all the participants are satisfactory.

The selection of IA clusters to maximize the number of satisfactory users consists of two steps. First, each IA cluster is preselected by the analysis of the local network. Second, all the preselected IA clusters are selected based on the conflict relationships among them.

Property 3: In a chordal interference graph G, if

$\exists! C^* \in \mathcal{C}, CL^* > K$,

the maximum number of satisfactory vertices will be equal to

$$\left|V\left(G\right)\right| - CL^{*} + K ,$$

and all the dissatisfactory vertices are in C^{*} in the optimal solution (Meng et al., 2015).

Property 4: In a chordal interference graph G, if

$$\exists! C^{*} \in \mathcal{C}_{1}\left(g\right), CL^{*} > K \geq 2$$

and

$$\forall C_{n} \in \mathcal{C}_{2}\left(g\right), CL_{n} - \left|C_{n} \cap g\right| + 2 \leq K$$

for the feasible IA cluster g, the use of g will increase the number of satisfactory FUEs in the partial network including g and the adjacent links (Meng et al., 2015).

We use the Property 4 as the preselection rule for all the feasible IA clusters. Then the result of preselection can be derived by the judgement of each IA cluster at tandem, the set of the preselected candidates is denoted as $\mathcal{S}_{P}$.

Property 5: In a chordal interference graph, if any two elements in the set of the preselected feasible IA clusters denoted as $\mathcal{S}_{P}$ are nonadjacent, the load of each overload conflict region will not be increased and there is no new overload maximal clique and by simultaneously using IA clusters in $\mathcal{S}_{P}$ (Meng et al., 2015).

We use the Property 5 as the selection rule for the preselected candidates in $\mathcal{S}_{P}$. According to the preselection rule, each preselected IA cluster can increase the number of the satisfactory UEs individually. Hence, to select the more IA clusters that satisfy the Property 5 is more beneficial for the performance improvement. Then we construct the conflict graph of the preselected IA clusters denoted as G_{IA}, where we can find the maximum independent set (MIS) in G_{IA}. Then we can get the set of selected feasible IA clusters $\mathcal{S}_{S}$.

Next, we give an example as shown in Figure 4 to illustrate the selection of IA cluster, where there are five feasible IA clusters. The intersection of the feasible IA clusters with the maximal cliques are listed in Table 1.

Assume the number of sub-channels is 3. Based on the judgement about each feasible IA cluster by the preselection rule though Table I, we can find that g_{1} is eliminated since the maximal clique in $\mathcal{C}_{1}\left(g_{1}\right)$ is no overloaded, while g_{2} is eliminated since the maximal clique in $\mathcal{C}_{2}\left(g_{2}\right)$ is overload by IA in g_{2}. Then the IA clusters g_{3}, g_{4} and g_{5} are preselected, where $\mathcal{S}_{P} = \left\{g_{3}, g_{4}, g_{5}\right\}$. Next, we can construct the conflict graph of preselected IA clusters as shown in Figure 5. Then the IA clusters g_{3} and g_{5} can be

Figure 4. The example of interference graph

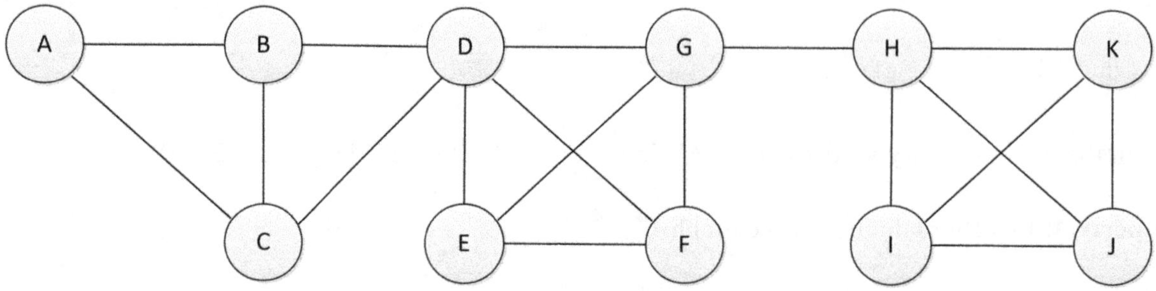

Table 1. The Intersection of a feasible IA cluster with a maximal clique in Figure 4

	$C_1 = \{A,B,C\}$	$C_2 = \{B,C,D\}$	$C_3 = \{D,E,F,G\}$	$C_4 = \{G,H\}$	$C_5 = \{H,I,J,K\}$
$g_1 = \{A,B,C\}$	$\{A,B,C\}$	$\{B,C\}$	$\varnothing$	$\varnothing$	$\varnothing$
$g_2 = \{D,E,F\}$	$\varnothing$	$\{D\}$	$\{D,E,F\}$	$\varnothing$	$\varnothing$
$g_3 = \{E,F,G\}$	$\varnothing$	$\varnothing$	$\{E,F,G\}$	$\{G\}$	$\varnothing$
$g_4 = \{H,I,K\}$	$\varnothing$	$\varnothing$	$\varnothing$	$\{H\}$	$\{H,I,K\}$
$g_5 = \{I,J,K\}$	$\varnothing$	$\varnothing$	$\varnothing$	$\varnothing$	$\{I,J,K\}$

Figure 5. The conflict graph of preselected IA clusters for the example in Figure 4

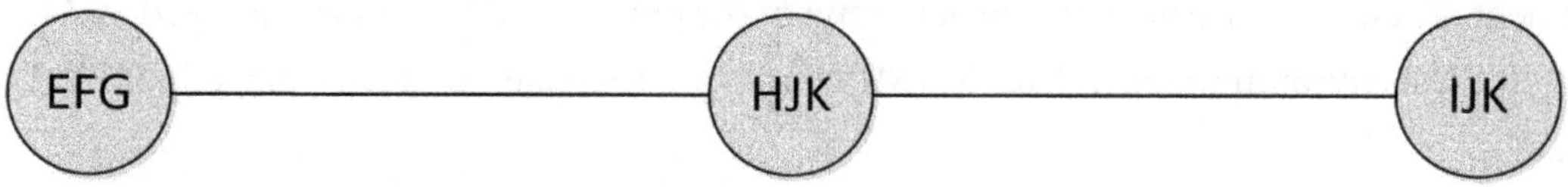

selected in $\mathcal{S}_s$ and the corresponding transformed conflict graph can be represented as shown in Figure 6. We can derived that the number of the satisfactory UEs can increase from 9 to 11 by executing IA in the illustrate example.

After the feasible IA clusters are selected, we need to allocate the sub-channels to vertices. The objective of resource allocation in this part is to maximize the number of satisfactory UEs based on the condition of the number of sub-channels is fixed. In graph theory, the definition of K-colorable is the graph

Figure 6. The transformed conflict graph of the example in Figure 4

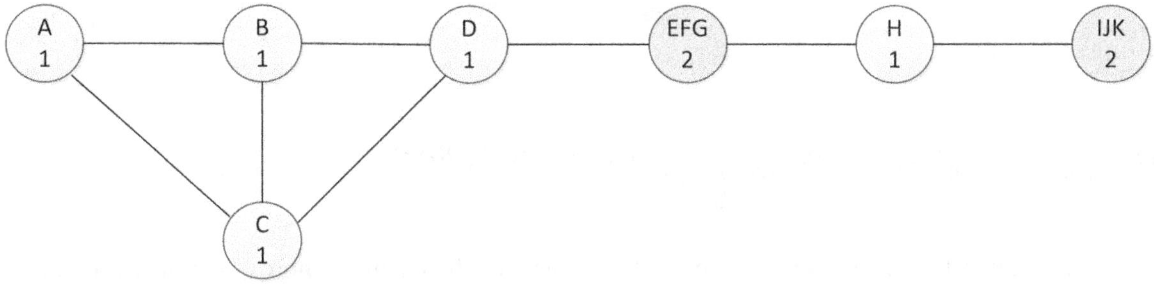

can be colored with K colors while each two adjacent vertices have the different colors. The maximum K-colorable subgraph problem is to find a maximum cardinality subset of vertices so that the induced subgraph of the subset is K-colorable. Therefore, it can be found that the resource allocation problem in this part is similar to the maximum K-colorable subgraph problem. In addition, the approximation algorithm for maximum K-colorable subgraph problem can be only used in chordal graph (Chakaravarthy & Roy, 2009) and the transformed conflict graph for a chordal graph is also chordal according to Property 1, hence we can allocate the sub-channels to UEs in the transformed conflict graph refer to the approximation algorithm. However, each vertex in the maximum K-colorable subgraph problem can only be allocated with one color, while the selected IA vertices need to be allocated with two colors. Therefore, we need to modify the approximation algorithm by considering both the non-IA UEs and the IA clusters.

Since the selected IA clusters in $\mathcal{S}_s$ can increase the satisfactory UEs of the network, the IA vertices in $\mathcal{S}_s$ are given the priority in the sub-channel assignment. Hence we give the larger weight to the IA vertices in $\mathcal{S}_s$ than the non-IA vertices and then we allocate the sub-channels to all the vertices in the transformed conflict graph as a perfect elimination ordering.

Since the triangulation from interference graph to chordal graph will bring about a conservative allocation, the sub-channel assignment need to be restored to further increase the number of satisfactory UEs. For each fill-in edge in the triangulation, we examine that whether both the two adjacent UEs are satisfactory. If a UE adjacent with a fill-in edge is not satisfactory and there exist the sub-channel that has not been allocated to its adjacent vertices, the remaining sub-channel will be allocated to the unsatisfactory UE. The scheme of Transformed conflict graph based Resource Allocation combining IA mentioned above is called TRAIA for short (Meng et al. 2015).

The Selection Criteria of IA Clusters and the Resource Allocation Algorithm for the Max-Min Fairness of Users

We use the number of data streams that can be transmitted simultaneously to measure the UE's performance in this part. To improve the max-min fairness of UEs, we need to increase the minimum number of data streams of UEs in network.

Property 6: In a chordal interference graph G with a feasible IA cluster g, if

$$CL_n^T < p, \forall C_n \in \mathcal{C}_2(g)$$

where

$p = \max_{C_n \in \mathcal{C}_1(g)} CL_n$,

$\max_{C_{n'}^T \in \mathcal{C}\left(G\left(v_g \cup adj_{G^g}(v_g)\right)\right)} CL_{n'}^T < \max_{C_n \in \mathcal{C}\left(G\left(g \cup adj_G(g)\right)\right)} CL_n$ (Li et al., 2015).

We use Property 6 as the selection criteria of IA clusters, which can to improve the max-min fairness of UEs by decreasing the maximum resource requirement of maximal cliques in G.

Next, we also give an example as shown in Figure 7 to illustrate the selection of IA cluster to improve the fairness of UEs, where there are two feasible IA clusters including $g_1 = \{B, C, E\}$ and $g_2 = \{C, D, E\}$. The intersection of a feasible IA cluster with a maximal clique is listed in Table 2.

Assume the number of sub-channels is 3. We judge each feasible IA cluster by the selection rule according Property 6, then we can find that g_2 is eliminated since the resource requirement of maximal

Figure 7. The example of interference graph

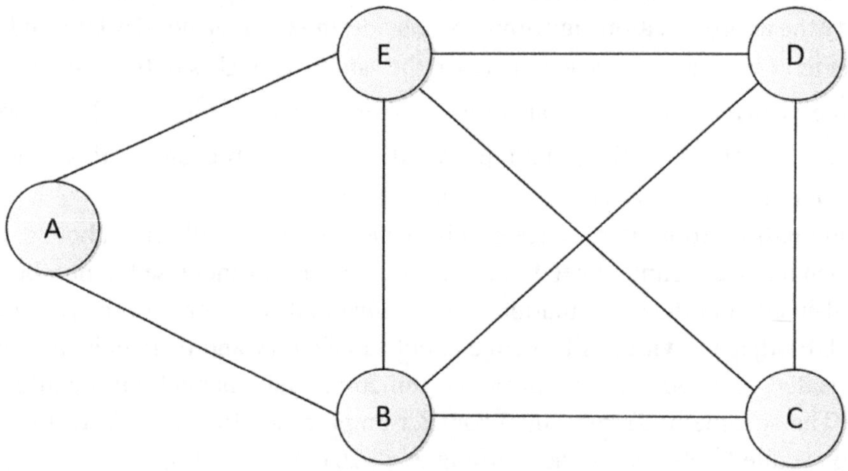

Table 2. The intersection of a feasible IA cluster with a maximal clique in Figure 7

	$C_1 = \{A, B, E\}$	$C_2 = \{B, C, D, E\}$
$g_1 = \{B, C, E\}$	$\{B, E\}$	$\{B, C, E\}$
$g_2 = \{C, D, E\}$	$\{E\}$	$\{C, D, E\}$

clique in $C_2(g_2)$ is not lower than that in $C_1(g_2)$. Then we can derived that the minimum number of data streams in the network can increase from 0 to 2 by executing IA in the illustrate example.

After the feasible IA clusters are selected, we need to allocate the sub-channels to UEs with the object of max-min fairness. The A^3 Algorithm in the works of Arslan et al. (2011,2013) can provide the max-min fairness for the UEs with different weights in a chordal graph. Since the transformed conflict graph for a chordal graph is also chordal according to Property 1, then we can allocate the sub-channels to UEs in the transformed conflict graph refer to the A^3 algorithm, which includes the restoration to update the sub-channels allocation. The Fair resource Allocation algorithm by combining IA mentioned above is called FAIA for short (Li et al., 2015).

SIMULATION RESULTS

The simulation results are shown in the following two parts. In the first part, we simulate the dependence of performance improvement on the network topology of IA in the number of satisfactory users and the users' fairness, which proves the necessity of selection of IA clusters in network. In the second part, we simulate the performance of the TRAIA scheme and FAIA scheme compared with other algorithms on the number of satisfactory users and the users' fairness, respectively.

The Dependence of Performance Improvement on the Network Topology of the IA

We use Z_{opt_IA} and Z_{opt_nonIA} to denote the maximum number of satisfactory users by executing at least one IA cluster and without IA respectively, similarly, we use $R_{\max\min\_IA}$ and $R_{\max\min\_nonIA}$ to denote the maximum value of the minimum number of data streams of UEs by executing at least one IA cluster and without IA respectively. To evaluate the impact of the network topology on the performance by executing IA, we compare Z_{opt_IA} ($R_{\max\min\_IA}$) and Z_{opt_nonIA} ($R_{\max\min\_nonIA}$) with different connection probabilities of vertices in interference graph as shown in Figure 8 (Figure 9). There are 5 vertices and 3 sub-channels in the network, and we assume that each clique of size three is a feasible IA cluster. The notations of p_1, p_2 and p_3 is used to represent the percentage of

$$Z_{opt\_IA} > Z_{opt\_nonIA}(R_{\max\min\_IA} > R_{\max\min\_nonIA}),$$

$$Z_{opt\_IA} = Z_{opt\_nonIA}(R_{\max\min\_IA} = R_{\max\min\_nonIA})$$

and

$$Z_{opt\_IA} < Z_{opt\_nonIA}(R_{\max\min\_IA} < R_{\max\min\_nonIA}),$$

Figure 8. The comparison between Z_{opt_IA} and Z_{opt_nonIA} with different connection probabilities of vertices

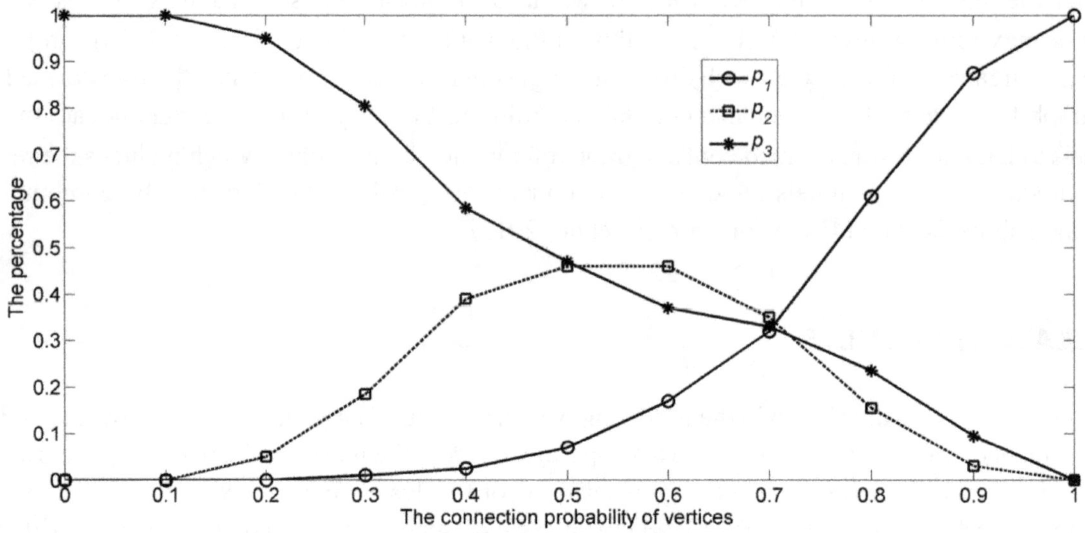

Figure 9. The comparison between $R_{\max\min\_IA}$ and $R_{\max\min\_nonIA}$ with different connection probabilities of vertices

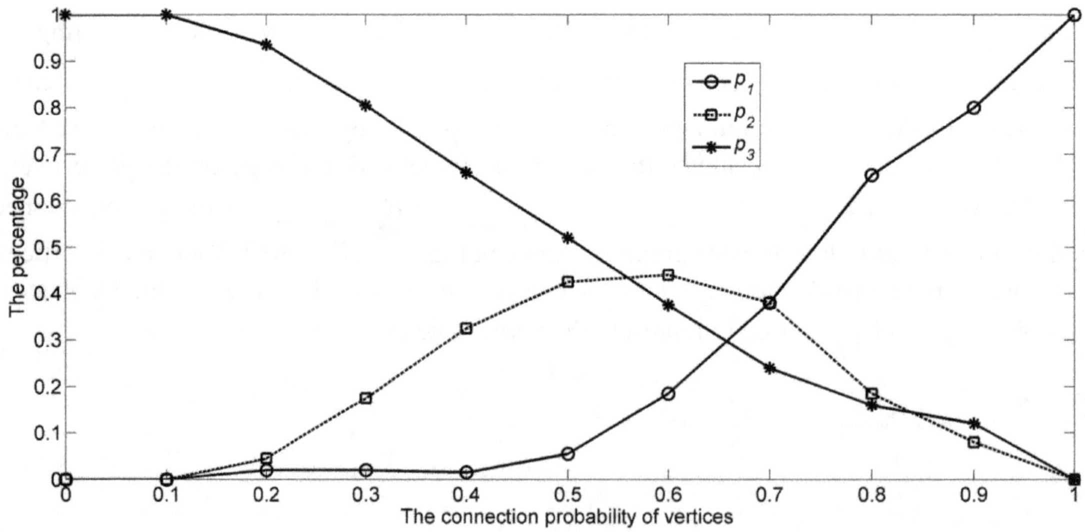

where Z_{opt_IA} ($R_{\max\min\_IA}$) and Z_{opt_nonIA} ($R_{\max\min\_nonIA}$) are found by using brute-force method. Each point is an average value of the results from 500 randomly generated network topologies.

From Figure 8 and Figure 9, we can find that Z_{opt_IA} ($R_{\max\min\_IA}$) is certainly higher than Z_{opt_nonIA} ($R_{\max\min\_nonIA}$) only when the connection probability of vertices equals to 1. In other cases, it is possible that the performance of executing IA is equal to or even lower than that without executing IA.

The Performance of the TRAIA Scheme and FAIA Scheme

In this part, the femtocells are randomly and uniformly distributed in a square area, where the UE is randomly and uniformly distributed in the coverage of its associated femtocell. The parameters used here are listed in Table 3.

Figure 10 shows the numbers of selected IA clusters with the increasing of the number of femtocells in the TRAIA scheme. From Figure 10, we can see that when the number of femtocells is smaller than 65, the numbers of selected IA clusters increase as the increase of the number of femtocells. The number of selected IA clusters of $K = 2$ is larger than that of $K = 3$ and $K = 4$ when the number of femtocells is smaller than 45. This is due to the fact that the resource conflict appears earlier in case of the number of sub-channels is smaller with the increase of the number of femtocells, which needs use IA to mitigate the interference in network. In addition, the number of selected IA clusters of $K = 4$ is

Table 3. Simulation parameters

Simulation Parameter	Value
Power of FAP (mW)	10
radius of femtocell(m)	20
Path loss exponent, α	3
Rayleigh fading, σ	1
Threshold to receive SIR (dB)	5
Number of UEs per femtocell	1
The length of the side of the square area, m	150

Figure 10. The number of selected IA clusters with the different number of femtocells in TRAIA

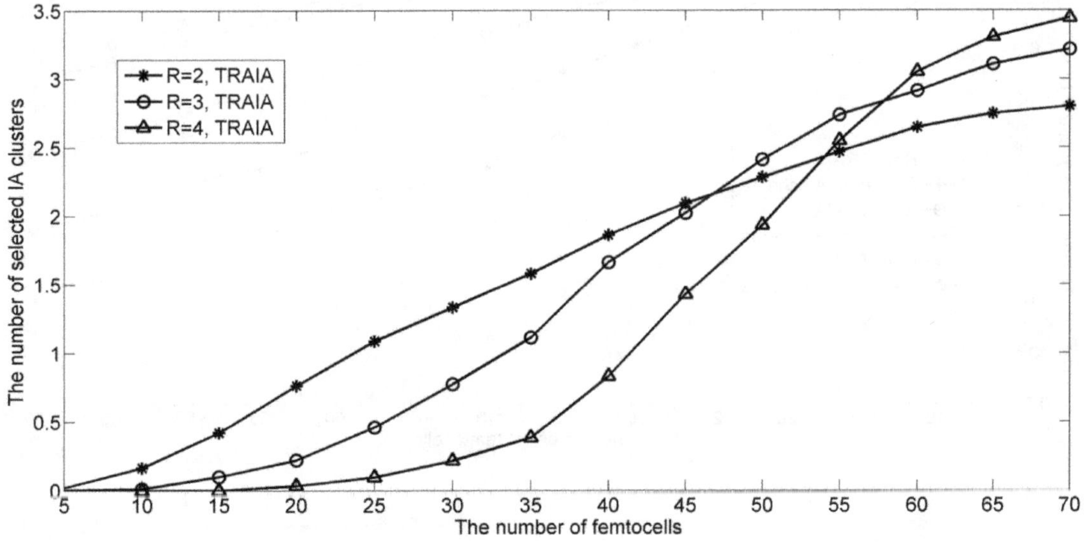

larger than that of $K = 3$ and $K = 2$ when the number of femtocells is larger than 60, the reason is that the constraint for the selection of IA clusters is more loose when the number of sub-channels is larger in dense deployment scenario. Moreover, the numbers of selected IA clusters increase with deceleration when the number of femtocells is larger than 55. This is due to the fact that when the interference becomes extremely serious, the multiple conflict regions have larger size and couple tighter with each other, and the feasible IA clusters are likely in the multiple overload maximal cliques that do not satisfy the selection criteria.

Figure 11 shows the performance of TRAIA scheme and the scheme without IA with different number of femtocells. From Figure 11, we can see that the satisfactory ratios of the UEs of the two schemes decrease as the increase of the number of femtocells. This is due to the fact that the interference is more and more serious as the increase of the number of femtocells. In addition, the performance improvement of the proposed TRAIA compared with the non-IA scheme is increasing as the increase of the number of femtocells when the number of femtocells is smaller than 55. The reason is that the numbers of the selected IA clusters increase with acceleration as shown in Figure 10. Moreover, when the number of femtocells is larger than 55, the performance improvement of the proposed TRAIA compared with the non-IA scheme is nearly invariable. This is due to the fact that the numbers of the selected IA clusters increase with deceleration in this stage as shown in Figure 10.

Figure 12 shows the number of the selected IA clusters with different number of femtocells in the FAIA scheme and the heuristic water-filling algorithm [21]. From Figure 11, we can find that the number of the selected feasible IA clusters increases as the increase of the number of femtocells. The reason is that the interference become serious where needs to perform IA to improve the system performance. In addition, the number of the selected IA clusters of the FAIA scheme increases with deceleration. The reason that is similar as Figure 10 is the feasible IA clusters are likely in multiple maximal cliques that couple with each other and do not satisfy the selection criteria.

Figure 11. The satisfactory ratios of FUEs with different number of femtocells

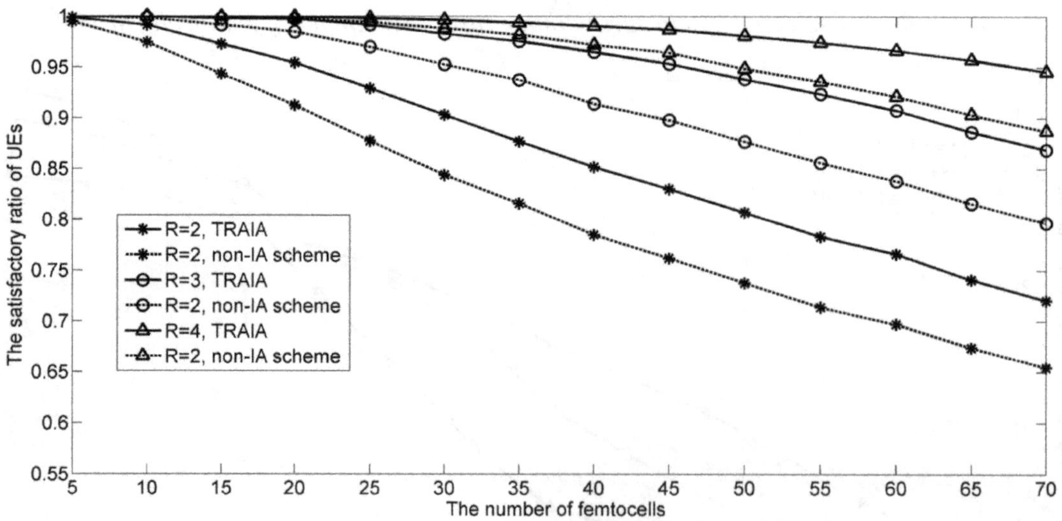

Figure 12. The number of selected feasible IA clusters with the different number of femtocells in FAIA

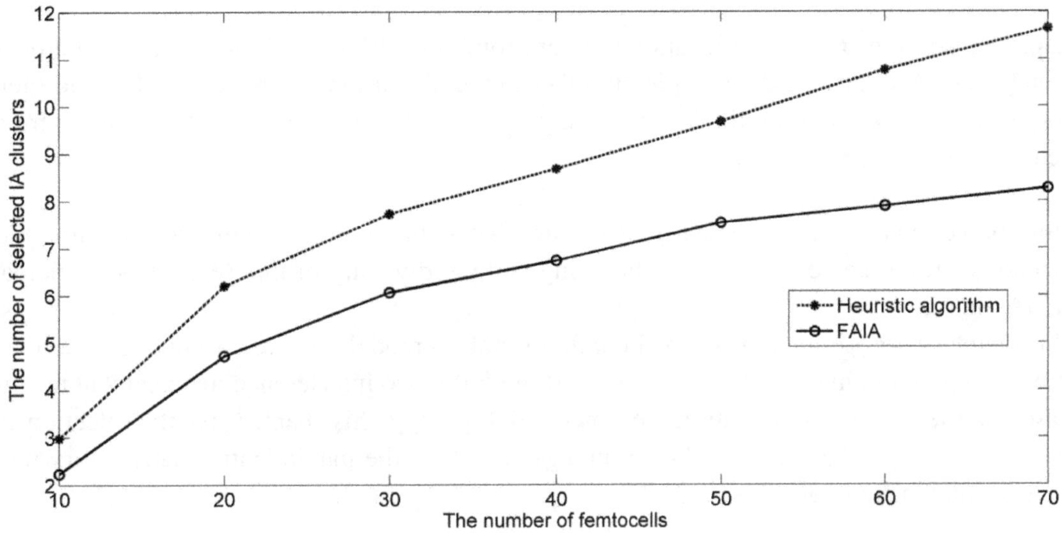

Figure 13 shows the fairness index of users with different schemes as the increase of the number of femtocells. In addition, we can see that the fairness index of users of the FAIA scheme is higher than the F-ALOHA-IA in the work of Nguyen et al. (2013) and the heuristic algorithm. This is due to that the interference of random spectrum allocation in F-ALOHA-IA can be not completely mitigated by IA due to the feasible constraint of IA in dense deployment scenario. Since the heuristic algorithm is lack of effective selection of feasible IA clusters, its performance is worse than the proposed scheme though the number of IA clusters is larger than the FAIA scheme.

Figure 13. The fairness index of users in different schemes

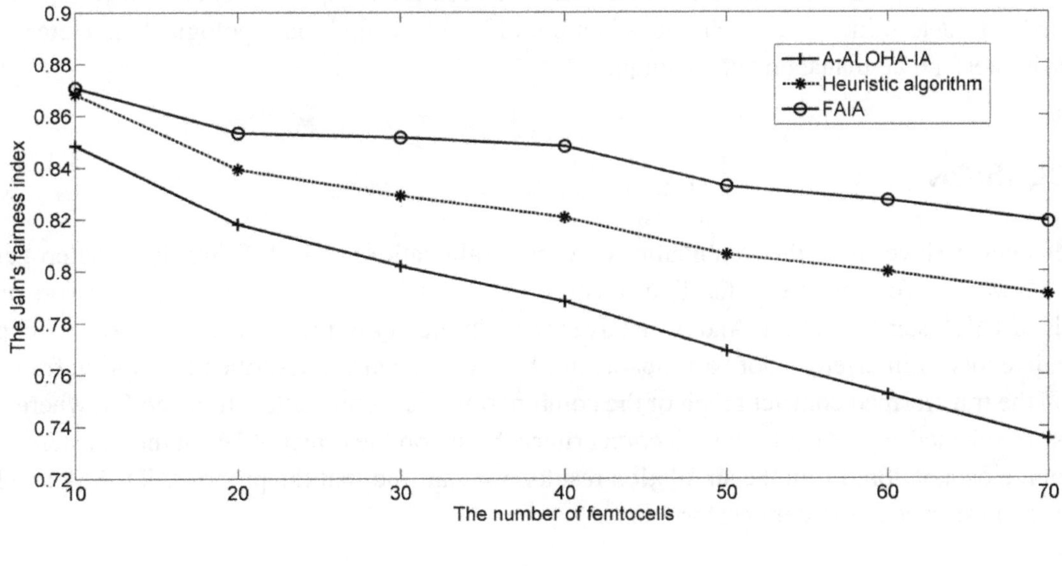

THE MAIN CONTRIBUTIONS OF THIS CHAPTER

This chapter presents the resource allocation problem combing with IA in the ultra-dense heterogeneous cellular networks. We analyze the complexity of resource allocation combing with IA, and then give the solution framework based on the interference graph with low-complexity. The main contributions of this chapter are listed as follows:

- This chapter analyze the complexity of resource allocation combing with IA, including the consistency of resource occupation in scheduling and the diversity of interference of participants in the IA cluster.
- This chapter propose the transformed conflict graph to model resource occupation and the interference expression in the interference network with the two interference management techniques.
- Based on the analysis of the interference network topology, this chapter give the selection criteria of IA clusters and the resource allocation algorithms for the maximization satisfactory ratio and the max-min fairness of users.

FUTURE RESEARCH DIRECTIONS

The interference problem is the key problem in the ultra-dense heterogeneous cellular networks, and as the analysis mentioned above, the combination of multiple interference management techniques is a promising solution. Based on the existing work, we give the possible future research directions as follows:

- The algorithms in this chapter is centralized. Then we can study the distributed schemes for the resource allocation combing with IA to decrease the signaling cost.
- Only the DoFs of users is considered in this chapter. Then the data rate can be studied for the maximization satisfactory ratio and the max-min fairness of users.
- The interference cancellation can mitigate the strong interference at under certain conditions, it is needed to determine which techniques is more efficient, or find the topological characteristic for each interference management technique.

CONCLUSION

In this chapter, we investigate the combination of resource allocation and IA for ultra-dense heterogeneous cellular networks. We show the IA feasible constraint, the classification of interference network and the methods of IA cluster formation. And we analyze the complexity induced by the consisten occupation and the diversity of interference of participants in IA cluster. Then we describe the solution framework based on the transformed conflict graph of the combination of resource allocation and IA, where the IA clusters are selected according to the selection criteria based on the effect of IA on the interference and the resource occupation. From the simulation results, we can find that the proposed TRAIA and FAIA schemes can improve the system performance.

REFERENCES

Andrews, J. G., Claussen, H., Dohler, M., Rangan, S., & Reed, M. (2012). Femtocells: Past, present, and future. *IEEE Journal on Selected Areas in Communications*, *30*(3), 497–508. doi:10.1109/JSAC.2012.120401

Ao, X., Yu, F. R., Yuan, H., & Jiang, S. (2014). Traffic-aware link scheduling with interference alignment for multi-user MIMO networks. In *Proceedings of IEEE International Conference on Communications*. Sydney: IEEE. doi:10.1109/ICC.2014.6883352

Arslan, M. Y., Yoon, J., Sundaresan, K., Krishnamurthy, S. V., & Banerjee, S. (2011). FERMI: A FEmtocell Resource Management System for Interference Mitigation in OFDMA networks. In *Proceedings of International Conference on Mobile Computing and Networking*. Las Vegas, NV: ACM. doi:10.1145/2030613.2030617

Arslan, M. Y., Yoon, J., Sundaresan, K., Krishnamurthy, S. V., & Banerjee, S. (2013). A resource management system for interference mitigation in enterprise OFDMA femtocells. *IEEE/ACM Transactions on Networking*, *21*(5), 1447–1460. doi:10.1109/TNET.2012.2226245

Blair, J. R. S., & Peyton, B. (1993). *An introduction to chordal graphs and clique trees, in graph theory and sparse matrix computation*. New York: Springer.

Cadambe, V. R., & Jafar, S. A. (2008). Interference alignment and degrees of freedom of the K-user interference channel. *IEEE Transactions on Information Theory*, *54*(8), 3425–3441. doi:10.1109/TIT.2008.926344

Chakaravarthy, V. T., & Roy, S. (2009). Approximating maximum weight K-colorable subgraphs in chordal graphs. *Information Processing Letters*, *109*(7), 365–368. doi:10.1016/j.ipl.2008.12.007

Chen, S., & Cheng, R. S. (2014). Clustering for interference alignment in multiuser interference network. *IEEE Transactions on Vehicular Technology*, *63*(6), 2613–2624. doi:10.1109/TVT.2013.2292897

Chen, S., & Cheng, R. S. (2014). Clustering for interference alignment in multiuser interference network. *IEEE Transactions on Vehicular Technology*, *63*(6), 2613–2624. doi:10.1109/TVT.2013.2292897

Da, B., & Zhang, R. (2011). Exploiting interference alignment in multi-cell cooperative OFDMA resource allocation. In *Proceedings of Global Telecommunications Conference*. Houston, TX: IEEE.

El-Absi, M., Shaat, M., Bader, F., & Kaiser, T. (2014). Interference alignment with frequency-clustering for efficient resource allocation in cognitive radio networks. In *Proceedings of Global Communications Conference*. Austin, TX: IEEE. doi:10.1109/GLOCOM.2014.7036936

Foschini, G. J., & Gans, M. J. (1998). On limits of wireless communications in a fading environment when using multiple antennas. *Wireless Personal Communications*, *6*(3), 311–335. doi:10.1023/A:1008889222784

Gomadam, K., Cadambe, V. R., & Jafar, S. A. (2011). A distributed numerical approach to interference alignment and applications to wireless interference networks. *IEEE Transactions on Information Theory*, *57*(6), 3309–3322. doi:10.1109/TIT.2011.2142270

Guler, B., & Yener, A. (2014). Uplink interference management for coexisting MIMO femtocell and macrocell networks: An interference alignment approach. *IEEE Transactions on Wireless Communications*, *13*(4), 2246–2257. doi:10.1109/TWC.2014.030314.131408

Jafar, S. (2011). Interference alignment: A new look at signal dimensions in a communication network. *Foundations and Trends in Communication and Information Theory*, *7*(1), 1–136. doi:10.1561/0100000047

Jaiswal, S., Ganz, A., & Mettu, R. (2009). An optimization framework for demand-based fair stream allocation in MIMO ad hoc Networks. *Mobile Networks and Applications*, *14*(4), 451–469. doi:10.1007/s11036-009-0161-x

Lertwiram, N., Popovski, P., & Sakaguchi, K. (2012). A study of trade-off between opportunistic resource allocation and interference alignment in femtocell scenarios. *IEEE Wireless Communications Letters*, *1*(4), 356–359. doi:10.1109/WCL.2012.051712.120268

Li, J. D., Meng, Y., Li, H. Y., & Suo, L. (2015). The Graph-based Fair Resource Allocation Scheme Combing Interference Alignment in Femtocell Networks. *IET Communications*, *9*(2), 211–218. doi:10.1049/iet-com.2014.0611

Meng, Y., Li, J. D., Li, H. Y., & Pan, M. (2015). A transformed conflict graph based resource allocation scheme combining interference alignment in OFDMA femtocell networks. *IEEE Transactions on Vehicular Technology*, *64*(10), 4728–4737. doi:10.1109/TVT.2014.2366747

Nguyen, T. M., Jeong, Y., Quek, T. Q. S., Tay, W. P., & Shin, H. (2013). Interference alignment in a Poisson field of MIMO femtocells. *IEEE Transactions on Wireless Communications*, *12*(6), 2633–2645. doi:10.1109/TWC.2013.040413.120024

Pantisano, F., Bennis, M., Saad, W., Debbah, M., & Latva-aho, M. (2013). Interference alignment for cooperative femtocell networks: A game-theoretic approach. *IEEE Transactions on Mobile Computing*, *12*(11), 2233–2246. doi:10.1109/TMC.2012.196

Park, S. H., Kim, M. J., & Ko, Y. C. (2014). Resource allocation for spatial-frequency domain based interference alignment scheme in MIMO-OFDM systems. In *Proceedings of International Conference on Communications*. Sydney: IEEE. doi:10.1109/ICC.2014.6884239

Peters, S. W., & Heath, R. W. (2009). Interference alignment via alternating minimization. In *Proceedings of International Conference on Acoustics, Speech and Signal Processing*. Taipei: IEEE.

Peters, S. W., & Heath, R. W. (2012). User partitioning for less overhead in MIMO interference channels. *IEEE Transactions on Wireless Communications*, *11*(2), 592–603. doi:10.1109/TWC.2011.122211.101188

Ronasi, K., Niu, B., Wong, V. W. S., Gopalakrishman, S., & Schober, R. (2014). Throughput-Efficient Scheduling and Interference Alignment for MIMO Wireless Systems. *IEEE Transactions on Wireless Communications*, *13*(4), 1779–1789. doi:10.1109/TWC.2014.031314.122040

Telatar, E. (1999). Capacity of multi-antenna Gaussian channels. *European Transactions on Telecommunications*, *10*(6), 585–595. doi:10.1002/ett.4460100604

Tresch, R., & Guillaud, M. (2009). Clustered interference alignment in large cellular networks. In *Proceedings of International Symposium on Personal, Indoor and Mobile Radio Communications*. Tokyo: IEEE. doi:10.1109/PIMRC.2009.5450107

Tresch, R., & Guillaud, M. (2010). Performance of interference alignment in clustered wireless ad hoc networks. In *Proceedings of International Symposium on Information Theory*. Austin, TX: IEEE. doi:10.1109/ISIT.2010.5513243

Tresch, R., Guillaud, M., & Riegler, E. (2009). A clustered alignment-based interference management approach for large OFDMA cellular networks. In *Joint NEWCOM++/COST2100 Workshop on Radio Resource Allocation for LTE.*

Weber, S., Andrews, J. G., & Jindal, N. (2007). The effect of fading, channel inversion, and threshold scheduling on ad hoc networks. *IEEE Transactions on Information Theory*, *53*(11), 4127–4149. doi:10.1109/TIT.2007.907482

West, D. (2001). *Introduction to Graph Theory* (2nd ed.). Upper Saddle River, NJ: Prentice Hall.

Zahir, T., Arshad, K., Nakata, A., & Moessner, K. (2013). Interference management in femtocells. *IEEE Communications Surveys and Tutorials*, *15*(1), 293–311. doi:10.1109/SURV.2012.020212.00101

KEY TERMS AND DEFINITIONS

Chordal Graph: The graph that each cycle longer than 3 has chord.

Clique: A set of vertices that each two vertices are adjacent in graph.

Interference Alignment (IA) Cluster: A set of users transmit data by IA technique. The interference among the participants in the IA cluster can be eliminated by IA.

K-Colorable: The graph can be colored with K colors while each two adjacent vertices have the different colors.

Maximal Clique: A clique and it cannot be a subset of other cliques.

Max-Min Fairness: A feasible allocation that each user can not increase the allocation unless to decrease the resource allocation of some other user with an equal or smaller resource allocation.

Subgraph: A subgraph of a graph that forms from the subsets of vertices and edges of the graph.

Chapter 7
Game Theory for Co-Tiered Interference Mitigation in 5G Small-Cell Networks

Ducheng Wu
PLA University of Science and Technology, China

Qihui Wu
Nanjing University of Aeronautics and Astronautics, China

Yuhua Xu
PLA University of Science and Technology, China

ABSTRACT

Small-cell technologies are seen as one of the most promising solutions for the rapid growth of wireless data services and 5G requirements. However, because most SBSs are deployed with minimum intervention from the end users and the service providers, it is hard to mitigate co-tiered interference. It is significant to study the self-organized distributed co-tiered interference mitigation and resource allocation. Game theory is an effective distributed approach towards handling the distributed co-tiered interference mitigation problem without a central controller. This chapter is to address the application of game theory and distributed learning solutions for distributed co-tiered interference mitigation. Two potential game models for static and dynamic co-tiered interference mitigation are presented and discussed for small-cell networks with fixed loads and dynamic loads separately. In addition, two distributed learning algorithms are presented and results are discussed. Finally, some future research directions are given.

INTRODUCTION

The drastic increase in the population of smart terminal devices and mobile data traffic with ubiquitous connectivity for the users will create a huge demand for wireless networks. However, there is a huge gap between the growth of wireless data traffic demand and the capacity growth rate of the fourth generation (4G) wireless technology which has been deployed in the world most recently. As a result, the 5G

DOI: 10.4018/978-1-5225-1712-2.ch007

mobile and wireless communication technologies are expected to resolve such a problem (Hu & Qian, 2014; Xu et al., 2014; Wang et al., 2014; Jo et al., 2014). In the future 5G wireless communication systems, the wireless capacity can be improved by node density increase, cooperative and collaborative radio technologies.

Therefore, the use of low-power, very dense, short-range, small-cell networks is seen as a promising option to handle the future data rate demands. With the ultra-dense small-cell deployments, the distance between the small-cell base stations (SBSs) and end users is reduced, leading to higher achievable data rates. On the other hand, the reuse of time-frequency resources across multiple cells can highly increase the spectrum efficiently. In this regard, SBSs, commonly known as femtocells, can improve the network coverage, especially in the interiors of houses and buildings and to provide ubiquitous high speed connectivity to the end users. Thus, a lot of SBSs are installed and managed by users in residential areas and small offices and provide better service performance for users in poor indoor coverage regions or dead zones.

On one hand, some SBSs can be connected to cellular core networks over digital subscriber line and exchange information with each other through the backhaul or gateway. In this way, SBSs can achieve an effectively and conveniently network deployment and radio resource management. SBSs can also exchange information over the X2 interface and allocate the radio resource with limited and time-insensitive information exchanged over the X2 interface with other SBSs. However, the information exchange overhead may be huge during network deployment and resource management. On the other hand, some other SBSs are deployed in an ad-hoc manner with minimum intervention from the end users and the service providers. In this way, SBSs deployment is difficult to perform centralized planning or be well controlled as macro-cell base stations (MBSs), and can't exchange information conveniently.

As more and more SBSs are randomly distributed in a surrounding area, users in the coverage of different cells overlapped may interfere with each other. There is a lot of necessary to mitigate interference to exploit the benefits promised by SBSs. Due to the random deployment, dynamic, flexible connection to cellular core networks and flat system architecture, it is difficult to perform central control to mitigate interference (Xu et al., 2014). It is very challenging and meaningful to develop efficient distributed algorithms for the radio resource allocation. Distributed self-organizing resource allocation and interference mitigation without information exchanging (or with limited information exchanging) is necessary for small-cell networks.

This chapter focuses on the mitigation of co-tiered interference with game theory approaches in 5G small-cell networks. The main contributions of this chapter are summarized as follows:

1. The motivation, issues and challenges for using game theory in co-tiered interference mitigation are presented and discussed. A summary of common game models for resource allocation in 5G small-cell networks is provided. Potential game is introduced which can be used for co-tiered interference mitigation without information exchanging in a non-cooperative manner.
2. A noncooperation interference minimization game for static co-tiered interference mitigation is presented and discussed in fixed load small-cell networks. Without central controllers and information exchange, fading memory joint strategy fictitious play with inertia with inertia is presented to achieve NE of the game.
3. The self-organized channel allocation with dynamic loads is analyzed and modeled as a sequential channel set selection problem. Then, a noncooperation dynamic interference graph game for

dynamic co-tiered interference mitigation is presented and discussed in dynamic load small-cell networks. A distributed dynamic learning algorithm is presented to achieve the NE of the dynamic interference graph game.

The rest of this chapter is organized as follows. In Section 2, it briefly reviews issues and solutions for interference mitigation in 5G heterogeneous cellular networks. In Sections 3, it will present and discuss motivation, issues and challenges for using game theory in co-tiered interference mitigation and resource allocation. Then, two potential game models for static and dynamic co-tiered interference mitigation are presented and discussed for fixed load and dynamic load small-cell networks separately. In Sections 4, two future research problems are listed. Conclusion is provided in Section 5.

BACKGROUND

The mass deployment of small-cells gives rise to several technical challenges. One of the major challenges is interference mitigation between neighboring small-cells and between small-cells and macro-cell (Chandrasekhar & Andrews, 2007, 2009). The small-cell interference can be specified as cross-tiered and co-tiered interference in the tiered networks.

Co-Tiered Interference Mitigation

The co-tiered interference is the interference among SBSs that belong to the same tier in the networks (Saquib, Hossain, Le, & Kim, 2012). As more and more SBSs are randomly distributed in a surrounding area, users in the coverage of different cells overlapped may interfere with each other. In case of a small-cell network, co-tiered interference occurs between neighboring small-cells. For example, a small-cell user (SU) causes uplink co-tiered interference to the neighboring SBSs. On the other hand, a SBS acts as a source of downlink co-tiered interference to the neighboring SUs in overlapping regions. When the number of SBSs is large, severe co-tiered interference is difficult to mitigate.

The problem of resource allocation and co-tiered interference mitigation in small-cell networks has been investigated widely (Kulkarni, Chin, & Farnham, 2010; Lien, Tseng, Chen, & Su, 2010; H. Lee, Oh & Y. Lee, 2010; Sundaresan & Rangarajan, 2009). In many existing works (Liang et al., 2012; Abdelnasser, Hossain, & Kim, 2014; Kang, Zhang, & Motani, 2012), a central management system or controller is necessary for handling resource allocation and interference mitigation, which leads to significant signaling overhead between the controller and SBSs. In Liang et al. (2012), the authors investigated a self-organization strategy for physical resource block (PRB) allocation with a small-cell management system to avoid the co-tiered interference. The authors utilized gateways to exchange information. In Abdelnasser et al. (2014), the authors proposed a semi-centralized interference management scheme, which was based on joint clustering and resource allocation for small-cells.

Some researchers applied different game models to solve this problem in a distributed manner, e.g., coalition games (Pantisano, Bennis, Saad, & Debbah, 2013), evolutionary games (Semasinghe, Hossain, & Zhu, 2015), and graphical games (Xu, Wu, Shen, Wang, & Anpalagan, 2013; Zheng, Cai, & Xu, 2014). In Xu et al. (2013), the authors applied a graph-based model to characterize the MAC-layer interference relationship among the cells. In Zheng et al. (2014), the authors investigated the problem of channel selection for MAC-layer interference mitigation with a proposed dynamic graphic game in the dynamic and distributed environment.

Cross-Tiered Interference Mitigation

The cross-tiered interference is the interference between different network tiers, e.g., MBSs and SBSs. For example, SUs and macro-cell users (MUs) act as a source of uplink cross-tiered interference to the serving MBS and the nearby SBSs, respectively. On the other hand, the serving MBS and SBSs cause downlink cross-tiered interference to the SUs and nearby MUs, respectively.

There are also many studies about the cross-tiered interference mitigation and both co-tiered and cross-tiered interference mitigation. In (Guruacharya, Niyato, Kim & Hossain, 2013; Kang, Zhang & Motani, 2012), the problem of resource allocation in the small-cell networks was formulated as a Stackelberg game, where the MBS acted as a leader and could put forward its own strategies and then the SBSs acted as followers and responded. In Li, Xu, Hu, & Qu (2010), a framework was presented to reduce both cross-tiered and co-tiered interference and enhanced the spectral efficiency for an OFDMA based closed access small-cell networks.

This chapter focuses on the mitigation of co-tiered interference and study the issue of distributed channel allocation in small-cell networks.

CO-TIERED INTERFERENCE MITIGATION GAMES

Motivation of Using Game Theory

Game theory is an applied mathematical tool which is used to model and analyze the mutual interactions of rational entities in a situation of conflict. It was primarily developed and used in Economics. It has also been used in other disciplines, e.g., biological systems, philosophy, human society and engineering. Nowadays, game theory has been successfully used in distributed wireless communication systems (Akkarajitsakul, Hossain, Niyato, & Dong, 2011; Felegyhazi & Hubaux, 2006; Mackenzie & Dasilva, 2006). In particular, it has been used to model and analyze resource allocation problems in complicated heterogeneous networks. Due to the distributed and autonomous decision process needed in 5G heterogeneous small-cell networks, game theory has been seen as one of the most important decision-theoretic solutions for 5G communication systems.

In 5G small-cell networks, SBSs share the limited radio spectrum resources. Therefore, one critical issue is how the SBSs allocate these resources to transmit data so that the interference can be mitigated and the optimal network performance can be achieved. Game theory is a useful decentralized approach toward handling this kind of issue. Game theory can be divided into two main groups, namely, cooperative game and non-cooperative game. In 5G small-cell networks, SBSs can either compete or cooperate to achieve their objectives, e.g., throughput, quality-of-service (QoS) and quality-of- experience (QoE). A summary of common game models for resource allocation in 5G small-cell networks is provided in Table 1.

The motivations of using game theory for interference mitigation and resource allocation in 5G small-cell networks are summarized as follows:

Table 1. Summary of game models for resource allocation in 5G small-cell networks

Game Model	Player	Player Behavior Type	Information Exchanges?	Solution Concept	Reference
Evolutionary game	SBSs	Non-cooperative	Yes	Evolutionary stable strategy	Semasinghe et al. (2015)
Potential game	SBSs	Cooperative/Non-cooperative	Yes/No	Nash equilibrium	Xu et al. (2013); Zheng et al. (2014)
Stackelberg game	MBS (leader) and SBSs (followers)	Non-cooperative	Yes	Stackelberg equilibrium	Guruacharya et al. (2013); Kang et al. (2012)
Auction game	MBS (auctioneer) and SBSs (bidders)	Non-cooperative	Yes	Nash equilibrium	Chen, Lin, & Zhang, (2014)
Coalitional game	SBSs	Cooperative	Yes	Nash stable	Pantisano et al. (2013)
Matching game	BSs and SUs	Non-cooperative	Yes	Stable matching	Gu, Saad, Bennis, & Debbah, (2015)

- **Theoretical Foundation for Interference Mitigation in 5G Small-Cell Networks:** Because in 5G small-cell networks centralized resource allocation is difficult to perform, a self-organized interference mitigation strategy is needed for SBSs with which each SBS can make decisions on its own depending on the decisions of others. Game theory is a useful approach toward handling such issues with multi-player. The equilibrium solution of a game model is defined as the action profiles of the different SBSs for which the chosen performance objective is optimized.
- **Distributed Protocols:** In most scenarios of 5G small-cell networks, SBSs make their decisions in a distributed manner rather than in a centralized manner. Game theory can be applied to 5G small-cell networks in the context of SBSs' random deployment and distributed resource allocation.
- **Modeling Cooperative/Non-Cooperative Behavior of SBSs:** In 5G small-cell networks, each SBS may behave selfishly or altruistically to reap performance advantage over or together with other SBSs. Cooperation or competition among SBSs for channel allocation in 5G small-cell networks is a multiplayer decision problem, which can be modeled as games in different network scenarios. Game theory can be used to model and analyze the selfish/altruistic behavior of SBSs and design robust mechanisms for channel allocation in 5G small-cell networks.
- **Mechanism Design:** The parameters and utility functions of a game can be designed such that it leads the self-interested and independent SBSs toward a network-wide desirable outcome. *Potential function* is one technique that can be used for such mechanism design to regulate the usage of radio resources by associating individual utilities with the collective utility.

Issues and Challenges

In 5G small-cell networks, there are some new challenges with regard to information constraints should be addressed when game theory is applied in wireless environments. As most SBSs are deployed by the end users and the service providers, the self-organizing and self-optimizing capabilities are needed without the amount of manual intervention involved in network planning. Meanwhile, the information exchange overhead should be reduced or limited during resource management. It is well known from Table 1 that in potential games players can act in a non-cooperative manner without information exchanging (or with limited information exchanging). Thus, potential games are suitable for resource allocation in such self-organizing 5G small-cell networks. The purpose of this chapter is to address the application of potential game theory and distributed learning solutions for distributed co-tiered interference mitigation and resource allocation.

Some noncooperation potential game based distributed interference management technologies have been proposed in the small-cell networks to make SBSs self-organize distributedly and reduce the signaling overhead as mentioned in the above Xu et al. (2013), Zheng et al. (2014). However, there are still some drawbacks.

On one hand, in Xu et al. (2013), Zheng et al. (2014), they assumed that one SBS can only select one channel to transmit. In fact, the demands of different SBSs may be different, so the numbers of required channels of SBSs may be different and bigger than one. Moreover, the proposed MAC-layer interference is not suitable for the small-cell networks without the multiple access control mechanisms, such as CSMA and Aloha. Thus, co-tiered interference mitigation with multichannel allocation in 5G small-cell networks is challenging problem to be solved.

On the other hand, all the aforementioned works have paid less attention on the dynamic loads of small-cells. It should be pointed out that Wei & Collings (2013) have conducted some preliminary researches on load-aware spectrum access for small-cell networks. However, even in Wei & Collings, (2013), it is assumed that the loads of small-cells are static and unchanged. In practice, in small-cell networks, more than one user can associate with each SBS and may connect or disconnect with their corresponding SBSs according to their specific service demand at non-deterministic moments in time. Thus, the loads (number of active SUs) of small-cells may change dynamically and vary widely within certain (short) time periods. Co-tiered interference mitigation with dynamic loads in 5G small-cell networks is still a very critical and challenging problem.

Potential Game

Suppose that the objective of the global small-cell network is captured by a potential function $\Phi : \mathcal{A} \rightarrow \mathbb{R}$. A suitable mechanism design can make each player's objective function to be appropriately "aligned" with the objective of the global small-cell network. Potential games require perfect alignment between the global objective and the players' individual objective functions, which are defined as follow.

Definition 1 (Potential Game (Monderer & Shapley, 1996)): A strategic form game $\{\mathcal{N},(\mathcal{A}_n),(u_n)\}$ is an exact potential game if there exists a potential function $\Phi : \mathcal{A} \rightarrow \mathbb{R}$ such that for all $n \in \mathcal{N}$, all $a_n \in \mathcal{A}_n$ and $a_n' \in \mathcal{A}_n$,

$$u_n\left(a_n, a_{-n}\right) - u_n\left(a_n', a_{-n}\right) = \Phi\left(a_n, a_{-n}\right) - \Phi\left(a_n', a_{-n}\right), \tag{1}$$

where

$$a_{-n} = \left(a_1, \ldots, a_{n-1}, a_{n+1}, \ldots, a_N\right)$$

is the collection of actions of players other than player n.

From the above definition, it can be found that when a player unilaterally changed its action, the change in the potential function would be equal to the change in the individual objective function. The problem of distributed resource allocation and interference mitigation in small-cell networks has been formulated as potential games in Xu et al. (2013), Zheng et al. (2014).

Most of the learning algorithms for potential games, like SLA algorithm (Xu et al., 2013), FM-JSFP algorithm (Marden, Arslan, & Shamma, 2009), guarantee convergence to a (pure) Nash equilibrium.

Definition 2 (Nash Equilibrium): An action profile $a^* \in \mathcal{A}$ is a pure strategy Nash equilibrium if and only if no player can improve its utility function by deviating unilaterally, i.e.,

$$u_n\left(a_n^*, a_{-n}^*\right) \geq u_n\left(a_n, a_{-n}^*\right), \forall n \in \mathcal{N}, \forall a_n \in \mathcal{A}_n . \quad (2)$$

It is easy to find that, in potential games, the action profile maximizing or locally maximizing the potential function is a pure Nash equilibrium. Thus, every potential game has at least one such equilibrium.

Noncooperation Interference Minimization Game for Static Co-Tiered Interference Mitigation

In this section, a noncooperation interference minimization game is proposed to allocate channels among the SBSs, and make an effort to mitigate the static co-tiered interference in small-cell networks. In the proposed game, each player (SBS) can transmit in multi-channel and tries to minimize the aggregate received interference signal and noise power in its selected channels. Compared with the existing MAC-layer interference minimization game (Xu, Wu, Shen, Wang, & Anpalagan, 2013), the interference in the proposed game is defined as the received interference signal and noise power in the selected channels rather than the MAC-layer interference, which is not suitable for the small-cell networks without the multiple access control mechanisms used in Xu et al. (2013), such as CSMA and Aloha. Note that the main analysis and results in this section were presented in Wu, Zhang, & Zhou (2015).

System Model and Problem Formulation

Let's consider a small-cell network, there are N small-cells, i.e., $\mathcal{N} = \{1, 2, \ldots, N\}$. It is assumed that there are M dedicated channels for small-cells to transmit, i.e., $\mathcal{M} = \{1, \ldots, M\}$. Due to the difference numbers of SUs associated with small-cells, there is an assumption that small-cell n requires k_n channels to transmit with its associated k_n SUs. The selected channel set can be denoted as:

$$a_n = \{c_{n1}, c_{n2}, \ldots, c_{nk_n}\}, \quad (3)$$

where $c_{nk} \in \mathcal{M}$, $0 < c_{n1} < c_{n2} < \ldots < c_{nk_n} \leq M$, $0 < k_n < M$.

Since SUs are located very closely to their associated SBSs and could move around their associated SBSs, it can be assumed that the average channel gain between a SBS j and a SU e_i, which is served by another SBS i, can be approximated by the channel gain between these two SBSs (i and j), i.e.,

$\overline{g}^{m}_{e_i,j} = g^{m}_{i,j}$ (Abdelnasser, Hossain, & Kim, 2014). Following this assumption, it can be deduced that channel gains are symmetric for two SBSs i, j and two SUs e_i, e_j, i.e.,

$$\overline{g}^{m}_{e_i,j} = g^{m}_{i,j} = g^{m}_{j,i} = \overline{g}^{m}_{j,e_j}.$$

To facilitate analysis, it does not consider power control and assumes that the transmission powers of SBSs in all channels are the same. In each channel m, the received interference signal and noise power by SBS n can be described approximately as

$$I_{nm} = \sum_{i\in\{\mathcal{N}\backslash n\}} \delta_{im} p g^{m}_{in} + \sigma^2, \tag{4}$$

where p represents the transmission power of each SBS i in each channel m, σ^2 is background noise, δ_{im} is the following interference indicator function:

$$\delta_{im} = \begin{cases} 1, m \in a_i \\ 0, m \notin a_i \end{cases}. \tag{5}$$

The bandwidth of each channel is B_0. The throughput obtained by small-cell n in channel m can be described approximately as:

$$r_{nm} = B_0 \log_2\left(1 + \frac{p g^{m}_{n}}{I_{nm}}\right) = B_0 \log_2\left(1 + \frac{p g^{m}_{n}}{\sum_{i\in\{\mathcal{N}\backslash n\}} \delta_{im} p g^{m}_{in} + \sigma^2}\right), \tag{6}$$

where g^{m}_{n} represents the average channel gain between SBS n and its' served SU e_n.

In the following, it considers the distributed channel allocation problem from the perspective of co-tiered interference mitigation. The channel allocation profile of the small-cell network is $\left(a_n, a_{-n}\right)$, where a_{-n} is the sequential channel selection profile of other SBSs. It is clear that each SBS prefers to choose the channels with lower interference to increase its transmission capacity. Particularly, when the action profile of all other players (a_{-n}) is fixed, with the selected channel set with the minimum aggregate interference SBS n would achieve the maximum aggregate throughput.

Thus, in a noncooperation manner, the objective of small-cell n is set as minimizing the aggregate interference, which can be expressed as

$$v_n\left(a_n, a_{-n}\right) = \sum_{m\in a_n} I_{nm}\left(a_n, a_{-n}\right). \tag{7}$$

Accordingly, a quantitative characterization of the network interference level is expressed as follows

$$\mathcal{V}(a) = \sum_{n \in \mathcal{N}} v_n \left(a_n, a_{-n}\right) \tag{8}$$

The network aggregate interference level reflects the whole interference level in the network. The global objective is to find the optimal channel selection profile out to minimize the network aggregate interference $\mathcal{V}(a)$.

Noncooperation Interference Minimization Game and FM-JSFP Algorithm

This chapter calls

$$G_I = \left\{\mathcal{N}, (\mathcal{A}_n)_{n \in \mathcal{N}}, (u_n)_{n \in \mathcal{N}}\right\}$$

noncooperation interference minimization game, where $\mathcal{N} = \{1, 2, \ldots, N\}$ is the set of SBSs, $\mathcal{A}_n$ is the strategy space of SBS n. Additionally, it notes that the size of the strategy space of each player n is

$$\left|\mathcal{A}_n\right| = C_{K_n}^{k_n} = \frac{K_n!}{k_n!(K_n - k_n)!},$$

where $X!$ denote the factorials of X. The size of the strategy space $\mathcal{A}_n$ is greater than the number (K_n) of available channels. u_n is the utility function of each player n, which can be defined as the opposite of the aggregate interference of SBS n. The utility function is expressed as

$$u_n(a_n, a_{-n}) = -v_n(a_n, a_{-n}) = -\sum_{m \in a_n} I_{nm}(a_n, a_{-n}). \tag{9}$$

Then, the noncooperation interference minimization game is expressed as follows:

$$(G_I): \max_{a_n \in \mathcal{A}_n} u_n \left(a_n, a_{-n}\right) \forall n \in \mathcal{N}. \tag{10}$$

For noncooperation interference minimization game G_I, a channel selection profile

$$a_{NE} = (a_1^*, a_2^*, \ldots, a_N^*)$$

is a pure strategy NE if and only if no player can improve its utility by deviating unilaterally, i.e.,

$$u_n \left(a_n^*, a_{-n}^*\right) \geq u_n \left(a_n, a_{-n}^*\right), \forall n \in \mathcal{N}, \forall a_n \in \mathcal{A}_n, a_n \neq a_n^*. \tag{11}$$

Theorem 1: G_I is an exact potential game which has at least one pure strategy NE. The optimal solution of the network interference minimization problem is a pure strategy NE of G_I.

Proof: The potential function is constructed as follows

$$\Phi\left(a_n, a_{-n}\right) = -\frac{1}{2}\mathcal{V}\left(a_n, a_{-n}\right) = \frac{1}{2}\sum_{n=1}^{N} u_n\left(a_n, a_{-n}\right). \tag{12}$$

The rest of the specific proof process is in Wu et al. (2015).

Thus, it is known that G_I is an exact potential game according to the definition of exact potential game. The most two important properties of potential game are as follows: i) Potential game has one pure strategy NE at least. ii) Local or globe maxima of potential function constitutes a pure strategy NE.

Although G_I is an exact potential game and has at least one pure strategy NE, how the autonomous small-cells can reach equilibrium in a distributed fashion is a challenging question. Fading memory joint strategy fictitious play with inertia (FM-JSFP) (Marden et al., 2009) is a very effective distributed learning algorithm for the distributed multichannel selection problem. FM-JSFP with inertia ($\varepsilon \in (0,1)$) guarantees convergence to pure Nash equilibrium almost surely in any potential game and maintains computational tractability even in complex large-scale small-cell networks.

The discount factor λ reflects the importance degree of the historical information. It is easy to identify two limiting cases of the discount factor. When $\lambda = 1$, it can be gotten the belief that only the most recent information matters. When $\lambda = 1/t$ (λ is not a constant), it can be concluded that all information is given equal importance.

It is clear that in FM-JSFP algorithm, each SBS only requires to sense the power of interference signal and noise power in each selected channel.

The FM-JSFP algorithm is described in Algorithm 1.

Then, simulations are presented to conduct for the evaluation of the performance of the proposed learning algorithm. The number of dedicated channels is $M = 4$, and bandwidth of each available channel is 200 kHz. The transmission power of SBSs is 40mW in each channel, the noise power density is -174dBm/Hz. Small-cells are randomly located. The maximum number of the associated SUs in each small-cell is 3. To model the propagation environment, it only considers path loss for analysis simplicity like many of previous studies. The path-loss between two SBSs is assumed equal to

$$PL = 38.46 + 20log(d_{ij}) + qL_w,$$

where d_{ij} is the distance between SBSs i and j, qL_w accounts for loss due to walls ($L_w = 5$ dB, for simplicity, it sets $q = 1$).

This subsection considers a small-cell network with 8 small-cells randomly located in a region of $100\text{m} \times 50\text{m}$. The cumulative distribution functions (CDF) of the iterations needed for converging of Algorithms 1 with $\varepsilon = 0.1, 0.2, 0.4$ are shown in Figure 1. It is noted from Figure 1 that when the value of ε increases, the iterations needed for converging of the FM-JSFP algorithm significantly decrease.

Then it varies the number of SBSs of the small-cell network in the square region of $100\text{m} \times 100\text{m}$. The aggregate interference varying with the number of small-cells is shown in Figure 2.

Algorithm 1. Fading memory joint strategy fictitious play with inertia

Step 1: Initially, each small-cell i is assigned a random action $a_i(1) \in \mathcal{A}_i$ and communicates with its users.

Step 2: Each small-cell i senses all available channels, and calculates the utility $V_i^{a_i}(1) \in R$ for each $a_i \in \mathcal{A}_i$, where $V_i^{a_i}(1) = u_i(a_i, a_{-i}(1))$. $V_i^{a_i}(1)$ is small-cell i's initial belief about the utility it would receive for playing action a_i in the ensuing time step.

Step 3: At time $t \geq 1$, each small-cell i plays the following strategy:

(1) $a_i(t) \in \arg\max_{a_i \in \mathcal{A}_i} V_i^{a_i}(t)$ with probability $(1-\varepsilon)$,

(2) $a_i(t) = a_i(t-1)$ with probability ε, where $\varepsilon \in (0,1)$ is referred to as the small-cell's inertia.

Step 4: Each small-cell i updates his beliefs as:

$$V_i^{a_i}(t+1) = (1-\lambda)V_i^{a_i}(t) + \lambda u_i(a_i, a_{-i}(t)), \forall a_i \in \mathcal{A}_i, \quad (13)$$

where $\lambda \in (0,1]$ is the discount factor. Return to Step 3 and repeat.

Figure 1. The convergence speed comparison of the FM-JSFP algorithm

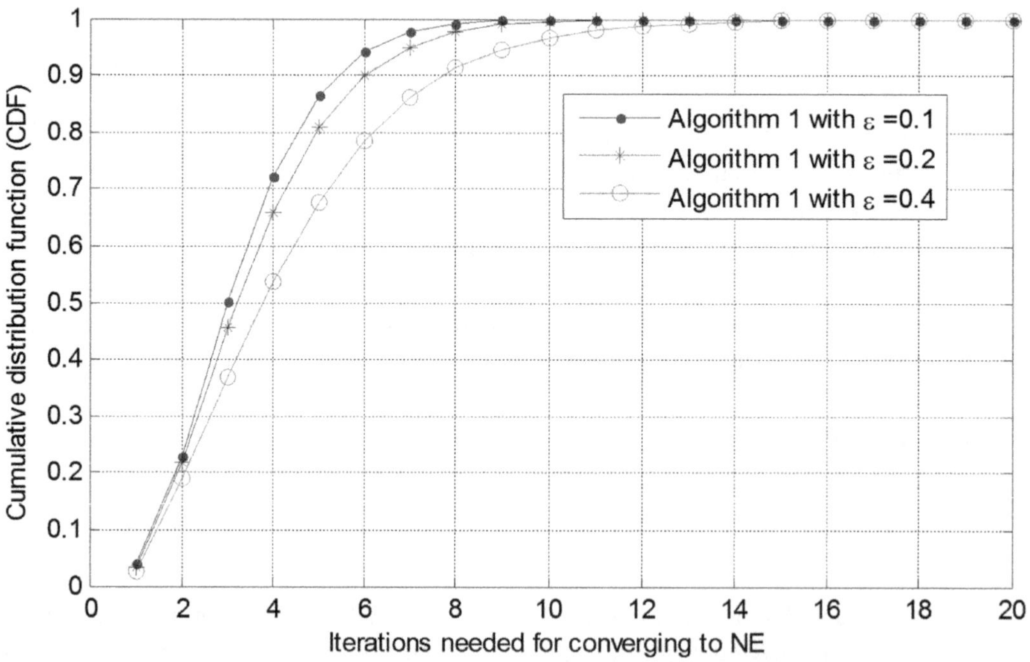

It is observed from the figure that when the network scale increases, the aggregate interference increases. More importantly, it is noted that NE solution of the noncooperation interference minimization game significantly outperforms the random selection.

Figure 2. The average aggregate interference levels

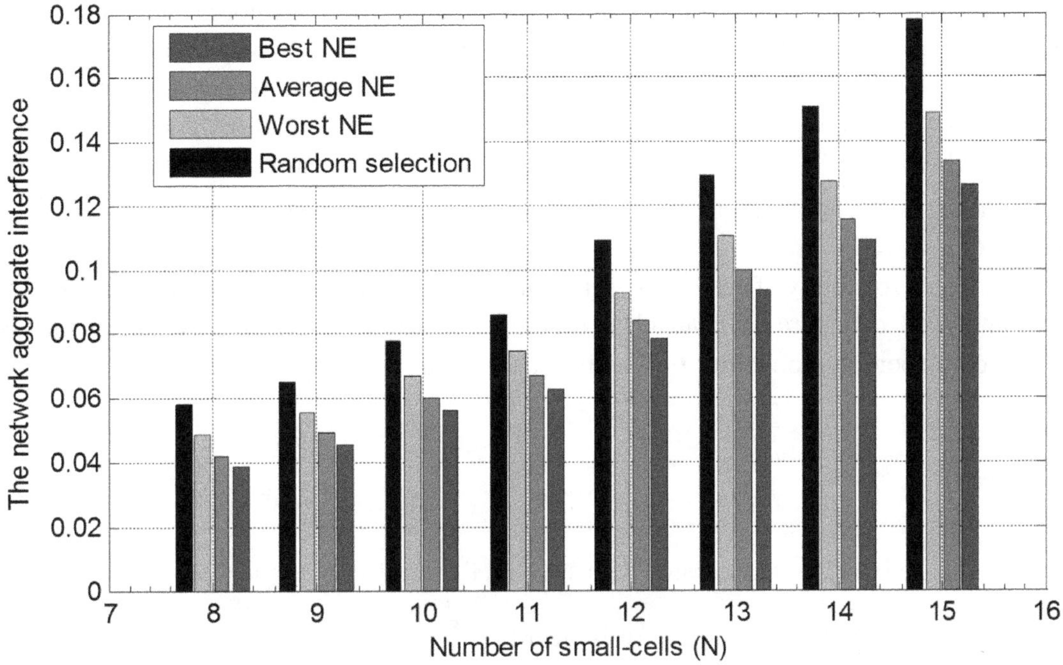

Noncooperation Dynamic Interference Graph Game for Dynamic Co-Tiered Interference Mitigation

In this subsection, it resorts to a game theory-based self-organizing approach to allocate channels for different and dynamic loads of small-cells. In the existing fixed-channel allocation method, each SU is allocated a fixed channel to transmit in each small-cell. Due to the dynamics of the loads, such fixed-channel allocation method is not applicable. What's more, due to the dynamics of SU loads, channels with low interference will be used inefficiently when SUs who have occupied these channels are inactive. Thus, a novel distributed channel allocation method is proposed, in which for each small-cell channels have different priorities to be used. The proposed distributed channel allocation method can be applied in overload situations when takes no account of serving queues and buffers. In the proposed channel allocation method, channels are not fixedly allocated to specific SUs, and the channels with lower interference are given a higher priority for transmission by active SUs. With such approach, the self-organized channel allocation with dynamic loads can be modeled as a sequential channel set selection problem. Note that the main analysis and results in this section were presented in D. Wu, Q. Wu, & Xu (2016).

System Model and Problem Formulation

Let us consider a small-cell network with N small-cells, i.e., $\mathcal{N} = \{1, 2, \ldots, N\}$, and M dedicated channels, i.e., $\mathcal{M} = \{1, \ldots, M\}$. Each cell n offers traffic services to K_n small-cell users (SUs). It is assumed that the transmission link between each SBS n and a SU only occupies one channel at each (short) time period. In this subsection, it only considers the downlink transmission.

There is an assumption that each SU can be in one of two states, namely "Sleep" (S) and "Active" (A). In the "Sleep" state, a SU has no wireless traffic demand and does not occupy any channels to transmit with the associated SBS, while in the "Active" state, the SU receives data from its associated SBS. The "Active-Sleep" model can be modeled as a simple two-state Markov chain. Taking an arbitrary chosen SU as an example, its two-state Markov chain is shown in Figure 3. It is assumed that the state transition probabilities (p^{AA}, p^{AS}, p^{SA}, p^{SS}) of each SU are different and fixed. Thus, it can be concluded that the active (α) probabilities and sleep (β) probabilities of SUs in each small-cell are fixed and different ($\alpha + \beta = 1$).

Due to the fixed probability of a SU being active, it can be concluded that the probability of the event, in which a small-cell n requires h channels to serve its h active SUs and meet the instantaneous load, is fixed. The corresponing probability θ_{nh} can be expressed as

$$\theta_{nh} = \sum_{\left|\mathcal{SU}_n^A\right|=h} \prod_{su_{ni} \in \mathcal{SU}_n^A} \alpha_{ni} \prod_{su_{nj} \in \mathcal{SU}_n^S} \beta_{nj} \tag{14}$$

where $\mathcal{SU}_n^A$ is the active SU set associated with small-cell n and $\mathcal{SU}_n^S$ is the sleeping SU set, and su_{ni} is the ith SU associated with small-cell n. It is clear that $\sum_{h=0}^{K_n} \theta_{nh} = 1$.

Then a distributed channel allocation method is proposed for small-cells with dynamic loads, which has taken into account that the load in each small-cell n changes dynamically according to $\{\theta_{nh}\}$. In the proposed method, each small-cell n selects a sequential channel set for active SU to transmit, which can be expressed as

$$C_n = \{c_{n1}, c_{n2}, \ldots, c_{nZ_n}\}, \tag{15}$$

where $Z_n = \min\{K_n, M\}$, and $c_{nk} \in \mathcal{M}$, $0 < k \leq Z_n$. These Z_n selected dedicated channels will been prioritized by small-cell n. Compared with existing fixed-channel allocation method, an active SU can

Figure 3. Two-state Markov chain of an arbitrary chosen SU

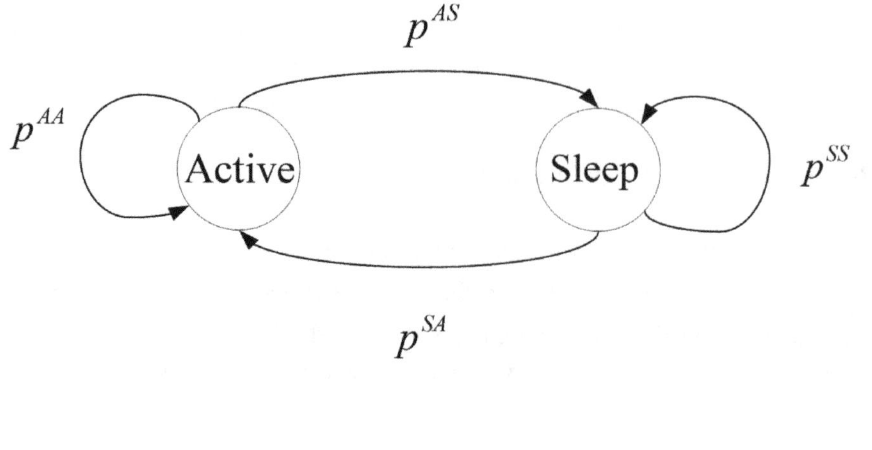

occupy an idle channel in the ordered channel set C_n instead of occupying a fixed channel in each time period.

More specifically, for channels k and k' in the selected channels set C_n of small-cell n, if $k < k'$, it says that c_{nk} has a higher priority to be used for transmission than $c_{nk'}$ for the cell n. In other words, when small-cell n has i active SUs to serve at a particular time period, the i channels with the highest priority, i.e., the first i channels in C_n, will be picked out for transmission. Thus, channels with lower interference should be given a higher transmission priority to achieve good performance. The distributed channel allocation needs temporary access control when the loads change which can be provided by SBSs.

Then, it can be concluded that the transmission probability of channel c_{nk}, i.e., the probability of channel c_{nk} being used to transmit by small-cell n, is given by $\vartheta_{nk} = \sum_{h=k}^{K_n} \theta_{nh}$. It is obvious that $1 \geq \vartheta_{n1} \geq \ldots \geq \vartheta_{nK_n}$. The ordering of the transmission probability of each channel is consistent with the corresponding priorities. In particular, when $c_{nk} = m \ (m \in \mathcal{M})$, it sets $\vartheta_{nk} = \vartheta_n^{c_{nk}} = \vartheta_n^m$. For a small-cell n, the transmission probability set of each channel c_{nk} ($c_{nk} \in C_n$) is represented as

$$\Theta_n = \{\vartheta_{n1}, \vartheta_{n2}, \ldots, \vartheta_{nZ_n}\}. \tag{16}$$

It can examine how the interference among the small-cells varies with the dynamic loads and the sequential channel set selected by each small-cell. A dynamic interference graph can be used to depict the statistical dynamic interference relationship. Such a graph of a small-cell network with three small-cells is shown in Figure 4. It is noted from Figure 4 that, small-cells 1 and 3 will interfere with each other only when they both transmit information in channel 3 at the same time. Thus, the probability that small-cell 1 and 3 interfering with each other in channel 3 is $\vartheta_1^3 \times \vartheta_3^3$.

To take a simple specific example, it is assumed that the distances between any two SBSs and the transmission powers of all SBSs in Figure 4 are the same. When any two small-cells transmit in the same channel at a same time period, the received interference in both small-cells is the same. Such a received interference can be denoted as I. Then, it is assumed that the active probabilities of SUs in these 3 small-cells are

$$\left\{\alpha_{11} = 0.6, \alpha_{12} = 0.4\right\},$$

$$\left\{\alpha_{21} = 0.6, \alpha_{22} = 0.5, \alpha_{23} = 0.4\right\}$$

and

$$\left\{\alpha_{31} = 0.8, \alpha_{32} = 0.5\right\}$$

Figure 4. A dynamic interference graph of the considered small-cell network with three small-cells

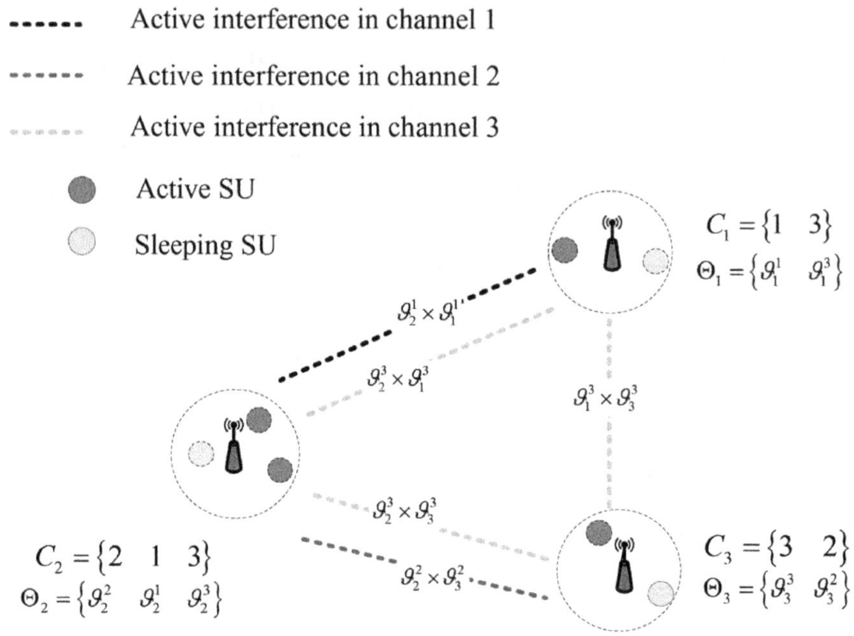

respectively. Then, based on (3), it can be calculated that

$$\Theta_1 = \{0.76, 0.24\},$$

$$\Theta_2 = \{0.88, 0.38, 0.12\}$$

and

$$\Theta_3 = \{0.9, 0.4\}.$$

With the channel sets for SBSs 1 and 3 being

$$\{c_{11} = 1, c_{12} = 3\}$$

and

$$\{c_{31} = 3, c_{32} = 2\}$$

respectively, the interference probability in channel 3 between SBS 1 and SBS 3 is 0.24×0.9=0.216. Then the average interference received by SBS 2 in channels 1, 2 and 3 are calculated as

$\{0.76I, 0.4I, 1.14I\}$.

Thus,

$$\{c_{21} = 2, c_{22} = 1, c_{23} = 3\}$$

should be the optimal sequential channel set of SBS 2.

A diagram (for small-cell 2) illustrating the dynamic load and the channel selection based on that load is shown in Figure 5. It is assumed that the loads of small-cell 2 in six time periods are $\{1,3,2,1,2,1\}$. As seen in the diagram, the utilization of channel 3 is the lowest, while the utilization of channel 2 is the highest. That is because the interference in channel 3 is the highest, while the interference in channel 2 is the lowest. Compared with allocating fixed channels to certain SUs, the proposed distributed channel allocation method with different channel priority, which has considered the dynamics of loads in small-cells, will make most use of the channels with low interference. The interference can be quantified in a statistical manner according to the dynamic interference graph. Then the average interference among small-cells can be calculated.

It is assumed that the channel subset being used to transmit by SBS n at time t is denoted as

$$C_n(t) = \{c_{n1}, c_{n2}, \ldots, c_{nl}\},\ C_n(t) \subset C_n.$$

Figure 5. A diagram illustrating dynamic load and channel selection based on that load

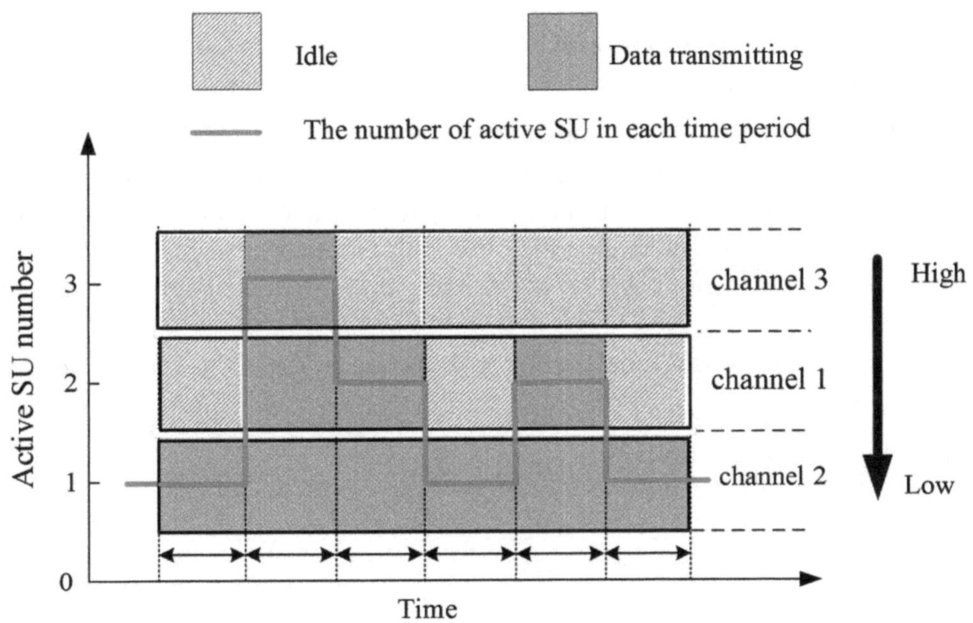

To facilitate the analysis, it does not consider power control and assume that the transmission powers of the SBSs in each channel are the same. The received instantaneous interference signal power[1] by SBS n in each channel m ($m \in C_n(t)$) can be approximated as:

$$I_{nm}(t) = \begin{cases} \sum\limits_{i \in \{\mathcal{N}-n\}} \delta_{im}(t) p g_{in}^{m}, & m \in C_n(t) \\ 0, & m \notin C_n(t) \end{cases} \tag{17}$$

where g_{in}^{m} represents the average channel gain between SBS i and n in channel m, p represents the transmission power of each SBS i in each channel, and $\delta_{im}(t)$ is the following interference indicator function:

$$\delta_{im}(t) = \begin{cases} 1, & m \in C_i(t) \\ 0, & m \notin C_i(t) \end{cases} \tag{18}$$

Following many previous studies which focus on small-cell networks, this chapter considers the average channel gain to be mainly determined by the transmission distance, in which case $g_{in}^{m} = g_{in}$.

The obtained instantaneous throughput of SBS n in channel m is

$$r_{nm} = B_0 \log_2 \left(1 + \frac{p g_n(t)}{I_{nm}(t) + \sigma^2} \right), \tag{19}$$

where $g_n(t)$ is the instantaneous channel gain between SBS n and the active SU at time t, σ^2 is the background noise, and B_0 is the bandwidth of each channel. Because of the mobility and activity of SUs, it is too hard to monitor and calculate the instantaneous channel gain $g_n(t)$. What's more, it is also difficult to calculate the instantaneous throughput or the average throughput considering the received instantaneous interference. On the other hand, it is easy to calculate the average interference in each channel for SBS n over a relatively long period of time. It is clear that each SBS will prefer to give higher priority to channels with lower average interference to transmit in order to increase its servicing capacity.

Suppose that (C_n, C_{-n}) is the sequential channel set selection profile of the small-cell network, where C_{-n} is the sequential channel set selection profile of other SBSs. The statistical average interference by SBS n in channel m can be described as:

$$\bar{I}_{nm} = \lim_{T \to +\infty} \frac{1}{T} \sum_{t=1}^{T} \sum_{i \in \{\mathcal{N}-n\}} \delta_{im}(t) p g_{in} \tag{20}$$

It can be assumed that for each SBS i with each channel $m' \in \{\mathcal{M} - C_i\}$, there will have $\delta_{im'}(t) = 0, \forall t$ and $\vartheta_i^{m'} = 0$. Thus, (20) can be rewritten as:

$$\overline{I}_{nm} = \sum_{i \in \{\mathcal{N} - n\}} \vartheta_i^m p g_{in}. \tag{21}$$

Because the probability that SBS n occupies channel $m = c_{nk}$ to communicate with an active SU is ϑ_{nk} ($\vartheta_{nk} = \vartheta_n^m$), when SBS n has a candidate channel set C_n, the average aggregate interference can be expressed as:

$$s_n\left(C_n, C_{-n}\right) = \sum_{c_{nk} \in C_n} \vartheta_{nk} \overline{I}_{nc_{nk}}. \tag{22}$$

The optimization goal for each small-cell is to minimize its received average aggregate interference. Accordingly, a quantitative expression of the average network interference level can be given by the following equation:

$$U(C) = \sum_{n=1}^{N} s_n\left(C_n, C_{-n}\right). \tag{23}$$

The average network interference level reflects the expectation of the aggregate dynamic co-tiered interference of the whole network. Thus, the global objective is expressed as follows:

$$(\mathcal{P}1) C_{opt} = \arg\min U(C). \tag{24}$$

Noncooperation Dynamic Interference Graph Game and Distributed Dynamic Learning Algorithm

Because centralized management is hard to achieve in most small-cell networks, this chapter propose a distributed channel allocation approach with incomplete information to solve this problem. The noncooperation dynamic interference graph game can be denoted as

$$G_d = \left\{\mathcal{N}, (\mathcal{A}_n)_{n \in \mathcal{N}}, (u_n)_{n \in \mathcal{N}}\right\},$$

where $\mathcal{N} = \{1, 2, \ldots, N\}$ is the set of SBSs, $\mathcal{A}_n$ is the strategy space of SBS n, with $C_n \in \mathcal{A}_n$. u_n is the utility function of each player n, which is the inverse of the received average aggregate interference of SBS n, and can be expressed as:

$$u_n(C_n, C_{-n}) = -s_n(C_n, C_{-n}). \tag{25}$$

Then the noncooperation dynamic interference graph game is expressed as follows:

$$(G_d): \max_{C_n \in \mathcal{A}_n} u_n\left(C_n, C_{-n}\right) \forall n \in \mathcal{N}. \tag{26}$$

It can be expected that each SBS prefers to choose the channels with the lowest average interference to maximize the capacity of the small-cell. Let us consider a scenario where the sequential channel set selection profile of other players C_{-n} is fixed, in which case it has the following theorem.

Theorem 2: For each SBS n with a fixed C_{-n}, for

$$C_n^* = \arg\max_{C_n} u_n(C_n, C_{-n}),$$

where

$$C_n^* = \left\{c_{n1}^*, c_{n2}^*, \dots, c_{nZ_n}^*\right\},$$

then it has

$$\bar{I}_{nc_{n1}^*} \leq \dots \leq \bar{I}_{nc_{nZ_n}^*} \leq \bar{I}_{nm'}, \forall m' \in \left\{\mathcal{M} - C_n^*\right\}.$$

Proof: The detailed proof lines can be found in Wu et al. (2016).

Therefore, each SBS n would get Z_n channels with the minimum average interference by minimizing the received average aggregate interference in all selected channels, i.e., solving (26). Moreover, the channel with the lowest interference would be selected with the highest priority for transmission.

Theorem 3: G_d is an exact potential game which has at least one pure strategy NE. The optimal solution of the network average interference minimization problem is a pure strategy NE of G_d.

Proof: The potential function is constructed as

$$\Phi(C_n, C_{-n}) = -\frac{1}{2}U(C_n, C_{-n}) = -\frac{1}{2}\sum_{n=1}^{N} u_n(C_n, C_{-n}) = \frac{1}{2}\sum_{n=1}^{N} s_n(C_n, C_{-n}). \tag{27}$$

It can be proofed that G_d is an exact potential game in accordance with the definition of exact potential games. The detailed proof lines can be found in Wu et al. (2016).

It is well known that many learning algorithms, such like the best response, fictitious play, spatial adaptive and log-linear learning algorithms, can converge towards pure strategy Nash equilibriums for an exact potential game. However, these algorithms are designed for static game models, in which the states and the utilities of players are static. Moreover, most of these algorithms need information exchange among players during learning and iterating. Therefore, these mentioned algorithms cannot be applied in the dynamic interference graph game.

Motivated by the imitative Boltzman-Gibbs algorithm with multiplicative weights (Tembine, 2012; Xu et al., 2015), a distributed dynamic learning algorithm is proposed to achieve the Nash equilibrium of the dynamic interference graph game. Without information exchange, in the proposed algorithm the utility of each player is considered to be the temporal aggregate interference which is varying with the dynamic load of each small-cell in each iteration. Moreover, the SBSs do not need to know the probability distributions of their dynamic loads. Algorithm 2 elaborates on this approach.

Algorithm 2. Distributed dynamic learning algorithm

Step 1: Initially, at time $t = 0$, each small-cell n sets the initial mixed strategy as $p^t_{nC_{nl}} = \frac{(M - Z_n)!}{M!}$ (the probability of action C_{nl} is $p^t_{nC_{nl}}$) and the initial estimation $Q_{nC_{nl}} = 0$, where $C_{nl} \in \mathcal{A}_n$, $1 \le l \le \frac{M!}{(M - Z_n)!} = |\mathcal{A}_n|$.

Step 2: At time $t \ge 1$, each small-cell n stochastically selects an action C^t_n according to its current sequential channel set selection probability vector p^t_n.

Step 3: All small-cells sense the interference of channels in C^t_n and transmit signal based on the current load. Then, using (17) and (23), they calculate the temporal aggregate interference, which serve as the temporal utilities, i.e.

$$u^t_n = -\sum_{c^t_{nk} \in C^t_n} \delta_{nc^t_{nk}}(t) I_{nc^t_{nk}}(t). \quad (28)$$

In order to facilitate the implementation of the algorithm, a constant is added to the utility of each small-cell to make the utility greater than zero, i.e.

$$r^t_n = Z_n I_{max} + u^t_n, \quad (29)$$

where I_{max} has a fixed value and is greater than or equal to the upper bound of interference detected in each channel.

Step 4: All the players update their estimation according to the following rules:

$$\begin{cases} Q^{t+1}_{nC_{nl}} = Q^t_{nC_{nl}} + \lambda_t \left(\frac{r^t_n}{Z_n I_{max}} - Q_{nC_{nl}} \right), & C_{nl} = C^t_n \\ Q^{t+1}_{nC_{nl}} = Q^t_{nC_{nl}}, & C_{nl} \ne C^t_n \end{cases} \quad (30)$$

where λ_t is the step factor. Finally, all the players update their mixed strategies according to the following rule:

$$p^{t+1}_{nC_{nl}} = \frac{p^t_{nC_{nl}} (1 + b)^{Q^t_{nC_{nl}}}}{\sum_{C_{nk} \in \mathcal{A}_n} p^t_{nC_{nk}} (1 + b)^{Q^t_{nC_{nk}}}}, \quad (31)$$

where b is the learning parameter.

Step 5: Return to Step 2 and repeat unless the stopping criterion has been met.

Because of the fixed value of $Z_n I_{max}$, $\max_{C_n \in \mathcal{A}_n} r_n(C_n, C_{-n})$ is equivalent to $\max_{C_n \in \mathcal{A}_n} u_n(C_n, C_{-n})$. Moreover, the properties of the game would not change even if replace the utility u_n by r_n. With sufficiently small values for λ and b, the distributed dynamic learning algorithm converges to the Nash equilibrium of the potential game G_d.

To evaluate the performance of the proposed learning algorithm in the dynamic and distributed environment, simulations are conducted in this section. The simulation scenario is set as follows. The transmission power of each SBS in each channel is 40mW, the background noise is -174dBm/Hz. The channel bandwidth is $B_0 = 500\text{kHz}$. The coverage radius of a small-cell is 10m. Like many previous studies, in order to simplify our analysis, it only considers path loss when modeling the propagation environment. The path loss between two SBSs is assumed equal to

$$PL = 38.46 + 20log(d_{ij}),$$

where d_{ij} is the distance between SBSs i and j.

For the sake of evaluating the interference performance and the fairness performance, a special scenario is considered, in which there have $M = 4$, $K_n = 3$, and $\alpha_{n1} = 0.6$, $\alpha_{n2} = 0.4$, $\alpha_{n3} = 0.2$, $n \in \mathcal{N}$. In the following section, the number of small-cells is varied from 15 to 25, and then the performance evaluation results are presented. In this scenario, the probability distributions of the SBSs' dynamic loads are set to the same. It is because that if the probability distributions are different, the demanded channels of small-cells are different, and fairness is hard to calculate under such conditions. The utilities of SBSs are calculated using (31). The interference performances for the best NE, the worst NE of the proposed game and random allocation are shown in Figure 6.

As shown in Figure 6, it is clear that the NE solutions are much better than random allocation which achieves the worst performance. In addition, it can be seen that the performance gap between the worst NE and the best NE is very small, i.e. the performance of the NE solution is relatively stable.

Then the simulation calculates the Jain's fairness index (JFI) (Jain, Chiu, & Haws, 1984) of the small-cell networks with the received aggregate interference (s_n in (22)) of each SBS. The result is shown in Figure 7. It can be noted from the figure that the JFI value achieved by the NE is also significantly higher than that of random allocation.

Then, the interference and the throughput performance in a more common scenario are evaluated. It set $M = 4$, K_n was selected from $\{1,2,3\}$ randomly, and the active probabilities of SUs were determined randomly from $[0,1]$ in each independent trial. Each SBS doesn't know the probability distributions of its own and other SBSs' dynamic loads. The utilities of SBSs were calculated using (28). Simulation simulated 1000 independent trials and then the average results were calculated. The average network interference levels varying with the number of small-cells are shown in Figure 8. It can be noted from Figure 8 that the proposed distributed dynamic channel allocation significantly outperforms the random allocation. It also can be concluded that in a given region, the number of small-cells increases, the average interference level increases either.

The average achieved throughput of the network varying with the number of SBSs is shown in Figure 9. In practice, due to restrictions on the capabilities (coding and modulation) of terminal devices, the signal to interference plus noise ratio (SINR) must be larger than a predetermined threshold. The

Figure 6. Average network interference levels with $M = 4$, $K_n = 3$, *and* $\alpha_{n1} = 0.6$, $\alpha_{n2} = 0.4$, $\alpha_{n3} = 0.2$, $n \in \mathcal{N}$

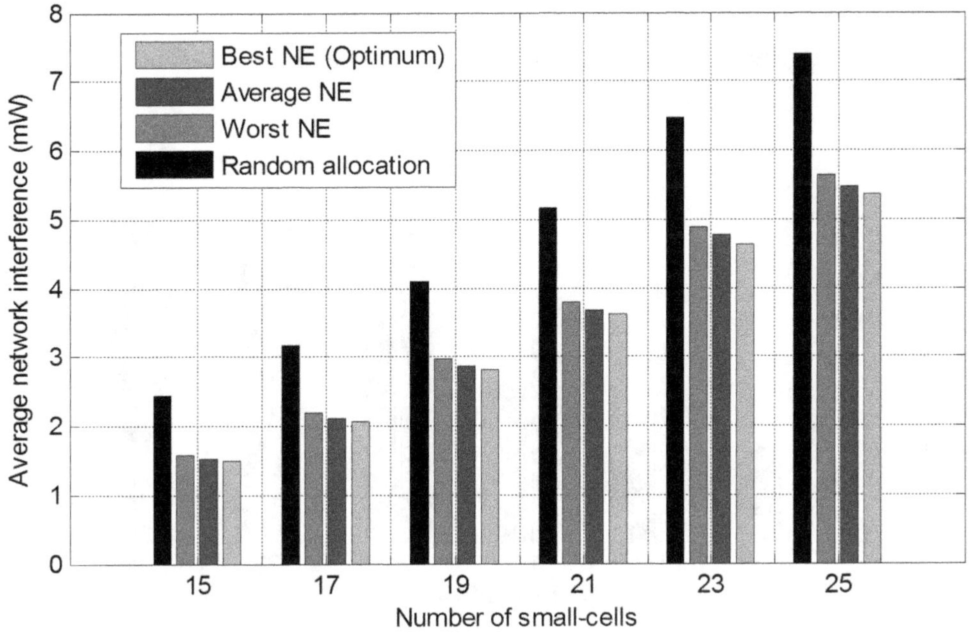

Figure 7. JFI of small-cell networks with $M = 4$, $K_n = 3$, *and* $\alpha_{n1} = 0.6$, $\alpha_{n2} = 0.4$, $\alpha_{n3} = 0.2$, $n \in \mathcal{N}$

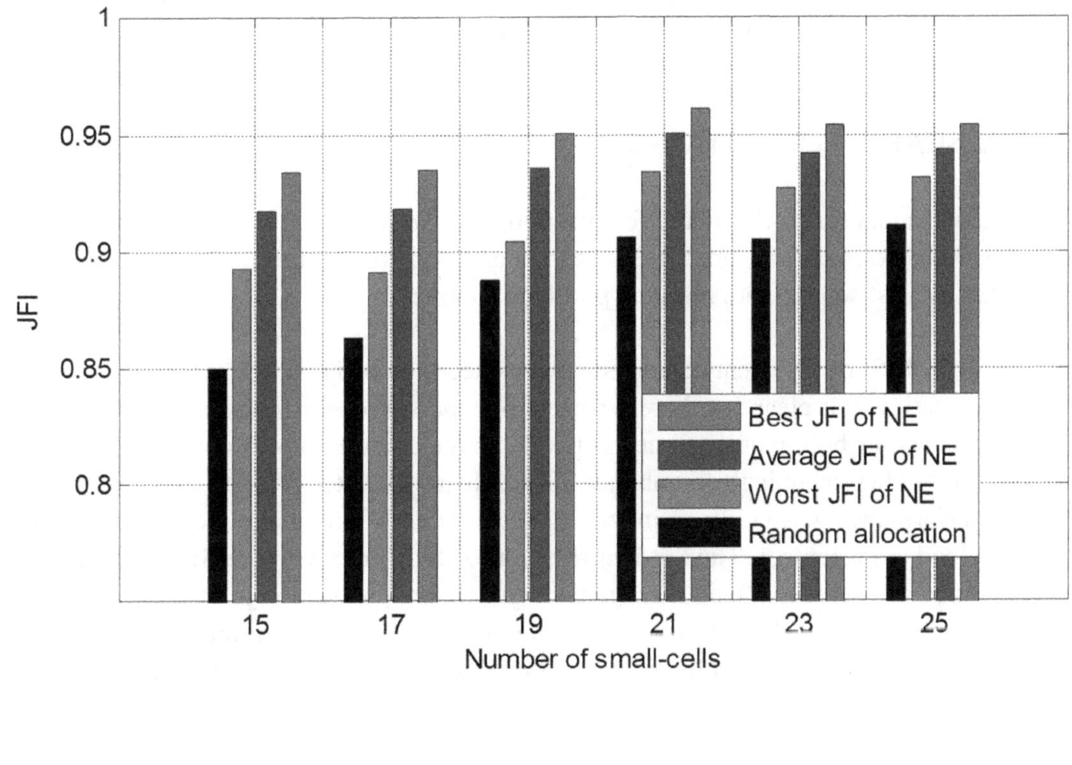

Figure 8. The average aggregate interference levels with $M = 4$, *randomly selected* K_n *and active probabilities* $(1 \leq K_n \leq 3, 0 \leq \alpha \leq 1)$

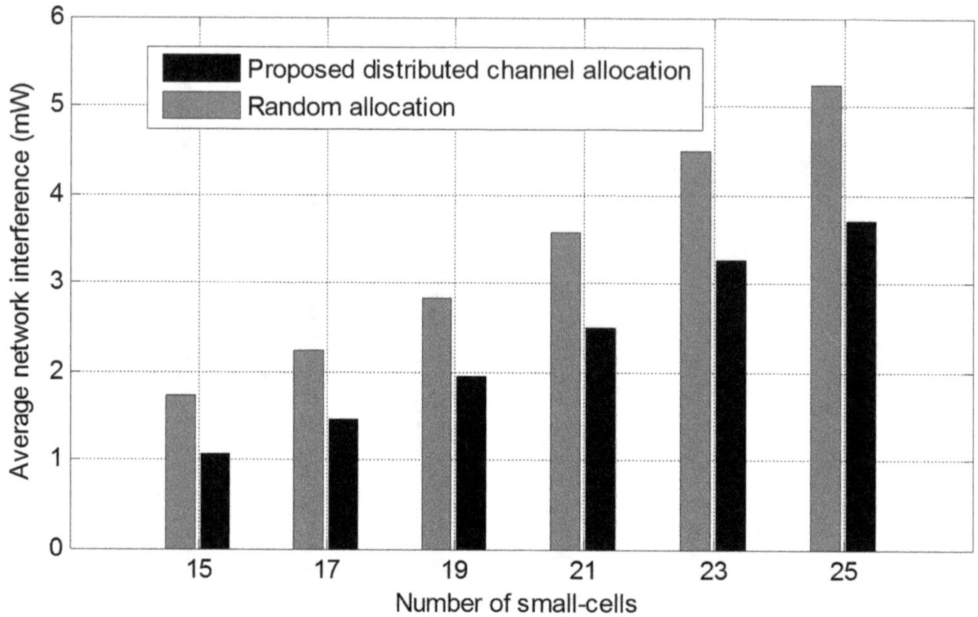

SINR threshold is set as 5dB, and the transmission rate of each available channel is 1Mbps. For a given channel, a SBS is allowed to transmit data with the transmission rate of 1 Mbps only when the temporal SINR is larger than 5dB. It can be found from the figure that the average throughput, which is achieved by the distributed channel allocation method, is much higher than the random allocation. Moreover, the average achieved throughput decreases when the number of SBSs increases.

FUTURE RESEARCH DIRECTIONS

It is seen that game theory for co-tiered interference mitigation in small-cell networks has been a research hotspot of 5G. However, consider practical scenarios and constraints, there are some problems need further research. We list some future research problems below:

1. Energy aware SBS management contained power selection, load management (SU allocation) and channel allocation. It is clear that there is a need to reduce the energy consumption of SCNs due to the increasing contribution of small-cell networks to the overall energy consumption of the world and the worldwide growth in the number of SBSs. Meanwhile, it is difficult to achieve the optimal network performance when only optimizes the channel allocation without considering the power selection, load management. Thus, it is important to study energy aware SBS management with considering power selection, load management and channel allocation jointly.
2. QoE-based multichannel allocation with cross-tiered interference constraints. It is clear that quality of experience (QoE) is a crucial indicator in 5G wireless communication system. In the underlay

Figure 9. The average achieved throughput of networks with $M = 4$, randomly selected K_n and active probabilities ($1 \le K_n \le 3, 0 \le \alpha \le 1$)

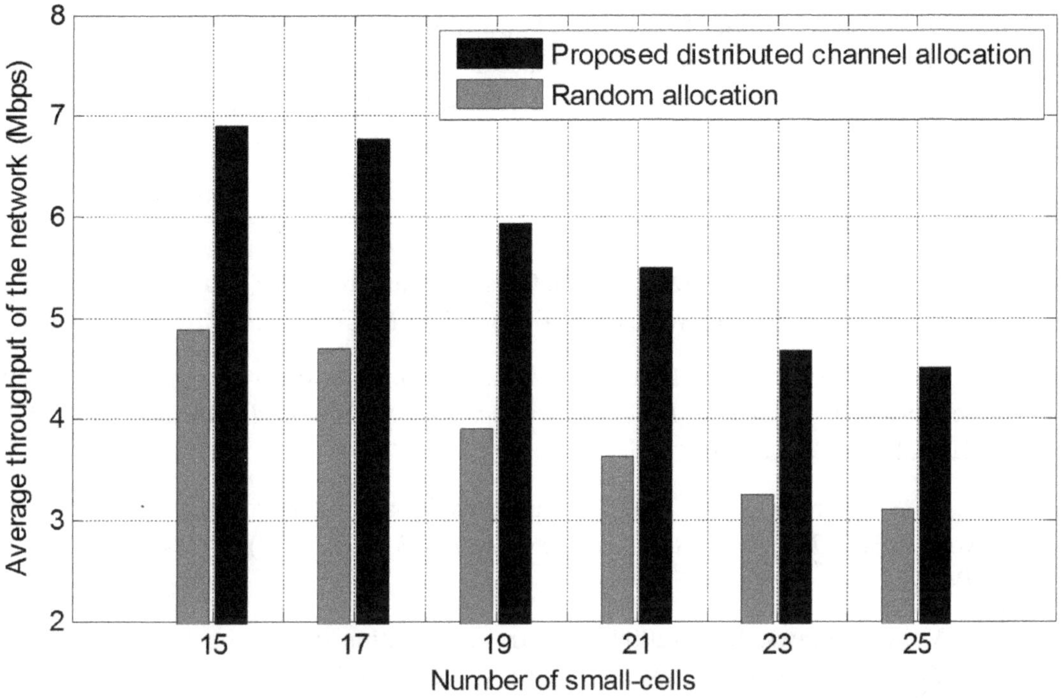

heterogeneous cellular networks, the cross-tiered interference caused by SBSs in each channel must be limited. However, each SBS is willing to occupy spectrum resource as much as possible to increase its QoE without cross-tiered interference constraints. In some existing works, cross-tiered interference was modeled into different utility functions of SUs, SBSs and macro-cell base stations without considering the individual influence to the satisfaction of each macro-cell user. There is a lot of necessary to mitigate co-tiered interference and increase users' QoE with cross-tiered interference constraints to exploit the benefits promised by SBSs

CONCLUSION

This chapter focused on the mitigation of co-tiered interference and studied the issue of distributed multichannel allocation in small-cell networks. Without central controller and information exchange, two game models for co-tiered interference mitigation and resource allocation games were presented and discussed for fixed load and dynamic load small-cell networks separately. For fixed load, a noncooperation interference minimization game was defined, in which the utility function of each small-cell was defined as the negative aggregate received interference signal and noise power in selected channels. For dynamic load, a noncooperation dynamic interference graph game was formulated based on a defined dynamic interference graph.

It proved that these games are potential games and have at least one pure strategy Nash equilibrium (NE) point. Moreover, the best pure strategy NE point minimizes the aggregate interference in the small-cell network. Two distributed dynamic learning algorithms were then designed to achieve the NE of these two games respectively, in which each SBS doesn't know the probability distributions of its own and the others' dynamic loads. Simulation results showed that the proposed approach could mitigate dynamic co-tiered interference effectively and outperform random selection significantly.

REFERENCES

Abdelnasser, A., Hossain, E., & Kim, D. I. (2014). Clustering and resource allocation for dense femtocells in a two-tier cellular OFDMA network. *IEEE Transactions on Communications*, *13*(3), 1628–1641.

Akkarajitsakul, K., Hossain, E., Niyato, D., & Dong, I. K. (2011). Game theoretic approaches for multiple access in wireless networks: A survey. *IEEE Communications Surveys and Tutorials*, *13*(3), 372–395. doi:10.1109/SURV.2011.122310.000119

Chandrasekhar, V., & Andrews, J. G. (2007). Uplink capacity and interference avoidance for two-tier femtocell networks. *IEEE Transactions on Wireless Communications*, *8*(7), 3498–3509. doi:10.1109/TWC.2009.070475

Chandrasekhar, V., & Andrews, J. G. (2009). Spectrum allocation in tiered cellular networks. *IEEE Transactions on Communications*, *57*(10), 3059–3068. doi:10.1109/TCOMM.2009.10.080529

Chen, Y., Lin, P., & Zhang, Q. (2014). Lotus: Location-aware online truthful double auction for dynamic spectrum access. *IEEE Transactions on Wireless Communications*, *14*(2), 510–518.

Felegyhazi, M., & Hubaux, J.-P. (2006). *Game Theory in Wireless Networks: A Tutorial.* EPFL Technical Report, LCA-REPORT-2006-002.

Gu, Y., Saad, W., Bennis, M., Debbah, M., & Han, Z. (2015). Matching theory for future wireless networks: Fundamentals and applications. *IEEE Communications Magazine*, *53*(5), 52–59. doi:10.1109/MCOM.2015.7105641

Guruacharya, S., Niyato, D., Kim, D. I., & Hossain, E. (2013). Hierarchical competition for downlink power allocation in OFDMA femtocell networks. *IEEE Transactions on Wireless Communications*, *12*(4), 1543–1553. doi:10.1109/TWC.2013.022213.120016

Hu, R., & Qian, Y. (2014). An energy efficient and spectrum efficient wireless heterogeneous network framework for 5G systems. *IEEE Communications Magazine*, *52*(5), 94–101. doi:10.1109/MCOM.2014.6815898

Jain, R., Chiu, D., & Hawe, W. (1998). A quantitative measure of fairness and discrimination for resource allocation in shared computer systems. *Computer Science*. Retrieved from cs.ni/9809099

Jo, M., Maksymyuk, T., Batista, R. L., Maciel, T. F., De Almeida, A. L. F., & Klymash, M. (2014). A survey of converging solutions for heterogeneous mobile networks. *IEEE Wireless Communications*, *21*(8), 54–62.

Kang, X., Zhang, R., & Motani, M. (2012). Price-based resource allocation for spectrum-sharing femtocell networks: A Stackelberg game approach. *IEEE Journal on Selected Areas in Communications*, *30*(3), 538–549. doi:10.1109/JSAC.2012.120404

Kulkarni, P., Chin, W. H., & Farnham, T. (2010). Radio resource management considerations for lte femto cells. *Computer Communication Review*, *40*(1), 26–30. doi:10.1145/1672308.1672314

Lee, H.-C., Oh, D.-C., & Lee, Y.-H. (2010). *Mitigation of Inter-Femtocell Interference with Adaptive Fractional Frequency Reuse.* Paper presented at the ICC Conference, Cape Town, South Africa. doi:10.1109/ICC.2010.5502298

Li, H., Xu, X., Hu, D., & Qu, X. (2010). *Graph Method Based Clustering Strategy for Femtocell Interference Management and Spectrum Efficiency Improvement.* Paper presented at the WiCOM Conference, Chengdu, China. doi:10.1109/WICOM.2010.5601236

Liang, Y. S., Chung, W. H., Ni, G. K., Chen, I. Y., Zhang, H., & Kuo, S. Y. (2012). Resource allocation with interference avoidance in ofdma femtocell networks. *IEEE Transactions on Vehicular Technology*, *61*(5), 2243–2255. doi:10.1109/TVT.2012.2191164

Lien, S. Y., Tseng, C. C., Chen, K. C., & Su, C. W. (2010). *Cognitive radio resource management for qos guarantees in autonomous femtocell networks.* Paper presented at the ICC Conference, Cape Town, South Africa. doi:10.1109/ICC.2010.5502784

Mackenzie, A., & Dasilva, L. (2006). Game theory for wireless engineers. *Synthesis Lectures on Communications,* (1), 1-86.

Marden, J. R., Arslan, G., & Shamma, J. S. (2009). Joint strategy fictitious play with inertia for potential games. *IEEE Transactions on Automatic Control*, *54*(2), 208–220. doi:10.1109/TAC.2008.2010885

Monderer, D., & Shapley, L. S. (1996). Potential games. *Games and Economic Behavior*, *14*(1), 124–143. doi:10.1006/game.1996.0044

Pantisano, F., Bennis, M., Saad, W., Debbah, M., & Latva-aho, M. (2013). Interference alignment for cooperative femtocell networks: A game-theoretic approach. *IEEE Transactions on Mobile Computing*, *12*(11), 2233–2246. doi:10.1109/TMC.2012.196

Pervaiz, H., Musavian, L., Ni, Q., & Ding, Z. (2015). Energy and spectrum efficient transmission techniques under qos constraints towards green heterogeneous networks. *IEEE Access*, *3*, 1655–1671. doi:10.1109/ACCESS.2015.2479406

Saquib, N., Hossain, E., Le, L. B., & Kim, D. I. (2012). Interference management in ofdma femtocell networks: Issues and approaches. *IEEE Wireless Communications*, *19*(3), 86–95. doi:10.1109/MWC.2012.6231163

Semasinghe, P., Hossain, E., & Zhu, K. (2015). An evolutionary game for distributed resource allocation in self-organizing small-cells. *IEEE Transactions on Mobile Computing*, *14*(2), 274–287. doi:10.1109/TMC.2014.2318700

Sundaresan, K., & Rangarajan, S. (2009). Efficient resource management in OFDMA Femto cells. *ACM Interational Symposium on Mobile Ad Hoc NETWORKING and Computing*, (pp. 33-42). doi:10.1145/1530748.1530754

Tembine, H. (2012). *Distributed strategic learning for wireless engineers*. CRC Press. doi:10.1201/b11896

Wang, C. X., Haider, F., Gao, X., You, X. H., Yang, Y., Yuan, D., & Hepsaydir, E. et al. (2014). Cellular architecture and key technologies for 5G wireless communication networks. *IEEE Communications Magazine*, *52*(2), 122–130. doi:10.1109/MCOM.2014.6736752

Wei, N., & Collings, I. B. (2013). A new adaptive small-cell architecture. *IEEE Journal on Selected Areas in Communications*, *31*(5), 829–839. doi:10.1109/JSAC.2013.130502

Wu, D., Wu, Q., & Xu, Y. (2016). Self-organized Spectrum Access in Small-cell Networks with Dynamic Loads. *Transactions on Internet and Information Systems (Seoul)*, *10*(5), 1976–1997.

Wu, D., Zhang, L., & Zhou, J. (2015). *Self-organized spectrum access in small-cell networks: A nonco-operation interference minimization game solution.* Paper presented at the WCSP Conference, Nanjing, China. doi:10.1109/WCSP.2015.7341204

Xu, J., Wang, J., Zhu, Y., Yang, Y., Zheng, X., & Wang, S. et al.. (2014). Cooperative distributed optimization for the hyper-dense small-cell deployment. *IEEE Communications Magazine*, *52*(5), 61–67. doi:10.1109/MCOM.2014.6815894

Xu, Y., Wang, J., Wu, Q., Zheng, J., Shen, L., & Anpalagan, A. (2015). Dynamic spectrum access with statistical QoS provisioning: A distributed learning approach beyond expectation optimization. *Computer Science*, (2): 174–175.

Xu, Y., Wu, Q., Shen, L., Wang, J., & Anpalagan, A. (2013). Opportunistic spectrum access with spatial reuse: Graphical game and uncoupled learning solutions. *IEEE Transactions on Wireless Communications*, *12*(10), 4814–4826. doi:10.1109/TWC.2013.092013.120862

Zheng, L., Cai, Y., & Xu, Y. (2014). *MAC-layer interference mitigation in dynamic and distributed environment: dynamic graphic game with stochastic learning*. Paper presented at the GAMENETS Conference, Beijing, China. doi:10.1109/GAMENETS.2014.7043692

KEY TERMS AND DEFINITIONS

Co-Tiered Interference: The interference among SBSs that belong to the same tier in 5G heterogeneous cellular networks.

Cross-Tiered Interference: The interference between different network tiers in 5G heterogeneous cellular networks.

Game Theory: A mathematical theory which is used to model and analyze the mutual interactions of rational entities in a situation of conflict. It was primarily developed and used in Economics.

Learning: A decision-maker updates its current action based on its (and other decision-makers') historical action-payoff information.

Potential Game: A class of cooperative/non-cooperative games which require perfect alignment between the players' local objective functions and the global objective. The unilateral improvement dynamics are guaranteed to converge in many practical cases.

Small-Cell Base Station: A class of low-power short-range base stations which provide wireless services to users.

ENDNOTE

1 The co-tiered interference from some SBSs, which are not close to SBS n, can be catered by assuming them as a part of thermal noise as shown in Pervaiz, Musavian, Ni, & Ding (2015)due to the low transmission power of small-cells. The remaining SBSs interfering with SBS n which can be defined as $J_n = \left\{ i : d_{in} < D_{\text{int}}, i \in \mathcal{N}, i \neq n \right\}$, where d_{in} is the distance between SBS i and SBS n, D_{int} is the defined interference distance. Then, it can replace $i \in \left\{ \mathcal{N} - n \right\}$ by $i \in J_n$ in (20), and the dynamic interferences can be more tractable with the following approach.

Chapter 8
Interference Management in Heterogeneous Networks

Yanxia Liang
Xi'an University of Posts and Telecommunications, China

ABSTRACT

In evolving fifth generation (5G), some demands that to be addressed are improved data rate, increased capacity, decreased latency and better quality of service. Heterogeneous networks (HetNets) is considered as an effective way to meet these prime objectives. Deployment of HetNets faces a number of challenges, among which interference management is of much importance. This chapter has summarized the major challenges and solutions of interference management. Interference cancellation, avoidance and coordination are discussed in detail, including enhanced intercell Interference Coordination (eICIC), Coordinated Multi-point Transmission (CoMP) et al. In addition, the future challenges of interference management have also been presented.

INTRODUCTION

There are many challenges in 5G mobile communication, such as higher capacity, higher data rate, low End to End latency, massive device connectivity, reduced cost and consistent Quality of Experience provisioning (Gupta & Jha, 2015), which were not effectively addressed in 4G . For the sake of meeting these challenges, some technologies are needed and crucial, which are massive Multi-Input and Multi-Output (MIMO), Interference Management (IM), Spectrum Sharing, Device to Device (D2D) communication System, Ultra Dense Networks, Multi-Radio Access Technology Association, A millimeter wave solution for 5G cellular Network, Cloud Technologies for flexible 5G radio Access Networks and Trends and quality of Service management. (Gupta & Jha, 2015)

Among all the schemes above in 5G, Interference Management is one of the biggest concerns. There have been many researches in this domain. In this chapter, the methods and algorithms for Interference Management will be summarized and analyzed in detail. The following three questions will be mainly talked about. What is Interference? Why should we deal with it? How can we solve or mitigate this problem? As it should be, we will talk about those methods and algorithms for interference elimination or mitigation under the condition of heterogeneous networks.

DOI: 10.4018/978-1-5225-1712-2.ch008

BACKGROUND

The 1st generation was announced in 1983 by Ameritech company in Chicago, America (Boucher, 2001;Frenkiel, 2010), which was analog system and called Advanced Mobile Phone System(AMPS). This was the first time that the thought of cells was applied.

The 2nd generation (2G) was a digital communication system. With the vigorous development of digital circuit technology, analog communication was gradually replaced by digital communication. Digital communication system has greater system capacity, lower equipment cost, volume and processing speed.Vital eminent technologies were GSM, Code Division Multiple Access (CDMA) and IS-95 (Halonen, Romero, & Melero, 2002; Santhi, Srivastava, Senthilkumaran, & Butare 2003). In order to offer data services, General Packet Radio Services(GPRS) was merged into 2G, which was called 2.5G.

The 3rd generation was established in late 2000. It imparts transmission rate up to 2Mbps. The International Telecommunication Union (ITU) chose three types of 3G in a variety of selection: wideband code division multiple access, which is WCDMA (Harte, 2004), cdma2000 (Vanghi, Damnjanovic, Harte, Vojcic, 2004) and Time Division-Synchronous Code Division Multiple Access, TD-SCDMA (Li, 2003). Compared with 2G, 3G has a faster transmission rate, and provides more convenient wireless applications. Long-Term Evolution technology (LTE) and Fixed Worldwide Interoperability for Microwave Access (WIMAX) are the future of mobile data services, which are called 3.75G and featured by better coverage with improved performance for less cost (Andrews, Ghosh, & Muhamed, 2007; Furht & Ahson, 2007; Sesia, Toufik & Baker, 2009).

3rd Generation Partnership Project (3GPP) is presently standardizing Long Term Evolution (LTE) Advanced as forthcoming 4G standard along with WIMAX. A 4G system improves the prevailing communication networks by imparting a complete and reliable solution based on IP.

The LTE we usually say is LTE release 8. OFDM and MIMO were introduced into LTE R8, and throughput on 100 Mbps is realized on 20 MHz bandwidth. But LTE R8 is not 4G standard. It is called 3.9G. The real 4G standard is LTE Release 10 and above 10, which is called LTE-Advanced and is the improved version of R8. It is adopted in R10 that enhanced MIMO, distributed antennas, repeaters, carrier aggregation and so on. And the maximum throughput can be 1Gbps on the bandwidth of 100MHz in LTE-A.

Interference has been a serious problem of wireless communication performance ever since 1G emerged. With the rapid development of wireless communication, controlling and reducing interference effectively becomes more and more urgent. Especially in recent years, more and more attention has been payed to it.

Inter-cell interference is controlled by frequency distribution in GSM (Mishra, 2006). The strategy is trying not to use same frequency in adjacent areas at early days. In addition, frequency hopping technology (Nielsen & Wigard, 2000) helps GSM make interference random. Moreover, frequency adjustment and frequency hopping pattern planning are used to manage interference. The thought of interference management in GSM is good reference to the future cellular network.

Pseudo code is used in CDMA system. Its imperfect orthogonalization causes intra-cell interference. Multi-user detection technology is used to reduce uplink interference (Michael, 2009). As well as equalization technology was utilized to combat multi-path interference in downlink way (Wu &Niu, 2005).

Orthogonal frequency division multiplexing (OFDM) technology is used in LTE system, which makes use of the orthogonal sub carriers to eliminate the intra cell interference, but inter cell interference still exists. In order to reduce the inter-cell interference, researchers have proposed a variety of techniques, such as interference coordination (Boudreau, Panicker, Guo, Chang, Wang, & Vrzic, 2009; Elayoubi, & Fourestié, 2006; Rahman, Yanikomeroglu, & Wong, 2009) technology based on frequency reuse, and multi-cell cooperative processing technology (Chang, Tao, Zhang, & Kuo, 2009; K. Kim, J. Lee, C. Lee, Jeon, & S. Kim, 2009).

INTERFERENCE MANAGEMENT

1. Definition of Interference

As is known to all, resources are limited in wireless communication, no matter in time, frequency, code or space. As for at least one of these resources being used effectively, reuse is the common concept in many wireless communication systems. We take frequency as an example to explain. Frequency is taken as a limited resource in cellular wireless communication systems. It is used in two or more cells which is distant from each other. And the distance between every two cells is same and called reuse distance, which determines the co-channel interference in conventional communication systems. This is the basic concept of interference in wireless communication systems, especially in earlier generations (1G and 2G).

Besides, thermal noise is included in interference as well. We've learned it since we were firstly exposed to this subject. It is common and unavoidable. If there is too much noise, it will lead to desired signal distortion and affect its transmission.

In addition, electromagnetic waves have nature of broadcast. As the medium of wireless communication, electromagnetic waves propagate in broadcasting way. That means one user can receive some information which is transmitted to other users. But the information is not useful for the former, so it is labeled interference.

It is indicated that, interference happens when the following three requirements satisfy no matter in a simple or complicated system. First, interference resource must exist. Second, there are channels which can propagate interference. Third, there is response in interfered receiver, which is called sensitive equipment. This is shown in Figure 1.

Figure 1. Requirements which make interference happen

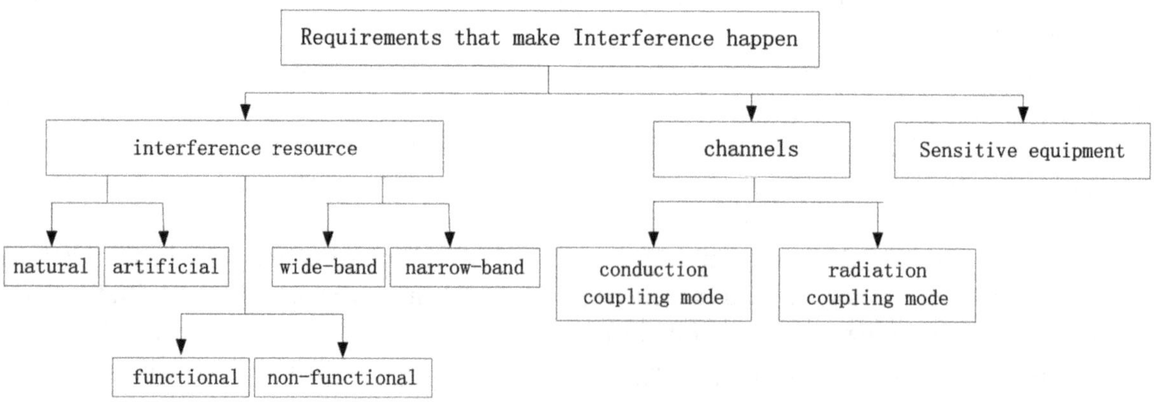

Interference resource can be divided into different types, such as natural and artificial resource, functional and non-functional resource, wide-band and narrow-band resource, and so on. The natural interference resource mainly comes from sky noise in atmosphere, and cosmic noise in outer space of the earth. Artificial resource is generated by electromechanical or other devices. Some of them are called intentional emission interference source, such as broadcasting, TV, radar, navigation, and so on. The others are called unintentional emission interference source, such as vehicle, power transmission line, lighting device and so on, which produce electromagnetic energy by the completion of its own function. Functional resources interfere other devices directly while they realize their own function. And non-functional resources function with side-effect of interference to other devices, for instance, arc discharge interference produced by switch closure. If the interference band width is larger than the designated receptor, it is called the wide-band interference resource, otherwise, called narrow-band interference resource.

The channels which propagate interference can be compartmentalized into conduction coupling mode and radiation coupling mode. There must be a complete circuit between interference resource and sensor in conduction coupling mode. Interference signals are transmitted to the sensor by this circuit. In radiation coupling mode, interference is propagated by electromagnetic wave. The interference energy is transmitted to the surrounding space according to the law of the electromagnetic field. Practically, the interference between two devices usually contains many kinds of ways. Because of this, there are repeated cross couplings, which together produce interference, so that the electromagnetic interference becomes difficult to control.

Sensitive equipment is a general term for interference objects, it can be a very small component or a circuit board component, and it can also be a separate power consumption equipment or even a large system.

Moreover, in heterogeneous cellular networks, interference can be divided into three types, which are noise, inter-system-interference and intra-system-interference. Noise includes inter noise and outer noise in the receiver. The inter noise is mainly considered in analysis, and it can be computed by the following formula:

$$Pn=KTB+NF$$

In which, K is Boltzmann constant, equals to $1.380662\times10^{-23}J/K$, and T is Kelvin absolute temperature, we usually use 290K as the normal temperature; B is the effective bandwidth; and NF is the noise coefficient at the receiver. 7 dB is usually taken as the base station noise coefficient in general. The thermal noise level of the channel will change with the change of the bandwidth.

The inner-system-interference is interference among the wireless network transceiver units in the mobile communication system. It includes co-channel interference and interference between up link and down link.

Intra-system-interference can be partitioned into Adjacent Channel Interference (ACI), spurious emissions, intermodulation interference in receiver and barrage jamming. All of these interference can be analyzed through Deterministic calculation method which is also called Minimum Coupling Loss (MCI) or simulation method.

To sum up, interference is some resource which is not desired to some user or in some system. And this resource could be frequency, time, code, space et al. Moreover, interference can be divided into different types and analyzed by different methods.

2. The Reasons to Deal with Interference

We can find some reasons for dealing with interference in last sub chapter. We need mitigate interference between users, so more users can be served (such as in CDMA). That is to say, interference management is used to improve throughput of the system. But this is just one of the reasons we should deal with interference, and there are many others.

Some breeds of interference are firstly introduced in conventional wireless communication system. They are co-channel interference, adjacent-channel interference, intermodulation interference and multi-access interference. Co-channel interference or CCI is crosstalk from two different radio transmitters using the same frequency. Adjacent-channel interference (ACI) is interference caused by extraneous power from a signal in an adjacent channel. Intermodulation (IM) or intermodulation distortion (IMD) is the amplitude modulation of signals containing two or more different frequencies, caused by nonlinearities in a system. We get these three definitions from wikipedia. And for multi-access interence, it appears in CDMA systems, because of the pseudo orthogonalization of the codes we use to distingwish users or channels.

The following discussion is about the reasons why interferecne happens and results which interference will lead to. It was recommended that the analysis should be divided into two portions, the up-link way and down-link way. And WCDMA in 3G will be taken into account as an example.

Up-link interference: if the Received Total Wide-band Power (RTWP) received by NodeB increases, the power of User Equipments (UEs) will increase as well because of power control, which will lead to inter-cell interference. And if the RTWP is too high, the quality of the up-link of UEs goes bad, which will cause the system to drop words.

Down-link interference: When the background noise of UEs increases, the Signal-to-Interference Ratio (SIR) decreases, then the Block Error Ratio (BLER) increases, which will lead to higher power by power control. After this, the quality of communication will go bad. And if the power is high enough and approach the threshold of the maximum value, dropping words will happen.

We can see that the mitigation of interference is beneficial to improving network performance, such as decreasing the drop-words ratio, higher call completing rate, and better speech quality. Besides, increasing power will shorten UEs' standby time.

There are 3 reasons which cause conventional up-link and down-link interference:

1. **Spurious Signals among Different Systems:** There are spurious signals in their frequency bands among systems like GSM, TD-SCDMA, WCDMA and so on. These signals might get into base stations or UEs and cause up-link and down-link interference.
2. **Intermodulation Interference:** So many communication systems exist together. Different bands will bring intermodulation interference inter- or intra- systems. And the imperfection of antenna systems would make signals in different frequency bands get into intermodulated ones, which lead to the 3-order and 5-order interference. This is very harmful.
3. **Block Interference:** The interference signals will block the receivers by making them not work normally because these signals are out of the working band of amplifiers and mixers. And block interference can be divided into two kinds, in-band and out-band, which will not be discussed in detail here.

The technical characteristics of 5G can be summed up in a few numbers: the capacity of 1000x, the connection of 100 billion +, the maximum speed of 10GB/s, the delay of 1ms. The characteristics of 5G are summarized as four aspects: continuous wide area coverage, hot spots of high capacity, low power consumption and low delay with high reliability. And these four aspects are able to meet the complex needs of different users, different industries for communications. Continuous wide area coverage means that 5G can make people access to high-speed internet in remote areas with high-speed mobility and other harsh environments. Hot spots of high capacity will enable the 5G to meet people's high network speed requirements in high-density, high-traffic-demand region. The low power consumption of 5G can play a role in smart city, environmental monitoring, intelligent agriculture, forest fire prevention and other applications. And for the car networking, unmanned driving, industrial control and other areas of high demand for delay and reliability, low delay and high reliability of the 5G characteristics can respond freely.

New situations in 5G will bring new challenges of interference management, which bring about interference dynamics in the up-link and down-link of the network. According to the work done by Hossain, Rasti, Tabassum and Abdelnasser (2014), there are following reasons which bring about interference management challenges.

- The network is heterogeneous, which has multi-tiers. And the wireless devices in network are arranged densely.
- The traffic load and the coverage are not balanced, because the transmit powers of different base stations are varying in the down-link.
- Public or private access are restricted in different tiers, which will lead to diverse interference levels.
- There are different priorities in accessing channels of different frequencies, and resource is allocated in different strategies. These cause complexity of interference management.
- It needs cooperation between base stations in some strategy, and users can communicate with each other directly. These will make interference management more complicated.

Besides, Zahir, Arshad, Nakata and Moessner (2013) discussed two-tiers network including conventional macrocellular network and femtocell network, who gives us some reasons of new challenges in interference management. In Zahir, Arshad, Nakata and Moessner's research (2013), three kinds of interference are talked about by co- or cross- tier and up- or down- link. We show them as Figure 2 and 3. Power of entities in system is the essential reason.

Figures 2 and 3 are both from the work of Zahir, Arshad, Nakata and Moessner (2013).

The up-link and down-link interference of co-tier are showed in Figure 2. The up-link co-tier interference is caused by the femtocell user equipment (FUE). If a UE in the neighboring femtocell transmits at a higher power, it will affect the victim femtocell and the performance will be degraded. The down-link co-tier interference is caused by an femtocell access point (FAP), where the FAP is the source of interference as it causes interference to the neighboring femtocell.

The up- and down- link interference of cross-tier are showed in Figure 3. The cross-tier up-link interference can take place when a FUE acts as a source of interference to the macrocell BS. When a macrocell UE (MUE) transmits at high power near a femtocell, it causes interference at FAP. The down-link cross-tier interference can be caused by an FAP to a close by MUE. And in the same way, the macrocell base station (MBS) could be a interference to FUE as shown in Figure 3.

Figure 2. A scenario showing co-tier interference between neighboring femtocells

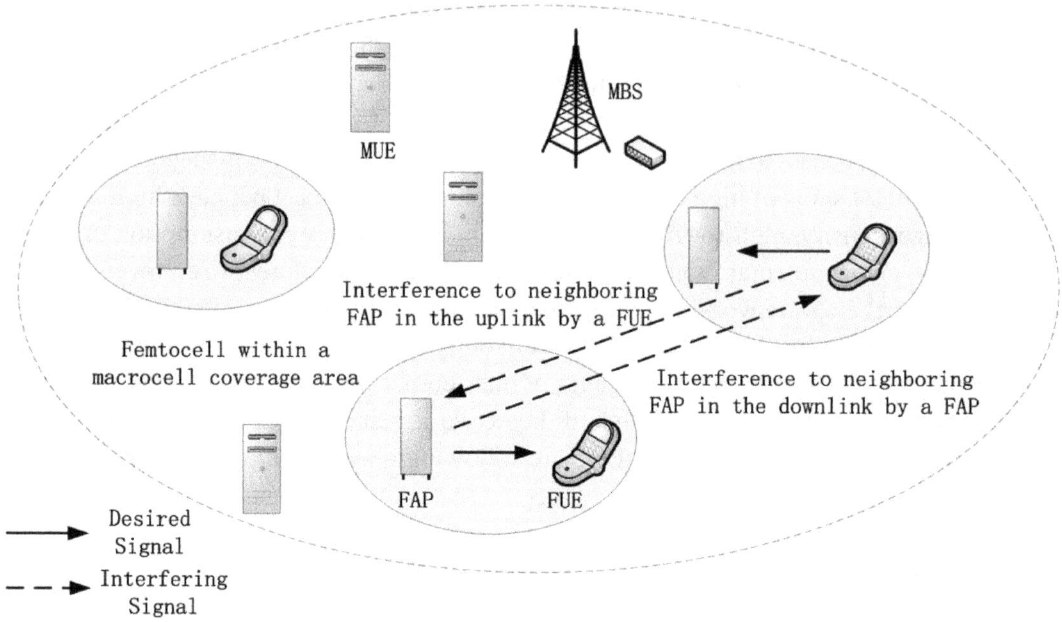

Figure 3. A scenario showing cross-tier interference between femtocell and macrocell

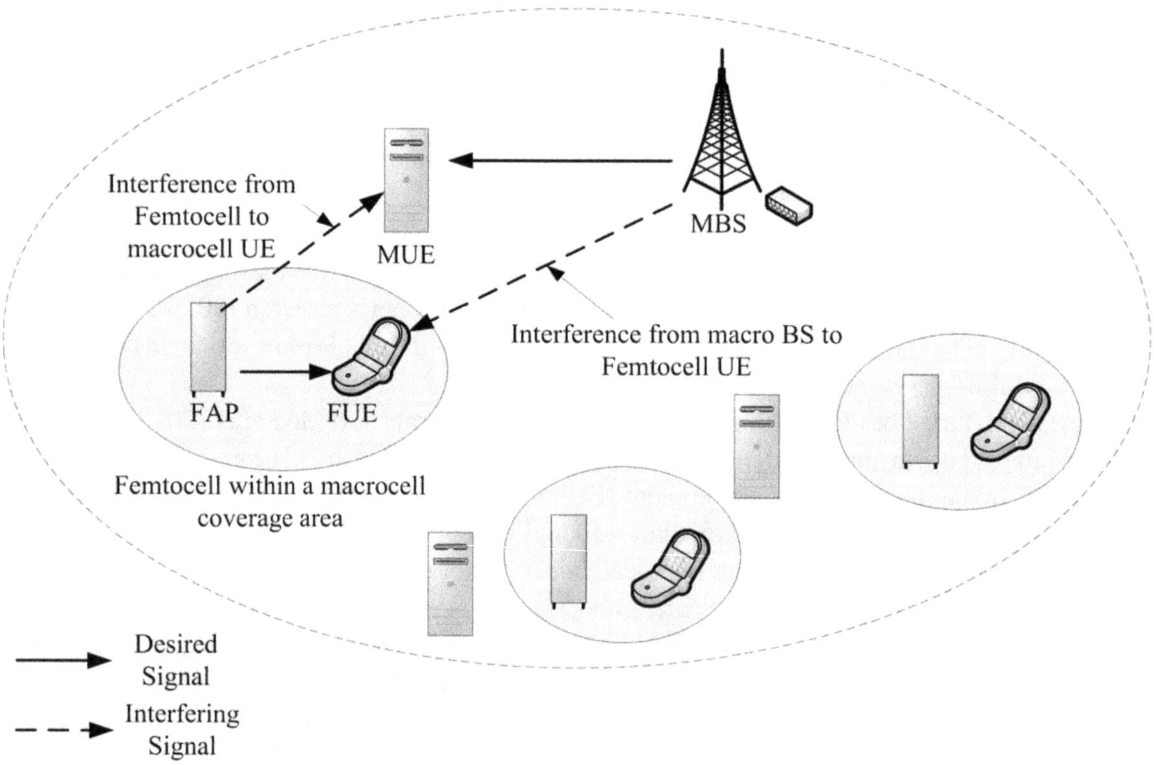

In brief, there are many reasons that interference takes place. So it is essential to understand and manage interference to achieve the highest network performance. (Lee & Heath, 2014)

3. Techniques to Eliminate or Mitigate Interference

As we have known the causes of interference, the final purpose is to solve or mitigate interference. Conventionally the approaches to attack interference are:

1. Increasing the value of reuse distance, the interference from the cells which use the same frequency will decrease. But the frequency is reused less, so the system throughput would not be more than before.
2. Utilizing the orthogonalization of the shared time/frequency resource (Lee & Heath, 2014). For example, frames are defined in time domain, which are distributed to different users. This method mitigates the interference between subscribers.
3. Treating other transmitters' signals as noise (Lee & Heath, 2014)

Other transmitters' signals are not the desired signal for the user we are supervising. But they are as intense as the desired signal, and don't seem like noise. So treating them as noise is not a proper way.

We need more effective interference mitigation methods in 5G, in order to get higher capacity and data rate, less end to end latency, massive number of connections, as well as sustainable cost and consistent quality of experience (QoE).(Gupta & Jha, 2015)We don't like interference at all. It is negative portion in communication systems. So we'd like to cancel it from the system, or avoid it from our desired signals, or take it as useful signal to help rebuild desired signals, or coordinate quantities in the network to mitigate it.

Some main schemes would be detailed in this section. The types of interference are Interference Cancellation, Interference Avoidance and Interference Coordination, which includes Inter Cell Interference Coordination (ICIC), Enhanced ICIC (eICIC), and Coordinated Multi-point Transmission (CoMP) technologies. And at last, the challenges in interference management are discussed.

Due to the importance of this portion, it is organized as flowing, Figure 4 shows the block diagram of all the techniques explained.

- Successive Interference cancellation (SIC), Parallel Interference Cancellation (PIC), Multi-stage SIC (MSIC) and Multi-User Detection (MUD) are focused on Interference Cancellation.
- Spectrum Splitting (SP), Power Control (PC), Time Hopping or Joint Hopping (TH or JH) and Interference Alignment (IA) are proposed in Interference Avoidance.
- ICIC, eICIC, Coordinated Scheduling/ Beamforming and Joint Processing (Dynamic Point Selection)/ Transmission (JP(DPS)/JP) in CoMP are also discussed.

Interference Cancellation

Interference has to be handled. And cancellation of interference can be a way. It can be canceled at the receiver end.The interference cancellation had been defined by Andrews.(Andrews, 2005) In this technique, desired information is got by demodulation or decoding. Then received interference is canceled from the received signal by using this information along with channel estimates. Interference cancella-

Figure 4. Different interference management techniques

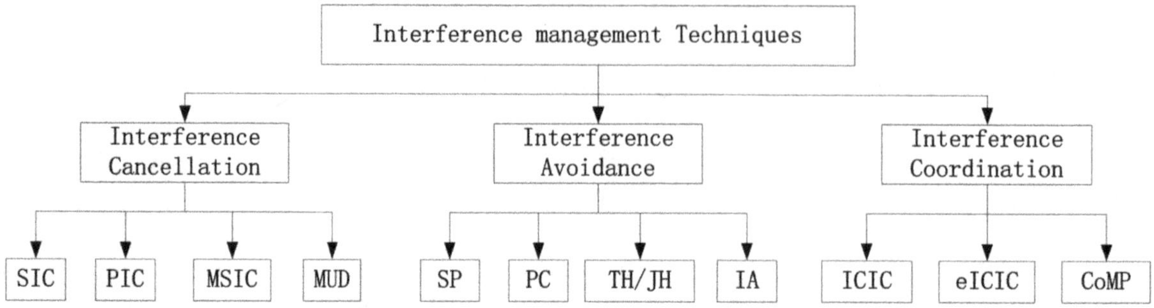

tion is extensively utilized in two classic ways, Successive Interference cancellation (SIC) and Parallel Interference Cancellation (PIC).

The obvious feature of PIC is that it detects all users at the same time. There are two steps, initial estimate and parallel detection. Initial estimate would be used to cancel interference in near future, and parallel detection is repeated all the time. This process is repeated over several stages. Multi-stage interference cancellation is another name of PIC (Frenger, Orten, & Ottosson, 2000). The performance of it is in two sides. It decreases latency as well as makes overall complexity higher. Let us make N be the number of users and P be the cancellation stages. N must be detected in parallel. We see that latency is proportional to P, but complexity is proportional to PN (Frenger, Orten, & Ottosson, 2000).

SIC has almost the same principles with PIC. But it detects one user per stage, and begins from the strongest signal, then the next strongest, and so on. The complexity and latency of SIC is proportional to N. So the latency N is much bigger than the number P of PIC. This latency is too big, so it may be unaffordable if N is too big with real time data, which means there are many users (Patel & Holtzman, 2010).

We use Multistage SIC to have better scheme than SIC and PIC. It is a tradeoff between PIC and SIC. The basic concept is as follows. Users are divided into several groups. Users are detected in parallel in each group, then the aggregate interference of each group is subtracted from the composite received signal. This is done one group followed by another.

If interference cancellation techniques are used, it is required that the knowledge of the characteristics of the interfering signal. As is known to all, the Multi-Input Multi-Output technique can mitigate interference, so antenna arrays need to be known at the receiver systems, in order to get any interference canceled. And there can be antenna arrays in base stations, so this technique is less suitable for UEs and more likely to implement in base stations (which is BS in macrocell, or FAP in femtocell) and hence mostly used for the up-link interference management. (Van, Janssen, & Prasad, 1995)

There is also another interference cancellation technique called Multi-user detection (MUD), which works on the same principle. In this technique, the signature of each user is identified. MUD observes signals to get their characteristics, then selects the most probable signal out of K number of users along with channel state information. From its principle we can see that it is something with multiple access technologies. And it is well suited for the CDMA systems. Different code is distributed to each user, and each user has a unique wave form. This will lead to the near far effect, which should be eliminated. The different CDMA users can be detected by MUD, and all the users can benefit from joint detection. (Duel-Hallen, Holtzman, & Zvonar, 1995) Decorrelating detector and linear minimum mean square error (MMSE) detectors (Honig, Madhow, & Verdu, 2010) are the common detectors, which is used to counter the multi-user interference in CDMA channels. But these detectors have their own shortages, such as they require prior knowledge about the system, such as signature wave form, received amplitudes, or some training data. And these algorithms are complex, and they are difficult to implement.

Interference Avoidance

There has been multi-tiers in heterogeneous network. As we talked before, there are two tiers in femtocell, we will take it as an example of small cell network, and to analyze these technologies. Femtocells have properties as plug and play device. So it has a nature of ad hoc. It is difficult to manage femtocells from a centralized controller as the cell planning, so femtocell need to cope with interference without any centralized controller. The FAP should be able to cope with most of the complex environment and make sure good quality of service to users. Therefore, intelligent interference avoidance technique is very important. In this section, some interference avoidance schemes proposed in recent research are summarized.

Spectrum Splitting (SP)

There are co-tier and cross-tier interference in heterogeneous network. And Spectrum Splitting (SP) is used to cope with the cross-tier interference. First, A spectrum band is divided into two portions. One portion is allocated for MUEs and the other portion is used for FUEs. But because of high cost and scarcity of spectrum, this technique would lead to reduced efficiency.

There might be very dense deployment of femtocells, in this case, it is hard to manage the cross-tier interference. it is suggested that separate portion of spectrum should be used for the femtocell operations (Karlsson, 1999). If it's OFDMA system, separate sub channels are allocated to the femtocells to mitigate the cross-tier interference. And some researchers suggested another approach (Zahir, Arshad, Nakata, & Moessner, 2013) to mitigate this kind of interference by limiting the maximum number of femtocell users in a macrocell. Given the properties of femtocells as plug and play device, so the number of femtocells in a macrocell can not be settled. And at a given time, it is hard to know the exact number of femtocells as well, so this approach is impractical.

Power Control (PC)

Power Control is an important technique in the interference avoidance. The purpose of PC is to overcome near-far effect (uplink), shadow fading and rapid shading (uplink/downlink), and get better quality of service, such as lower interference, higher capacity, call-dropping rate and congestion. So required link quality is sustained by reducing interference from other cells and UEs. In the case of dense femtocells, if the femtocells' transmit power is controlled properly, those MUEs can be protected sufficiently. A method is proposed in Han, Choi, Kim, Na, Choi and Han's work (2009), in which the indoor coverage is restricted, and the interference caused by femtocell BS is constrained too. So there should be a balance of indoor coverage of femtocell and the interference caused by femtocell BS as well. This problem turns into a mixed integer problem. As a result, it obtains maximum transmit power and operational frequency of femtocell. And the femtocell has enough indoor coverage simultaneously the interference to MUE is not sufficient. But the complexity is high and time is consuming, so it would not be feasible for implementation in a FAP. There are other researchers who give a simple technique which provides power control to femtocells to avoid causing interference to nearby macrocell users. It is based on the assumption that the femtocell takes information about macrocell users from the macrocell base station. Once the macrocell user is under interference, the nearby femtocell reduces its transmit power.

Power control plays a vital role in the interference avoidance in CDMA systems. There are two main and common power control schemes, open loop power control and closed loop power control, which can be used to mitigate uplink and downlink interference.

An uplink power control technique is taken as an example, and the downlink is an analogue. The purpose of uplink power control is to make the received signals at BS from UEs have comparable power.

In the open loop way, in CDMA system as an example, all BSs in one area send pilot signals at the same frequency. Each UE tests the powers of the pilot signal, estimates the path loss according to the test result, and determine its transmission power. Open loop power control is simple, fast and low-cost. In the range of a cell, signal fading caused by fast fading is usually more serious than that caused by propagation loss. Therefore, open loop power control is applied only at the beginning of connection setup, generally in setting the initial power value, but it's still inaccurate.For WCDMA-FDD system, the uplink fading is not related to the downlink one because of the big frequency interval of them. Therefore, the path loss and interference estimated by downlink cannot reflect the one in uplink completely. So open loop power control is suitable for mitigation of slow fading, but not of fast fading, which needs to be revised by closed loop power control.

Figure 5 shows the principle of closed loop power control, which can convergence the transmission power of uplink and downlink very fast, and decrease interference in system with higher quality of service. Outer loop power control is used to maintain the quality of communication at the level of bearer service quality requirement, while using as low power as possible. Inner loop power control, which is also called fast closed loop power control, is used to adjust UEs' uplink Power every one slot. Inner loop power control frequency is 1500Hz in WCDMA system.

In WCDMA system, NodeB calculates the Signal-to-Interference Ratio (SIR) by estimating the power strengthen and the current interference, then compare it with SIR target. If the value is less than SIR target, it needs to increase UE transmission power, otherwise to decrease transmission power. The receiver will adjust the transmission power by algorithms.

Figure 5. Inner and outer power control in uplink

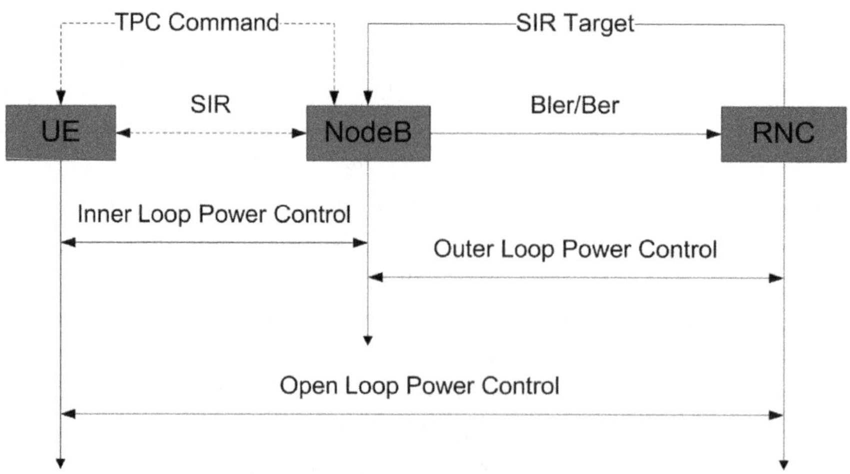

The purpose of inner loop power control of the WCDMA system is to maintain a certain signal-to-interference ratio of transmission signal power when the signals reach the receiving end. But the outer loop power control is to determine the SIR target by information from network. SIR target should be satisfied with the requirement of decoding correctly. But different multiple path radio environment request different SIR. Therefore, the outer-loop power control can adjust the SIR target to get a stable Block Error Rate(BLER) in the changeable radio environment.

In a two-tier scenario as we talked above, there is cross-tier interference. In the open loop power control (Jo, Mun, Moon, &Yook, 2009), a femtocell UE estimates the maximum allowable transmit power by estimating its cross-tier interference to the macrocell BS. The femtocell UE finds out the cross-tier interference which will cause to the macrocell BS, and adjusts its power in such a way that the cross-tier interference which is causing to the macrocell BS should be less than a predefined threshold. In the closed loop power control, femtocell UEs adjust their maximum transmit power based on two things. First is the estimate of additional cross-tier interference at the macrocell BS due to the femtocell UE, which is similar to the open loop technique. Second, the Noise and Interference (NI) level is also taken into account at the macrocell BS. The NI level is broadcast to a femtocell UE by the FAP. This NI level is calculated by macrocell BS, which informs the FAP about this through the backhaul network. It is also shown that the closed loop power control technique performs better than the open loop technique. The closed loop power control performs better at the low expense of macrocell throughput. Another solution is provided in Li, Qian and Kataria's research (2009), where a joint power control, channel management and admission control algorithm is proposed to cope with the co-channel interference in the down link. This algorithm gives higher priority to the macro UEs and provides power control solutions in both distributed and centralized Manner.

Time Hopping or Joint Hopping (TH or JH)

Time hopping is one major scheme proposed to reduce the cross-tier up-link interference. In the 3G system, Time Hopped CDMA (TH-CDMA) is used to mitigate this kind of interference. In the scheme, the transmission period is subdivided into small portions.A user transmits in one portion, and remains silent for the other portions (Elhakeem, Girolamo, Bdira, &Talla, 1993). If there is no synchronization between the two network tiers, then each network tier independently chooses its periods. For example, a CDMA transmission period P can be subdivided into N portions or hops. Each user then selects one of the hops (having length P/N) to transmit. This scheme reduces the co-tier and cross-tier interference by a factor of N (Chandrasekhar & Andrews, 2007).

A joint hopping scheme can also be implemented by all the users of one femtocell transmitting in the same time slot. Due to the averaging of aggregate interference in a CDMA system, the femtocell users would not cause interference to each other. The neighboring femtocells independently select time slots, which reduces the co-tier interference by a factor of N as well. The cross-tier uplink interference at the FAP is reduced due to the same reason of independently selecting time slots (Chandrasekhar &Andrews, 2007).

Interference Alignment (IA)

Interference Alignment(IA) (Jafar, 2010) is an effective scheme of interference management. In this technique, signal space is divided into two portions, which are desired signal subspace and interference signal subspace. Interference is overlapped at the receiver by precoding, which is compressed in its capacity as well. So the effect to desired signal is eliminated and the channel capacity increases.

IA is representative interference shaping technique. It offers better performance in the interference-limited communication regime than traditional methods. In this section, the basic mathematics thought of IA is introduced first. Then challenges and solutions of IA will be presented. At last, we talk about the application of IA in heterogeneous network.

- **Mathematics Thoughts of Interference Alignment:** IA comes from linear algebra. It can be explained by an example as follows. These are several equations.

$$\begin{aligned} y_1 &= h_{11}x_1 + h_{12}x_2 + \cdots + h_{1K}x_k \\ y_2 &= h_{21}x_1 + h_{22}x_2 + \cdots + h_{2K}x_k \\ &\vdots\vdots\vdots \\ y_B &= h_{B1}x_1 + h_{B2}x_2 + \cdots + h_{BK}x_k \end{aligned}$$

B is the number of receivers. So $y_1, y_2, \cdots y_B$ present the receiving signals. *K* is the number of transmitters. It is supposed that each transmitter just send on signal, signified as $x_1, x_2, \cdots x_K$. And h_{ij} represents the effective channel coefficients from transmitter *j* to receiver *i*.

If these equations are generic, that is to say, if these channel coefficients come from a continuous distribution, all the signals can be rebuilt with the fact that the dimension the channel could afford is at least the same with the dimension of receivers. This means that these linear equations have solutions only if the rank of channel matrix is at least the same with the number of receivers. Therefore, if all of the signals are desired signals to each receiver, *K* number of signals are needed. However, if just one subset is needed in a receiver, all the other subsets turn to be interference. For example, desired signals for some receiver are x_1, x_2, x_3, then $x_4, x_5, \cdots x_K$ will be the interference signals for this receiver.

Without loss of generality, it is supposed that one receiver only desires one signal. So the problem turns into that how we get the desired signal from K-1 interference signals. In general, we solve this kind of problem by using orthogonality of signals, i.e. each signal is in one dimension and every two dimensions are orthogonal. So as we have supposed, if there are *K* number of receivers and each receiver just has one desired signal, there will be *B* (number of receivers) number of equations with *K* number of coefficients after signals transmitted through linear channel. The aim is that we can get solutions of these equations and each receiver gets the desired signal. Thus, *K* dimensions of signals is needed to reach this purpose. This method is called cake cutting in wireless interference network. And the signal dimensions are distributed into *K* number, so each UE can only use 1/*K* channel capacity to transmit. If *K* is big, i.e. there are enormous transmitters in the system, the quality of communication would reduce.

Cake cutting is not optimal, there is possibility of get solutions in the condition that the number of linear equations is less than the number of unknown quantities, which will be talked as follows.

If x_1 is the desired signal, what should be the conditions to get it? We rewrite those equations as follow by matrix:

$$\mathrm{Y} = \mathrm{H}_{*1}x_1 + \mathrm{H}_{*2}x_2 + \cdots + \mathrm{H}_{*K}x_K$$

In which,

$$\mathbf{Y} = \begin{pmatrix} y_1 \\ y_2 \\ \vdots \\ y_B \end{pmatrix}, \mathbf{H}_{*k} = \begin{pmatrix} h_{1k} \\ h_{2k} \\ \vdots \\ h_{Bk} \end{pmatrix},$$

which present the vector of receiving signals and vector of the channel coefficients along with $x_k, k = 1, 2, \cdots K$.

Vector $\mathbf{H}_{*k}$ is called the receiving beam of x_k. That is to say, there is 1 antenna at each transmitter, and one transmitter just send 1 signal. So there are *K* number of signals received by receivers, including *K*-1 interference signals. If x_1 is what we want, $\mathbf{H}_{*1}$ would not intersect the space spanned by

$$\mathrm{H}_{*2}x_2, \mathrm{H}_{*3}x_3, \cdots, \mathrm{H}_{*K}x_K,$$

so zero forcing criteria can be used to eliminate interference and just get x_1 left. In other words, if x_1 can be solved at the receiver, if and only if

$$\mathrm{H}_{*2} \notin \mathrm{span}(\mathrm{H}_{*2}x_2, \mathrm{H}_{*3}x_3, \cdots, \mathrm{H}_{*K}x_K)$$

is established.

(*K*-1) interference beams

$$\mathrm{H}_{*2}x_2, \mathrm{H}_{*3}x_3, \cdots, \mathrm{H}_{*K}x_K$$

is contained in space of dimension of min (*B*, *K*-1). So when *B*<*K*, the interference space is in dimension of *B*. Since *B* is the number of receivers, so under the assured condition of independent receivers, the rank of coefficients matrix is *B*, which equals to the dimension of interference space. As a result, the desired signal is in the interference space. However, if interference can be merged into a smaller subset, and this subset is small enough to have not any intersection with the space spanned by desired signals, the receiver can recover its desired signal. This is the thought of IA. An example is given as follows to explain in detail. These are some equations,

$$\begin{aligned} y_1 &= 3x_1 + 2x_2 + 3x_3 + x_4 + 5x_5 \\ y_2 &= 2x_1 + 4x_2 + x_3 - 3x_4 + 5x_5 \\ y_3 &= 4x_1 + 3x_2 + 5x_3 + 2x_4 + 8x_5 \end{aligned}$$

In this example, K=5, B=3, the dimension is min(B,K-1)=3, and there are 5 unknown quantities. We need 5 number of independent equations to solve these five unknowns. But for these particular equations, x_1 can be solved even the number of equations is less than the number of unknowns.

The dimension of interference space is just 2. The reason is as follows.

$$\mathbf{H}_{*5} = \begin{pmatrix} 5 \\ 5 \\ 8 \end{pmatrix} \mathbf{H}_{*4} = \begin{pmatrix} 1 \\ -3 \\ 2 \end{pmatrix} \mathbf{H}_{*3} = \begin{pmatrix} 3 \\ 1 \\ 5 \end{pmatrix} \mathbf{H}_{*2} = \begin{pmatrix} 2 \\ 4 \\ 3 \end{pmatrix}$$

So $\mathbf{H}_{*4} = \mathbf{H}_{*3} - \mathbf{H}_{*2}$, $\mathbf{H}_{*5} = \mathbf{H}_{*3} + \mathbf{H}_{*2}$. We get $Rank(\mathbf{H}_{*1} \quad \mathbf{H}_{*2} \quad \mathbf{H}_{*3}) = 3$. This is to say, they are linear independent, $\mathbf{H}_{*2}$ and $\mathbf{H}_{*3}$ don't interfere $\mathbf{H}_{*1}$. So desired signal x_1 can be projected into this interference-free dimension to transmit and solve at last. It is noticed that $\mathrm{U} = \begin{bmatrix} 17 & -1 & -10 \end{bmatrix}^{\mathrm{T}}$ is orthogonal to all of the interference signals, so $\mathbf{U}^{\mathrm{T}}Y$ can eliminate interference and $17y_1 - y_2 - 10y_3 = 9x_1$ is got. So the x_1 is solved. This can be explained in linear algebra. If B<K, x_1 is in one dimension. (B-1) signals are in independent other (B-1) dimensions, which are part the remaining (K-1) interference signals. And the last (K-B) signals are in the spanned space by the (B-1) signals.

- **Challenges and Solutions of Interference Alignment:** This subsection explores the challenges faced by IA schemes and the ingenious solutions that get around those challenges. Two main issues faced by interference alignment schemes are:

First, when the number of interfering users is increased, the number of alignment constraints would grow very rapidly. For example, in a K user interference channel, each of the K receivers needs an alignment of (K−1) interfering signal spaces, for a total of O(K^2) signal space alignment constraints. Since there are only K signal spaces (one at each transmitter) to be chosen in order to satisfy O(K^2) signal space alignment constraints, the problem can quickly appear infeasible.

Second, The diversity of channels which enables the relativity of alignment is often a limiting factor. Further issues to be dealt with by interference alignment schemes include the imperfect, noisy, localized and possibly delayed nature of channel knowledge feedback to the transmitters where such knowledge is crucial to achieve interference alignment. These challenges have given rise to ingenious solutions in the form of new interference alignment schemes, invariably providing new and surprising insights into the nontrivial interplay of signaling dimensions in an interference network.

The earliest application of interference alignment (Birk & Kol, 2006), that we are aware of, appears in Example 7 of the 1998 INFOCOM paper of Birk and Kol (1998) in the context of the Informed Source Coding on Demand (ISCOD) problem (also known as the *index coding* problem). The problem formulation, introduced in the same paper, can correspond to a variety of settings.

- **Linear Interference Alignment:** I.e., alignment of signal spaces based on linear precoding (beamforming) schemes is the simplest form of interference alignment. Each space of receiving signal space can be divided into desired signal space and interference signal space by beamforming on the space spanned by antennas of transmitters and receivers. E.g., this purpose can not be achieved if there's only one antenna on each node because the space is not big enough to execute IA schemes. The maximum degree of freedom (DoF) is an integer by beamforming. Since beamforming schemes are common in existing MIMO point to point, broadcast (BC) and multiple access (MAC) networks, the maximum DoFs are all integers. In the condition of two-user MIMO MAC channel,the DoF is min(M,N), in which M and N are numbers of antennas at the transmitter and receiver respectively. The DoF is going to be min(M1+M2,N), if there are M1 and M2 antennas at transmitter with others staying equal. And DoF will change to min(M,N1+N2), if there are N1 and N2 antennas at the receiver and M antennas at the transmitter. In last condition, if M1 and M2 are antennas at the transmitter, and N1 and N2 are the numbers of antennas at the receiver, the DoF turns to min(M1+M2,N1+N2,max(M1,N2),max(M2,N1)).

The idea of using symbol extensions, i.e., beamforming across multiple channel uses, was introduced in Jafar and Shamai's paper (2006) for the two user MIMO X channel where all nodes are equipped with $M > 1$ antennas. Symbol extensions (Jafar & Shamai, 2006) enlarged the achievable DoF of the two user MIMO X channel from the $\lfloor (4/3)M \rfloor$ DoF, which is previously achieved by Maddah-Ali, Motahari and Khandaniv (2008) to 4M/3, which also matched the outer bound for this class of channels.

Since channel fades are drawn from a continuous distribution which makes the probability of any given channel state equal to zero, Nazer, Gastpar, Jafar and Vishwanath (2011) first quantize the channel states and then use a strong typicality argument to show that if the channel fading process is ergodic, then almost all quantized states are matched with complementary states, i.e., the fraction of unmatched states approaches zero. Because of the reliance on the ergodic fading property, this approach is called ergodic interference alignment.

The ergodic interference alignment scheme is an opportunistic scheme that exploits the existence of complementary channel states in equal proportions to achieve linear interference alignment. While this assumption applies to a broad variety of channel distributions, including widely studied Rayleigh fading models, it is not universally applicable. Arbitrary channel distributions, or even standard ones like Rician fading, do not satisfy the symmetric phase assumptions made by ergodic interference alignment.

The idea of rationally independent dimensions is intuitively quite similar to the idea of linear independence of vectors. A set of vectors is considered linearly independent if no nontrivial linear combination of the vectors can produce the zero vector. Similarly, a set of scalars is considered to be rationally independent if no nontrivial rational combination of the scalars. —i.e., a linear combination with only rational coefficients at least one of which must be nonzero — can add up to zero. Just as signals received along linearly independent vectors are separable, Motahari et al. showed that superimposed lattices scaled by rationally independent scalars were also separable.

Recent theoretical advances advocate highly sophisticated structured codes while the achievable rates considered practical draw largely on basic random coding arguments. Perhaps the strongest case for structured codes is the idea of lattice alignment. The idea of lattice alignment is introduced by Bresler, Parekh and Tse (2010) for the many-to-one interference channel where only one receiver experiences interference, and is shown to achieve the capacity of this channel within a constant number of bits/sec/Hz (regardless of SNR and channel parameters). The layering of information in signal levels is an essential aspect of this problem because of the different strengths of signals arriving from each interferer to the interfered receiver.

Blind interference alignment is introduced in Jafar's paper (2010) as the idea that even if the actual channel coefficient values are entirely unknown to the transmitters, interference alignment may be possible based on the knowledge of the distinct autocorrelation properties of the channels seen by different receivers.

Retrospective interference alignment is introduced by Maddah-Ali and Tse (2010) and refers to interference alignment schemes that exploit only delayed channel state information at the transmitter (CSIT). The delayed CSIT is assumed independent of the current channel state. The general intuition prior to the Maddah-Ali and Tse's work (2010) was that channel knowledge is only useful to the transmitter to the extent that it helps learn the current channel state. It is therefore surprising that even delayed CSIT that is independent of the current channel state is, not only useful, but so useful that it can increase the available DoF.

- **Interference Align Application in Heterogeneous Networks (HetNets):** The interference alignment (IA) is a revolutionary cooperative technique to reduce interference by aligning multiple interfering signals within a reduced dimensional subspace at each receiver and leaving the desired signal to transmit in an interference-free subspace. (Peng, Wang, Li, Xiang, &Lau, 2015) It was firstly proposed and proven to achieve the optimal degrees of freedom at high SINR in Cadambe and Jafar's (2008) paper. Furthermore, another paper (Gomadam, Cadambe, &Jafar, 2011) studied the distributed iterative solutions to find IA precoders and proposed a max-SINR algorithm which approaches almost the same performance gains with regards to baselines as the high SINR regime at the low SINR regime.The underlay HetNet consists of one High Power Node (HPN) and two Low Power Nodes (LPNs) with four and two antennas, respectively, and each UE is equipped with two antennas. The LPN serves one UE per cell and the HPN serves two UEs simultaneously. To mitigate both the intertier interference and the inra-tier interference by aligning the corresponding beamforming matrices to the nullspace of each other's effective channels, the beamforming matrices for LPNs and HPN are successively determined.

The involved signal dimensions in IA design are provided by symbol extension (time or frequency domain) or by multiple antennas (spatial domain) (Zhao, Peng, Ding, &Wang, 2014). However, IA requires perfect Channel State Information (CSI) knowledge and tight synchronization between cooperative transmitters, which are non-trivial, especially in the context of underlay HetNets. Substantial gains by IA over the non-cooperative transmission strategies, such as time division multiple access, are observed at low speeds, which is benefited by the relatively accurate CSI acquired with a low overhead cost at the low mobility speed. It is also indicated that in a high mobility configuration, where accurate CSI acquisition is costly, a lower overhead strategy or coordinate within a smaller UE group may be preferable.

Apparently, the challenge of obtaining accurate CSI is critical to the application of IA. With the consideration of overwhelming signaling overhead and the accuracy of required CSI, several types of solutions are considered in the published literature. The iterative IA algorithm in Gomadam, Cadambe and Jafar's paper (2011) is realized with the reciprocity of time division duplex (TDD). Nevertheless, the reciprocity may incur non-negligible pilot transmission overhead and does not hold in the frequency division duplex (FDD) mode. An analog feedback for IA, which directly transmits the channel coefficients as uncoded quadrature amplitude modulation(QAM) symbols, was researched by Ayach and Heath (2012). However, the analog CSI feedback quality is naturally influenced by the SINR of the reverse links.

Motivated by the interesting fact that the IA algorithm may not always require full knowledge of CSI to achieve cooperative IA interference nulling, a dynamic feedback profile was proposed, a novel IA precoder/decorrelator design was developed, and new IA feasibility conditions were established by Rao, Ruan and Lau (2013a). Both analytical and simulation results showed that the proposed solution could significantly reduce costs of the CSI feedback for the IA under the same DoF requirement.

Another interesting fact is that the general heterogeneity in terms of power, pathloss, and the spatial correlations of multiple antennas will make the interference topology asymmetric. Therefore, not all interfering links should be treated equally. To this end, Rao, Ruan and Lau (2013b) tried to exploit the potential benefits brought by the asymmetric interference topology to improve the SE performance with limited feedback.

Inspired by the substantial theoretical gain in interference channels, there have been increasing efforts recently on making IA a feasible cooperative interference control technique for underlay HetNets to deal with the rising severe inter-tier and intra-tier interference (Shin, Noh, Jang, &Choi, 2012; Nguyen, Jeong, Quek, Tay, &Shin, 2013; Guler &Yener, 2014). Hierarchical interference alignment (HIA) with multiple antennas was firstly proposed by Shin, Noh, Jang and Choi (2012) to mitigate the inter-tier interference in underlay HetNets under the closed and open access modes. Motivated by the fact that LPNs are usually configured with much fewer transmit antennas than HPNs, the closed-form IA precoder/decorrelator was calculated with the strategy of a two-stage beamforming design for LPNs and HPNs instead of iterative computations, like in Gomadam, Cadambe and Jafar's paper (2011). The multi-cell simulations by Shin, Noh, Jang and Choi (2012) indicated that an appropriate HIA mode should be chosen adaptively according to the bias value to maximize the HIA gain for underlay HetNets.

An opportunistic interference alignment was proposed by Nguyen, Jeong, Quek, Tay and Shin (2013) to mitigate the intra-tier and inter-tier interference in a two-tier MIMO network, where a random spectrum allocation for LPNs was applied on top of the opportunistic interference alignment. To further revaluate the effect of challenges met by underlay HetNets on the IA performance, Nguyen, Jeong, Quek, Tay and Shin (2013) utilized stochastic geometry to analyze the proposed IA scheme for underlay HetNets in terms of the distribution of received SINR, spatial average capacity, and Energy Efficiency (EE) performance in the presence of imperfect CSI configurations. With the merits of exploiting cognition and coordination between tiers, a novel inter-tier interference control solution called selective IA was presented by Guler and Yener (2014) for two-tier underlay HetNets, where the cognitive LPNs sense their environment to judiciously choose the subset of interfered receivers and design their precoders using IA to avoid inter-tier interference causing performance degradation to nearby receivers.

Lots of challenges still remain for the application and development of IA in underlay HetNets, including the general precoding algorithm design and the robustness guarantee for the arbitrary number of UEs whenever non-ideal CSI and limited feedback exist.

Interference Coordination

Inter Cell Interference Coordination (ICIC)

Inter Cell Interference Coordination (ICIC) (Academic group of Baidu, 2016) is used to mitigate inter cell Interference in co frequency network. OFDM is used in LTE system, in which high-speed data is modulated to orthogonal sub-channels to reduce Inter Channel Interference (ICI) effectively. But this orthogonality is limited to UEs in current cell. Users of different cells would interfere with each other, especially for those ones at the edge of a cell in a co frequency network. ICIC technique is utilized to solve this problem.

The purpose of ICIC is to improve the spectrum efficiency of UEs at the edge of a cell with unchangeable system throughput. The essential thought of ICIC is to control inter cell interference by radio resource management, which is restricted in each cell and includes used time/frequency restriction or sending power restriction.

There are Fractional Frequency Reuse (Fractional FR), Soft Frequency Reuse(SFR), Full Frequency Reuse (Full FR) in static ICIC and dynamic ICIC.

- **Fractional Frequency Reuse (Fractional FR):** The frequency band is divided into two reuse portions. One portion is a frequency assemble with Frequency Reuse Factor(FRF) 1, which is for centric users. And another assemble with FRF greater than 1, which is for cell-edge users.

The system band is divided into 4 portions which is shown in Figure 5. The FRF of centric UE is 1, and FRF of cell-edge UE is 3. Different patterns represent different cell edges. So UEs use different frequencies in different area of the cell, which is shown in Figure 6. So it is sure that UEs at the edge use different frequency to avoid inter-cell interference.

Figure 6. Diagram of FFR

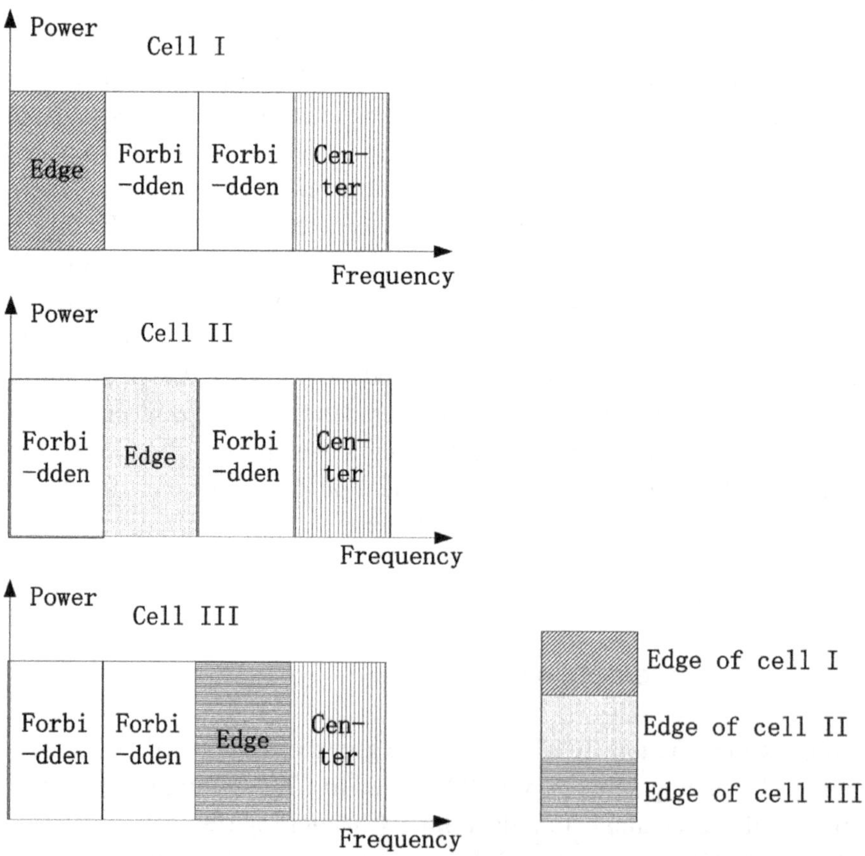

Figure 7. Frequency allocation in one cell of FFR

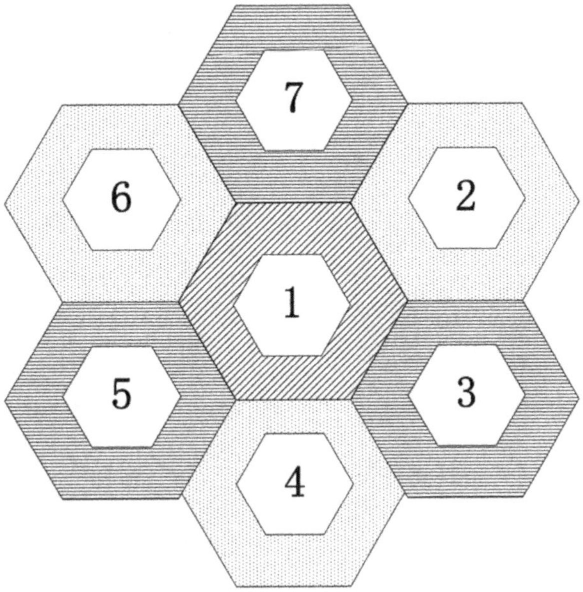

Uplink and downlink Fractional FR is the same by BS scheduling. But one sub-band is used by all cells in the same power (this is to say FRF is 1), meanwhile other power of sub-bands is coordinated in adjacent cells, so there is a low-interference sub-band in each cell, which is used for cell-edge UEs. Fractional FR scheme can be classified into Fractional FR Full Illustration (Fractional FRFI) and Part Illustration (Fractional FRPI).

Fractional FRFI scheme suggests no data sending at the edge of adjacent cells, which can mitigate inter-cell interference by reducing available resources and be shown in Figure 8. However all the cells can use all the bandwidth in Fractional FRPI. So the available resources is the same with the situation of FRF 1, and this is similar to the Soft FR(SFR) which will be discussed later.

In Fractional FRPI scheme, the general sending power is the same, but the quality is guaranteed by raising the power in the sub-band for cell-edge UEs and reducing the power in other sub-bands. This is demonstrated in Figure 8. The proportion of fractional frequency reuse must be proper, even so some frequency resource is not intelligently used. Thus different cells have different proportions of fractional frequency reuse according to their situation. But services change by time, the proportion should be changed according to services. If user load is high, the problem becomes more difficult.

- **Soft Frequency Reuse:** The band is divided into 3 portions in Soft Frequency Reuse(SFR), as is shown in Figure 9. In a scenario of 3 cells, the cell-edge UEs use one portion, and centric UEs use the left 2 portions. So the FRF is 3/2 for centric UEs as well as 3 for cell-edge UEs.

For example, we emblem the 3 portions with A,B and C. In Cell I, A+B is allocated to centric UEs. Some part of C is for cell-edge UEs, and others of C can be assigned flexibly. SFR combined with power control can mitigate interference by adjusting powers of some sub-bands, which is shown in Figure 10.

Figure 8. Fractional FR

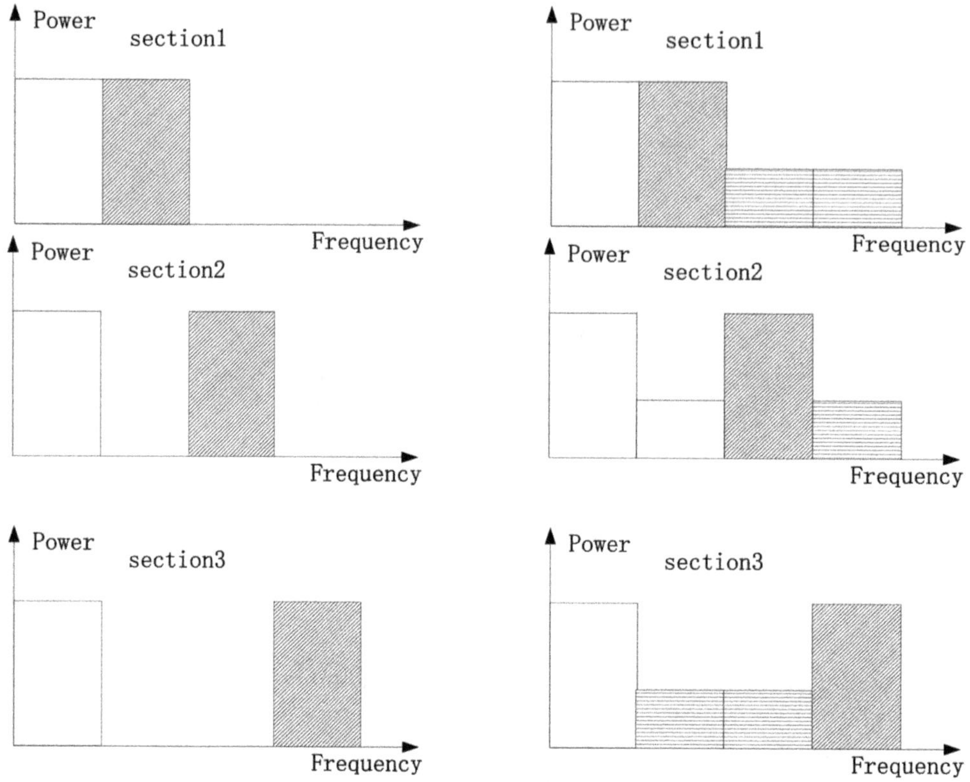

Figure 9. Diagram of SFR

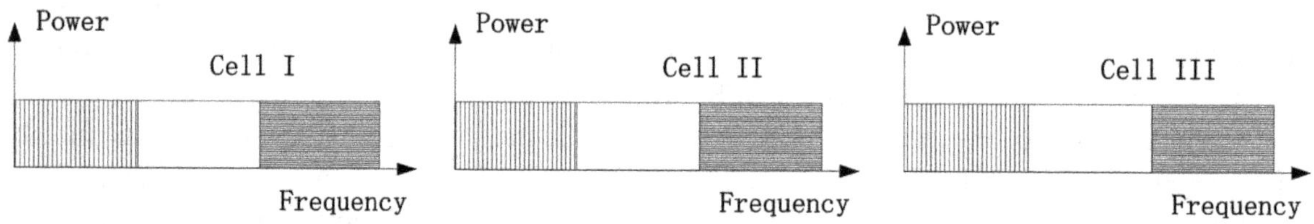

Figure 10. Band division of SFR I

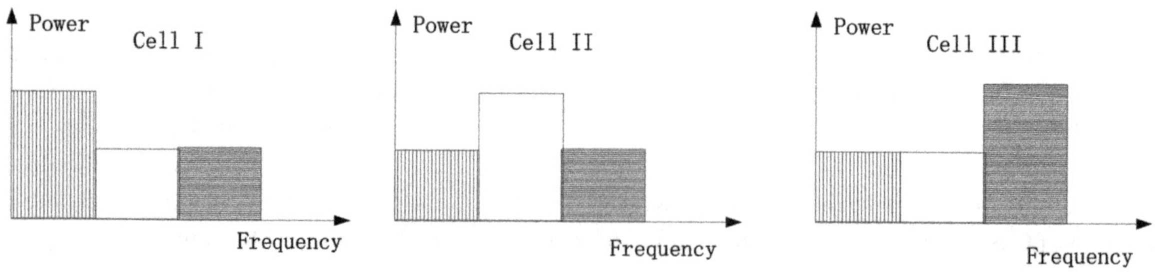

The power is high in the edge of cell, because of higher path loss for cell-edge UEs. And the sub-bands for adjacent cells are orthogonal, which can mitigate interference. The power is low in the center of cell, because of less path loss for centric UEs. The interference is low even they use the same sub-band. The power can be high in the center of cell to increase transmit rate by high-order modulation. Because centric UEs are far away from adjacent cell, and the sub-band is orthogonal to each other which leads to lower interference to adjacent cell. If the cell-edge sub-bands can be used, they can be assigned to centric UEs. This is shown in Figure 11. SFR combined with power control, it is sure that the cell-edge UEs at adjacent cells use different frequencies, in order to avoid inter-cell interference.

Compared to normal FR schemes, the power of some sub-bands just partially reduces in SFR, but is not totally restricted. So the sub-band power needs to adjust. To sum up, there are two different power assignments in SFR scheme: one is that sending power reduces on some given sub-bands with equal full power on other sub-bands; the other is that full power is sent on some given sub-bands with reduced power on other sub-bands. Both of them are shown in Figure 12. In general, SFR is similar to the scheme with FRF equaling 1, both of them can use all the available frequency resources. But the average Signal to Interference plus Noise Ratio (SINR) is not totally the same on different sub-bands of SFR. The average SINR stays still or increases if full power is sent on sub-band, otherwise SINR will reduce if reduced power is sent on sub-band.

- **Full Frequency Reuse:** Physical Resource Block (PRB) is the essential unit in Full Frequency Reuse (Full FR), which is different from Fractional FR and SFR. Fractional FR and SFR A group of successive PRB is used to allocate resource and restrict sending power in. But it is different in Fractional FR and SFR. PRB is scheduled and power restricted separately to keep cell-edge UEs away from severe bad impact caused by high power interference in Full FR. In Full FR scheme, there is no difference of IC and OC, but high-interference PRB indication is measured to coordinately schedule PRB in eNodeB. The system can use frequency resource of a cell, which means the FRF is 1.
- **Dynamic ICIC:** Dynamic ICIC is realized by sharing information between cells. In this scheme, indications are sent between adjacent cells to control inter-cell interference.

There are two schemes of dynamic ICIC in uplink of LTE. The first one, cells send High Interference Indication (HII) to adjacent cells, so adjacent cells can use proper resource scheduling to avoid interference. The second one,when the adjacent cells receive Overload Indication(OI), some kind of mitigation of interference will be executed, such as reduction of sending power. This is shown in Figure 13.

Figure 11. Band allocation in cells

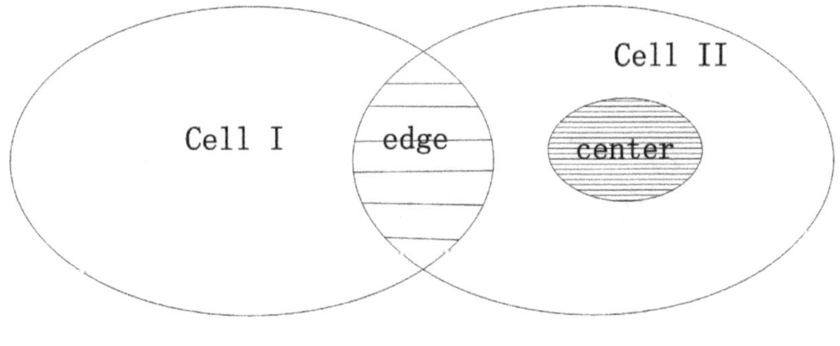

Figure 12. diagram of SFR II

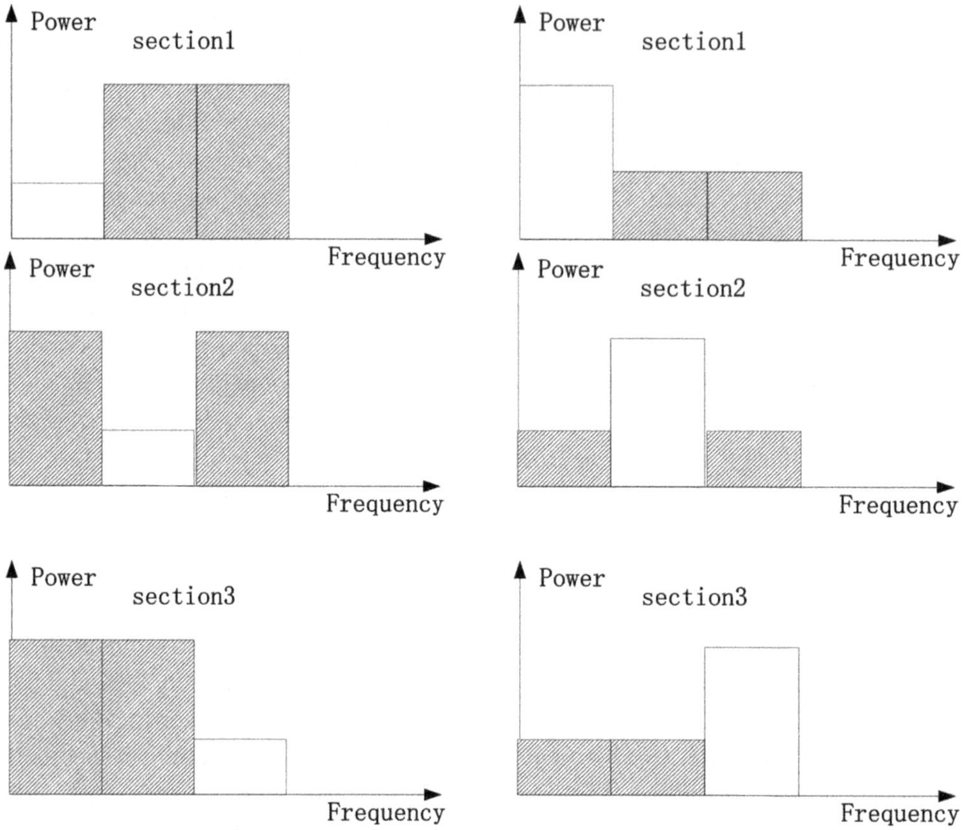

HII is scheduling information from cell-edge UEs of source cell. It used to indicate the RBs which will be allocated to cell-edge UEs. These RBs will interfere adjacent eNodeB seriously. And HII is very sensitive to the interference from adjacent eNodeB.The cell which receives HII will know the adjacent cell and will produce strong interference by monitoring HII from it. The receiver will schedule RBs to avoid interference or adjust power. In the low-load scenario, avoidance can be performed.

Figure 13. Sending OI through interface X2

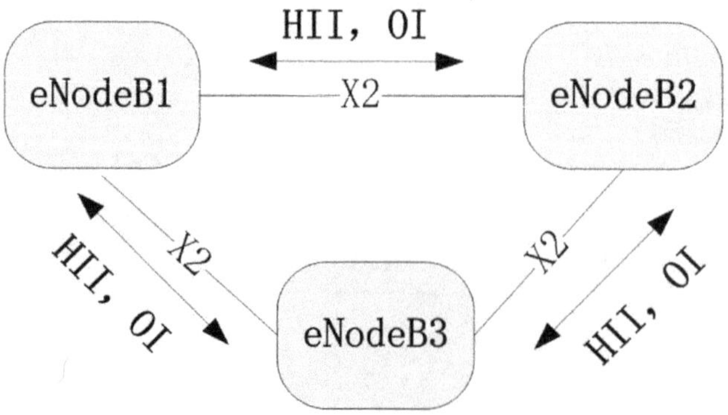

OI is based on measuring information, which is used to indicate serious interference on some sub-bands in the source cell. In the cell which receives OI, sending power will reduce according to the hierarchy of OI to mitigate inter-cell interference.

In the downlink transmission of LTE, the downlink interference coordination is realized by sending Relative Narrowband Transmission Power(RNTP) through X2 interface. RNTP is used to tell adjacent cell if the send power of RB is lower than a fixed upper threshold. In this way, adjacent cell will take interference intensity into consideration while scheduling UEs.

Enhanced ICIC (eICIC)

Enhanced ICIC (eICIC) is in time domain for co frequency inter-cell interference coordination, while ICIC is in frequency domain. In ICIC, frequency is divided into sub-bands to reduce inter-cell interference. This is not suitable for heterogeneous network.

The main problem in Heterogeneous Networks (HetNets) is the co-frequency interference between different access points (APs) in the situation of same coverage of different networks. Because APs in piconets and Macronets use same carrier, and the downlink power of Macro BS is much bigger than that of Local Packet switched Network(LPN), which will seriously interfere the downlink reception of cell-edge UEs in LPN. And the big power of UEs at the edge of Macro BS will interfere the uplink transmission of nearby LPN.

eICIC technique can effectively solve the interference problem between different types of base stations (Macro BS, fetmo BS, pico BS) in HetNets. It can increase LTE-A system throughput and the effectiveness of the whole Hetnet. Time is considered in eICIC, and some UEs of different cells are orthogonal to each other in time domain. Compared to ICIC technique, eICIC can reduce interference of control channels in different cells, not just of traffic channels. That is to say, ICIC coordinates inter-cell interference in frequency domain with power control for traffic channels, as well as eICIC coordinates inter-cell interference in both time and frequency domains with power control for traffic channels and control channels.

There are two key techniques in eICIC. One is called Almost Blank Subframe (ABS), and the other is named Cell Range Expansion(CRE). The former will be emphatically introduced later.

- **Almost Blank Subframe (ABS):** Almost Blank Subframe (ABS) (Cierny, Wang, Wichman, Ding, &Wijting, 2013; Koutlia, Perez-Romero, &Agusti, 2015) technique is mainly used to mitigate interference in control channels. If there is signal on Physical Downlink Control Channel (PDCCH) of interference cell in every sub-frame. These signals will cause words-drop of UEs in the interfered cell. If ABS sub-frames are deployed first, power of signals will reduce or even to be zero on some physical channel of ABS sub-frames in interference cell. So the interference of UEs in interfered cell can be avoided on PDCCH and Physical Downlink Shared Channel(PDSCH). These UEs will decode PDCCH and transmit data on those ABS sub-frames deployed by interference cell. So interference is avoided effectively. Some researches of interference mitigation have been done in LTE-A. And these schemes includes:
    - First, total different frequencies, in which macro BS and LPN are networking in totally different frequencies, so there is hardly any interference between adjacent cells.
    - Second, based on Carrier Aggregation (CA), in which interference of control signals from different cells can be avoided by cross carrier scheduling.

    ◦ Third, based on non-CA, in which control channels from two different types of APs are orthogonalized by time/frequency division, or are partially orthogonalized by other methods, in order to solve the inter-cell interference.

- **eICIC Based on Carrier Aggregation:** Cross carrier scheduling is introduced in LTE-A, the inter-cell interference of control information problem can be solved by this feature in Hetnet. Different carriers are adopted to carry control information in macro and pico cells. Traffics are sent on other carriers which is deployed by cross carrier scheduling. Thus the interference is avoided for control information in different cells. As is shown in Figure 14, f_1 and f_2 are aggregated, f_1 is taken as a main carrier to carry control information in macro cells, and deploys traffic in assistant carrier f_2 by cross carrier scheduling. The PDDCH interference problem can be abstained because carriers carrying control information in macro cells are different to that in pico cells.
- **eICIC based on non-CA:** In the scenario of non-CA, the inter-cell interference can be solved in time, frequency or space domain. In time domain, interference of downlink control channels can be mitigated by transmitting information at different time slot in different types of cells. This can at least reduce interference from one cell to another one. But this method requires time synchronization. In frequency domain, control signals are sent in different frequency bands for different cells to mitigate interference. This needs time synchronization as well. Macro cells deploy one or a few sub-frames as ABS. Micro/Femto/Pico cells offer services in ABS sub-frames to cell-edge UEs. The interference from macro cells are avoided, so the speed of cell-edge UEs increases, as shown in Figure 15.
- **Cell Range Expansion(CRE):** The idea of CRE has emerged as a remedy to the problem of load imbalance in the downlink. It aims to increase the downlink coverage footprint of low power BSs by adding a positive bias to their signal strengths. Such BSs are referred to as biased BSs. This biasing allows more users to associate with low power or biased BSs and thereby achieve a better cell load balancing. Nevertheless, such off-loaded users may experience unfavorable channel from the biased BSs and strong interference from the unbiased high-power BSs. (Guvenc, 2011)

Figure 14. Cross carrier scheduling in HetNet

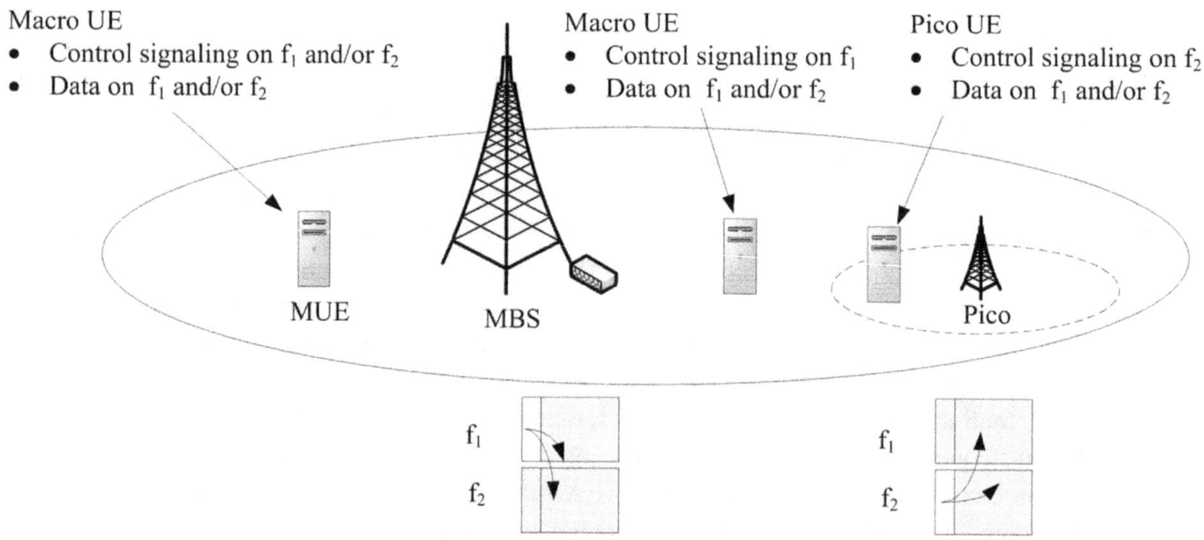

Figure 15. eICIC based on non-CA

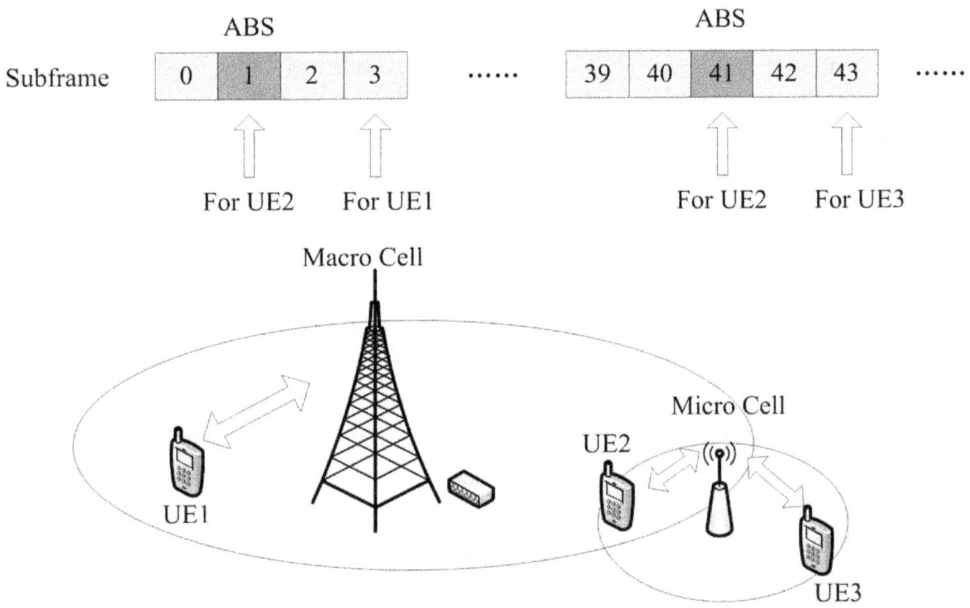

Cell Range Expansion (CRE) is one kind of techniques of eICIC as well. System sets a Bias for UE to access to macro BS when UE is choosing macro or micro BS to access. UE gets into macro BS only when the difference of the signal intensity of macro BS from that of micro BS is more than the bias. In this way the system can extend the coverage of micro BS. CRE aims at UEs accessing to micro BS as far as possible. Therefore the trade-off between cell load balancing and system throughput strictly depends on the bias values, which need to be optimized in order to maximize the system utility, obtain maximum CRE gain and get better network coverage and flow performance combing with ABS interference elimination technique.

Coordinated Multi-Point Transmission (CoMP)

For efficient utilization of limited resources, reuse is one of the concept that is being used by many specifications of cellular wireless communication systems. Along with this, densification of the network is one of the key aspect for improving traffic capacity and user throughput. So with the introduction of reuse and densification concept, there will be an additional enhancement in terms of efficient load sharing between macro cells and local access networks. But all these advantages have come up with a problem that the density and load of the network have increased considerably, correspondingly receiver terminals in the network suffer from increased co-channel interference, mainly at the boundaries of cells.

In 3GPP LTE-Advanced, Release 10 and 11, through probability readings, it was realized that there was a space for additional performance improvement at the cell edges with the help of synchronized transmission among multiple transmitters dispersed over different cell sites. For calibrating the development, some typical coordinated multipoint schemes, like coordinate scheduling, coordinated beam forming, dynamic point selection, and joint transmission were normally conferred (Nam, Bai, Lee, &Kang, 2014).

Coordinated Scheduling/ Beamforming

Coordinated Scheduling (CS) is a Coordinated Multi-Point Transmission and Reception (CoMP) technique that allows several eNodeBs (eNBs) to coordinate service to a set of UEs: by deciding who addresses whom and using which Resource Blocks (RB), pairs of cell-UEs transmissions can be scheduled concurrently with a tolerable increase in interference, thus maximizing the benefits of spatial spectrum reuse (Haile, Mutafungwa, & Hämäläinen, 2015); Lee et al., 2012). Coordinated eNBs share the scheduling information to serve their UEs in Coordinated Scheduling/beamforming. Resource will not be assigned to two UEs simultaneously which have severe interference in Coordinated Scheduling by proper scheduling algorithm.

Cells can be coordinated in both a distributed and a centralized architecture. The former relies on eNBs running independent algorithms and sharing information through peer-to-peer inter-eNB connections. This approach may suffer from limited state visibility (i.e., each eNB only possesses partial information on the state of the network, and especially of neighboring cells, hence makes sub-optimal decisions) and may entail higher inter-eNB communication latencies. Centralized coordination, instead, can leverage cloud-based architectures, such as Cloud Radio Access Network (C-RAN) (China Mobile Research Institute, 2011). This makes it possible to exploit network-wide information to make better decisions, provided that the computational overhead does not become a bottleneck itself.

Coordinating a (possibly large) number of cells entails deciding which cells are active on which RB, to target which UE, so as to minimize the interference and increase the overall throughput. In order to do this, the system needs to be able to assess the effect of interference of subsets of cells on single UEs. The main problem with this approach is scale: UE channel reporting is limited in practice, and an eNB can only be expected to be made aware of the interference of a few (e.g., one or two) neighboring cells by each UE. Even though increasing the coordination scale is likely to yield diminishing returns in the long run, the scale at which coordinated scheduling is beneficial goes beyond these figures.

In Coordinated Beamforming scheme, UEs use the same resource with each other, but they are in different beams to avoid interference. Beamforming is a kind of signal pre-processing technique based on antenna aerial array. Array gain is obtained by weighting coefficients of each array element getting adjusted. It has obvious superiority at extending coverage, improving cell-edge UEs' throughput and mitigating interference. (Peng, Wang, Li, Xiang, & Lau, 2015)

As shown in Figure 16, the data traffic for CS/CB is only transmitted by the serving BS, but the corresponding beamformer is jointly calculated in a coordination cluster to control the interference to all other UEs within the CoMP cluster. Numerical results suggested that the coverage and throughput of CoMP with zero-forcing beamformers (ZFBF) fall almost linearly with increasing average backhaul delay. When the time delay grows to over 60% of the fading coherence time, the proposed scheme does not bring any gain (Peng, Wang, Li, Xiang, & Lau, 2015).

Joint Processing (Dynamic Point Selection and Joint Transmission)

Data, which are for one UE and transmitted from several cells, are shared in each CoMP cooperation set. Interference becomes part of the desired signals, so the receiving SINR increases for the cell-edge UEs. But the cooperated cells need to share UEs' channel and data information. Joint Transmission and Dynamic Point Selection (DPS, which is also called Dynamic Points Selection DPS) are both included in Joint Processing, which demands shared UEs' information and short time delay. So JP is suitable for those scenarios of low time delay with high capacity. The data traffic is jointly transmitted for JT by

Figure 16. Diagram of CS/CB

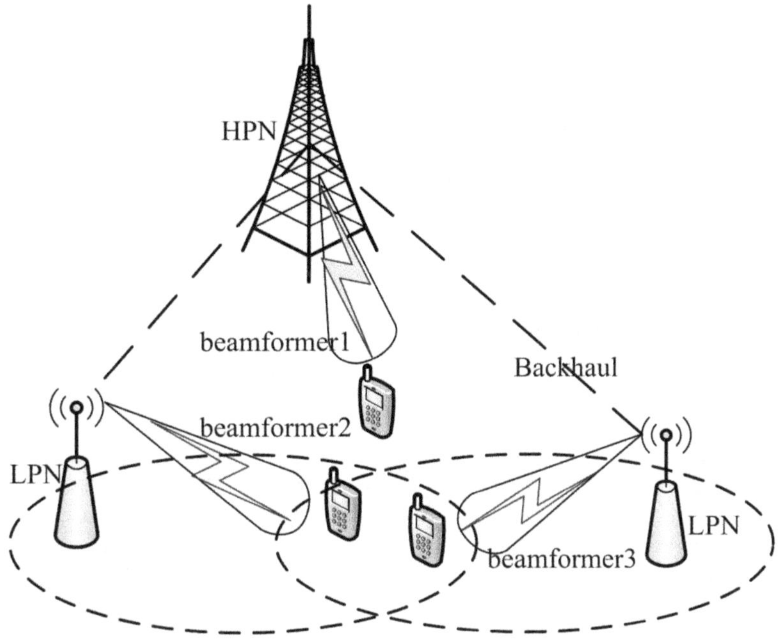

all transmission nodes within the same CoMP cluster; therefore the traffic payload is available at each distributed BS through capacity-limited backhaul links. The diagram of JP is shown in Figure 16. (Peng, Wang, Li, Xiang, & Lau, 2015)

When the JT technique is applied in downlink transmission, the user data for each UE needs to be shared among multiple BSs through the capacity-limited backhaul. If the number of cooperative BSs is large, this data exchange will result in a huge backhaul signaling overhead. (Peng, Wang, Li, Xiang, & Lau, 2015)

Joint of JT and CS/CB

To minimize the backhaul consumption without the loss of exploiting the strong interference and maintain the required QoS for the desired UEs, it is crucial to jointly design the transmit beamformers and resource allocation, which can be categorized as a non-deterministic polynomial time hard (NP-hard) problem. By increasing the number of antennas configured in each cell, the backhaul overhead consumption can be reduced by 66.7% compared to the full cooperation scenario, which is contributed by more spatial degrees of freedom with more antennas at LPNs and a proper choice of beamformers to reduce the inter-cell interference.

The performance gain of inter-CoMP highly depends on perfect knowledge of the channel state information (CSI) and tight synchronization, both of which pose strict restrictions on the backhaul. The non-coherent JT (Peng & Wang, 2010), without prior phase-mismatch correction and tight synchronization across cells, is particularly appealing due to the low complexity and small overhead. Nevertheless, a general analysis of the non-coherent JT is difficult, mostly due to the lack of tractable models. Tanbourgi, Singh, Andrews and Jondral (2014) addressed this gap and presented a tractable model to analyze the

Figure 17. Diagram of JT

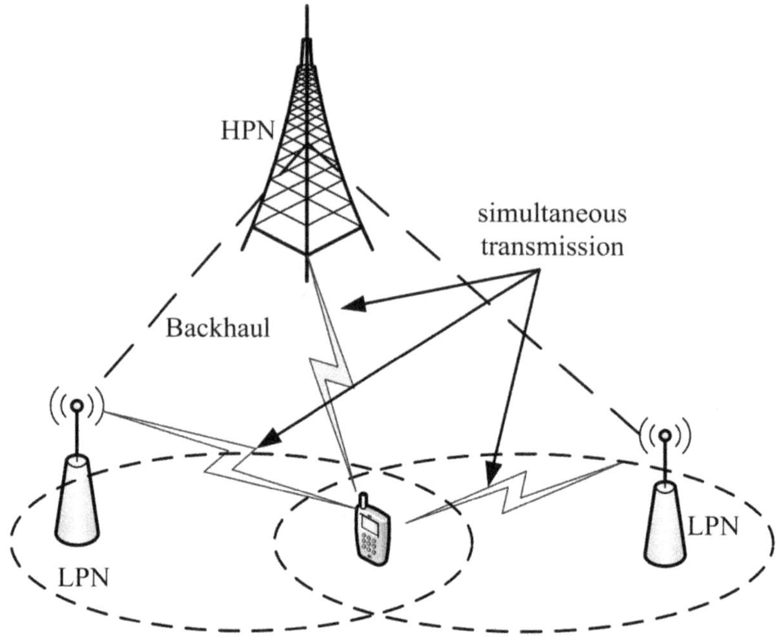

non-coherent JT, incorporating cooperation aspects of practical importance, such as user-centric clustering and channel-dependent cooperation activation.

To handle the limited backhaul impact on the information exchange of the JT scheme, the inter-CoMP CS/CB scheme has adopted, in which the Channel Direction information (CDI) and codebook are essential for the beamforming in cells. Considering that the fixed size of the codebook cannot be adaptive to the time-varied radio channels, the CDI and codebook adaptive strategy was proposed by Yuan, Yang, Wang and Lei (2013), where the overall number of feedback bits for each UE and the codebook sizes for quantizing the per-cell CDIs of each UE are optimized jointly to maximize the average sum rate. Through the way of characterizing the degree of freedom (DoF) achieved by each UE, the results showed that the proposed adaptive feedback scheme can thoroughly avoid the inter-cell interference with imperfect CDI. Substantial performance gain of the proposed feedback scheme was observed over the non-CoMP and CoMP CS/CB schemes with a fixed codebook size, and was only slightly less than that of the optimal feedback configuration. In addition, considering the huge gains from a large scale antenna (Larsson, Edfors, Tufvesson, & Marzetta, 2013),it is challenging that to propose a low-complexity inter-CoMP solution with high spectral efficiency and energy efficiency performances, where the limited backhaul and non-ideal CSI should be taken into account.

4. Comparison of Interference Mitigation Methods

Interference cancellation, avoidance and coordination have been introduced. All of them are able to mitigate interference in certain extent and scope. Prior knowledge is used in interference cancellation about the system or channel estimation to get interfering signals, which is then canceled from the receiving signals. But if there are too many users, it is too complex to implement. Interference avoidance utilizes resource allocation to avoid interference in SP, PC and TH/JH, through splitting spectrum for different usages, controlling power for different items and allocating different users in different time slots. IA is an effective mechanism for interference management. It makes the interference at the receiving end overlap through precoding techniques, in order to completely eliminate the influence of interference to the desired signal. Great channel capacity is obtained by IA strategy through compressing interference dimension. But global channel state information is hard to get though it's important in IA. With users increasing, IA constraints will increase dramatically, which will lead to failure of controlling interference. ICIC and eICIC mitigate interference in frequency and time area respectively as interference coordination methods. Several base stations allocate same frequency to the cell-edge user in CoMP. These base stations serve the user at the same time. These methods are used to improve the services to cell-edge users. A comparison of these interference mitigation methods is shown in Table 1.

OPEN CHALLENGES IN INTERFERENCE MANAGEMENT

The efficiency of an interference management scheme depends on the specific small cell scenario. We take a small cell as femtocell an example. In case of less dense deployment, simple schemes can be helpful and provide sufficient interference management for both the macrocell and femtocell network layers. The main aim is to enable femtocell to be deployed densely and hence, efficient interference management schemes should be developed that can cope with the ad hoc nature of femtocells and provide QoS to both macrocell and femtocell users. It should also be kept in mind that the FAP is a small low powered device and it should be able to handle the complexity of these schemes. (Zahir, Arshad, Nakata, & Moessner, 2013)

The limitations of a femtocell should be studied and the schemes should be able to overcome these limitations. For example, the femtocell may have a limited bandwidth on the backhaul connection and may not be able to handle too much signalling through the backhaul. In case of cooperation among neighbouring femtocells, the overall process should be evaluated and studied. (Zahir, Arshad, Nakata, & Moessner, 2013)

Interference management schemes can be mainly centralised or distributed. In the centralised schemes, there is a need of a central entity which can provide information to a particular cell about the global scenario. There is also a need of communication link between femtocells and the central entity, which can create extra burden and delay. On the other hand, this can also save the femtocell a lot of processing overhead required for sensing the environment and using this information. On the other hand, distributed techniques provide much flexibility to the femtocells. The femtocells can take their own decisions based on their local environment and also exchange information with its neighbouring femtocells. As compared to the centralised schemes, the distributed schemes are more complex and require self organisation and self optimisation capabilities. It needs to have specific knowledge about the macrocell users in order to avoid causing any harmful interference to it. For example, in a fully distributed algorithm, the femtocell needs to know the location of any nearby macrocell users using the same spectrum. A lot of schemes are proposed as discussed in the previous section, but there is still a need to have an efficient and less complex Technique. (Zahir, Arshad, Nakata, & Moessner, 2013)

Any interference management scheme strongly depends on the radio access technology (e.g CDMA or OFDMA) and access mode (CSG, open access or hybrid access) being used. Techniques like adaptive power control, resource allocation, interference cancellation and beamforming for multiple antenna

Table 1. Comparison of interference mitigation methods

Catalog	Technologies	Principles	Shortages	Advantages
Interference Cancellation	PIC	• Detect all users at the same time • Cancel the interference by steps	Make overall complexity higher	Decrease latency
	SIC	• Detects one user per stage • Begins from the strongest signal	Complexity and latency is big when there are many users	Easy to realize
	MSIC	• Users are divided into groups and detected in parallel in each group • The aggregate interference of each group is subtracted from the composite received signal	Trade-off of PIC and SIC	
	MUD	• Observes signals to get their characteristics • Selects the most probable signal out of K number of users along with channel state information	• Requires prior knowledge about the system • Algorithms are to complex to implement	• Greatly reduce multipath • Multiple access interference
Interference Avoidance	SP	Spectrum band is divided into two portions for MUEs and FUEs separately.	• High cost of spectrum • Reduced efficiency	Cope with the cross-tier interference
	PC	UEs get sufficient performance through power control, which are divided into open-loop and close-loop PC.	Other factors which effect performance should be considered	• Overcome near-far effect, shadow fading and rapid shading • Get better quality of service
	TH/JH	• Transmission period is subdivided into small portions • A user transmits in one portion, and remains silent for the other portions.	Need no synchronization between users in selecting time slots	Effectively avoiding interference
	IA	• Makes the interference at the receiving end overlap through precoding techniques • In order to completely eliminate the influence of interference to the desired signal	• Global channel state information is hard to get • Doesn't work with increasing users	• Completely eliminate interference • Significantly improve the system capacity
Interference Coordination	ICIC	The inter-cell interference is controlled by managing radio resources, especially in frequency area	Low frequency resource utilization	Maintain the system throughput Improve the spectral efficiency of edge users
	eICIC	Inter cell interference of traffic and control channels are coordinated in time and frequency domain, as well as power control	Cannot completely solve the problem of control channel interference, especially for the TDD system	• Signals are orthogonal in time domain • Interference is reduced both in traffic and control channels
	CoMP	• Several base stations allocate same frequency to the cell-edge user • These base stations serve the user at the same time	• Complex • Need much interactive information which is big overhead	• Reduce inter-cell interference • Enhance the spectral efficiency of cell-edge users

transceivers can be of help in having an efficient interference management scheme. Furthermore, hybrid schemes such as joint power control and resource allocation can be very helpful. These schemes have more flexibility and can result in good performance of femtocell and macrocell. A very good and detailed study of the open challenges in femtocell deployment is provided by Saquib, Hossain, Le and Kim (2012).

CONCLUSION

Interference problem is getting more and more serious with the development of communication technology, which must be solved in mobile communication. This problem is particularly prominent especially in the presence of dense cells and heterogeneous network. This chapter outlines and summarizes interference mitigation strategies, which are interference cancellation, avoidance and coordination.

Both desired signals and interference are decoded or demodulated at the receiver along with channel information, then interference is canceled to get desired signals. This is the basic thought of interference cancellation. The defect of this scheme is that the channel information has to be known to the receiver.

In interference avoidance, the cross-tier interference can be coped with by spectrum splitting. Power control can overcome uplink near-far effect, shadow fading and rapid shading, and get better quality of service, such as lower interference, higher capacity, call-dropping rate and congestion. Time hopping is one major scheme proposed to reduce the cross-tier uplink interference. And cross-tier uplink interference at the FAP is reduced by joint hopping scheme due to independently selected time slots.

Interference alignment is the most important technique of interference cancellation. The basic mathematical thought has been introduced. The interference signals are compressed into a space which doesn't include desired signal. And many novel methods are discussed as well, such as linear interference alignment, symbol extension, ergodic, lattice alignment, blind and retrospective interference alignment. The application of IA in HetNets has been presented too.

ICIC, eICIC and CoMP are also of great importance interference coordination schemes. The spectrum efficiency of UEs at the edge of a cell has been improved by ICIC technology. The interference problem between different types of base stations can be solved by eICIC technique effectively in HetNets. ABS and CRE are two most important techniques of eICIC. CoMP technology helps to get additional performance improvement at the cell edges with synchronized transmission among multiple transmitters dispersed over different cell sites.

And there are more challenges in interference management. It is of great importance and is worth to research on this theme in order to get better system performance.

REFERENCES

Academic group of Baidu. (n.d.). *ICIC*. Retrieved from http://baike.baidu.com

Ali, M., & Tse, D. (2010). *On the degrees of freedom of miso broadcast channels with delayed feedback*. Available from http://www.researchgate.net/profile/Mohammad_Maddah-Ali/publication/264869062_On_the_Degrees_of_Freedom_of_MISO_Broadcast_Channels_with_Delayed_Feedback/links/54c7b3800cf238bb7d0b0a81.pdf

Andrews, J. G. (2005). Interference cancellation for cellular systems: A contemporary overview. *IEEE Wireless Commun.*, *2*(3), 19–29. doi:10.1109/MWC.2005.1421925

Andrews, J. G., Ghosh, A., & Muhamed, R. (2007). *Fundamentals of wimax: Understanding broadband wireless networking*.Prentice Hall Communications Engineering & Emerging Technologies.

Ayach, O. E., & Heath, R. W. (2012). Interference alignment with analog channel state feedback. *Wireless Communications IEEE Transactions on*, *11*(2), 626–636. doi:10.1109/TWC.2011.122211.101814

Birk, Y., & Kol, T. (1998). Informed-source coding-on-demand (ISCOD) over broadcast channels. In *Proceedings of the conference on computer communications: seventeenth annual joint conference of the IEEE Computer and Communications Societies, Gateway to the 21st century*. IEEE.

Birk, Y., & Kol, T. (2006). Coding on demand by an informed source (iscod) for efficient broadcast of different supplemental data to caching clients. *IEEE Transactions on Information Theory*, *52*(6), 2825–2830. doi:10.1109/TIT.2006.874540

Boucher, N. J. (2001). *The cellular radio handbook: a reference for cellular system operation*. Retrieved from http://www.gbv.de/dms/ilmenau/toc/321833627.PDF

Boudreau, G., Panicker, J., Guo, N., Chang, R., Wang, N., & Vrzic, S. (2009). Interference coordination and cancellation for 4g networks. *IEEE Communications Magazine*, *47*(4), 74–81. doi:10.1109/MCOM.2009.4907410

Bresler, G., Parekh, A., & Tse, D. N. C. (2010). The approximate capacity of the many-to-one and one-to-many gaussian interference channels. *IEEE Transactions on Information Theory*, *56*(9), 4566–4592. doi:10.1109/TIT.2010.2054590

Cadambe, V., & Jafar, S. (2008). Interference alignment and degrees of freedom of the K-user interference channel. *IEEE Transactions on Information Theory*, *54*(8), 3425–3441. doi:10.1109/TIT.2008.926344

Chandrasekhar, V., & Andrews, J. G. (2007). Uplink capacity and interference avoidance for two-tier femtocell networks. *IEEE Transactions on Wireless Communications*, *8*(7), 3498–3509. doi:10.1109/TWC.2009.070475

Chang, R. Y., Tao, Z., Zhang, J., & Kuo, C. C. J. (2009). Multicell ofdma downlink resource allocation using a graphic framework. *IEEE Transactions on Vehicular Technology*, *58*(7), 3494–3507. doi:10.1109/TVT.2009.2014384

China Mobile Research Institute. (2011). *C-RAN- The Road Towards Green RAN White Paper (Version 2.5)* [PDF document]. Retrieved from Lecture Notes Online Web site: http://labs.chinamobile.com/report/view_59826

Cierny, M., Wang, H., Wichman, R., Ding, Z., & Wijting, C. (2013). On number of almost blank subframes in heterogeneous cellular networks. *IEEE Transactions on Wireless Communications*, *12*(10), 5061–5073. doi:10.1109/TWC.2013.090513.121756

Duel-Hallen, A., Holtzman, J., & Zvonar, Z. (1995). Multiuser detection for cdma systems. *Personal Communications IEEE*, *2*(2), 46–58. doi:10.1109/98.382531

Elayoubi, S. E., & Fourestié, B. (2006). On frequency allocation in 3G LTE systems. In *Proceedings of the IEEE 17th International Symposium on Personal, Indoor and Mobile Radio Communications (PIMRC '2006)*. Helsiniki, Finland: IEEE. doi:10.1109/PIMRC.2006.254118

Elhakeem, A. K., Girolamo, R. D., Bdira, I. B., & Talla, M. (1993). Delay and throughput characteristics of TH, CDMA, TDMA, and hybrids for multipath faded data transmission channels. *Universal Personal Communications, 1993. Personal Communications: Gateway to the 21st Century. Conference Record., 2nd International Conference on.*

Frenger, P. K., Orten, P., & Ottosson, T. (2000). Code-spread cdma with interference cancellation. *IEEE Journal on Selected Areas in Communications*, *17*(12), 2090–2095. doi:10.1109/49.814815

Frenkiel, R., & Schwartz, M. (2010). Creating cellular: A history of the amps project (1971-1983). *Communications Magazine IEEE*, *48*(9), 14–24. doi:10.1109/MCOM.2010.5560579

Furht, B., & Ahson, S. A. (2007). *Long Term Evolution: 3GPP LTE Radio and Cellular Technology*. Prentice-Hall.

Gomadam, K., Cadambe, V. R., & Jafar, S. A. (2011). A distributed numerical approach to interference alignment and applications to wireless interference networks. *IEEE Transactions on Information Theory*, *57*(6), 3309–3322. doi:10.1109/TIT.2011.2142270

Guler, B., & Yener, A. (2014). Selective interference alignment for mimo cognitive femtocell networks. *IEEE Journal on Selected Areas in Communications*, *32*(3), 439–450. doi:10.1109/JSAC.2014.140306

Gupta, A., & Jha, R. K. (2015). A survey of 5G network: Architecture and emerging technologies. *Access IEEE*, *3*, 1206–1232. doi:10.1109/ACCESS.2015.2461602

Guvenc, I. (2011). Capacity and fairness analysis of heterogeneous networks with range expansion and interference coordination. *IEEE Communications Letters*, *15*(10), 1084–1087. doi:10.1109/LCOMM.2011.082611.111387

Haile, B. B., Mutafungwa, E., & Hämäläinen, J. (2015). Coordinated multi-point transmission for relaxation of self-backhauling bottlenecks in heterogeneous networks. *EURASIP Journal on Wireless Communications and Networking*, (1): 1–17.

Halonen, T., Romero, J., & Melero, J. (2002). *GSM, GPRS and EDGE performance: evolution towards 3G/UMTS*. New York: Wiley.

Han, K., Choi, Y., Kim, D., Na, M., Choi, S., & Han, K. (2009). Optimization of femtocell network configuration under interference constraints.*Proceedings of the 7th international conference on Modeling and Optimization in Mobile, Ad Hoc, and Wireless Networks*. IEEE Press.

Harte, L. (2004). *Introduction to Wcdma, Physical Channels, Logical Channels, Network, and Operation*. Retrieved from https://www.land.ufrj.br/laboratory/repository/upfiles/book/introduction_to_evdo_book.pdf

Honig, M., Madhow, U., & Verdu, S. (2010). Blind adaptive multiuser detection. *Information Theory IEEE Transactions on*, *6*(4).

Honig, M. L. (2009). *Advances in Multiuser Detection*. Available from http://www.ece.northwestern.edu/~mh/Advances/Peacock_Collings_Honig.pdf

Hossain, E., Rasti, M., Tabassum, H., & Abdelnasser, A. (2014). Evolution toward 5G multi-tier cellular wireless networks: An interference management perspective. *IEEE Wireless Communications*, *21*(3), 118–127. doi:10.1109/MWC.2014.6845056

Jafar, S. A. (2010). Interference alignment — a new look at signal dimensions in a communication network. *Foundations & Trends® in Communications & Information Theory*, *1*(1), 1-136.

Jafar, S. A. (2010). *Exploiting channel correlations - simple interference alignment schemes with no csit*. Retrieved from http://newport.eecs.uci.edu/~syed/papers/nocsitalignconf.pdf

Jafar, S. A., & Shamai, S. (2006). Degrees of freedom region for the mimo x channel. *Eprint Arxiv Cs*, *54*(1), 151–170.

Jo, H. S., Mun, C., Moon, J., & Yook, J. G. (2009). Interference mitigation using uplink power control for two-tier femtocell networks. *IEEE Transactions on Wireless Communications*, *8*(10), 4906–4910. doi:10.1109/TWC.2009.080457

Karlsson, R. S. (1999). Radio resource sharing and capacity of some multiple access methods in hierarchical cell structures. *Vehicular Technology Conference*. *5*, 2825-2829. doi:10.1109/VETECF.1999.800302

Kim, K. H., Lee, J. H., Lee, C. H., Jeon, N. R., & Kim, S. C. (2009). Coordinated Beamforming with Limited BS Cooperation for Multicell Multiuser MIMO Broadcast Channel. In *Proceedings of Vehicular Technology Conference (VTC '2009 Spring)*. Barcelona, Spain: IEEE. doi:10.1109/VETECS.2009.5073473

Koutlia, K., Perez-Romero, J., & Agusti, R. (2015). On Enhancing Almost Blank Subframes Management for Efficient eICIC in HetNets. *Vehicular Technology Conference (VTC '2015 Spring)*. Glasgow, UK: IEEE. doi:10.1109/VTCSpring.2015.7145960

Larsson, E., Edfors, O., Tufvesson, F., & Marzetta, T. (2013). Massive mimo for next generation wireless systems. *IEEE Communications Magazine*, *52*(2), 186–195. doi:10.1109/MCOM.2014.6736761

Lee, D., Seo, H., Clerckx, B., Hardouin, E., Mazzarese, D., Nagata, S., & Syana, K. (2012). Coordinated multipoint transmission and reception in lte-advanced: Deployment scenarios and operational challenges. *IEEE Communications Magazine*, *50*(2), 148–155. doi:10.1109/MCOM.2012.6146494

Lee, N., & Heath, R. W. Jr. (2014). Advanced Interference Management Technique: Potential and Limitations. *IEEE Wireless Communications*, *23*(3), 30–38. doi:10.1109/MWC.2016.7498072

Li, S. H. (2003). *The standard of the third generation wireless communication. TD-SCDMA*. Beijing, China: Posts & Telecom Press.

Li, X., Qian, L., & Kataria, D. (2009). Downlink power control in co-channel macrocell femtocell overlay. In *Proceedings of 43rd Annual Conference on Information Sciences and Systems, CISS 2009*. IEEE. doi:10.1109/CISS.2009.5054750

Maddah-Ali, M. A., Motahari, A. S., & Khandani, A. K. (2008). Communication over mimo x channels: Interference alignment, decomposition, and performance analysis. *Information Theory IEEE Transactions on*, *54*(8), 3457–3470. doi:10.1109/TIT.2008.926460

Mishra, A. R. (2006). *Advanced Cellular Network Planning and Optimisation*. Available from http://download.e-bookshelf.de/download/0000/5683/51/L-G-0000568351-0002356592.pdf

Nam, W., Bai, D., Lee, J., & Kang, I. (2014). Advanced interference management for 5g cellular networks. *IEEE Communications Magazine*, *52*(5), 52–60. doi:10.1109/MCOM.2014.6815893

Nazer, B., Gastpar, M., Jafar, S. A., & Vishwanath, S. (2011). Ergodic interference alignment. *IEEE Transactions on Wireless Communications*, *11*(2), 842–851.

Nguyen, T. M., Jeong, Y., Quek, T. Q. S., Tay, W. P., & Shin, H. (2013). Interference alignment in a poisson field of mimo femtocells. *IEEE Transactions on Wireless Communications*, *12*(6), 2633–2645. doi:10.1109/TWC.2013.040413.120024

Nielsen, T. T., & Wigard, J. (2000). *Performance Enhancements in a Frequency Hopping GSM Network*. Available from http://www.openisbn.org/download/0792378199.pdf

Patel, P., & Holtzman, J. (2010). Analysis of a simple successive interference cancellation scheme in a ds/cdma system. *IEEE Journal on Selected Areas in Communications*, *12*(5), 796–807. doi:10.1109/49.298053

Peng, M., Wang, C., Li, J., Xiang, H., & Lau, V. (2015). Recent advances in underlay heterogeneous networks: Interference control, resource allocation, and self-organization. *IEEE Communications Surveys and Tutorials*, *17*(2), 700–729. doi:10.1109/COMST.2015.2416772

Peng, M., & Wang, W. (2010). Technologies and standards fortd-scdma evolutions to imt-advanced. *IEEE Communications Magazine*, *47*(12), 50–58. doi:10.1109/MCOM.2009.5350368

Rahman, M., Yanikomeroglu, H., & Wong, W. (2009). Interference Avoidance with Dynamic Inter-Cell Coordination for Downlink LTE System. In *Proceedings of IEEE Wireless Communications & Networking Conference*. Budapest, Hungary: IEEE. doi:10.1109/WCNC.2009.4917761

Rao, X., Ruan, L., & Lau, V. K. N. (2013a). Csi feedback reduction for mimo interference alignment. *IEEE Transactions on Signal Processing*, *61*(18), 4428–4437. doi:10.1109/TSP.2013.2269902

Rao, X., Ruan, L., & Lau, V. K. N. (2013b). Limited feedback design for interference alignment on mimo interference networks with heterogeneous path loss and spatial correlations. *IEEE Transactions on Signal Processing*, *61*(10), 2598–2607. doi:10.1109/TSP.2013.2252168

Santhi, K. R., Srivastava, V. K., Senthilkumaran, G., & Butare, A. (2003). Goals Of True Broad band's Wireless Next Wave (4G-5G). In *Proceedings of IEEE Vehicular Technology Conference*. Orlando, FL: IEEE.

Saquib, N., Hossain, E., Le, L. B., & Kim, D. I. (2012). Interference management in ofdma femtocell networks: Issues and approaches. *IEEE Wireless Communications*, *19*(3), 86–95. doi:10.1109/MWC.2012.6231163

Sesia, S., Toufik, I., & Baker, M. (2009). *LTE: The UMTS Long Term Evolution: From Theory to Practice*. New York: Wiley Publishing. doi:10.1002/9780470742891

Shin, W., Noh, W., Jang, K., & Choi, H. H. (2012). Hierarchical interference alignment for downlink heterogeneous networks. *Wireless Communications IEEE Transactions on*, *11*(12), 4549–4559. doi:10.1109/TWC.2012.101912.120421

Tanbourgi, R., Singh, S., Andrews, J. G., & Jondral, F. K. (2014). Analysis of Non-Coherent Joint-Transmission Cooperation in Heterogeneous Cellular Networks. In *Proceedings of IEEE International Conference on Communications (ICC '2014)*. Sydney, Australia: IEEE. doi:10.1109/ICC.2014.6884140

Van, W. F., Janssen, G. M. J., & Prasad, R. (1995). Groupwise successive interference cancellation in a DS/CDMA system. In *Proceedings of Personal, Indoor and Mobile Radio Communications (PIMRC '95),Sixth IEEE International Symposium on Wireless: Merging onto the Information Superhighway*. Toronto, Canada: IEEE.

Vanghi, V., Damnjanovic, A., Harte, L., & Vojcic, B. (2004). *The cdma2000 System for Mobile Communications: 3G Wireless Evolution*. Retrieved from http://www.openisbn.org/download/0131416014.pdf?name=The_Cdma2000_System_For_Mobile_Communications_3G_W

Wu, W. L., & Niu, K. (2005). *Principle of Mobile communication*. Beijing, China: Electronics Industry Press.

Yuan, F., Yang, C., Wang, G., & Lei, M. (2013). Adaptive channel feedback for coordinated beamforming in heterogeneous networks. *IEEE Transactions on Wireless Communications*, *12*(8), 3980–3994. doi:10.1109/TWC.2013.071913.121553

Zahir, T., Arshad, K., Nakata, A., & Moessner, K. (2013). Interference management in femtocells. *IEEE Communications Surveys and Tutorials*, *15*(1), 293–311. doi:10.1109/SURV.2012.020212.00101

Zahir, T., Arshad, K., Nakata, A., & Moessner, K. (2013). Interference management in femtocells. *IEEE Communications Surveys and Tutorials*, *15*(1), 293–311. doi:10.1109/SURV.2012.020212.00101

Zhao, Z., Peng, M., Ding, Z., Wang, W., & Chen, H.-H. (2014). Denoise-and-forward network coding for two-way relay mimo systems. *IEEE Transactions on Vehicular Technology*, *63*(63), 775–788. doi:10.1109/TVT.2013.2281330

Section 3

Novel Mathematical Frameworks for Interference and Energy Management

Chapter 9
Geometric Programming Based Resource Allocation for 5G High-Speed Mobile Networks

Shaoyi Xu
Beijing Jiaotong University, China

Tianhang Fu
Beijing Jiaotong University, China

ABSTRACT

The high-speed railway (HSR) is a typical application case in 5G systems. Mobile relay stations (MRSs) which are mounted in a high-speed train (HST) is popular system architecture for high-speed mobile communications. However, sharing spectrums between the macro cell and the MRS cell, interference exists in this hybrid system. In this chapter, we investigate the downlink of a multi-cellular decode and forward (DF) relayed orthogonal frequency division multiple access (OFDMA) system and formulate the problem to maximize the system sum rate of all cells subject to a total power constraint and a new proposed time delay constraint. An effective resource allocation scheme combined by greedy sub-carriers allocation and geometric programming (GP) based power allocation algorithm is proposed to optimize subcarrier allocation and power allocation. Numerical experiments verify that the proposed resource allocation scheme outperforms the other traditional approaches and the necessity of introducing the time delay constraint.

INTRODUCTION

In the near future, to achieve the objectives of increased capacities, improved data rates, decreased latencies, and better quality of services, key technologies and drastic improvements need to be made for the beyond 4G cellular networks which is called fifth generation (5G) mobile communications system. The main two drivers of 5G are mobile internet and the Internet of Things (IoT) and it will penetrate into every element of the future society and create an all-dimensional, user-centered information ecosystem. 5G will be able to deliver a consistent experience across a variety of scenarios including the case of

DOI: 10.4018/978-1-5225-1712-2.ch009

an ultra-high traffic volume density, ultra-high connection density, and ultra-high mobility (IMT-2020 (5G) Promotion Group, 2014).

Compared with previous generations of mobile communications, 5G needs to meet extremely high performance requirements in more diverse scenarios. It is expected that the upcoming 5G systems should fulfill the following key performance indicators (KPIs) (IMT-2020 (5G) Promotion Group, 2014).

- 1000 times mobile data traffic compared to current traffic volume: tens of Gbps per square kilometer.
- 100 times connections: 1 million connections per square kilometer.
- 0.1~1Gbps user experienced data rate.
- Lower end-to-end latency: millisecond level.
- Higher spectrum efficiency.
- Lower cost.
- Higher energy efficiency.

To meet the above KPIs, many emerging techniques are identified as the key enabling technologies of 5G. Among them, it is expected that the Ultra Dense Hetnets (UDHs) architecture will play an important role in supporting such a 5G network, especially for hotspot and blind wireless areas. But for achieving ultra-dense, heterogeneous networks (Hetnets) will play an important role.

Heterogeneous networks have received much attention these days as a new paradigm for increasing cellular capacity and are under standardization for the current 3rd Generation Partnership (3GPP). In heterogeneous networks, some low power nodes (LPNs) are introduced to the traditional macro cellular deployment, which include micro eNBs, pico eNBs (PeNBs), home eNBs (HeNBs), relay nodes (RNs), etc (Gupta, 2015; Xu, 2012). Figure 1 illustrates the basic concept of heterogeneous networks in an LTE-Advanced (LTE-A) system.

As a specific application scenario, 5G is required to support higher than 500Km/h mobility. Recently, the high-speed railway (HSR) has been rapidly developed in the world within these years. Providing a safe and comfortable environment for long distance traveling, high-speed trains (HSTs) have been the good choice for people. In USA, for example, a high-speed rail layout which includes 10 regional routes has been outlined. In China, high-speed railways will reach 18,000 km by 2020 (Zhou, Pan, Hu, Shi, & Mo,2011), among which the highest velocity will reach to more than 350km/h. Catering for such development and requirements of the travelers, the technologies of broadband wireless access for HSR communications are needed.

Among the existing architectures for HSR communications, deploying the mobile relay station (MRS) will potentially benefit cellular networks and is supported by 3GPP. However, because of the huge number of users on the train, the mobile relay stations must reuse the frequency resources with the base station (BS). So there will be co-channel interference (CCI) between user equipments (UEs) in the high speed train (HUEs) and UEs in the macro cell zone (MUEs) both in uplink (UL) and downlink (DL) traffic. Moreover, if we deploy more than one MRS in the train (placing one MRS on each carriage is a common case), they will interfere with each other too. (Fu, Xu, & Kwak, 2014).

With this respect, effective resource allocation mechanisms are required to coordinate the interference for the heterogeneous networks in the high-speed environments.

Figure 1. Illustration of heterogeneous networks in an LTE-Advanced system

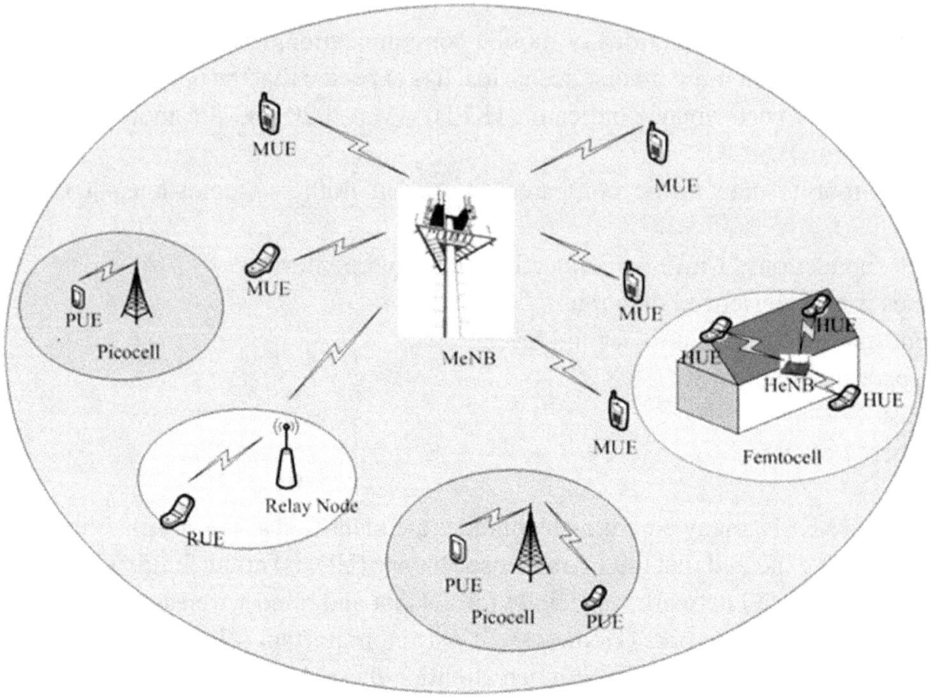

BACKGROUND

In the near future, there will be a growing demand for mobile services in vehicles, trains and even aircrafts. Providing a safe and comfortable environment for long distance traveling, high-speed trains have been the good choice for people. High speed trains with a speed of more than 300km per hour are being deployed rapidly in many countries around the world such as China, USA, Germany, France, Japan and so on, having attracted a lot of attention in recent years. Nowadays more and more people choose HSTs for travelling as a fast, safe and green public transportation vehicle. During the travel passengers may request varied wireless communication demands, including high data rates services such as video meetings and online movies. Yet, these demands cannot be guaranteed by the existing global system for mobile (GSM) for rail (GSM-R).

Actually, GSM-R is the dominant wireless communication system for current railways. Unfortunately, GSM-R presents the same features to the second generation mobile GSM and provides the data transmission rate of 9.6kbps. Such a data rate cannot satisfy the current and future applications and is unaffordable for the HST deployment (Aguado, Onandi, Agustin, Higuero, & Taquet, 2008). Some alternative or complementary solutions have been proposed in (Aguado, Onandi, Agustin, Higuero, & Taquet, 2008; Yamada, et al., 2010; Kolios, Friderikos, & Papadaki, 2011; Xiao, Fuja, & Costello, 2010; Balakrishnan, Yang, Venkatachalam, & Akyildiz, 2011), including using WiMax (Aguado, Onandi, Agustin, Higuero, & Taquet, 2008), WiFi(Yamada, et al., 2010) and mobile relays (Kolios, Friderikos, & Papadaki, 2011; Xiao, Fuja, & Costello, 2010; Balakrishnan, Yang, Venkatachalam, & Akyildiz, 2011). Among the existing architectures, deploying the MR will potentially benefit cellular networks and is supported by

3rd Generation Partnership Project (3GPP) in Long Term Evolution (LTE) Release 10 (Haider, Dianati, & Tafazolli, 2011;Bhat, et al., 2012). Firstly, an MR can provide higher spectral efficiency and improve the coverage. Secondly, such deployment contributes to the reduction of signaling overhead of the network operation especially greatly decreasing the signaling of fast handover. Thirdly, the battery life of the user equipments associated to the MRS can be prolonged considering the transmission distance is delimited within the carriage. Figure 2 illustrates the system architecture by using a mobile relay for a high-speed mobile environment based on an LTE network. From this figure we can observe that under such an architecture, there are two subsystems. One is the macro cell which includes the macro eNB (MeNB) and its connected macro UEs (MUEs). The other is the high-speed railway cell which consists of the MR and its HST UEs (HUEs).

Based on the above description, in our study, we focus on the MRS architecture and use LTE as the wireless access technology. However, considering the scarcity of the spectrum nowadays, it is difficult to allocate dedicated spectra for the communication between the MRS and its connected UEs. Consequently, the spectra reusing between the two cells is inevitable and such resource sharing will lead to the interference between these two networks.

Figure 3 illustrates the system architecture and interference scenario by using multiple mobile relays (MRs) for a high-speed mobile environment based on an LTE network. From this figure we can observe that under such an architecture, there are two subsystems. One is the macro cell which includes the macro BS and its connected macro UEs. The other is the high-speed railway cell which consists of the MRS and its HST UEs. By sharing resources between the macro network and the mobile relay subsystem, interference is incurred in both uplink and downlink communication. In the UL transmission, the

Figure 2. System architecture by using a mobile relay for a high-speed mobile environment

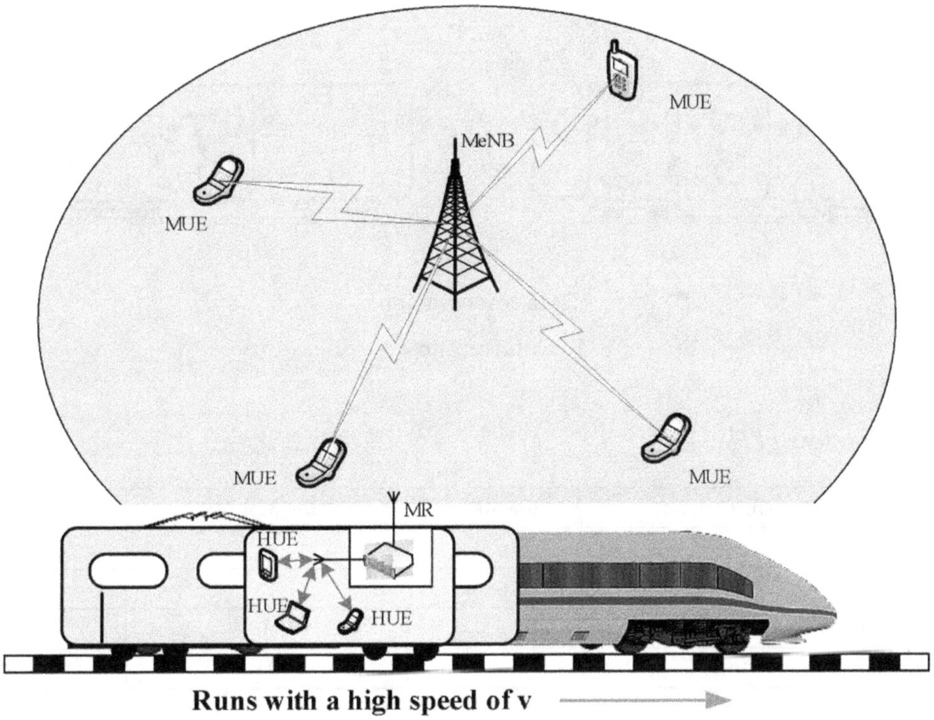

interference includes two parts: from the HUEs to the macro BS and from the MUEs to its nearby MR. Similarly, for the DL traffic, there are also two kinds of interference, namely, from the macro BS to the HUEs and from the MRS to its close MUEs.

Although the signals sent by the macro users are attenuated by the train body, this interference can not be neglected. For instance, a penetration loss of 8.3 dB is proposed in (Yamada, et al., 2010) for Japanese HSTs, and Table 1 shows the measurement results of penetration loss for the Chinese HSTs (Aguado, Onandi, Agustin, Higuero, & Taquet, 2008). From Table 1 we can see that the average penetration loss of an HST is usually below 20dB which is similar to the scenario from the outdoor to the indoor. To this end, the interference between these two subsystems must be suppressed to guarantee the system performance.

Some researchers have discussed a number of algorithms on resource allocation previously. In (Mao, Xu, Fu, & Haung, 2010) a scheme based on effective capacity and Nash equilibrium (NE) was proposed. Researchers of (Jin & Vandendorpe, 2011) presented a conditioned subcarrier allocation and mode selection approach which is actually a conditioned water filling power allocation (CWFPA) algorithm. (Hao

Figure 3. Illustration of DL and UL interference scenario

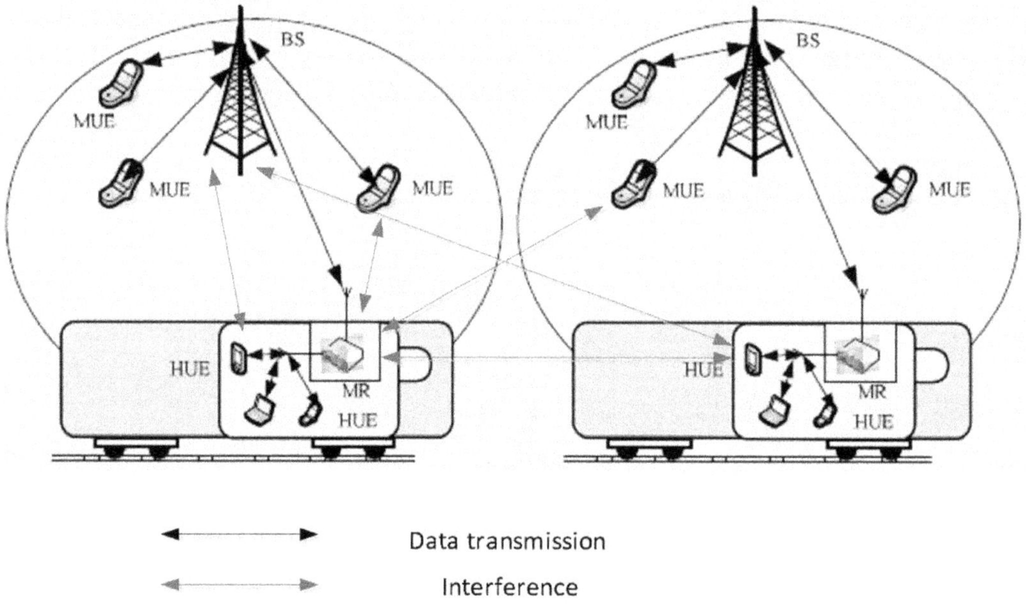

Table 1. Penetration loss of HST

HST	Sleeping Car (dB)	Aisle (dB)	Sitting Car (dB)	Average (dB)
K-type	14	16	13	14
T-type	_	16	12	12
CRH2	_	_	10	10
CRH380	_	_	24	24

& Zhuang, 2012) introduced a single-machine preemptive scheduling scheme based on a time-capacity mapping. A novel resource allocation approach by considering multiple bit error rate (BER) requirements for different types of packets in one data stream was studied in (Zhu, 2012). Besides, researchers of (Li, Vucetic, Zhou, & Dohler, 2007) proposed an adaptive sub-optimum power allocation solution based on the SNR upper bound.

All above papers focus on negligible CCI or fixed CCI while designing algorithms of the sub-carriers and transmission power allocation. However, in our opinions, this is impractical for the realistic scenarios. In order to make the algorithm closer to reality and obtain a more appropriate allocation solution, we propose a scheme with the consideration of dynamic CCI in this chapter. However, the formulation of such dynamic resource allocation makes it infeasible to solve the optimization goal function by previously known methods such as the Lagrangian multiplier method. So we propose a new geometric programming based algorithm to calculate the solution of the resource allocation. What's more, we further take the consideration of time delay constraints causing by the high travelling speed of the train. As can be seen from the simulation results, all our above work is meaningful to improve the system sum throughput.

The rest of this chapter is organized as follows. At first, we present the system model. Then, the problem formulation is discussed. After that, we propose our scheme of resource allocation. Numerical experiments results are demonstrated and finally conclusions are drawn.

SYSTEM MODEL AND PROBLEM FORMULATION

System Model

We consider a multi-cell scenario and there are totally N cells, and K orthogonal frequency division multiple access (OFDMA) sub-carriers are reused in each cell. In cell n there are J_n nodes, including one BS and $J_n - 1$ mobile relay stations set on railway carriages. For simplicity we denote $j = 1$ as the BS and $j = 2, 3, \ldots J_n$ as mobile relay stations. In the zone of node j in cell n, there are $I_{n,j}$ users, who transmit and receive data using K orthogonal sub-carriers so they do not interfere with each other. The channel gain of subcarrier k from node j' in cell n' to UE i of node j in cell n is denoted by $G^{k,i}_{n',j',n,j}$.

We define the following resource allocation related notations for the considered system. $P^{k,i}_{n,j}$ denotes the transmission power of node j in cell n to user i, on sub-carrier k. And the binary variable $b^{k,i}_{n,j}$ is introduced. $b^{k,i}_{n,j} = 1$ indicates that sub-carrier k is allocated to user i of node j in cell n, otherwise $b^{k,i}_{n,j} = 0$. We further assume that in one node of a certain cell, each sub-carrier can be used by only one user.

The rate of user i on subcarrier at node j in cell n is expressed as

$$R^{k,i}_{n,j} = \log\left(1 + SINR^{k,i}_{n,j}\right), \tag{1}$$

where $SINR^{k,i}_{n,j}$ denotes the signal to interference plus noise ratio (SINR) of each channel. And its expression is

$$SINR_{n,j}^{k,i} = P_{n,j}^{k,i} G_{n,j,n,j}^{k,i} \left(\sigma^2 + \sum_{\substack{j'=1 \\ j' \neq j}}^{J_n} \sum_{i'=1}^{I_{n,j'}} P_{n,j'}^{k,i'} G_{n,j',n,j}^{k,i} + \sum_{\substack{n'=1 \\ n' \neq n}}^{N} \sum_{j'=1}^{J_{n'}} \sum_{i'=1}^{I_{n',j'}} P_{n',j'}^{k,i'} G_{n',j',n,j}^{k,i} \right)^{-1}, \tag{2}$$

where σ^2 denotes the Additive White Gaussian Noise (AWGN) variance and we assume that it is fixed on each channel. The second and third items in the brackets represent the CCI of other nodes in the same cell and nodes in other cells, respectively.

Problem Formulation

Our considered resource allocation scheme is to maximize the total rate of all users in all cells subject to the total power constraint and considering the time delay in the high-speed environments. Specifically, given the formulation in last Section, the optimization goal function can be expressed as

$$\underset{\mathbf{P},\mathbf{B}}{\text{maximize}} \ R\left(\mathbf{P},\mathbf{B}\right) = \sum_{k=1}^{K} \sum_{n=1}^{N} \sum_{j=1}^{J_n} \sum_{i=1}^{I_{n,j}} R_{n,j}^{k,i} b_{n,j}^{k,i}, \tag{3}$$

s.t.

$$\text{C1: } \sum_{k=1}^{K} \sum_{n=1}^{N} \sum_{j=1}^{J_n} \sum_{i=1}^{I_{n,j}} P_{n,j}^{k,i} \leq P_{total}, \tag{4}$$

$$\text{C2: } \begin{array}{l} \sum_{i=1}^{I_{n,j}} b_{n,j}^{k,i} \leq 1, \forall k, n, j \\ b_{n,j}^{k,i} = \left\{0,1\right\}, \forall k, n, j, i \end{array}. \tag{5}$$

$$\text{C3: } \sum_{k}^{K} R_{n,j}^{k,i} b_{n,j}^{k,i} \geq \psi\left(T_{n,j}^{i}, \lambda_{n,j}^{i}, F\right), \forall n, j, i, \tag{6}$$

where $\mathbf{P} = \left\{P_{n,j}^{k,i}\right\}$ and $\mathbf{B} = \left\{b_{n,j}^{k,i}\right\}$ denote the vectors of transmission power and base station plus MRSs. C1 is the total power constraint of all users in all cells. C2 means that each subcarrier can be used by only one user of one node in one certain cell.

In another side, a UE demands not only a high bits rate but also a delay low enough to guarantee its Quality of Service (QoS). So we introduce the constraint of C3 as follows. We assume that each UE has a delay constraint, denoted by $T_{n,j}^{i}$, and if the delay of a UE is higher than this constraint its service will be considered as a failure. It is obvious that users on a train are experiencing faster channel changing and shall have a stricter delay constraint than other ordinary users on ground if they get the same service. Furthermore, the faster the train travels, the smaller the delay constraint should be. Based on the above analysis, the maximal delay constraint can be converted into a minimum bits rate constraint of each user as C3 (Shi, Wang, Wang, & Zhang, 2011) and we rewrite it as

$$\text{C3: } \sum_{k}^{K} R_{n,j}^{k,i} b_{n,j}^{k,i} \geq \psi\left(T_{n,j}^{i}, \lambda_{n,j}^{i}, F\right), \forall n, j, i,$$

where $\lambda_{n,j}^{i}$ and F denote the packet arrival rate of Poisson process index and packet size respectively, and

$$\psi\left(T_{n,j}^{i}, \lambda_{n,j}^{i}, F\right) = \frac{\left(T_{n,j}^{i}\lambda_{n,j}^{i} + 1\right) + \sqrt{\left(T_{n,j}^{i}\lambda_{n,j}^{i} + 1\right)^{2} - 2T_{n,j}^{i}\lambda_{n,j}^{i}}}{2T_{n,j}^{i}} E\left\{F\right\}. \tag{7}$$

GEOMETRIC PROGRAMMING BASED RESOURCE ALLOCATION SCHEME

In the previous part, we have formulated the optimization goal function. In this section, we further discuss how to solve it. As we can see from the above analysis, the problem in (3) is a mixed integer non-linear programming (MINLP), which is a NP-hard problem (Yao, Zhang, & Niu, 2010). That is to say that we cannot solve the optimization problem easily with existing mathematic methods because of the high computation complexity. Therefore, to reduce the complexity, we propose a new scheme that can be separated into two stages as sub-carrier allocation and transmission power allocation.

Greedy Sub-Carrier Allocation

We first propose a greedy sub-carrier allocation that allows each node (a BS or an MRS) to allocate its sub-carriers autonomously. At the beginning of the optimization scheme, each node updates its sub-carrier allocation matrix

$$b_{n,j}^{k,i} = \begin{cases} 1, i = \left(i_{opt}\right)_{k,n,j} \\ 0, \text{otherwise} \end{cases}, \tag{8}$$

where

$$i_{opt} = \arg\max_{i} G_{n,j,n,j}^{k,i}, \forall k, n, j. \tag{9}$$

The expression of (8) and (9) mean that each node chooses a best user for each sub-carrier and then allocates the sub-carrier to its best user. In other words, on each sub-carrier the best user has a higher channel gain than all other users. This algorithm can be summarized as follows.

```
lgorithm: Greedy Sub-carrier Allocation
Input: $G^{k,i}_{n,j,n,j}$, $K$, $N$, $J_n$, $I_{n,j}$
Output: $\boldsymbol{B} = \{b^{k,i}_{n,j}\}$
Initialize $\{b^{k,i}_{n,j}\}=\mathbf{0}$
for $k$ = 1 to $K$ do
    for $n$ = 1 to $N$ do
        for $j$ = 1 to $J_n$ do
            Initialize $a$ = $b$ = 0;
            for $i$ = 1 to $I_{n,j}$ do
                if $G^{k,i}_{n,j,n,j} \geq b$ then
                    $a = i, b = G^{k,i}_{n,j,n,j}$;
                end if
            end for
            $b^{k,a}_{n,j} = 1$;
        end for
    end for
end for
```

Geometric Programming Based Power Allocation

As the sub-carrier allocation result is given in the former part, the optimization goal function (3) can be denoted as

$$\underset{\mathbf{P}}{\text{maximize}}\ R(\mathbf{P}) = \sum_{k=1}^{K}\sum_{n=1}^{N}\sum_{j=1}^{J_n}\sum_{i=1}^{I_{n,j}} R^{k,i}_{n,j} b^{k,i}_{n,j}, \tag{10}$$

s.t.

$$\text{C1: } \sum_{k=1}^{K}\sum_{n=1}^{N}\sum_{j=1}^{J_n}\sum_{i=1}^{I_{n,j}} P^{k,i}_{n,j} \leq P_{total}$$

$$\text{C3: } \sum_{k}^{K} R^{k,i}_{n,j} b^{k,i}_{n,j} \geq \psi\left(T^{i}_{n,j}, \lambda^{i}_{n,j}, F\right), \forall n, j, i$$

Unlike present usual optimization programming, our proposed problem is a non-linear and non-convex function which results in that this optimization problem cannot be solved by using the traditional method like Lagrangian multiplier easily. Therefore, we introduce the geometric programming (GP) to pursue the result.

GP is a class of non-linear, non-convex optimization programming with many useful theoretical and computational properties. Since GP can be turned into a convex optimization problem, a local optimum is also a global optimum, Lagrange duality gap is zero under mild conditions, and a global optimum can always be computed very efficiently (Chiang, Tan, Palomar, O'neill, & Julian, 2007). Numerical efficiency holds both in theory and in practice, for example, interior point methods applied to GP have provably polynomial time complexity (Nesterov, Nemirovskii, & Ye, 1994), and are very fast in practice with high-quality software (e.g., the MOSEK package). GP in a standard form is not a convex optimization problem, because posynomials are not convex functions (Chiang, Tan, Palomar, O'neill, & Julian, 2007). With a logarithmic change of the variables and multiplicative constants, we can turn the standard form into the convex form, which is a convex optimization problem. And then we can solve the new convex optimization problem with common mathematical methods such as the interior point method.

Back to our optimization problem, since the allocation of $b_{n,j}^{k,i}$ has been already determined in the former part, the optimization goal function (10) can be written as

$$\underset{\mathbf{P}}{\text{maximize}} \sum_{\substack{k,n,j,i \\ b_{n,j}^{k,i}=1}} \log\left(1+SINR_{n,j}^{k,i}\right), \tag{11}$$

and we rewrite (11) as

$$\underset{\mathbf{P}}{\text{minimize}} \prod_{\substack{k,n,j,i \\ b_{n,j}^{k,i}=1}} \left(\frac{1}{1+SINR_{n,j}^{k,i}}\right). \tag{12}$$

Here we define two new functions as follows:

$$F_{n,j}^{k,i}\left(\mathbf{P}\right) = \sigma^2 + \sum_{\substack{j'=1 \\ j'\neq j}}^{J_n} \sum_{i'=1}^{I_{n,j'}} P_{n,j'}^{k,i'} G_{n,j',n,j}^{k,i} + \sum_{\substack{n'=1 \\ n'\neq n}}^{N} \sum_{j'=1}^{J_{n'}} \sum_{i'=1}^{I_{n',j'}} P_{n',j'}^{k,i'} G_{n',j',n,j}^{k,i}, \tag{13}$$

$$G_{n,j}^{k,i}\left(\mathbf{P}\right) = \sigma^2 + \sum_{n'=1}^{N} \sum_{j'=1}^{J_{n'}} \sum_{i'=1}^{I_{n',j'}} P_{n',j'}^{k,i'} G_{n',j',n,j}^{k,i}, \tag{14}$$

so $1+\frac{1}{SINR_{n,j}^{k,i}} = \frac{F_{n,j}^{k,i}\left(\mathbf{P}\right)}{G_{n,j}^{k,i}\left(\mathbf{P}\right)}$, and the optimization goal function (12) can be rewritten as

$$\underset{\mathbf{P}}{\text{minimize}} \prod_{\substack{k,n,j,i \\ b_{n,j}^{k,i}=1}} \frac{F_{n,j}^{k,i}\left(\mathbf{P}\right)}{G_{n,j}^{k,i}\left(\mathbf{P}\right)}. \tag{15}$$

In addition, we rewrite C1 and C3 for standardization as

$$\text{C4: } \frac{1}{P_{total}} \sum_{\substack{k,n,j,i \\ b_{n,j}^{k,i}=1}} P_{n,j}^{k,i} \leq 1,$$

and

$$\text{C5: } \frac{1}{\exp\left(\psi\left(T_{n,j}^{i}, \lambda_{n,j}^{i}, F\right)\right)} \prod_{\substack{k \\ b_{n,j}^{k,i}=1}} \frac{F_{n,j}^{k,i}\left(\mathbf{P}\right)}{G_{n,j}^{k,i}\left(\mathbf{P}\right)} \leq 1, \forall n, j, i.$$

Obviously, our optimization goal function (15) is not a GP in the standard form because the denominator $\prod_{\substack{k,n,j,i \\ b_{n,j}^{k,i}=1}} G_{n,j}^{k,i}\left(\mathbf{P}\right)$ in (15) is not a monomial but a posynomial. In fact, such a power allocation optimization problem by minimizing a ratio between two posynomials like (15) belongs to a non-convex class of problems known as complementary geometric programming which is an intractable NP-hard problem, as revealed in (Chiang, Tan, Palomar, O'neill, & Julian, 2007).

To solve this complementary GP, we propose an iterative algorithm consisting of two parts which will be shown in the following part. Firstly we convert the complementary GP (15) into a GP in the standard form approximated by the Single Condensation Method (SCM). Then we solve the GP in a standard form by the Dual Programming Method (DPM). The proposed GP based algorithm can be summarized as follows.

Step 1: Start with an initial power vector **P**.
Step 2: Convert the complementary GP (15) into a GP in standard form approximated by the SCM.

Algorithm of SCM

1: Calculate $G_{n,j}^{k,i}\left(\mathbf{P}\right)$ with the given **P.**
2: Compute the weight for each term of $G_{n,j}^{k,i}\left(\mathbf{P}\right)$ in the above PA formulation as

$$a_{xn,j}^{k,i} = \frac{g_{xn,j}^{k,i}\left(\mathbf{P}\right)}{G_{n,j}^{k,i}\left(\mathbf{P}\right)}, \forall k, n, j, i, \tag{16}$$

where $g_{xn,j}^{k,i}\left(\mathbf{P}\right)$ is a monomial, denoting the *x*th term of the posynomial as $G_{n,j}^{k,i}\left(\mathbf{P}\right)$.

3: Convert each denominator posynomial $G_{n,j}^{k,i}\left(\mathbf{P}\right)$ into a monomial with the weight index $a_{xn,j}^{k,i}$

$$G'^{k,i}_{n,j}\left(\mathbf{P}\right)=\prod_{x}\frac{g_{xn,j}^{k,i}\left(\mathbf{P}\right)^{a_{xn,j}^{k,i}}}{a_{xn,j}^{k,i}},\forall k,n,j,i\,. \tag{17}$$

In other words, substitute the monomial $G'^{k,i}_{n,j}\left(\mathbf{P}\right)$ of (17) into (15) for $G_{n,j}^{k,i}\left(\mathbf{P}\right)$ approximately. So the optimization goal function (15) is converted to

$$\underset{\mathbf{P}}{\text{minimize}}\prod_{\substack{k,n,j,i\\ b_{n,j}^{k,i}=1}}\frac{F_{n,j}^{k,i}\left(\mathbf{P}\right)}{G'^{k,i}_{n,j}\left(\mathbf{P}\right)}. \tag{18}$$

4: Because the denominator $\prod_{\substack{k,n,j,i\\ b_{n,j}^{k,i}=1}}G_{n,j}^{k,i}\left(\mathbf{P}\right)$ in (15) is converted to a monomial $\prod_{\substack{k,n,j,i\\ b_{n,j}^{k,i}=1}}G'^{k,i}_{n,j}\left(\mathbf{P}\right)$, the optimization goal function is now a standard GP problem. In addition, we rewrite the optimization goal function as

$$\underset{\mathbf{P}}{\text{minimize}}\; h_0\left(\mathbf{P}\right), \tag{19}$$

s.t.

$h_m\left(\mathbf{P}\right)\leq 1$, m=1,2,...,M+1,

where

$$h_0\left(\mathbf{P}\right)=\prod_{\substack{k,n,j,i\\ b_{n,j}^{k,i}=1}}\frac{F_{n,j}^{k,i}\left(\mathbf{P}\right)}{G'^{k,i}_{n,j}\left(\mathbf{P}\right)}, \tag{20}$$

$$h_m\left(\mathbf{P}\right)=\frac{1}{\exp\left(\psi\left(T_{n,j}^{i},\lambda_{n,j}^{i},F\right)\right)}\prod_{\substack{k\\ b_{n,j}^{k,i}=1}}\frac{F_{n,j}^{k,i}\left(\mathbf{P}\right)}{G'^{k,i}_{n,j}\left(\mathbf{P}\right)},\; m=1,2,...,M \tag{21}$$

$$h_{M+1}\left(\mathbf{P}\right)=\frac{1}{P_{total}}\sum_{\substack{k,n,j,i\\ b_{n,j}^{k,i}=1}}P_{n,j}^{k,i}, \tag{22}$$

and M denotes the number of UEs allocated with sub-carriers.

Step 3: Solve the resulting GP in the standard form using DPM as follows.

Algorithm of DPM

1: Define T_m as the number of monomials in the posynomial $h_m(\mathbf{P})$, and $\beta_{m,t}, m = 0,\ldots,M+1, t = 1,\ldots,T_m$ as the constant multiplier in $h_m(\mathbf{P})$.
2: Construct a dual vector $\omega = \left\{\omega_{m,t}\right\}, m = 0,\ldots,M+1, t = 1,\ldots,T_m$.
3: Construct dual programming as

$$\underset{\omega}{\text{maximize}} \prod_{m=0}^{M+1}\prod_{t=1}^{T_m}\left(\frac{\beta_{m,t}}{\omega_{m,t}}\right)^{\omega_{m,t}} \bullet \prod_{m=1}^{M+1}\left(\sum_{t=1}^{T_m}\omega_{m,t}\right)^{\sum_{t=1}^{T_m}\omega_{m,t}}, \tag{23}$$

with the constraints of

$$\begin{cases} \sum_{t=1}^{T_0}\omega_{0,t} = 1 \\ \sum_{m=0}^{M+1}\sum_{t=1}^{T_m} c_{m,t}^{1}\omega_{m,t} = 0 \\ \sum_{m=0}^{M+1}\sum_{t=1}^{T_m} c_{m,t}^{2}\omega_{m,t} = 0, \\ \vdots \\ \sum_{m=0}^{M+1}\sum_{t=1}^{T_m} c_{m,t}^{M}\omega_{m,t} = 0 \end{cases}$$

where $c_{m,t}^{y}, y = 1,\ldots,M$ indicates the index of the yth UE transmission power in the tth monomial of $h_m(\mathbf{P})$.

4: Solve the linear equations to obtain the result **P**.

Step 4: Return to *Step 1* with the refreshed **P** until the error between two iterations is less than the fixed tolerance.

Now we can solve the optimization problem in our proposed scheme, which is summarized as follows.

Step 1: Determine the sub-carrier allocation $\left\{b_{n,j}^{k,i}\right\}$ matrix by using our greedy sub-carrier allocation algorithm.

Step 2: Substitute the sub-carrier allocation result in **Step 1** into the optimization goal function, and to obtain the optimal power allocation by using the geometric programming based iterative algorithm which is summarized above.

So far we solve the problem and the numerical experiment results will be discussed in the next section.

NUMERICAL EXPERIMENTS

Simulation Parameters

We consider the downlink of a multi-cellular decode-and-forward (DF) relayed OFDMA system, with $N = 3$ cells and $K = 8$ sub-carriers in each cell. Each cell has $J = 3$ nodes, namely, one base station and two mobile relay stations. The radius of each micro cell and relay cell are set as 500m and 50m, respectively. The channel impulse response (CIR) is generated in Rayleigh distribution for each channel in our channel matrix. And the AWGN power spectral density is set to -174dBm/Hz. In the large scale, users are distributed uniformly in each macro cell or mobile relay station cell.

We give each user a random time delay parameter, then we can obtain the bit rate constraint of each UE based on the time delay parameter and (17). Specifically, each HUE has a random bit rate constraint in uniform distribution range in (0, 5) bits/symbol, and each MUE has a random bit rate constraint in uniform distribution range in (0, 2) bits/symbol. As we can see, users on the train have tighter delay parameters than those on the ground communicating directly with the BS. That is to say, the HUE has a higher bit rate constraint than the MUE while they have the same service.

To verify the effectiveness of the proposed resource allocation scheme, we present other three schemes as comparisons which are defined as

1. Our greedy sub-carrier assignment and geometric programming based power allocation scheme with constraints of the time delay.
2. Our proposed scheme without time delay constraints.
3. Compared scheme 1: greedy sub-carriers allocation and fixed equal power allocation scheme.
4. Compared scheme 2: random sub-carriers and fixed equal power allocation scheme.

As an ideal upper bound, we introduce a benchmark using orthogonal sub-carriers allocation and water-filling power allocation in addition.

Simulation Results

In order to evaluate the performance of these schemes, we set up different user numbers of each node (BS or relay node) ranging from 8 to 40. Users are distributed uniformly in each macro cell or mobile relay station cell. Figure 4 shows sum system data rates of all schemes with different user numbers.

We can make three observations from Figure 2. Firstly, our proposed scheme has the performance close to the ideal upper bound which verifies the effectiveness of our proposed scheme. Secondly, our proposed scheme outperforms the same scheme without delay constraints and this proves that the time delay constraint is necessary in such a high-speed environment. Specifically, our proposed scheme has a

Figure 4. Comparison of total rates for three schemes with different user numbers

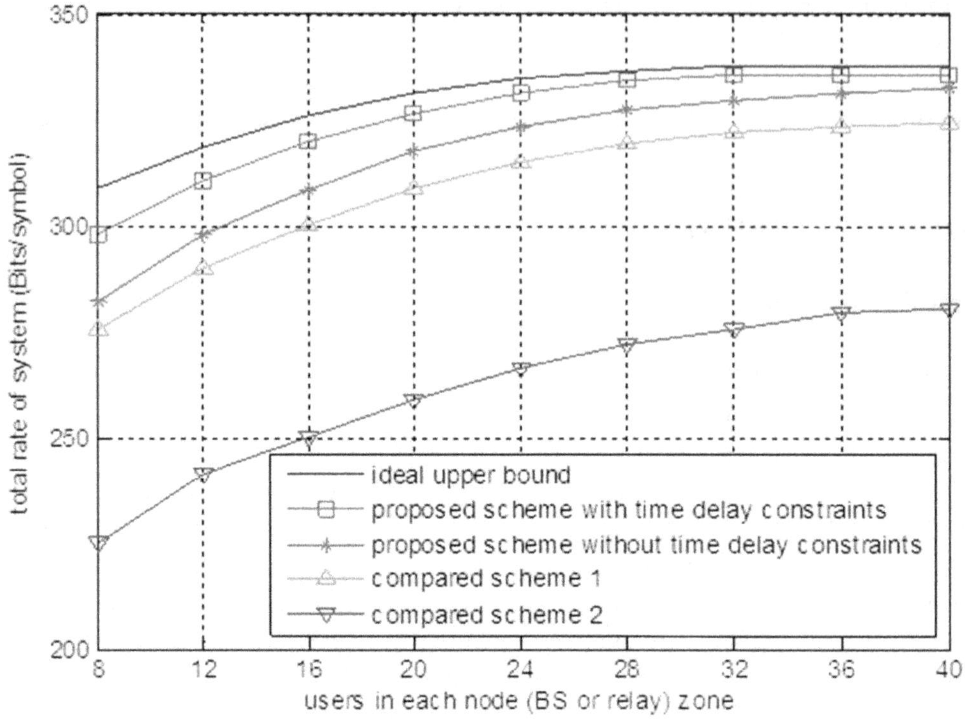

system data rate increase about 6% at most compared to the same scheme without delay constraints. Finally, our proposed scheme has an average performance improve about 8% and 30% than the compared scheme 1 and compared scheme 2 respectively, caused by better sub-carriers and power allocation algorithms.

Furthermore, to investigate the algorithm performance under different signal-to-noise ratio (SNR), we set up different SNRs ranging from 10dB to 35dB. Interestingly, as showed in Figure 5, all 4 schemes perform similarly at low SNR regimes. This is because that for low transmit power, noise is the dominant factor affecting the system performance. On the other hand, the performance are improved as the constrained total power increases, and our proposed scheme achieves much better performance improvement than other three schemes at high SNR regimes.

We also simulate the average throughput for each user. From Figure 6 we can observe that the average throughput for each user is reduced when the total number of users raises. This can be explained that each user can obtain sufficient sub-carrier when the total users are few. However, with the raising of the number of users, it is impossible for all users to get the enough frequencies to transmit data so that the average throughput decreases. With considering the coordination of CCI, our proposed algorithm outperforms the other three schemes.

CONCLUSION

In this chapter, we consider the DL of a multi-cellular DF relayed OFDMA system with multiple BSs and MRs in a high-speed environment. We propose a resource allocation scheme combined by greedy

Figure 5. Comparison of three schemes with different values of channel SNRs

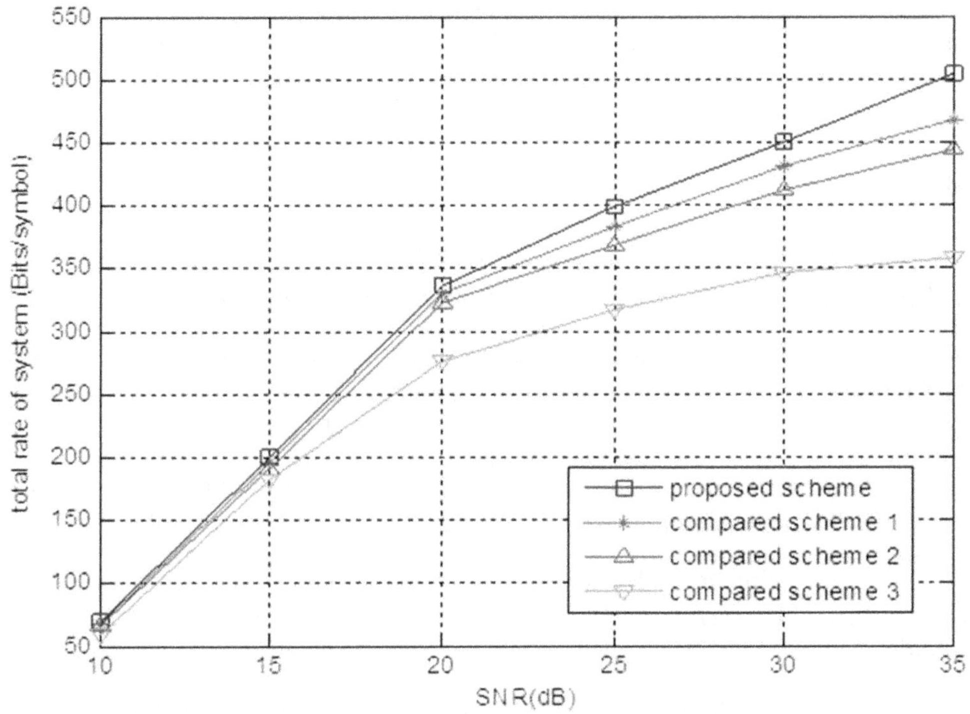

Figure 6. Comparison of average throughput for three schemes with different values of channel SNRs

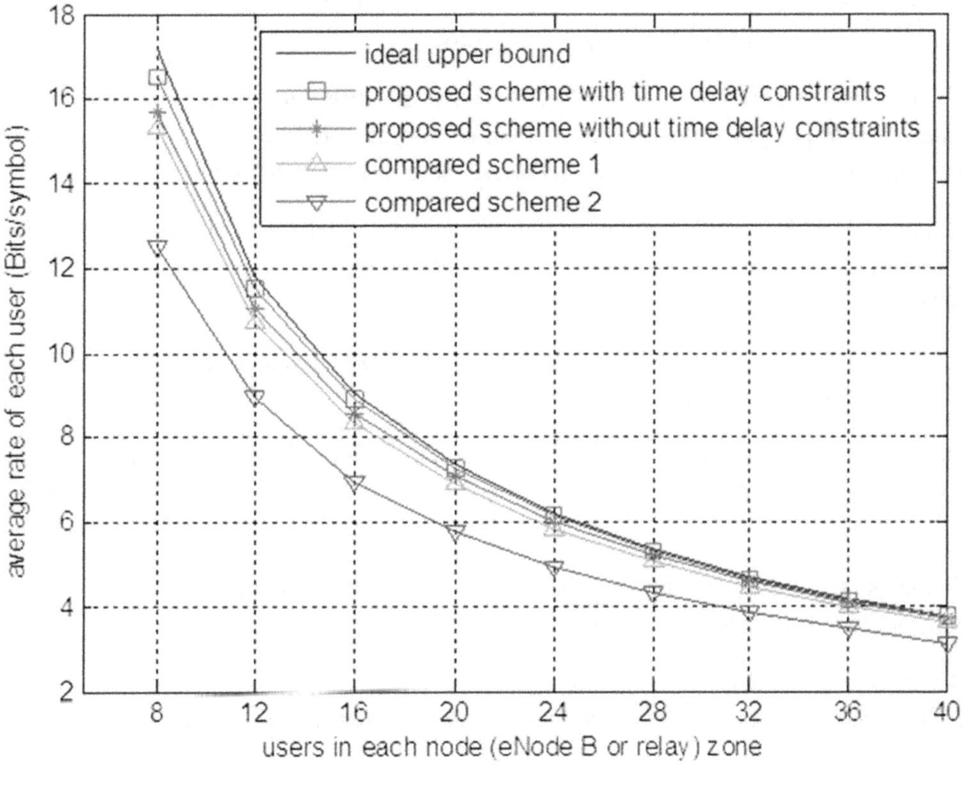

sub-carriers allocation and geometric programming based power allocation to maximize the system sum rate subject to a total power constraint for all cells. Furthermore, we add a new time delay constraint into the optimization goal function, which is necessary for the high-speed mobile environment. Simulations verify our scheme has better performance than other schemes in varied situations.

REFERENCES

Aguado, M., Onandi, O., Agustin, P. S., Higuero, M., & Taquet, E. J. (2008). WiMax on Rails. *IEEE Vehicular Technology Magazine*, *3*(3), 47–56. doi:10.1109/MVT.2008.927489

Balakrishnan, R., Yang, X., Venkatachalam, M., & Akyildiz, I. F. (2011, March). *Mobile relay and group mobility for 4G WiMAX networks*. Paper presented at the meeting of the IEEE Wireless Communications and Networking Conference, Cancun, Mexico. doi:10.1109/WCNC.2011.5779334

Bhat, P., Nagata, S., Campoy, L., Berberana, I., Derham, T., & Liu, G. et al. (2012). LTE-advanced: An operator perspective. *IEEE Communications Magazine*, *50*(2), 104–114. doi:10.1109/MCOM.2012.6146489

Chiang, M., Tan, C. W., Palomar, D. P., O'neill, D., & Julian, D. (2007). Power control by geometric programming. *IEEE Transactions on Wireless Communications*, *6*(7), 2640–2651. doi:10.1109/TWC.2007.05960

Fu, T. H., Xu, S. Y., & Kwak, K. S. (2014). *Geometric Programming Based Resource Allocation for Mobile Relays on Vehicles*. Paper presented at the meeting of the IEEE 79st Vehicular Technology Conference, Seoul, South Korea. doi:10.1109/VTCSpring.2014.7022930

Gupta, A., & Jha, R. K. (2015). A survey of 5G network: Architecture and emerging technologies. *IEEE Access*, *3*, 1206–1232. doi:10.1109/ACCESS.2015.2461602

Haider, F., Dianati, M., & Tafazolli, R. (2011, July). *A simulation based study of Mobile Femtocell assisted LTE networks*. Paper presented at the meeting of the 7th International Wireless Communications and Mobile Computing Conference, Istanbul, Turkey. doi:10.1109/IWCMC.2011.5982771

Hao, L., & Zhuang, W. (2012). Efficient on-demand data service delivery to high-speed trains in cellular/infostation integrated networks. *IEEE Journal on Selected Areas in Communications*, *30*(4), 780–791. doi:10.1109/JSAC.2012.120512

IMT-2020. (5G) Promotion Group. (2014). *White paper on 5G vision and requirements*. Retrieved February 20, 2015, from http://www.imt-2020.cn/zh

Jin, Z., & Vandendorpe, L. *Resource allocation in multi-cellular DF relayed OFDMA systems*. (2011, December). Paper presented at the meeting of the IEEE Global Telecommunications Conference, Houston, TX.

Kolios, P., Friderikos, V., & Papadaki, K. (2011). Future Wireless Mobile Networks. *IEEE Vehicular Technology Magazine*, *6*(1), 24–30. doi:10.1109/MVT.2010.939905

Li, Y., Vucetic, B., Zhou, Z., & Dohler, M. (2007). Distributed adaptive power allocation for wireless relay networks. *IEEE Transactions on Wireless Communications*, *6*(3), 948–958. doi:10.1109/TWC.2007.05256

Mao, L., Xu, S. Y., Fu, T. H., & Huang, Q. *Game theory based power allocation algorithm in high-speed mobile environment.* (2012, September). Paper presented at the meeting of the IEEE 72nd Vehicular Technology Conference, Ottawa, Canada. doi:10.1109/VTCFall.2012.6399136

Nesterov, Y., Nemirovskii, A. S., & Ye, Y. (Eds.). (1994). *Interior-point polynomial algorithms in convex programming*. Philadelphia: Society for Industrial and Applied Mathematics. doi:10.1137/1.9781611970791

Shi, C., Wang, Y., Wang, T., & Zhang, T. (2011). *Cross-layer design for interference-limited spectrum sharing systems with heterogeneous QoS*. Paper presented at the meeting of the IEEE Global Telecommunications Conference, Houston, TX.

Xiao, L., Fuja, T. E., & Costello, D. J. (2010). Mobile Relaying: Coverage Extension and Throughput Enhancement. *IEEE Transactions on Communications*, *58*(9), 2709–271. doi:10.1109/TCOMM.2010.080310.090199

Xu, S. Y., Han, J., & Chen, T. (May, 2012). *Enhanced Inter-Cell Interference Coordination in Heterogeneous Networks for LTE-Advanced.* Paper presented at the meeting of the Vehicular Technology Conference, Yokohama, Japan. doi:10.1109/VETECS.2012.6240203

Yamada, K., Sakai, Y., Suzuki, T., Kawahara, Y., Asami, T., & Aida, H. (2010, September). *A Communication System with a Fast Handover under a High Speed Mobile Environment.* Paper presented at the meeting of the IEEE 72nd Vehicular Technology Conference, Ottawa, Canada. doi:10.1109/VETECF.2010.5594584

Yao, H., Zhang, Q., & Niu, Z. (2010, March). *Resource allocation in multi-cell OFDMA-based relay networks*. Paper presented at the meeting of the IEEE INFOCOM, San Diego, CA.

Zhou, Y., Pan, Z., Hu, J., Shi, J., & Mo, X. (2011, April). *Broadband wireless communications on high speed trains*. Paper presented at the meeting of the 20th Annual Wireless and Optical Communications Conference, Newark, NJ. doi:10.1109/WOCC.2011.5872303

Zhu, H. (2012). Radio resource allocation for OFDMA systems in high speed environments. *IEEE Journal on Selected Areas in Communications*, *30*(4), 748–759. doi:10.1109/JSAC.2012.120509

Chapter 10
Self-Organization and Optimization in Heterogenous Networks

Aradhana Misra
Gauhati University, India

Kandarpa Kumar Sarma
Gauhati University, India

ABSTRACT

Driven by the continuous growth of wireless and mobile communication users, Long Term Evolution (LTE) and Long Term evolution-Advanced (LTE-A) have taken up densification of network as a new paradigm to meet the growing capacity demands. Small cells come with the advantage of enhanced coverage in indoor and hard reach areas and offer traffic offloading capacity in hotspots. However, there are challenges of interference management and self-adaptability with the overlaying macro cellular network since most small cell base stations are user deployed and do not have centralized control on their configuration and operation. The purpose of this chapter is to elaborate the concept of heterogeneous network and Self Organization Network (SON) in LTE-A. The various use cases of SON that can benefit the heterogeneous network have been discussed laying emphasis on interference management use case. Further the current trend of research in this field has been highlighted. It provides a holistic picture of the heterogeneous network and SON in LTE-A and upcoming mobile communication generations.

INTRODUCTION

With an exponential increase in mobile traffic in recent times, the need to enhance cellular capacity has been a major issue of focus for mobile operators as well as in the field of research. Literature has suggested three conventional ways to address the need for increased cellular capacity. The first is to utilize more frequency bandwidth, the second is by making use of spectrally efficient transmission techniques through air interface and the last is by spatial frequency reuse which employs more number of transmitters and receivers at closely spaced distance. It has been established that increasing spectral reuse

DOI: 10.4018/978-1-5225-1712-2.ch010

increases the capacity more than the first two methods combined (Chandrashekhar, 2008). Densification of network is the new paradigm of Long Term Evolution-A (LTE-A) to meet the high traffic demand and need for seamless service in indoor environments such as homes, offices, and dense hotspots like airports or stadium and isolated areas. Network densification means under laying the existing macro cellular network with small cells like micro cells, pico cells and femto cells thus resulting in a heterogenous topology. Traffic load from macro cellular network can be offloaded to these small cells to increase the network capacity. Also, such small cells can give better coverage in indoor environments as well as hotspots. Another advantage of small cells is that they use low cost, low power base station as the coverage radius is limited up to a few meters. Thus they offer energy saving as well. However, there are certain issues that have to be handled in dense network deployment. Due to the close proximity of femto cells in a dense deployment and spectrum sharing between cells belonging to different tiers, there are serious issues of interference in a heterogenous network. The interference management techniques devised for heterogenous networks are called the enhanced Inter Cell Interference Coordination (eICIC) and methods include algorithms for resource allocation, controlling transmit power or beamforming.

Deployment and operation of multi-tier cellular network is a complicated task and needs to consider multiple stages of planning, dimensioning, testing, installation and failure maintenance and correction and pre and post launch optimization. In a dense mobile network, all these activities may become labour extensive as well as expensive. Often these small cells are user deployed and network planning and management of such cells cannot be accomplished centrally. Another major challenge is the delay that may be incorporated due to human intervention in these manual maintenance activities. Consequently, the network becomes inefficient in terms of throughput and reliability of operation. This in turn lowers the Quality of Service (QoS) and increases dissatisfaction at the customer end. Therefore, with increasing densification of cellular networks, the mobile operators are facing the challenge of providing high quality services to users while reducing its CAPital EXpenditure (CAPEX) and OPerational EXpenditures (OPEX). These challenges prompt the need for an automated system wherein, maintenance and operation of the system may be managed by the network itself according to the system dynamics.

The human intervention in network operation and maintenance can be reduced to a great extent by incorporating self-configuration and self-optimization mechanisms in the network. With this objective, the Self Organizing Network (SON) has been introduced as part of Third Generation Partnership Project (3GPP) Long Term Evolution (LTE) to operate with automated features like self-configuration, self-healing and self-optimization (3G Americas, 2009; 4G Americas, 2011). While self-configuration and optimization are mainly useful in the initial deployment phase of a new or extended cellular network, self-healing helps in automatic detection and adjustment of parameters in the later operational phase. A major source of SON development is the industry forum Next Generations Mobile Networks (NGMN). NGMN had established the initial set of requirements of SON and several use cases were later identified to cover multiple aspects of network operation such as planning, deployment, optimization and maintenance. SON concepts have been included in the LTE standards from the first release of the technology (3GPP Release 8) and its scope has been expanding in the subsequent releases (Release 9, 10 and 11). While the first release mainly includes the initial use cases of deployment phase like Automatic Software Download, Automatic Neighbour Relation (ANR), and Automatic Physical Cell Identity (PCI) assignment, the subsequent releases include optimization use cases like coverage and capacity optimization, mobility optimization, random access channel optimization and load balancing optimization and self-healing use cases. Thus with the expanding mobile networks and inclusion of heterogeneous services, SON and its use cases have become a core area of research and it has abundant scope to evolve as a promising technology to meet the requirements of fourth and fifth generations of mobile networks.

The purpose of this chapter is to elaborate the concept of interference management in LTE-A heterogeneous network and the viability of SON for such cases. The basic theory behind the working of the heterogeneous network, interference issues in a heterogenous network, SON as part of LTE-A, the impact of SON use cases in heterogenous networks, its challenges, review of the solutions available and suggested, benefits to be derived and system design trends have been discussed in this chapter. It is likely to provide a holistic picture of the heterogeneous network and SON in LTE-A and upcoming mobile communication generations.

HETEROGENOUS NETWORKS

Initially LTE networks were deployed as homogenous networks with macro base stations providing the basic coverage in an area. Such networks are planned prior to deployment and can be optimized to some extent after roll out. Although LTE has been successful in meeting the user requirements of higher data rate and capacity, the performance of such networks is limited by the growing demand for data and unequal distribution of user traffic. With the explosion in data traffic, capacity increase in homogenous networks is primarily achieved by increase in node density through cell splitting or by addition of carriers to the existing network. However, these methods are limited by increase in intercell interference due to high density deployment and also the high capital expenditure associated with the installation and operation of high power Macro Base Stations (MBS). In order to increase coverage and capacity cost-effectively, an alternative strategy considered by LTE is the deployment of low power nodes within a macro base station cellular area to form a heterogeneous network. As compared to a fully macro network, a heterogenous layout has improved throughput and interference management as shown in the report by Qualcomm (2011). In this section, an insight is provided into heterogenous networks and its deployment in LTE and the type of interferences that may be seen in such networks. Further in this chapter, issues related to SON and interference management has been discussed keeping heterogenous network structure as the preferred network layout.

A heterogenous network, also called a HetNet, is a layered Radio Access Network (RAN) topology that uses a macro cell and some smaller (micro, pico and femto) cells to provide wireless services. MBS power typically varies between 5 and 40W. Pico base stations are normal evolved NodeB (eNB) with low transmit power, particularly used for outdoor deployment. The power of pico base stations normally ranges from 250 mW to 2W. Femto base stations are for indoor use and have power 100mW or less. With the mobile data traffic expected to increase by ten folds by the end of 2021 and indoor data traffic growing 20% faster than outdoor traffic, femto cell deployment is almost inevitable in the future generations of mobile communication. Femto cells are low power, short range and low cost small base stations typically designed to provide coverage in indoor environment (Chandrashekhar, 2008). They operate in the licensed spectrum and are connected to the network operator through broadband connection like Digital Subscriber Line (DSL), cable modem or separate radio frequency backhaul channel. With a transmit power of less than 100 mW, they have a short range of 30-40m. The low power requirement also guarantees the advantage of added battery life. A heterogenous network layout is illustrated in Figure 1.

Figure 1. A Heterogenous network

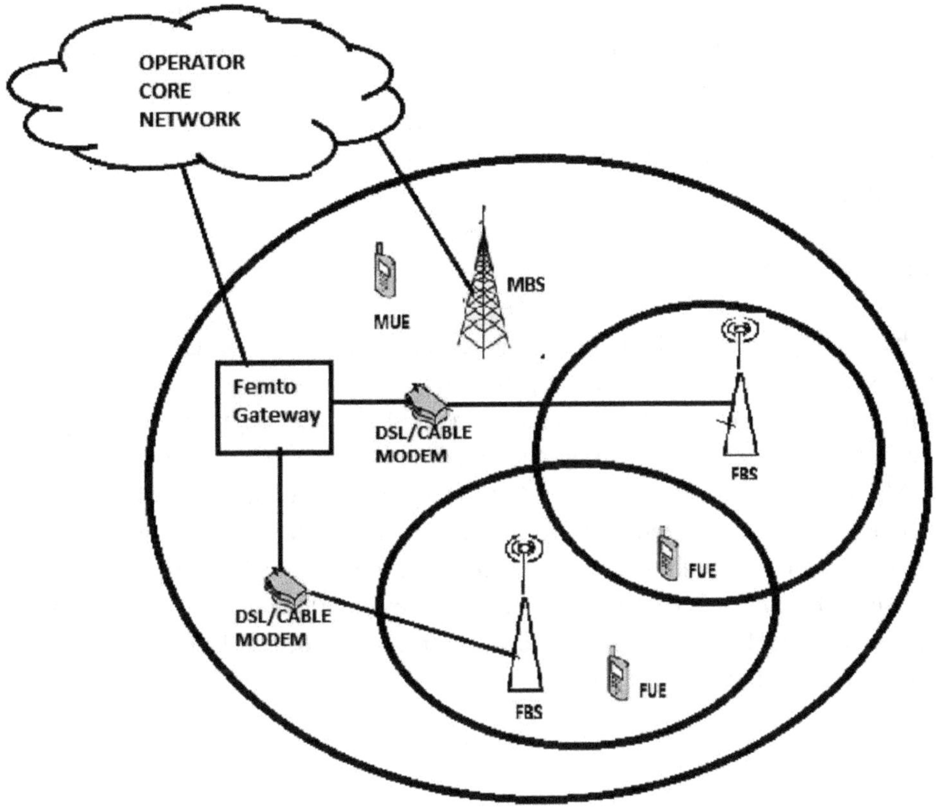

Radio Resource Allocation

There are three types of radio resource allocation in femto cells. The first kind is orthogonal resource allocation where independent segments of the spectrum are allotted to the macro cell and the femto cell. The second type is co-channel deployment where the spectrum is shared by the macro and femto cells. The second approach has efficient spectrum utilization but results in inter-cell interference problem. In the third type of deployment i.e. partial co-channel deployment, the whole spectrum is divided into two parts and one part is allotted as dedicated spectrum to the macrocell while the remaining part is shared by both the macro cell and the femto cell (Damnjanovic et al., 2011).

Access Modes

Femtocells have access control modes to withstand potential network intrusion and hacking. Femtocells have a list of subscribers to prevent unauthorized users from accessing the network services through the femtocell. The access methods can be classified as open, closed and hybrid access methods (Claussen et al., 2008). In closed access, the owner of the femtocell can restrict the usage to a closed group of members. A rejection message is sent to unauthorized mobile users that try to access the femtocell with restricted access and the mobile device is switched to another radio channel of the macrocell network to avoid interference. This kind of access can be used in homes and offices for private use. In open access femtocells, the access to the femtocell is open for all subscribers in the vicinity. Such cells are mostly useful in public places like malls, restaurants, airports etc. Open access femtocells increase capacity as macrocell users can connect to a neighboring femtocell if macrocell coverage decreases. The drawback with open access is that there are security issues and also the number of handoff increases. Hybrid access is a kind of open access but subscribed users are given priority of access as compared to non-subscribed users (Jo et al., 2012).

Interference in HetNets

HetNets have been adopted by LTE-A for its related advantages of spectrum utilization and energy efficiency. However, the coverage area of the pico and femto cells being small, Hetnets have high cell density and this lead to serious inter-cell interference problem. Deployments of femtocells, however, pose a lot of technical challenges like timing and synchronization, resource allocation and interference management. Allotment of separate carriers to femtocells may turn out useful in mitigation of interference, but it is not a practical solution considering the constraint on the available spectrum for usage. Even if a dedicated spectrum is allotted to femtocell operation, intercell interference remains a serious problem, particularly because of the un-coordinated deployment of femtocells. There are two types of interference that is seen in Hetnets: co-tier and cross-tier interference. A brief discussion of these interferences are given in the following sections.

Co-Tier Interference

Co-tier interference occurs among network elements of the same tier in the network. In a femtocell network, interference between neighboring femtocells will be termed as co-tier interference. In the uplink, a Femto User Equipment (FUE) may cause co-tier interference to a neighboring femtocell, whereas in the downlink, a Femto Base Station (FBS) may cause interference to a FUE of the neighbouring base station. In Orthogonal Frequency Division Multiple Access (OFDMA) based heterogenous network, co-tier interference will exist only if the source of interference and the victim are allotted the same sub-channels. Management of this kind of interference is possible if channel allocation to the neighboring femtocells is carefully planned. Co-tier interference mitigation schemes have been discussed by Lee et al., (2010), Kurda et al., (2014) and Huang et al., (2015).

Cross-Tier Interference

Cross-tier interference occurs between network elements of separate tiers of the network. For example, a FUE can cause cross- tier interference to the MBS and vice versa. This kind of interference occurs due to the spectrum sharing by the macro and femto cell and causes severe performance degradation if the density of the network increases. Cross-tier interference and suitable mitigation schemes have been discussed in the works of Zhenwei et al., (2009), Choi et al., (2011) and Bouras et al., (2012) to name a few.

Heterogenous networks offer throughput increase and improved coverage and capacity when the existing single tier network resources are exhausted to meet the demands of increased data traffic. However, there are complexities involved in the planning, deployment and maintenance of such layered networks.

Certain automation of installation and maintenance activities became necessary and this motivated the development of SON which in the recent times has come out successful in increasing the performance as well as reducing the deployment complexities of layered networks. The characteristics of SON and their usefulness have been discussed in the following section of the chapter.

SELF ORGANIZING NETWORKS

The SON is an LTE network that has self-ability to change and configure, optimize the network coverage, capacity, cell size, topology and frequency allocation and bandwidth based on dynamic changes in traffic pattern, interference and signal strength and can recover and heal itself in case of system or network failures. The main purpose of SON is to minimize the operational cost of cellular network by eliminating human intervention during planning, deployment, operational and maintenance phases of the network. This also has the added benefit of reduced delay in operational and maintenance issues. Therefore, apart from reducing operational expenditure and capital expenditure, SON provides efficiency in terms of time, accuracy and self-adaptability to changing network environment. In this section, the SON functionalities, its use cases and the practical challenges facing the SON network are discussed.

Functionalities of SON

Self-organization is the key factor for the future evolution of mobile networks. A number of use cases and the related research results in each use case have been highlighted in (Barth and Kuehn, 2010). Some potential optimization parameters and the use cases served by these parameters are shown by 4G Americas, (2011). As depicted, most of the parameters serve more than one use case. As such there is high chance of getting contradictory objectives. It is therefore important to focus on the issue of minimizing interdependencies of use cases. In an attempt to obtain an automatic intelligent network, 3GPP has defined the main functionalities of SON as self-configuration, self-optimization and self-healing capabilities (4G Americas, 2009). These three use cases of SON have been discussed in the following sections and some important literatures related to these cases have been summarized in each section.

Self-Configuration

Self-configuring is the process where newly deployed nodes are configured through an automatic installation process. It includes basic set up and the initial radio related configurations in the pre-operational stage. The major aspects of self-configuration are automatic configuration of initial radio transmission parameters, automatic connectivity management and self-test and inventory check before proceeding with the connection. Self-configuration of eNBs reduces the amount of manual process in network planning, integration and configuration and therefore results in faster deployment at reduced capital expenditure.

Some of the use cases of SON pertaining to self-configuration functionality are Physical Cell Identity (PCI) and Automatic Neighbour Relation (ANR). In order for the User Equipment (UE) to easily identify the source of the receiving signal, each eNB is given a unique signature sequence known as the PCI. To assign PCI it is necessary to maintain a list of neighbor cells. The ANR function maintains the list of minimum number of cells required for handover. Some of the literatures related use cases under self-configuration category are summarized in Table 1.

Table 1. Summary of selected literatures on SON self-configuration use cases in heterogenous networks

Use Case	Reference	Salient Feature
Automated configuration of PCI	Wu et al., 2010	• PCI assignment and reuse based on geolocation information • About 58% saving in femtocell transmit power
	Ahmed et al., 2013	Based on UE measurements
Automatic Neighbour Relation	Mueller et al., 2011	• Convergence time of ANR function • Proposed method has lower measurement overhead

Self-Optimization

Self-optimization functions auto tune the network using performance measures from the UEs and eNBs. It is initiated after the eNBs are in the operational state to get the best level of efficiency under conditions of dynamic changes in traffic pattern, propagation characteristics and deployment patterns in the network. The objective functions to be optimized may vary and depend on a number of criteria like energy consumption, interference conditions, resource utilization, mobility optimization, etc. Self-optimization reduces work load of site survey, analysis of network performances and therefore results in reduced operational expenditure.

Self-optimization use cases in SON include energy saving optimization, Coverage and Capacity Optimization (CCO), Random Access Channel (RACH) Optimization, Mobility Robustness or Handover Optimization and Mobility Load Balancing (MLB) optimization (Feng and Seidel, 2008). In heterogenous networks, energy saving is mainly obtained by turning off eNBs with low traffic while coverage can be optimized by transmit power control and adjustment of antenna down tilt angle. The objective of Mobility Robustness Optimization (MRO) is to dynamically improve the network performance of handover in order to provide improved end-user experience as well as increased network capacity. The process where the network traffic load is shared by similar network elements is called load balancing. The SON algorithms that are used for load balancing may initiate traffic offloading from one element and may use inter carrier, intra carrier or inter technology resources. RACH is an uplink, unsynchronized channel that is used for initial access or uplink synchronization. Automation of RACH configuration through SON can reduce synchronization time; call setup time and handover delays. Detection of probable interferers and building self-optimizing mechanism to reduce interference through antenna tilting or resource allocation is classified as another use case of Inter Cell Interference Coordination (ICIC).

Table 2 gives a summary of few literatures under the use cases pertaining to self-optimization in heterogenous networks.

Self-Healing

Every system develops faults and this causes inconvenience to users as well as network operators. The self-healing process aims at automatic detection and localization of failures and correction of such cases with self-healing mechanism such a self-recovery of software, self-healing of board faults, cell outage detection, recovery and compensation. Some benefits of self-healing are reduction of manual work for failures, automatic detection and removal of failures and reduction of failure possibilities.

Table 2. Summary of selected literatures on SON self-optimization use cases in heterogenous networks

Use Case	Reference	Salient Feature
Coverage and Capacity optimization	Dudnikova and Panno, 2013	Transmit power control based on traffic
Energy Saving	Zhao and Chen, 2014; Hou and Chen, 2013	Automatic turning on and off of base stations if UE is out of coverage area
	Dudnikova and Panno, 2013	Switching off cells with low traffic
Mobility Robustness Optimization	Changati et al., 2013	• Based on estimated user position • Almost 30% reduction in handover
	Zheng et al., 2013	• Based on gradient algorithm in femtocell and cost Function • Reduced radio link failure and unnecessary handover
	Abdullah et al., 2014	• Handover optimization based onTime to Trigger(TTT) parameter • Reducing Ping Pong rate, lower handover rate
Mobility Load balancing	Aguilar-Garcia et al., 2015	• Low complexity method based on fuzzy logic controller • Temporary resize of overloaded cells by adjusting transmit power
ICIC	Zhou et al., 2013	• Game theoretic approach for frequency reuse • Self-organizing clustering and optimal frequency allocation to cluster
	Sathya et al., 2013	• Variable radius algorithm to increase or decrease cell edge region • Reduction of co-tier interference
	Halder et al., 2013	• Identification of interfering femtocells •Allocation of different Physical Resource Blocks to interfering cells •14% decrease in Inter Cell Interference to femto users

Outage detection and outage compensation are important use cases related to self-healing functionality. A cooperative outage detection scheme has been described in (Wang et al., 2014) where statistical characteristic of the Reference Signal Received Power (RSRP) is observed by FBSs over a time period to detect outage. The detections are triggered in the Femto Access Points and intercell communication is greatly reduced. A local cooperative architecture of femtocell network and outage detection with distributed trigger mechanism is explored in (Wang and Zhang. 2014) and (Fan and Tian, 2015). The work proposes a coalition game based resource allocation algorithm to compensate for outage detection in a small cell network and facilitate self-healing. When cell outage is detected, small cells form a coalition and serve the users of the faulty cell cooperatively with optimized power allocation. Some literatures related to the self-healing use case are summarized in Table 3.

Table 3. Summary of selected literatures on SON self-optimization use cases in heterogenous networks

Use Case	Reference	Salient Feature
Outage Detection and compensation	Wang et at., 2014	• Cooperative scheme to detect outage • Statistical characteristics of RSRP is observed by FBS over span of time
	Wang and Zhang, 2014; Fan and Tian, 2015	• Coalition game based resource allocation to compensate outage • Small cells form a coalition to serve the faulty cell cooperatively with optimized power.

SON Architectures

The Operation and Management (OAM) unit of the SON network has a Self-Organization Subsystem which helps in the optimization process. Depending on the location of this subsystem SON architectures are classified into three types: centralized, distributed and hybrid SON (Feng and Siedel, 2008). In centralized SON, all SON functions are located in the OAM system. It is easy to deploy but since different vendors have their own OAM, it offers low optimization for different vendors. In distributed SON, the SON functionalities are executed in the eNBs. Since there are many locations of SON functionality it becomes difficult to deploy them. At times for complex optimization, coordination of several eNBs may be required. It solves the difficulty faced by centralized SON in multivendor environment. Optimization algorithms are executed both in OAM and eNBs in case of hybrid SON. Simple and quick optimizations are carried out in the eNBs while the complex optimizations are carried out in the OAM. The three types of architectures are illustrated in Figure 2.

INTERFERENCE MANAGEMENT IN HETEROGENOUS NETWORKS

Interference mitigation schemes in LTE can be broadly classified into two major categories such as multi –antenna techniques (for interference alignment) and interference avoidance through intercell interference coordination. While both techniques are suitable for macro cell networks, the former method is not preferable in dense network deployments. Since small cell networks like femtocells use internet for backhaul connection, it can experience delay in information exchange and so antenna technique which requires channel state feedback is not preferred. Interference coordination through resource allocation is preferred over such femto cell networks. The interference mitigation techniques standardized by 3GPP for heterogenous networks is terms as enhanced Inter Cell Interference Coordination. Some of the interference management schemes popularly studied for femtocell deployment scenarios have been discussed in this section.

Figure 2. (a) Centralized, (b) distributed and (c) hybrid SON network

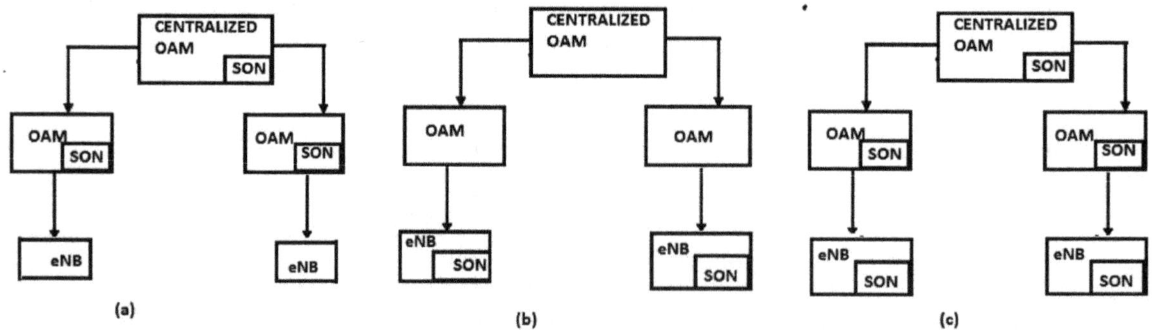

Transmit Power Control

Preconfiguration of the transmit power of the femto base station leads to simple deployment strategy but may result in sub-optimal performance. If the fixed power of the FBS is too low, interference received by the femto user from the neighbouring macro station will be too high, while very high power in the FBS may cause interference to any macro user at the cell edge. As such, proper control of transmit power of FBS is essential to manage the interference that may be caused either to the femto or macro user. Power control algorithms can be roughly categorized as centralized, decentralized or assistance based (Kurda et al., 2015).

Wang et al., (2011) have proposed a power control based interference mitigation scheme, where FBS transmit power is adjusted dynamically based on the information of RSRP measured at the victim Macro UE (MUE). It has been shown that interference at the victim MUE is lowered considerably at the expense of throughput degradation of the femto user. Interference mitigation through downlink power control has also been considered in the work by Xianfang et al. (2009).

In their work, Kurda et al., (2014) have considered a MBS assisted power adjustment algorithm which provides throughput improvement of both macro and femto user. Measured Signal to Interference Noise Ratio (SINR) and location information of both MUE and FUE is used to mitigate cross tier interference caused to victim MUE by neighbouring FBSs.

Choi et al., (2011) have proposed a new Time Division Duplexing (TDD) frame called the Listening-TDD frame (LTDDF) to acquire Channel Quality Information (CQI) from the MBS. During the downlink period of macro cell, the femtocell has both uplink and downlink periods. The uplink period of macrocell is used by the femtocell as the listening period to obtain CQI based on which the transmit power of the femto BS is adjusted. There is improved throughput of macro users, but due to the reduced frame time of femto user, its capacity is reduced by 70%.

Adaptive Frequency Reuse

In LTE/LTE-A, OFDMA is used in downlink communication where frequency reuse factor of one is supported to increase the capacity of the cellular system. Frequency reuse of one means that all the cells in the network can make use of all available channels. However, the drawback associated with the frequency reuse of one is that the users at the cell edge may have degraded signal quality due to interference from adjacent cells. In OFDMA, signal transmission occurs on the sub-channels which form only a fraction of the total available bandwidth. So to tackle the interference problem, proper allocation of sub-channels to the cell edges and cell centres is done which is known as Fractional Frequency Reuse (FFR). In FFR, the cell centres are allowed to operate using all of the available channels. At the cell edge, the users are allocated a separate set or fraction of frequencies. Interference is avoided by allotting the set of subchannels to the edges in such a way that adjacent cells do not share the same set of subchannels. Signal strength and throughput at the cell edge is improved. Partial Frequency reuse (PFR) and Soft Frequency Reuse (SFR) are two variations of the FFR scheme. The FFR schemes are popular in macrocells. However, they require prior planning of frequency and so FFR schemes are not feasible in femtocell networks which are mostly user deployed ad hoc networks. So another version of frequency planning technique is gaining popularity in heterogenous networks. This is known as adaptive frequency reuse and considers dynamic assignment of frequency based on availability, load conditions and interference caused to neighbouring cells.

An adaptive fractional frequency reuse scheme, where frequency reuse factor is determined based on the location information of femtocell is presented in the work by Lee et al., (2010). The work aims at mitigation of co-tier interference, ensures higher spectral efficiency and better QoS at cell boundary. A position based adaptive frequency reuse algorithm has been presented by Bouras et al., (2012) with increase in overall system performance.

Huang and Li (2015) in their work have proposed two algorithms, Primary Sub-Channel Self Configuration (PSC-SC) for inter cell interference coordination and Interference Aware Resource Allocation (IARA) for intra-cell interference management in a femtocell network. The entire set of sub-channels is divided into primary and secondary channels and the optimum number of primary sub-channels allotted to each FUE has been determined. The FUEs can access the secondary channels in case the traffic load demand greater bandwidth. As compared to frequency reuse of one, the adaptive frequency reuse algorithms proposed in this work have been able to show a gain of 100% in terms of spectral efficiency of the cell edge users. The proposed algorithm has low signaling overhead which is a desirable trait in capacity limited backhaul connection of femtocells.

Transmit Beamforming

Beamforming is a signal processing technique that is used in array antennas for directional transmission and reception of signals. The process of beamforming is accomplished by adjusting the complex weights of the antenna elements so that the beam is focused in the direction of the user. It has the capability to focus a beam in a desired direction, so most of the energy is concentrated in the main lobe. This reduces wastage of energy in redundant directions and also increases the SINR in a multipath fading environment. Interference control through coordinated beamforming in a Multiple Input Multiple Output (MIMO) femtocell network that requires sharing of beam index and angle information between heterogenous cells has been discussed in the work of Zhu (2011). Park et al., (2011) have proposed a codebook based beamforming for cross-tier interference reduction with the advantage that it does not require information exchange through backhaul. A null steering beamforming based on the angle of arrival information is presented in the work by Bartoli et al., (2014).

A summary of the various interference mitigation techniques and some related literatures are highlighted in Table 4.

SON AND INTERFERENCE MANAGEMENT

Co-tier and cross-tier interference problem as seen in heterogenous networks can be managed using various interference coordination techniques. A distributed and efficient self-organizing strategy is needed to implement the interference mitigation processes in a cost effective way (Andrews et al., 2012). Interference management schemes in 3GPP are termed as eICIC. Self-optimization of eICIC parameters can greatly enhance the QoS of small cells in a heterogenous network. Two parameters of eICIC that need to be optimized are the Cell Range Extension (CRE) and the Almost Blank Subframe (ABS). CRE optimization based on self-optimized load balancing between the macro and small cell has been discussed in (Tall et al., 2014). The ABS ratio defines the mute ratio for macrocell, i.e. when macrocells are muted for certain portions of the subframe so as to avoid interference to the users in the

Table 4. Summary of selected literatures on intereference mitigation in heterogenous networks

ICI Mitigation Technique	Reference	Features
Transmit power control	Wang et al.,2011	• Centralized power control algorithm • Power adjustment of femto BS based on RSRP measured at victim MUE • Cross-tier interference considered • Limitation of degradation of femto user throughput
	Choi et al., 2011	• Channel Quality information is acquired by listening to the surrounding environment using LTDDF • Based on CSI, femto base stations adjust transmit power •mitigates cross tier interference •30% higher MUE throughput • Decrease in femtocell capacity due to shorter downlink period
	Kurda et al., 2014	• Macro BS assisted power control algorithm • Power adjustment of both macro and femto BS • Both co-tier and cross-tier interference • Based on SINR and location of macro and femto user
Adaptive frequency reuse (ICIC)	Lee et al., 2010	• Based on location information of femtocell • Mitigation of co-tier interference • High spectral efficiency and QoS at cell boundary
	Bouras et al., 2012	• Reuse factor determination based on location of femto user • Throughput maximization via various combinations of inner cell radius and resource allocation • Cross-tier interference considered
	Huang et. al., 2015	• Primary sub-channel self-configuration for cross tier interference mitigation • Co-tier interference mitigation by Interference aware resource allocation algorithm •FUEs can use secondary channels in case of heavy traffic load
Transmit beamforming	Bartoli et al., 2014	• Reduction of cross-tier interference from femtocell on macrocell downlink communications •Based on estimation of direction of arrival •Uses zero forcing algorithm places nulls of femtocell transmission path in direction of macrocell user
	Park et al., 2011	• Codebook based beamforming • Eliminates the need of information exchange through limited backhaul • Reduction of cross-tier interference and improvement of throughput
	Zhu and Yang 2011	• Coordinated beamforming in MIMO femtocell network • Beamforming based on beam index and angle information • Cross-tier interference •Low complexity, reduced overhead signaling
eICIC: Resource allocation and ABS	Barbeiri et al., 2012	• Joint resource partitioning and power allocation •Closed access femtocell
	Yaacoub, 2015	• Sensing of interference (channel sensing) on subcarriers • Prioritize allocation of sub-bands with lower interference
	Zhang et al., 2013	• Resource allocation through graph coloring based clustering of cells to avoid co-tier interference •Orthogonal spectrum sharing between macro and femto cell to reduce cross tier interference
	Kamel and Elsayed, 2012	• Almost Blank Subframe framework (ABSF) with a procedure to mark and unmark the macrocell user victim •ABSF triggering at femto node

CRE area. Self-optimization of these two algorithms can lead to a cooperative functioning with proper load balancing and reduced interference between small and macro BS.

In (Guo et al., 2013), a self-deploying algorithm has been proposed for proper placement of FBS in indoor environment. The interference aware SON algorithm that indicates the best suitable location with minimum interference and maximum throughput depends only on three static network parameters:

the direction of strongest interference, propagation characteristics of medium and building material parameters and dimension. The work eliminates the need of coordination between femto BSs and hence is suitable for closed subscriber group femtocells. Also the need of interface between base stations is eliminated. As compared to random deployment, a throughput gain between 20%-50% has been reported using this SON algorithm.

Most of the ICIC techniques depend on partitioning of resources among the nodes to reduce the intercell interference problem. Further improvements can be noticed when the ICIC scheme is associated with autonomous power control algorithms. SON based resource allocation and scheduling has been reported in Liang et al., (2010), Bernardo et al., (2011), Lopez-Perez et al., (2013) and Shahid et al., (2014) among many. The contributions from these researches have been summarized in Table 5.

GAME THEORY FOR SON BASED ICIC IN HETNETS

Game theory is a mathematical tool that can be used for modeling and analyzing the interaction between the decision makers in a strategic manner. A game theoretic model consists of a set of players, a set of available actions or strategies for each player and a utility function that maps the action of each player to a real value. The set of players are the decision makers of the game, who choose the strategies depending on the state of the game. The preference of each player under the current strategy profile defines the payoff and can be modeled as a cost function or utility function. The choice of payoff depends not only on the player's own strategy but also on the strategies of other players. In recent years, game theory has found research importance in the field of wireless communication to optimize various issues and also for resource allocation problems. A game approach can be particularly useful in heterogenous network where each small base station is deployed in unplanned manner and hence needs to adapt to the network environment by interacting with the other neighboring base stations. By considering the FBSs as potential players, a decentralized framework can be modeled to serve the purpose of self-adaptation and configuration in a heterogenous network. A game theory process meets the basic idea of a SON to learn from the environment and autonomously adapt to reach equilibrium with emergent behavior.

Game theoretic models are broadly classified into two classes: cooperative games and non-cooperative games. In non-cooperative games, each player is considered rational and self-interested and selects strategies to optimize its own payoff function. Most non-cooperative games are based on Nash equilib-

Table 5. Summary of selected literatures on SON based interference management

Reference	Technique Used
Nazir et al., 2010	Self-organized interference mitigation through Game theory based power control in subcarriers, performance comparison of RL and evolutionary learning methods
Liang et al., 2011	Interference avoidance through self-organized resource allocation
Lopez-Perez et al., 2013	• Interference mitigation through downlink power control on each Resource Block • Power control through coordinated resource allocation
Shahid et al., 2014	• SON algorithm for joint component carrier selection and scheduling in the downlink • Makes use of CA concept to increase bandwidth. • Particle swarm optimization based selection and scheduling
Bernardo et al, 2011	• Self-organized frequency resource allocation using machine learning • Decentralized approach

rium. A non-cooperative game theoretic approach for interference mitigation in an OFDM network by transmit power adaptation of each co-channel user is presented in Al-Zahrani and Yu (2011). Another self-organized clustering of femtocells and non-cooperative game theoretic resource allocation to the clusters to minimize the co-channel interference is proposed in the work by Zhou et al., (2013). In co-operative game approach, the players cooperate and form coalitions and the optimization of coalition payoff is targeted rather than the individual payoff. A cooperative game theoretic approach for transmit power control in base stations is presented in Dirani and Altman (2010).

Learning Mechanisms for Game Theory in SON

In game theory based SONs, learning is an important paradigm through which the femtocell interacts with the environment and neighbouring interferers or learn from local information to self-organize itself for interference mitigation in a heterogenous network layout. Two strategic learning mechanisms viz. the evolutionary game approach and the Q-learning based approach for interference mitigation in a macro-femtocell network have been discussed in the following sections (Nazir et al., 2010). While evolutionary game approach depends on the behavior of other base stations to converge to an equilibrium state, in Reinforcement Learning (RL), the FBS self-organize based on local information and does not depend on other players' strategies. The two learning mechanisms and how they have been proposed as SON solutions in heterogenous networks are highlighted in this section.

Evolutionary Based Approach

In the Evolutionary game based approach, each FBS chooses its strategy against other FBSs within the same network. In this approach, there is exchange of information through a local controller between small cell BSs. The FBS observes the behavior of other BSs, learns from the behavior and makes decisions based on the instantaneous payoff as well as average payoff of all other base stations. At equilibrium, the instantaneous payoff is equal to the average payoff.

The works in Jiang et al., (2013) and Niyato and Hossain (2009), are based on formulating resource allocation and network selection in a heterogenous network as an evolutionary game, where the femto BSs are the game players and the evolutionary equilibrium is considered to be the solution to this game. A local combiner of the estimates from the neighboring femto BSs is used for the learning and strategy selection. Evolutionary game theory is used for subcarrier and power allocation to small cells in a heterogenous network in the work by Semasinghe et al., (2013). In their work, replicator dynamics is used to model the strategy adaptation and solution is reached when equilibrium is obtained. A summary of these literatures is provided in Table 6.

Table 6. Summary of selected literatures on Evolutionary based learning approach

Reference	Technique Used
Niyato and Hossain, 2009	Evolutionary game based resource allocation and network selection.
Semasinghe et al., 2013	• Evolutionary game theory for subcarrier and power allocation to small cells in a Hetnet • Replicator dynamics to model the strategy adaptation

While evolutionary game approaches require information exchange, this method is not so efficient in a femto network where information exchange between stations occur through backhaul connections thus introducing latency and partial information. As such, a reinforcement learning method, called the Q-learning method which requires no exchange of information is best suited for a heterogenous network considering the limitations of the backhaul network.

Reinforcement Learning

In reinforcement learning, an agent interacts with its environment and in the process the agent learns the optimal behaviour of states (Bennis et al., 2013). The learning problem is usually modeled as a Markov Decision Process (MDP) which is then solved using dynamic programming techniques. Most of these techniques require an accurate model for solving the problem and this is typically difficult in a dynamic femtocell network. A model free RL technique is the Q-learning which uses Temporal Difference (TD) to solve problems even without models. The Q-Learning method consists of three elements: state space, action space and the reward. The state space represents the various situations of the environment. The action space defines the possible values of the control parameters and rewards define how accurately the actions define the desired objective. The reward is defined as the difference between the previous state and the state reached after the current action has had its effect on the environment. In this learning method, a Q function $Q(S,A)$ is estimated where S is the set of states and A is the set of actions. The Q function maps the state $s \in S$ and the action $a \in A$, to a utility value called the Q value $Q(s,a)$. An optimal policy that returns the action a against state s so that the long term reward is maximized is given by

$$\pi = \arg\max_a Q\left(s,a\right)$$

Interference mitigation using a distributed Q- learning algorithm has been presented by Bennis and Niyato (2010), where the FBS learns to adapt to a channel selection strategy based on only local available information by a trial and error method until it reaches convergence. In this work, a self-organized macro-femto layout with spectrum sharing between the two tiers has been considered. With the proposed Q learning algorithm, the femtocells are able to self-organize and reduce cross-tier interference from femtocell to macrocell. While most game theoretic ICIC methods optimize a single parameter, Sallakh et al., (2014) have proposed a multiple parameter optimization method using Q-learning to reduce interference in a LTE SON. In Simsek and Czylwik (2012), a decentralized cross layer interference mitigation technique in a OFDMA based HetNet deployment is proposed where small cells optimize their downlink transmission through SON methods. A RL framework, based on an improved decentralized Q-learning algorithm for small cells sharing the macrocell spectrum with an improved effective initialization procedure has been discussed by the authors. In the work by Dirani and Altman (2010), Reinforcement Learning is used to optimize a Fuzzy Inference System (FIS), which is the controlled system. The approach is cooperative in the sense that the base stations which are the agents, share their learned experience and in turn the other agents learn about the power allocation strategy to reduce interference. The literatures discussed in this section are summarized in Table 7.

Table 7. Summary of selected literatures on SON based reinforcement learning

Reference	Technique Used
Bennis and Niyato, 2010	• FBS learns to adapt to channel selection strategy based on local information • Trial and error method is used during reinforcement learning process until convergence • Work is tested on a two tier femto-macro network with spectrum sharing
Sallakh et al., 2014	Q-learning to optimize multiple parameters to reduce interference
Simsek and Czylwik (2012)	• A RL framework based on improved decentralized Q-learning • Cross layer interference mitigation in OFDMA based Hetnets • Small cells optimize downlink transmission through SON methods.
Dirani and Altman, 2010	• RL to optimize Fuzzy Inference system • Interference reduction in a cooperative sense as base stations share their learning experience with other base stations.

CHALLENGES OF SON IMPLEMENTATION IN HETEROGENOUS NETWORKS

Some of the technical challenges seen in the implementation of SON algorithms in a dense network are discussed in this section.

- **Network Deployment:** The deployment of small cells is often random and dynamic as they are mostly plug and play devices installed by customers themselves without prior network planning. The number of active cells at any instant is therefore time varying. SON algorithms that optimize small cell parameters should therefore be adaptive to the random and dynamic nature of the network. Further, SON algorithms depend on information exchange and interaction with neighboring small cells and users. In an ultra-dense network, such complicated interaction of small cells can be an added bottleneck in the precise implementation of algorithms.
- **Limited Backhaul:** Since small cell networks like femtocell use internet for backhaul connection, it can experience delay in information exchange. SON algorithms which are dependent on environment sensing and neighbourhood information face challenge of the limited capacity and latency introduced by the backhaul connection. Therefore, feedback with large signaling overhead needs to be avoided leading to partial information exchange.
- **Imperfect Information:** SON optimization algorithms depend on information from other small cells or users. Due to the dynamic nature of the network, the information is time varying and the network and spectrum conditions may vary from the observed period to the decision period. Hence, the information available at the base station is not the true state of the network at that instant. Also, due to constraints in hardware and effect of noise, the information sensed by the network is most often corrupted. These issues pose a problem in information gathering from distant base stations and have to compromise with optimization using limited information from only the neighbouring base stations.

RECENT TRENDS IN SON NETWORKS

With the growth of mobile data and introduction of new technologies, there comes the challenges of increased network complexities and SON plays a vital role in simplifying network planning, configuration, optimization and self-healing of such complicated cellular networks.

Some of the major players in SON technology include Amdocs Ltd., Ericsson, Huawei Technologies Co. Ltd., Nokia Solutions and Networks, Cellwize Wireless Technologies Pte. Ltd., Cisco Systems Inc., Reverb Networks and Qualcomm Incorporated among others. Amdocs Ltd., launched its SON solutions in 2014, to manage heterogenous networks efficiently. Amdocs SON supports 2G, 3G and LTE services to optimize usage of existing network resources. Amdocs SON which has already been deployed with leading service providers globally has delivered 10% faster mobile broadband speeds, 20% increase in capacity utilization and 40% fewer dropped calls ("Self-Optimizing Networks,"). Ericson's SON optimization manager is a network management system that offers multi-vendor and multi-technology integration to mediate centralized access to networks ("Ericsson SON Optimization Manager", 2012). Reverb Networks launched the InteliSON 4.0 in 2015 to provide SON solutions in Middle East, Africa and Asia.

Huawei, a leading service provider in information and communication technology has been in SON research since 2006 and in 2011 it launched 'SingleSON', worlds first multimode, multi-layer SON ("SingleSON"). The first test in of SingleSON on a commercial LTE network in Hongkong was done in 2012 and the test reported a 30% increase in throughput for cell edge users by using Huawei's Adaptive ICIC technology. The first commercial launch of Huawei's ANR feature of SingleSON was announced on a LTE network in Germany.

UltraSON, a product of Qualcomm Technologies Inc., was developed for small cell to achieve 1000 times the capacity of conventional networks. UltraSON equipped small cells have reduced deployment time and cost, improved coverage and throughput performance improvement by 2 to 4 times for indoor users ("Cost effective Enterprise small cells", 2016). Qualcomm Wi-Fi SON, another major development by Qualcomm Atheros Inc., was launched in January 2016 and it has made its way to make deployment of Wifi networks in homes and offices and also providing cloud managed services in a cost efficient way. Qualcomm Wi-Fi claims to have easy installation of Wi-Fi routers and home gateways and also makes addition of new devices and range extenders very simple and hassle free. It can optimize its own performance as well as detect and rectify interference issues. This network has a capability to allot the best Wi-Fi link and bandwidth to each connected device to provide optimal performance and at the same time it builds a security mechanism on detection of threats.

The "SON (Self-Organizing Networks) Ecosystem: 2015 – 2030 – Opportunities, Challenges, Strategies & Forecasts" report published in 2015 by Signals and Systems Telecom presents an in-depth assessment of the SON and its optimization, OpEx and CapEx saving potentials, use cases, revenues forecasts, SON deployment case studies and key market drivers and challenges. According to the report, there shall be an expected growth of $4 Billion in SON revenue by end of 2017 in spite of implementation complexities. This exceeds the conventional mobile network optimization revenue by 60%. Dynamic activation and deactivation of nodes in the network according to the traffic conditions can lead to a power saving of 40%. Further mobile operators with SON implementations have reported 50% lesser dropped calls and 20% higher data rates.

CONCLUSION

In this chapter, the importance, issues and challenges of heterogenous network deployment have been elaborated. The background and the evolution of the technologies requiring SON deployment have been outlined. The issues involving the heterogenous networks, interferences threatening the QoS, SON and its related aspects have been discussed. The functionalities of SON have been outlined along with the various use cases defined by 3GPP and NGNM. The importance of SON in a dense network and their implementation algorithms has been discussed. Emphasis is laid on how interference mitigation has been dealt with in research and what are the solutions provided by various studies. Certain reported literatures related to SON have been covered and the key contributions have been summarized. This review has enabled to identify the challenges related to heterogenous networks and network densification. The SON approach in HetNets using game theory concept has been elaborated and various studies based on evolutionary and Q learning based approaches have been discussed.

Enhancements proposed in interference mitigation techniques in Release 12 focus more on the role of receiver in interference cancellation with or without network assistance and use of non-linear receivers. Further, with the increasing density of small cells, the role of SON and its functions shall continuously grow to provide robustness and automation in massive deployment of small cells. As interference is bound to increase with density of small cells, there is a need to develop more autonomous schemes to tackle the interference problem in a decentralized manner. Game theoretic approaches for interference mitigation through resource allocation which requires minimum information exchange through backhaul and can stand as a greener opportunity to resolve interference issues.

REFERENCES

3G Americas. (2009). *The benefits of SON in LTE: Self optimizing and self organizing networks.* Whitepaper.

4G Americas. (2011). *Self optimizing networks: The benefits of SON in LTE.* Whitepaper.

Abdullah, L. M., Baba, M. D., & Ali, S. G. A. (2014). Parameters optimization for handover between femtocell and macrocell in LTE-based network.*2014 IEEE International Conference on Control System, Computing and Engineering (ICCSCE).* doi:10.1109/ICCSCE.2014.7072796

Aguilar-Garcia, A., Barco, R., Fortes, S., & Munoz, P. (2015). Load balancing mechanisms for indoor temporarily overloaded heterogeneous femtocell networks. *EURASIP Journal on Wireless Communications and Networking*, *2015*(29).

Ahmed, M., Yoon, S.-G., Jeong, S., & Bahk, S. (2013). Geo-location based self-organization scheme for femtocell networks. *2013International Conference on ICT Convergence (ICTC).* doi:10.1109/ICTC.2013.6675415

Al-Zahrani, A. Y., & Yu, F. R. (2011). A game theory approach for inter-cell interference management in OFDM networks.*IEEE International Conference on Communications (ICC).* doi:10.1109/icc.2011.5962512

Andrews, J. G., Claussen, H., Dohler, M., Rangan, S., & Reed, M. C. (2012). Femtocells: Past, present, and future. *IEEE Journal on Selected Areas in Communications*, *30*(3), 497–508. doi:10.1109/JSAC.2012.120401

Barbieri, A., Damnjanovic, A., Tingfang, J., Montojo, J., Yongbin, W., Malladi, D., & Horn, G. et al. (2012). LTE femtocells: System design and performance analysis. *IEEE Journal on Selected Areas in Communications*, *30*(3), 586–594. doi:10.1109/JSAC.2012.120408

Barth, U., & Kuehn, E. (2010). Self-organization in 4G mobile networks: motivation and vision. *2010 7th International Symposium on Wireless Communication Systems (ISWCS)*.

Bartoli, G., Fantacci, R., Marabissi, D., & Pucci, M. (2014). Angular interference suppression in cognitive LTE-A femtocells.*2014 International Wireless Communications and Mobile Computing Conference (IWCMC)*. doi:10.1109/IWCMC.2014.6906488

Bennis, M., & Niyato, D. A. (2010) Q-learning based approach to interference avoidance in self-organized femtocell networks. *IEEE GLOBECOM Workshops (GC Wkshps)*.

Bennis, M., Perlaza, S. M., Blasco, P., & Han, Z. (2013). Self-organization in small cell networks: A reinforcement learning approach. *IEEE Transactions on Wireless Communications*, *1*(1), 1–11.

Bernardo, F. F., Agusti, R. R., Perez-Romero, J. J., & Sallent, O. (2011). Intercell interference management in OFDMA networks: A decentralized approach based on reinforcement learning. *IEEE Transactions on Systems, Man and Cybernetics. Part C, Applications and Reviews*, *41*(6), 968–976. doi:10.1109/TSMCC.2010.2099654

Bouras, C., Kavourgias, G., Kokkinos, V., & Papazois, A. (2012). Interference management in LTE femtocell systems using an adaptive frequency reuse scheme.*Wireless Telecommunications Symposium (WTS)*. doi:10.1109/WTS.2012.6266120

Chaganti, R., Sathya, V., Ahammed, S. A., Rex, R. A., & Tamma, B. R. (2013). Efficient SON handover scheme for enterprise Femtocell networks.*2013 IEEE International Conference on Advanced Networks and Telecommunications Systems (ANTS)*. doi:10.1109/ANTS.2013.6802870

Chandrasekhar, V., Andrews, J. G., & Gatherer, A. (2008). Femtocell networks: A survey. *IEEE Communications Magazine*, *46*(9), 59–67. doi:10.1109/MCOM.2008.4623708

Choi, B.-G., Cho, E. S., Chung, M. Y., Cheon, K.-Y., & Park, A.-S. (2011). A femtocell power control scheme to mitigate interference using listening TDD frame.*2011 International Conference on Information Networking (ICOIN)*. doi:10.1109/ICOIN.2011.5723186

Claussen, H., Ho, L. T. W., & Samuel, L. G. (2008). An overview of femtocell concept. *Bell Labs Technical Journal.*, *13*(1), 221–246. doi:10.1002/bltj.20292

Cost effective Enterprise small cell deployment with UltraSON. (2016). Qualcomm Technologies Inc. Retrieved June 26, 2016, from https://www.qualcomm.com/documents/whitepaper-cost-effective-enterprise-small-cell-deployment-ultrason

Damnjanovic, A., Montojo, J., Wei, Y., Ji, T., Luo, T., Vajapeyam, M., & Malladi, D. et al. (2011). A survey on 3GPP heterogenous networks. *IEEE Wireless Communications.*, *18*(3), 10–21. doi:10.1109/MWC.2011.5876496

Dirani, M., & Altman, Z. (2010). *A cooperative reinforcement learning approach for inter-cell interference coordination in ofdma cellular networks*. Avignon, France: WiOpt.

Dudnikova, A., & Panno, D. (2013). An energy saving approach for femtocell coverage function. *2013 22nd ITC Specialist Seminar on Energy Efficient and Green Networking (SSEEGN).*

Ericsson SON Optimization Manager. (2012). Retrieved June 26, 2016, from http://archive.ericsson.net/service/internet/picov/get?DocNo=28701-FGC1011634&Lang=EN&HighestFree=Y

Fan, S., & Tian, H. (2015). Cooperative resource allocation for self healing in small cell networks. *IEEE Communications Letters*, *19*(7), 1221–1224. doi:10.1109/LCOMM.2015.2433268

Feng, S., & Seidel, E. (2008). *Self organizing networks (SON) in 3GPP Long Term Evolution*. Nomor Research GmbH. White Paper.

Guo, W., Wang, S., Chu, X., Zhang, X., Chen, J., & Song, H. (2013). Automated small-cell deployment for heterogeneous cellular networks. *IEEE Communications Magazine*, *51*(5), 46–53. doi:10.1109/MCOM.2013.6515046

Haldar, K. L., Li, H., & Agrawal, D. P. (2013). A cluster-aware soft frequency reuse scheme for inter-cell interference mitigation in LTE based femtocell networks. *2013IEEE 14th International Symposium and Workshops on World of Wireless, Mobile and Multimedia Networks (WoWMoM)*. doi:10.1109/WoWMoM.2013.6583436

Hou, I.-H., & Chen, C. S. (2013). An energy-aware protocol for self-organizing heterogeneous LTE systems. *IEEE Journal on Selected Areas in Communications*, *31*(5), 937–946. doi:10.1109/JSAC.2013.130512

Huang, G., & Li, J. (2015). Interference mitigation for femtocell networks via adaptive frequency reuse. *IEEE Transactions on Vehicular Technology*, (99). 1-1.

Jiang, C., Chen, Y., & Liu, K. J. R. (2013, November). Distributed adaptive networks: A graphical evolutionary game-theoretic view. *IEEE Transactions on Signal Processing*, *61*(22), 5675–5688. doi:10.1109/TSP.2013.2280444

Jo, H.-S., Xia, P., & Andrews, J. G. (2012). Open, closed, and shared access femtocells in the downlink. *EURASIP Journal on Wireless Communications and Networking*, *2012*(1), 363. doi:10.1186/1687-1499-2012-363

Kamel, M. I., & Elsayed, K. M. F. (2012). Performance evaluation of a coordinated time-domain eICIC framework based on ABSF in heterogeneous LTE-Advanced networks.*2012 IEEE Global Communications Conference (GLOBECOM)*. doi:10.1109/GLOCOM.2012.6503967

Kurda, R., Boukhatem, L., & Kaneko, M. (2015). Femtocell power control methods based on users' context information in two-tier heterogeneous networks. *EURASIP Journal on Wireless Communications and Networking*, *2015*(1), 129. doi:10.1186/s13638-015-0328-z

Kurda, R., Boukhatem, L., & Yahiya, T. A. (2014). Interference mitigation in mobile environment through power adjustment in macro and femto cell systems.*2014 IEEE Wireless Communications and Networking Conference (WCNC)*. doi:10.1109/WCNC.2014.6952491

Lee, H.-C., Oh, D.-C., & Lee, Y.-H. (2010). Mitigation of inter-femtocell interference with adaptive fractional frequency reuse.*2010 IEEE International Conference on Communications (ICC)*. doi:10.1109/ICC.2010.5502298

Liang, Y. S., Chung, W. H., Zhang, H., & Kuo, S. Y. (2011). Interference mitigation through self-organization in OFDMA femtocells. *IEEE 22nd International Symposium on Personal Indoor and Mobile Radio Communications (PIMRC)*.

Lopez-Perez, D., Chu, X., Vasilakos, A. V., & Claussen, H. (2013). On distributed and coordinated resource allocation for interference mitigation in self-organizing LTE networks. *IEEE/ACM Transactions on Networking*, *21*(4), 1145–1158. doi:10.1109/TNET.2012.2218124

Mueller, C. M., Bakker, H., & Ewe, L. (2011). Evaluation of the automatic neighbor relation function in a dense urban scenario. *IEEE 73rd Vehicular Technology Conference (VTC Spring)*.

Nazir, M., Bennis, M., Ghaboosi, K., MacKenzie, A. B., & Latva-aho, M. (2010). Learning based mechanisms for interference mitigation in self-organized femtocell networks.*Conference Record of the Forty Fourth Asilomar Conference on Signals, Systems and Computers (ASILOMAR)*. doi:10.1109/ACSSC.2010.5757866

Niyato, D., & Hossain, E. (2009). Dynamics of network selection in heterogeneous wireless networks: An evolutionary game approach. *IEEE Transactions on Vehicular Technology*, *58*(4), 2008–2017. doi:10.1109/TVT.2008.2004588

Park, S., Seo, W., Choi, S., & Hong, D. (n.d.). A beamforming codebook restriction for cross-tier interference coordination in two-tier femtocell networks. *IEEE Transaction on Vehicular Technology, 60*(4), 1651-1663.

Qualcomm. (2011). *LTE advanced: heterogenous networks*. Author.

Sallakh, U., Mwanje, S. S., & Mitschele-Thiel, A. (2014). Multi-parameter Q-Learning for downlink inter-cell interference coordination in LTE SON.*IEEE Symposium on Computers and Communication (ISCC)*. doi:10.1109/ISCC.2014.6912599

Sathya, V., Gudivada, H. V., Narayanam, H., Krishna, B. M., & Tamma, B. R. (2013). Enhanced distributed resource allocation and interference management in LTE femtocell networks. *2013 IEEE 9th International Conference on Wireless and Mobile Computing, Networking and Communications (WiMob)*.

Self Optimizing Networks. (n.d.). Retrieved June 26, 2016, from http://www.amdocs.com/products/centralized-management-optimization-solution/pages/self-optimizing-networks.aspx

Semasinghe, P., Zhu, K., & Hossain, E. (2013). Distributed resource allocation for self-organizing small cell networks: An evolutionary game approach. *IEEE Globecom Workshops (GC Wkshps)*.

Shahid, A., Aslam, S., Sohaib, S., Kim, H. S., & Lee, K.-G. (2014). A self-organized metaheuristic approach towards inter-cell interference management for LTE-Advanced. *EURASIP Journal on Wireless Communications and Networking*, 2–15.

Simsek, M., & Czylwik, A. (2012). Improved decentralized fuzzy Q-learning for interference reduction in heterogeneous LTE-networks. *17th Proceedings of International OFDM Workshop 2012 (InOWo'12).*

Single, S. O. N. (n.d.). Retrieved June 26, 2016, from http://www.huawei.com/ilink/en/solutions/broader-smarter/HW_104774

SON (Self-Organizing Networks) Ecosystem. (2015). *2015 – 2030 – Opportunities, Challenges, Strategies & Forecasts. Market Research Report. Signals and Systems Telecom.* Retrieved June 26, 2016, from http://www.snstelecom.com/son

Tall, A., Altman, Z., & Altman, E. (2014). Self organizing strategies for enhanced ICIC (eICIC).*12th International Symposium on Modeling and Optimization in Mobile, Ad Hoc, and Wireless Networks (WiOpt).* doi:10.1109/WIOPT.2014.6850315

Wang, W., Liao, Q., & Zhang, Q. (2014). COD: A cooperative cell outage detection architecture for self organizing femtocell networks. *IEEE Transactions on Wireless Communications*, *13*(11), 6007–6014. doi:10.1109/TWC.2014.2360865

Wang, W., & Zhang, Q. (2014). Local cooperation architecture for self healing femtocell networks. *IEEE Wireless Communications.*, *21*(2), 42–49. doi:10.1109/MWC.2014.6812290

Wang, Z., Xiong, W., Dong, C., Wang, J., & Li, S. (2011). A novel downlink power control scheme in LTE heterogeneous network.*International Conference on Computational Problem-Solving (ICCP).* doi:10.1109/ICCPS.2011.6089792

Wu, Y., Jiang, H., Wu, H., & Zhang, D. (2010). Physical cell identity self-organization for home eNodeB deployment in LTE. *2010 6th International Conference on Wireless Communications Networking and Mobile Computing (WiCOM).*

Xiangfang, L., Lijun, Q., & Kataria, D. (2009). Downlink power control in co-channel macrocell femtocell overlay.*Annual Conference on Information Sciences and Systems (CISS).*

Yaacoub, E. (2015). Interference mitigation in femtocell networks with joint channel sensing and resource allocation.*2015 IEEE Wireless Communications and Networking Conference (WCNC).* doi:10.1109/WCNC.2015.7127569

Zhang, Q., Zhu, X., Wu, L., & Sandrasegaran, K. (2013). A coloring-based resource allocation for OFDMA femtocell networks.*2013 IEEE Wireless Communications and Networking Conference (WCNC).* doi:10.1109/WCNC.2013.6554644

Zhao, X., & Chen, P. (2014). Improving UE SINR and networks energy efficiency based on femtocell self-optimization capability.*2014 IEEE Wireless Communications and Networking Conference Workshops (WCNCW).*

Zheng, W., Zhang, H., Chu, X., & Wen, X. (2013). Mobility robustness optimization in self-organizing LTE femtocell networks. *EURASIP Journal on Wireless Communications and Networking*, *2013*(27).

Zhou, X., Feng, S., Ding, Y., Liu, Y., Wang, W., & Zhang, Y. (2013). Game-theoretical frequency reuse method for complex cognitive femto-cell network.*8th International ICST Conference on Communications and Networking in China (CHINACOM)*.

Zhu, J., & Yang, H.-C. (2011). Interference control with beamforming coordination for two-tier femtocell networks and its performance analysis.*IEEE International Conference on Communications (ICC)*. doi:10.1109/icc.2011.5963130

Chapter 11
Stackelberg Game Theoretic Framework in Cognitive Green Heterogeneous Networks

Chungang Yang
Xidian University, China

Jia Xiao
Xidian University, China

Pengyu Huang
Xidian University, China

Lingxia Wang
Xidian University, China

Jiandong Li
Xidian University, China

ABSTRACT

Game theory has found an extensive application in wireless communication networks including cognitive radio networks, heterogeneous cellular networks, cooperative relay networks. Also, cognitive radio networks, green communications and heterogeneous cellular networks have attracted a wide attention on improve the spectrum efficiency and energy efficiency; therefore, the capacity, the coverage and the energy consumption. However, interference problem and energy consumption are critical for these networks. Introducing hierarchy among different decision-making players in cognitive, heterogeneous, green, cooperative cellular networks can both save energy and mitigate interference, thus enhance throughput. Stackelberg game suits to model, analyze and design the distributed algorithms in these hierarchical decision-making networking scenarios. In this chapter, we introduce basics of Stackelberg game and survey the extensive applications of Stackelberg game in cognitive, heterogeneous, cooperative cellular networks with the emphasis on resource management, green commutations design and interference management. This chapter highlights the potentials and applications with the promising vision of Stackelberg game theoretic framework for future cognitive green heterogeneous cellular networks.

INTRODUCTION

Game theory is a branch of mathematics modeling, analyzing and designing the strategic interactions among autonomous decision makers. These players are usually assumed to be selfish and rational, which mean that they are always pursuing the utility maximization without considering the others' strategy,

DOI: 10.4018/978-1-5225-1712-2.ch011

utility the context. However, these actions have mutual, probably conflicting, consequences. Game theory has developed for decades to model problems in the field of economics, and it has recently been applied to wireless network problems, in most cases to solve the resource allocation problems in a competitive environment. Certainly, with its promising applications in particular for the decade, game theory including both strategic non-cooperative game and cooperative game is also used in access control, scheduling, network security, interference management and mitigation, even in the field of network information theory.

The reason that game theory is an appropriate choice for studying various problems in communications is multi-fold. First, nodes in the network are autonomous agents, making decisions only for their own interests. Game theory provides us sufficient theoretical tools to analyze the network users' behaviors and actions. Second, game theory primarily deals with distributed optimization, which often requires local information only. Thus, it enables us to design distributed algorithms.

Games can be classified into two categories, strategic form game and extensive form game. The strategic form game is a one-shot game. In this game, the players make their decisions simultaneously without knowing what others will do. On the contrary, the extensive form game represents the structure of interactions between players and defines possible orders of moves. The repeat game is the most simple dynamic game as an example of the extensive form game, in which each stage is a repetition of the same strategic game. At the beginning of each stage, players observe the past history of strategies before making decisions. The number of stages may be finite or infinite. The utility of each player is the accumulated utility through all the stages. Therefore, players care not only the current utility but also the future utilities. The Stackelberg game is another extensive form game, which is used to model the competition between one player, called the leader, and a set of players, called the followers. In this game, the leader takes action first and then the followers take actions. The leader knows that the followers observe its action and take actions accordingly. The Nash equilibrium in the Stackelberg game is called Stackelberg equilibrium. Although non-cooperative game also has found extensive application in wireless networks, we concentrate on the survey of Stackelberg game theoretic applications in the recent promising wireless communication networks including cognitive radio networks, heterogeneous cellular networks, cooperative relay networks.

The existing iterative power water-filling algorithm is regarded as one type of non-cooperative game application has found extensive use to allocate power among competing users in wireless communication networks. As we known, strategic non-cooperative games have been well explored in wireless community. However, it always leads to a non-social optimal solution, let alone the fairness among selfish and rational players. For instance, if selfish users choose their transmission probabilities independently without any coordination mechanism with other players, then it results in a network collapse. In (Park & Schaar, 2009), a methodology that transforms the non-cooperative game into a Stackelberg game is proposed. Stackelberg equilibrium solution of the Stackelberg game can overcome the deficiency of the Nash equilibrium solution of the original non-cooperative game. A particular type of Stackelberg intervention is constructed to show that any positive payoff profile feasible with independent transmission probabilities can be achieved as a Stackelberg equilibrium payoff profile. In (Lasaulce et al., 2009), hierarchy in wireless networks is modeled by a decentralized multiple access channel and for which energy-efficiency is the main performance index. In these networks users are free to choose their power control strategy to selfishly maximize their energy-efficiency. Two kind of hierarchies are included: first, assuming single-user decoding at the receiver, a Stackelberg formulation of the game is investigated where one user is the leader whereas the other users are assumed to be able to react to the leader's decisions; and then assuming neither leader nor followers among the users, introducing hierarchy by assuming successive interference cancellation at the receiver. It is shown that introducing a certain degree of hierarchy in non-cooperative power control games not only improves the individual energy efficiency of all the users but can also be a way of insuring the existence of a non-saturated equilibrium and reaching a desired trade-off between the global network performance at the equilibrium and the requested amount of signaling. In (Su & Schaar, 2009), the Stackelberg game was introduced to share a frequency-selective interference channel, where the interaction among selfish users is modeled as the non-cooperative Stackelberg game and analyze their strategic behavior via Stackelberg equilibrium in the framework of bi-level optimization. In the same scenario, the authors in (Su & Schaar, 2009) studied how a foresighted user, who knows the channel state information and response strategies of its competing users, should optimize its own transmission strategy. To characterize this multiuser interaction, the Stackelberg equilibrium is introduced. A simple two-user scenario was analyzed in detail, where the foresighted user can determine its optimal transmission strategy by solving a bi-level program which allows him to account for the myopic user's response strategies. Therefore, the competition among users is transformed into a cooperative competition since the foresighted user will avoid interfering the myopic user. Since the optimal solution is computationally prohibitive, we propose a low-complexity algorithm based on Lagrangian duality theory.

In addition to the dynamic access and power control problems, interference management is also critical Due to the increasingly tense situation of spectrum scarcity. An adaptive inter-cell interference management scheme is formulated as a game theoretic framework in (Chen et al., 2008) for an OFDMA-based WiMAX/3GPP-LTE system. Here, the users is categorized into two kinds including the ones who are, or are not, subject to potential interference, and different optimization opportunities for heterogeneous BSs are identified and analyzed. The rational interactions among them are modeled by a Stackelberg game where BSs can make intelligent decisions to reduce interference with minimum optimization cost. In the upper layer of the communication protocol, Stackelberg game also finds uses, in (Yang et al., 2007) a network flow routing problem is viewed from the Stackelberg game perspective, and then it is formulated as a mathematical program with equilibrium constraints. Using a marginal function approach, the problem is transformed into an equivalent, continuously differentiable single-level optimization problem. A Stackelberg game based cooperative user relay assisted load balancing scheme to tackle signal to noise ratio degradation problem of shifted cell-edge users which commonly occurs in a conventional direct handover load balancing scheme. In the proposed scheme of (Liu et al., 2013), users from a lightly loaded cell can be selected as cooperative user-relays and will be paid by the shifted cell edge users. Stackelberg game theory is applied to optimize the strategies of both the user relays and shifted cell-edge users, in order to maximize both of their utilities in terms of signal to noise ratio and payment. Therefore, we can conclude that Stackelberg game has been well investigated for various problems from the communication protocol perspective. It across multiple technologies and problems including the medium access control, power control, routing and relay selection and so on. In this chapter, cognitive radio networks, green communications and heterogeneous cellular networks have attracted a wide attention on improve the spectrum efficiency and energy efficiency; therefore, the capacity, the coverage and the energy consumption. However, interference problem and energy consumption are critical for these networks. Introducing hierarchy among different decision-making players in cognitive, heterogeneous, green, cooperative cellular networks can both save energy and mitigate interference, thus enhance throughput. Stackelberg game suits to model, analyze and design the distributed algorithms in these hierarchical decision-making networking scenarios. In this chapter, we introduce basics of Stackelberg game and survey the extensive applications of Stackelberg game in cognitive, heterogeneous, cooperative cellular networks with the emphasis on resource management, green communications design and interference management, as shown in Figure 1.

BASICS OF STACKELBERG GAME

We should first introduce the basics of Stackelberg games including definitions, equilibrium concepts, equilibrium solution analysis, and distributed algorithm design. In accordance with dynamic behaviors of the network users, there are always different game models for modeling the same problem. Different game models will lead to different concepts of the equilibrium solutions, for instance, the Nash equilibrium solution and correlated equilibrium solution cause different overall performance and need high or low implementation complexity. So far, for multiple users with short-sighted behaviors, we model it as a one-shot game model with a short-sighted utility function. Meanwhile, a more rational player will adopt a long-term expected utility, considering not only the current value of the utility function but also the future situation of available utility.

We first differentiate two types of selfish users according to the different degree of rationality and interests based on their response strategies to the dynamic environment.

1. **Myopic User:** A network user that always acts to maximize its immediate achievable utility. It is myopic, in the sense that it treats other users' actions as fixed, ignores the dependence between its competitors' actions and its own action, and determines its response so as to maximize its immediate payoff.
2. **Foresighted User:** A network user that selects its transmission action by considering the long-term impact on its performance. It anticipates how the others will react, and maximizes its performance

Figure 1. Extensive applications in emerging wireless networks

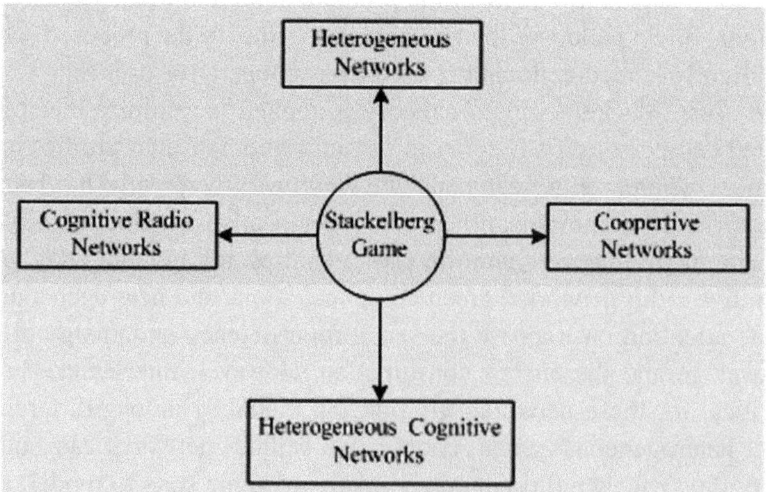

by considering their reactions. It should be highlighted that additional information is required to assist the foresighted user in its decision making.

At this time, we illustrate the basics of a Stackelberg game in Figure 2. There are two kinds of players in a Stackelberg game including the leader and follower players, always the myopic players represent the leaders, and foresighted players are the followers. Leaders can be the base station in the cellular networks, primary users in cognitive radio networks; and followers can be common subscribers, or secondary users, relays and smallcells. Usually, leaders has the first-move advantage, and followers can make much smarter decision supporting with history information. The two-tier interaction between leaders and followers make the Stackelberg game, and that is why Stackelberg game is also referred to hierarchical game in some published works.

Stackelberg equilibrium solution is always derived solution of a Stackelberg game, and always the optimization analysis via backwards induction methods is used to analyze such a game.

COGNITIVE RADIO NETWORKS

Cognitive radio networks contain two kinds of players, the secondary players and the primary users, where the secondary players are always the users do not have spectrum resources, however, they are with super spectrum sensing capabilities. They can find out the spectrum holes in the white or gray spaces, then analyze them in detail and opportunistically access of them with the required interference introduced to primary users below a threshold. We can find that most of the current works concentrate on spectrum sensing and dynamic spectrum access from the secondary users perspective, where no primary users participate the cognitive radio games. Certainly, there is some published works involving the primary users as the players in the gaming processing formulations.

Cognitive radio technology and networks enable the reuse of the available spectrum resources. The basic limiting factor for spectrum reuse is interference across the two kinds of players and among the secondary users. Around these issues such spectrum sensing and access, power control, spectrum pricing and leasing, and multiple antennas in the cognitive radio systems, we summarize the Stackelberg game theoretic formulation, analysis, design and simulation, as shown in Figure 3.

Figure 2. Basics of a Stackelberg game

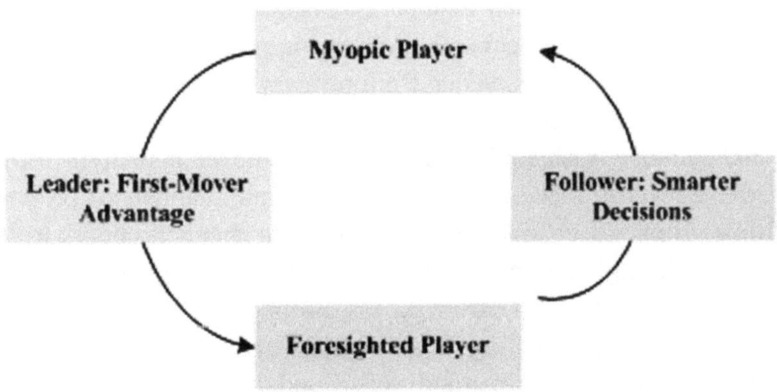

Figure 3. Stackelberg game in cognitive radio networks

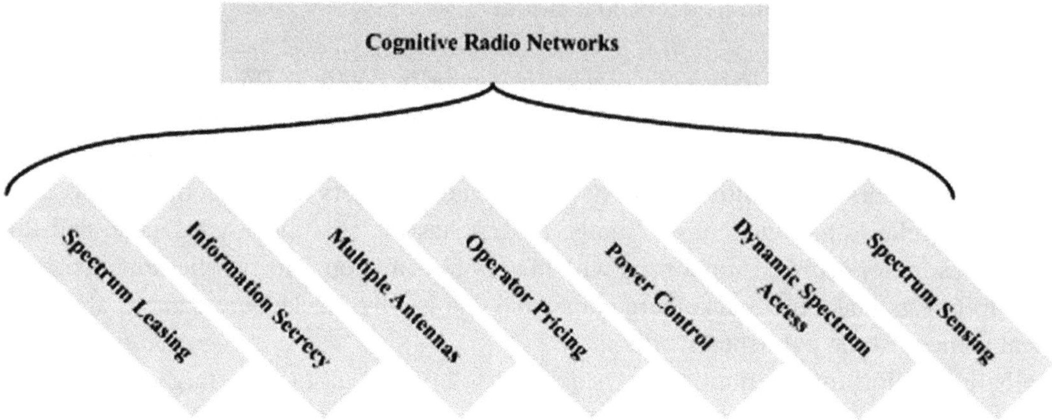

SPECTRUM SENSING

Packet delay compromises sensing energy consumption during spectrum sensing process, and to overcome the optimal trade-off problem, both packet delay and energy consumption are concerned in (Oussama et al., 2013). The authors find out that if the licensed channels are more occupied by primary users, then secondary users improve the spectrum utilization. Based on this observation, a Stackelberg game-theoretic formulation is proposed where the network manager increases the occupation of licensed channels in order to improve the secondary users' average throughput.

Dynamic Spectrum Access

With the development of cognitive radio technologies, dynamic spectrum access becomes a promising approach to increase the efficiency of spectrum utilization and solve spectrum scarcity problem. Under dynamic spectrum access, secondary users can dynamically access the licensed bands from legacy spectrum holders on an opportunistic basis. Currently, most primary users assume secondary transmissions as negative interference and don't actively involve them into the primary transmission. A cooperative cognitive radio framework is proposed in (Zhang et al., 2009), where primary users, aware of the existence of secondary users, may select some of them to be the cooperative relay, and in return lease portion of the channel access time to them for their own data transmission. Secondary users cooperating with primary transmissions have the right to decide their payment made for primary user in order to achieve a proportional access time to the wireless media. Both primary and secondary users target at maximizing their utilities in terms of their transmission rate and revenue/payment. This model is formulated as a Stackelberg game and a unique Nash equilibrium point is achieved in analytic format.

Risk-aware cooperative spectrum access schemes with multiple channels are proposed in (Zhang et al., 2014), whereby multiple primary users operating over different channels choose trustworthy secondary users as relays to improve throughput, and in return secondary users gain transmission opportunities. To study the multi-channel cooperative spectrum access, cooperation over single channel is investigated first, which involves a primary users selecting the suitable SU and granting a period of access time to the selected SU as a reward, considering trustworthiness of secondary users. The above procedure is modeled as a Stackelberg game, through which access time allocation and power allocation are obtained. Based on the above results, cooperation over multiple channels is studied from the perspectives of the primary network and secondary network, respectively. The problem of distributed throughput maximization with the multi-channel ALOHA medium access protocol is studied in (Cohen et al., 2013). The problem of choosing the access probability is modeled as a leader-followers Stackelberg game, where a single user is chosen to be the leader to manage the network.

Pricing is an effective approach for spectrum access control in cognitive radio networks. In (Do et al., 2014), the pricing effect on the equilibrium behaviors of selfish secondary users' data packets which are served by a CR base station (BS). From the secondary users' point of view, a spectrum access decision on whether to join the queue of the BS or not is characterized through an individual optimal strategy that is joining the queue with a joining probability. A price competition between one opportunistic-spectrum BS and one dedicated-spectrum BS as a Stackelberg game via the dual decomposition algorithms. (Yang et al., 2013) studied the pricing-based spectrum access control, where primary users sell the temporarily unused spectrum and secondary users compete via random access for such spectrum opportunities. Compared to existing market-based approaches with centralized scheduling, pricing-based spectrum management with random access provides a platform for secondary users contending for spectrum access and is amenable to decentralized implementation due to its low complexity. For the multiple-primary users market model, we cast the competition among Primary users as a three-stage Stackelberg game, where each SU selects a primary users's channel to maximize its throughput.

POWER CONTROL

Cognitive and non-cognitive transmitters are regarded as two kinds of players, where each transmitter aims at maximizing his energy efficiency is considered in the framework of Stackelberg game in (Treust et al., 2013). A Stackelberg equilibrium analysis of the presented two-level Stackelberg game is conducted, which allows to better understand the effects of cognition on energy efficiency. Authors in (Zheng et al., 2013) investigate a price-based power control problem in a spectrum sharing networks, where the base stations work as the primary users, and they permit the secondary users to access if the introduced interference power is under the predefined interference power constraint. The secondary users as followers should pay for the interference power, and the base stations as leaders first decides the price for each secondary user to maximize its revenue. Then, each secondary user controls its transmit power to maximize its revenue. Such an interaction is modeled as a Stackelberg game between the leader BS and the secondary user. Furthermore, the revenue function of the BS is expressed as a function of the transmit power of the secondary users via backward induction. Authors in (Xiao et al., 2012) focus on maximizing both secondary users and primary users' payoff by jointly optimizing transmit powers of secondary users, sub-band allocations, and the prices charged by primary users in the OFDMA-based cognitive radio networks. To further improve the performance of secondary users, secondary users are allowed to share the same sub-band to cooperate with each other to send and receive signals. To help understand the interaction among secondary users and primary users, the proposed network model is investigated from coalition formation and hierarchical Stackelberg game theoretic perspectives.

The primary user allow secondary users to access by pricing their interference power under the interference power constraint. The interaction between the primary users and the secondary users is formulated as a Stackelberg game in (Wang et al., 2014). Similar to (Zheng et al., 2013), The revenue function of the primary users is expressed as a non-concave function of SU's transmit power by backward induction. In (Zhang et al., 2014), a Stackelberg game is built to model the joint power allocation of the primary user network and secondary user network. With the analysis of the Stackelberg equilibrium of the proposed Stackelberg game model, an analytic hierarchic power allocation method is proposed when the primary users can acquire the additional information to anticipate secondary users' reactions.

(Elias et al., 2013) addresses the joint pricing and network selection problem in cognitive radio networks., which is formulated as a Stackelberg game. In this game, the primary and secondary operators first set the network subscription price to maximize their own revenue. Then, users perform the network selection process, deciding whether to pay more for a guaranteed service or to use a cheaper best-effort secondary network, where congestion and low throughput may be experienced. (Tran et al., 2013) studies a price-based spectrum access control scheme in cognitive radio networks, which characterizes network operators' service provisions to delay-sensitive secondary users via game pricing. A price competition between two operators employing shared-use and exclusive-use DSA, is respectively modeled as a two-stage Stackelberg game. Using a backward induction method, a unique equilibrium for the game is achieved and the authors also investigate the equilibrium convergence.

MULTIPLE ANTENNAS

Multiple-input multiple-output (MIMO) enables transmission of multiple independent data streams in the spatial domain to improve capacity. However, taking advantage of the MIMO technique in cognitive radio networks is an unexplored area. A novel MIMO-based cognitive radio networking framework is proposed in (Hua et al., 2014), which enables the secondary users to utilize MIMO techniques to cooperatively relay the traffic for the primary users, while concurrently accessing the same channel to transmit their own traffic.

In (Xu & Mao, 2014), a cooperative spectrum leasing scheme for primary and secondary users is considered incorporating two advanced physical-layer technologies, i.e., multiple-input multiple-output and distributed interference alignment. A Stackelberg game is also formulated, where the primary user is the leader, and secondary users are followers. With backward induction, the optimal strategies for primary and secondary users achieve the unique Stackelberg equilibrium, where no player can gain by a unilaterally changing strategy. In (Wang et al., 2012), the authors propose a pricing-based spectrum leasing framework between one primary user and multiple secondary users. In this scenario, the primary user attempts to maximize its utility by setting the price of spectrum, the selected secondary users have the right to decide their power levels to help the primary user's transmission, aiming to obtain corresponding spectrum access time. Then, the spectrum leasing problem can be cast into a Stackelberg game, which jointly maximizes the utilities of primary user and secondary users. In (Taranto et al., 2010), a primary user leases part of its resources to independent secondary terminals in exchange for a tariff in dollars per bit, under the constraint that secondary transmissions do not cause excessive interference at the primary receiver. The primary user's spectrum leasing and the secondary user's transmission scenario gives rise to a Stackelberg game.

Spectrum pooling has been shown to have a great potential to improve the spectrum utilization, especially when primary users and secondary users are allowed to utilize a common spectrum pool. (Xiao et al., 2014) studies the joint optimization problem for a spectrum pooling system with both primary users and secondary users. A novel hierarchical game theoretic model which consists of an overlapped coalition formation game model to analyze the pricing cooperation/competition strategy among primary users and a non-cooperative game model to investigate the resource competition among secondary users. These two game models are interrelated in a hierarchical game structure, in which we also study the interaction between secondary users and primary users. Our model does not require secondary users to have information about spectrum access scheduling of primary users.

OPERATOR PRICING

(Elias et al., 2013) addresses the joint pricing and network selection problem in cognitive radio networks. The problem is formulated as a Stackelberg game, where the primary and secondary operators first set the network subscription price to maximize their revenue. Then, users perform the network selection process, deciding whether to pay more for a guaranteed service or to use a cheaper best-effort secondary network, where congestion and low throughput may be experienced. (Wu et al., 2011) study the joint pricing and power allocation via a Stackelberg game. In our model, primary users is the game leader and jointly determines its power allocation (to guarantee its QoS requirement) and the interference price charged to Secondary User (SU) (to reap revenue). Meanwhile, SU is the game follower and determines its power demand in response to primary users's decisions. We quantify primary users's and SU's benefit from the channel sharing model by deriving the Stackelberg equilibrium.

INFORMATION SECRECY

(Gabry et al., 2014) investigates cooperation for secrecy in cognitive radio networks. In particular, we consider a four-node cognitive scenario where the secondary receiver is treated as a potential eavesdropper with respect to the primary transmission. Although cognitive radio technology improves efficiency of spectrum utilization, primary users usually do not gain from opening up the spectrum in the opportunistic spectrum access, and sometimes even suffer from collisions due to secondary users' imperfect spectrum sensing. However, in (Wu & Liu, 2011), information secrecy is a concern, primary users could actually be better off by allowing secondary users to cooperatively share the spectrum. Specifically, a new cooperation paradigm in cognitive radio networks that primary users improve secrecy with the help of trustworthy secondary users, in the presence of an intelligent and passive eavesdropper attempting to decode primary users' messages. After deriving the achievable pair of primary users' secrecy rate and secondary users' transmission rate under various circumstances, we model the interaction between primary users and secondary users as a Stackelberg game in which transmission power levels are the key to maximize data rates.

HETEROGENEOUS CELLULAR NETWORKS

Heterogeneous wireless cellular networks are considered as promising technologies to improve both energy efficiency and spectral efficiency. In heterogeneous cellular networks, interference management is also important due to spectrum reuse and cell splitting can significantly improve the overall performance. In the existing works, various resource allocation methods are proposed to either improve energy efficiency or mitigate interference. It is worth mentioning that Femtocell technology addresses the problem of poor indoor coverage, benefiting both wireless service provider and end users. With the introduction of femtocell, the cross-tier interference between macro link and femto link becomes a major factor which greatly impacts the network performance. Different access control approaches, by generating different interference patterns, also severely affect the overall throughput of the network and need to be carefully investigated. Among all the access control mechanisms, hybrid access is the most promising one, which allows roaming unregistered users to access the nearby femto base station while reserving certain resource for registered femto users, improving overall network capacity. However, to successfully leverage hybrid access is challenging because the femto holders are selfish, unwilling to share their femto facilities and spectrum resource with macro users without any incentive mechanism. Therefore, game theory can play a huge role in heterogeneous cellular networks with macrocells underlaid with various kinds of smallcells, for instance, the femtocells, pico cells and even radio remote heads. Since there exists the nature hierarchies in such a heterogeneous networks, Stackelberg game also find the suitable places as shown in Table 1.

In (Guruacharya et al., 2013), a downlink power allocation is investigated in a heterogeneous cellular network with macrocells underlaid with femtocells. The femto-access points and the macro-base stations in the heterogeneous network are assumed to compete with each other to maximize their throughput with the individual power constraint. A Stackelberg game is formulated with the macro-base stations as the leaders and the femto-access points as the followers. The leaders are foresight enough to consider the responses of the followers while formulating their own strategies. Typical Stackelberg equilibrium is recognized as the solution of the formulated Stackelberg game, and it is shown to exist under some mild assumptions.

Authors in (Bu et al., 2013) jointly consider both the energy efficiency and the interference control via interference power constraint strategies. A game-theoretical scheme using energy-efficient resource

Table 1.

Player		Strategy	Utility	Reference
Leader	Follower			
Macro-base stations	Femto-access points	Downlink power allocation	Maximize their throughput	(Guruacharya et al., 2013)
Macrocells	Small-cells	Interference power constraint strategies	Energy-efficient resource allocation and interference pricing	(Bu et al., 2013)
Macrocells	Small-cells	Power adaptation policies	Energy efficiency by "Revenue per Cost"	(Chen et al., 2013)
Mobile network operator	Small cell network	Optimal differentiated subsidies	Maximize their throughput	(Yang & Quek, 2014) (Rose et al., 2014)
Macrocell	Femtocells	Spectrum sharing approach	Expand the cell range	(Hamouda et al., 2014)
Macrocell	Femtocells	Utility-aware refunding framework	Connectivity	(Chen et al., 2012)
Cellular operator	Users	Economic incentive	Resource utilization	(Duan et al., 2013)

allocation and interference pricing for an interference-limited environment in heterogeneous networks. The problems of resource allocation and interference management is modeled as a Stackelberg game with incomplete information. A backward induction method is used to analyze the proposed game. A closed-form expression of the Stackelberg equilibrium is achieved for the proposed game with different interference power constraints.

A Stackelberg learning game is formulated in (Chen et al., 2013), in which the macrocells behave as the leaders and the small-cells are followers. First, the leaders update their power adaptation policies by knowing the best-responses of all followers, while the followers compete against each other in each time slot with only the leaders' action information. A combined learning framework is established, through combining the Stackelberg learning formulation and the technique of transfer learning, to tell players how to plan the action decisions. A low-complexity game theoretic approach is presented in (Wang et al., 2014) for energy-efficient resource allocation in a two-tier heterogeneous network. The energy efficiency by "Revenue per Cost" in the utility domain considering both power allocation and cross-tier interference coordination. Since global optimization for maximizing the overall energy efficiency is computationally expensive, we transform this non-convex problem to a two-stage Stackelberg game inspired by fractional programming.

(Yang & Quek, 2014) study the optimal differentiated subsidies from a mobile network operator to form a shared small cell network using social network information. It is a win-win strategy for the mobile network operator to offload data and for small cell users to receive enhanced QoS. We formulate the problem as a Stackelberg Game: in Stage I, the mobile network operator sets differentiated subsidies and maximizes a subsidy elasticity of sharing metric, which captures the goal of using the least subsidies to obtain the most shared network; in Stage II, each secondary users decide the degree of sharing by maximizing a utility function, which is a tradeoff between proportional-fair capacity and subsidy. In the model, at subgame perfect equilibrium, the shared network can be maintained with the termination of subsidies. A novel game-theoretic model of the complex interactions between network service providers and users in heterogeneous small-cell networks is investigated in (Rose et al., 2014). In this game, the service providers selfishly aim at maximizing their profit while, simultaneously, the users seek to optimize their chosen service's quality-price trade-off. A Stackelberg formulation in which the service providers act as leaders and the users as followers is proposed.

Authors in (Hamouda et al., 2014) propose a new spectrum sharing approach, where macrocell and femtocells can simultaneously share the available bandwidth, which is formulated as a Stackelberg game. The macrocell sells a specific bandwidth to femtocells in exchange of some victim macro-users to serve. The presented approach avoid the intra-tier interference and help the macrocell to offload by expanding the cell range of some femtocells. In (Chen et al., 2012), a novel utility-aware refunding framework is proposed to motivate hybrid access in femtocell. Wireless service provider provides certain refunding to motivate femtocell to open their resource for macro users. Femtocells decide the resource allocation among femto and macro users according to the amount of refunding wireless service provider offers via a Stackelberg game. Femtocells can effectively resolve the poor connectivity issue of indoor cellular users. (Duan et al., 2013) investigates the economic incentive for a cellular operator to add femtocell service on top of its existing macrocell service. They model the interactions between a cellular operator and users as a Stackelberg game: The operator first determines spectrum allocations and pricing of femtocell and macrocell services, and then heterogeneous users choose between the two services and the amount of resource to request.

COOPERATIVE NETWORKS

Cooperative communication has been proposed to improve the channel capacity in wireless networks. It takes advantage of the broadcast nature of wireless transmission and utilizes the antennas on wireless nodes to achieve spacial diversity. Depending on how the relay node processes the overheard signal, two primary cooperative communication modes are widely used: amplify-and-forward and decode-and-forward. In the AF mode, the relay node amplifies the signal before forwarding it to the destination node. In the DF mode, the relay node decodes and encodes the signal before forwarding it to the destination node. Cooperative communication has potential applications in many different networks, including cellular networks, ad-hoc networks, and cognitive networks. Extensive research in recent years has shown the benefits of cooperative relaying in wireless networks, where nodes overhear and cooperatively forward packets transmitted between their neighbors. Most existing studies focus on physical-layer optimization of the effective channel capacity for a given transmitter-receiver link; however, the interaction among simultaneous flows between different endpoint pairs, and the conflicts arising from their competition for a shared pool of relay nodes, are not yet well understood. Cooperative communication has great potential to improve the wireless channel capacity by exploiting the antennas on wireless devices for spatial diversity. However, applications of cooperative communication are barely seen in reality. A main obstacle blocking its wide application is the lack of incentives for wireless nodes to participate in cooperative communication. Three primary mechanisms designed to provide cooperation incentives are pointed out in (Yang et al., 2012), which include reputation based mechanism, resource-exchange-based mechanism, and pricing-based mechanism.

Pricing as Cooperation Incentive

An AF cooperative scheme is modeled using Stackelberg market framework in (Tous & Barhumi, 2013), where a relay can sell its resources, power and bandwidth to multiple users to maximize its net revenue. Joint optimization problem including pricing of the relay and optimal power and bandwidth allocation of users is investigated as complex interaction problem. In (Cao et al., 2012), a single-relay and multiple users scenario is considered, where the relay gets paid for forwarding users' signals, and the users have to pay for the relay service. The relay power allocation and pricing problems are investigated and the interaction between the users and the relay is modeled as a two-level Stackelberg game in (Cao et al., 2012), where the relay is modeled as the service provider and the leader of the game, sets the relay price to maximize its revenue, whereas the users are modeled as customers and followers who buy power from the relay for higher transmission rates. In addition, a two level Stackelberg game is employed in (Wang et al., 2009) to jointly consider the benefits of the source and relay nodes in which the source node is modeled as a buyer and the relay nodes are modeled as the sellers, respectively. To improve the system performance and stimulate cooperation, a Stackelberg game theoretic framework is formulated in (Wang et al., 2006) to distribute resource allocation over multiuser cooperative communication networks. Two questions of who should relay and how much power for relaying are answered, by employing a two-level Stackelberg game to jointly consider the benefits of source nodes as buyers and relay nodes as sellers.

Resource Leasing

In summary, there exist natural hierarchical resource property-rights, and the primary players own a given licensed bandwidth and can lease it for a fraction to secondary players in exchange for additional revenue. Here, the leased spectrum resource can be multidimensional including time, frequency, and space. For instance, (Simeone et al., 2008) analyze such an implementation where a primary link has the possibility to lease its owned spectrum to the secondary nodes working in an ad hoc network, in exchange for cooperation in the form of distributed space-time coding. The investigated model is conveniently cast in the framework of Stackelberg games. The problem of medium sharing is investigated in (Canzian et al., 2013), which is often seen as a case where game theory just proves the inefficiency of distributed access, without proposing any remedy. Instead, in this work, by properly modeling the agents involved in such a scenario, and enabling simple but effective incentives towards cooperation for the users, we obtain a resource allocation scheme which is meaningful from both perspectives of game theory and network engineering. Such a result is achieved by introducing throughput redistribution as a way to transfer utilities, which enables cooperation among the users. Finally, a Stackelberg formulation is proposed, involving the network access point as a further player. The cooperative transmission between primary user and secondary user has been regarded as an effective method in (Feng & Li, 2013) to promote harmonious coexistence between themselves where how to balance user's benefits and improve the system throughput is a key issue. A joint pricing and power allocation strategy is studied for the cognitive relay transmission by using the Stackelberg model which is a classical dynamic game model.

To improve network performance of all networks, authors in (Yi et al., 2011) propose a cooperative communication-aware spectrum leasing framework. In the proposed framework, the primary network leverages secondary APs as cooperative relays, and decides the optimal strategy on the relay selection and the price for spectrum leasing. Based on primary network's strategy, secondary networks determine the length of spectrum access time they purchase from the primary network. Finally, each network allocates the total spectrum access time of the network among its end users. The above sequential decision procedure is formulated as a Stackelberg game, with primary network acting as the leader and secondary networks as the followers.

COGNITIVE GREEN HETEROGENEOUS CELLULAR NETWORKS

Rapidly rising energy costs and increasingly rigid environmental standards have led to an emerging trend of addressing the energy efficiency aspect of wireless cellular networks. In cellular networks, base stations are power-hungry elements and absorbing them in the smart grid can significantly increase power efficiency. However, this important perspective has been ignored in the energy-efficiency research on cellular networks. Both cognitive radio and femtocell have been considered as promising techniques in wireless networks. However, most of previous works are focused on spectrum sharing and interference avoidance, and the energy efficiency aspect is largely ignored. Cognitive heterogeneous mobile networks are considered as important techniques to improve the energy efficiency.

A price-based resource allocation for heterogeneous cellular networks with a central macrocell underlaid with distributed femtocells is presented in (Kang et al., 2012), and they all operate over the same frequency band. A Stackelberg game is formulated to study the joint optimization of the macrocell and femtocells, where the macrocell base station protects itself by pricing the interference from femtocell users. In (Bu & Yu, 2012), heterogeneous networks, cognitive radios, and the smart grid are jointly studied to improve energy efficiency. The joint decision problems of electricity price of multiple retailers, energy-efficient power allocation and interference price of macrocell base station and femtocell base stations are formulated as a three-level Stackelberg game in (Bu & Yu, 2012).

The authors in (Xie et al., 2012) study the problem of energy efficiency during spectrum sharing and power allocation in heterogeneous cognitive radio networks with femtocells, where both the macrocell and the femtocell are assumed to be with the cognitive capability. The energy-efficient resource allocation problem with femtocells is formulated as a Stackelberg game. A gradient based iteration algorithm is proposed to solve the Stackelberg game-theoretic energy-efficient resource allocation problem. In (Bu & Yu, 2014), green cognitive mobile networks with small cells is studied in the smart grid environment. Unlike most existing studies on cognitive radio networks, where only the radio spectrum is sensed, in (Bu & Yu, 2014) the presented cognitive radio networks sense not only the radio spectrum environment but also the smart grid environment, based on which power allocation and interference management for multimedia communications are performed. We formulate the problems of electricity price decision, energy efficient power allocation and interference management as a three-stage Stackelberg game. Currently, the power grid is experiencing a significant shift from the traditional grid to the smart grid. In the smart grid environment, only considering energy efficiency may not be sufficient, since the dynamics of the smart grid will have significant impacts on mobile networks. In (Bu et al., 2013), cognitive heterogeneous mobile networks is studied in the smart grid environment. Unlike most existing studies on cognitive networks, where only the radio spectrum is sensed, the cognitive networks sense not only the radio spectrum environment but also the smart grid environment, based on which power allocation and interference management are performed. We formulate the problems of electricity price decision, energy-efficient power allocation and interference management as a three-level Stackelberg game.

CONCLUSION AND FUTURE WORK

In this chapter, we summarize the extensive applications of Stackelberg game in wireless communication networks, in particular, we concentrate on the Stackelberg game theoretic formulations in the cognitive radio networks, heterogeneous cellular networks, cooperative networks since there naturally exist various hierarchies between different players. There hierarchies contain secondary and primary users in cognitive radio networks, Macrocell and smallcell in heterogeneous cellular networks, direct link and relay link in cooperative communications. Finally, looking forward to the future vision of 5G wireless mobile communication, we further survey the current use of Stackelberg game in cognitive heterogeneous green cooperative networks. Although more research attention has been poured in this topic, and Stackelberg game can find more applications in the future.

1. Multi-tier Stackelberg game with less signaling overhead;
2. Distributed learning algorithm design for pursuing the Stackelberg equilibrium solutions;
3. Combine Stackelberg game with other games, for instance, the cooperative game;
4. More advanced analysis method in addition to backward induction method.

ACKNOWLEDGMENT

This work was supported in part by the National Science Foundation of China under Grant 61231008, 61201139; by the Natural Science Basic Research Plan in Shaanxi Province of China under Grant no. 2012JQ8012; the open research fund of National Mobile Communications Research Laboratory, Southeast University (No.2014D10); by ISN02080001; by the 111 Project under Grant B08038; and by Shaanxi Province Science and Technology Research and Development Program (2011KJXX-40).

REFERENCES

Al-Tous, H., & Barhumi, I. (2013). Joint Power and Bandwidth Allocation for Amplify-and-Forward Cooperative Communications Using Stackelberg Game. *Vehicular Technology. IEEE Transactions*, *62*(4), 1678–1691.

Bu, S. R., & Yu, F. R. (2012). Dynamic energy-efficient resource allocation in cognitive heterogeneous wireless networks with the smart grid.*Global Communications Conference*. doi:10.1109/GLOCOM.2012.6503579

Bu, S. R., & Yu, F. R. (2014). Green Cognitive Mobile Networks With Small Cells for Multimedia Communications in the Smart Grid Environment. *Vehicular Technology. IEEE Transactions*, *63*(5), 2115–2126.

Bu, S. R., Yu, F. R., & Qian, Y. (2013). Energy-efficient cognitive heterogeneous networks powered by the smart grid. *INFOCOM, 2013. Proceedings of the IEEE*, 980–988.

Bu, S. R., Yu, F. R., & Senarath, G. (2013). Interference-aware energy-efficient resource allocation for heterogeneous networks with incomplete channel state information. *Communications (ICC),2013 IEEE International Conference*.

Canzian, L., Badia, L., & Zorzi, M. (2013). Promoting Cooperation in Wireless Relay Networks Through Stackelberg Dynamic Scheduling. *Communications. IEEE Transactions*, *61*(2), 700–711. doi:10.1109/TCOMM.2012.120512.110654

Cao, Q., Zhao, H. V., & Jing, Y. D. (2012). Power Allocation and Pricing in Multiuser Relay Networks Using Stackelberg and Bargaining Games. *Vehicular Technology. IEEE Transactions*, *61*(7), 3177–3190.

Chen, L., Libman, L., & Leneutre, J. (2011). Conflicts and Incentives in Wireless Cooperative Relaying: A Distributed Market Pricing Framework. *Parallel and Distributed Systems. IEEE Transactions*, *22*(5), 758–772.

Chen, X. F., Zhang, H. G., & Lasanen, M. (2013). Combined learning for energy efficiency in heterogeneous cellular networks. *Personal, Indoor and Mobile Radio Communications (PIMRC Workshops), 2013 IEEE 24th International Symposium*.

Chen, Y. D., Hoo Teo, K., Kishore, S., & Zhang, J. Y. (2008). Inter-Cell Interference Management in WiMAX Downlinks by A Stackelberg Game between BSs. *ICC*, *08*, 3442–3446.

Chen, Y. J., Zhang, J., & Zhang, Q. (2012). Utility-Aware Refunding Framework for Hybrid Access Femtocell Network. *Wireless Communications. IEEE Transactions*, *11*(5), 1688–1697.

Cohen, K., Leshem, A., & Zehavi, E. (2013). Game Theoretic Aspects of the Multi-Channel ALOHA Protocol in Cognitive Radio Networks. *Selected Areas in Communications. IEEE Journal on*, *31*(11), 2276–2288.

Di Taranto, R., Popovski, P., Simeone, O., & Yomo, H. (2010). Efficient Spectrum Leasing via Randomized Silencing of Secondary Users. *Wireless Communications. IEEE Transactions*, *9*(12), 3739–3749.

Do, C. T., Tran, N. H., Han, Z., Le, L. B., Lee, S., & Hong, C. S. (2014). Optimal Pricing for Duopoly in Cognitive Radio Networks: Cooperate or not Cooperate? *Wireless Communications. IEEE Transactions*, *13*(5), 2574–2587.

Duan, L. J., Huang, J. W., & Shou, B. Y. (2013). Economics of Femtocell Service Provision. *Mobile Computing. IEEE Transactions*, *12*(11), 2261–2273.

Elias, J., Martignon, F., Chen, L., & Altman, E. (2013). Joint Operator Pricing and Network Selection Game in Cognitive Radio Networks: Equilibrium, System Dynamics and Price of Anarchy. *Vehicular Technology. IEEE Transactions*, *62*(9), 4576–4589.

Elias, J., Martignon, F., Chen, L., & Altman, E. (2013). Joint Operator Pricing and Network Selection Game in Cognitive Radio Networks: Equilibrium, System Dynamics and Price of Anarchy. *Vehicular Technology. IEEE Transactions*, *62*(9), 4576–4589.

Feng, L., & Li, W. (2013). A Dynamic Game Algorithm for Power Allocation in Cognitive Relay Transmission *Communications. IEEE Transactions*, *61*(11), 4440–4447. doi:10.1109/TCOMM.2013.101113.130034

Gabry, F., Li, N., Schrammar, N., Girnyk, M., Rasmussen, L. K., & Skoglund, M. (2014). On the Optimization of the Secondary Transmitter's Strategy in Cognitive Radio Channels with Secrecy. *Selected Areas in Communications. IEEE Journal*, *32*(3), 451–463.

Guruacharya, S., Niyato, D., Kim, & Hossain, E. (2013). Hierarchical Competition for Downlink Power Allocation in OFDMA Femtocell Networks. *Wireless Communications, IEEE Transactions*, *12*(4), 1543-1553.

Hamouda, S., Zitoun, M., & Tabbane, S. (2014). Win–win relationship between macrocell and femtocells for spectrum sharing in LTE-A. *Communications, IET*, *8*(7), 1109–1116. doi:10.1049/iet-com.2013.0845

Hua, S., Liu, H., Zhuo, X. J., Wu, M. Q., & Panwar, S. S. (2014). Exploiting Multiple Antennas in Cooperative Cognitive Radio Networks. *Vehicular Technology. IEEE Transactions*, *63*(7), 3318–3330.

Kang, X., Zhang, R., & Motani, M. (2012). Price-Based Resource Allocation for Spectrum-Sharing Femtocell Networks: A Stackelberg Game Approach. *Selected Areas in Communications. IEEE Journal*, *30*(3), 538–549.

Kang, X., Zhang, R., & Motani, M. (2012). Price-Based Resource Allocation for Spectrum-Sharing Femtocell Networks: A Stackelberg Game Approach. *Selected Areas in Communications. IEEE Journal*, *30*(3), 538–549.

Lasaulce, S., Hayel, Y., El Azouzi, R., & Debbah, M. (2009). Introducing hierarchy in energy games. *Wireless Communications. IEEE Transactions*, *8*(7), 3833–3843.

Le Treust, M., Lasaulce, S., Hayel, Y., & He, G. N. (2013). Green Power Control in Cognitive Wireless Networks. *Vehicular Technology, IEEE Transactions*, *62*(4), 1741-1754.

Liu, D. T., Chen, Y., Zhang, T. K., Keong Chai, K., Loo, J., & Vinel, A. (2013). Stackelberg Game Based Cooperative User Relay Assisted Load Balancing in Cellular Networks. *Communications Letters, IEEE*, *17*(2), 424–427. doi:10.1109/LCOMM.2012.122012.122377

Oussama, H., El-Azouzi, R., & Yezekael, H. (2013). A Stackelberg Model for Opportunistic Sensing in Cognitive Radio Networks. *Wireless Communications, IEEE Transactions*, *12*(5), 2148-2159.

Park, J., & Van der Schaar, M. (2009). *Stackelberg Contention Games in Multiuser Networks*. Computer Science and Game Theory.

Rose, L., Belmega, E. V., Saad, W., & Debbah, M. (2014). Pricing in Heterogeneous Wireless Networks: Hierarchical Games and Dynamics. *Wireless Communications. IEEE Transactions*, *13*(9), 4985–5001.

Simeone, O., Stanojev, I., Savazzi, S., Bar-Ness, Y., Spagnolini, U., & Pickholtz, R. (2008). Spectrum Leasing to Cooperating Secondary Ad Hoc Networks. *Selected Areas in Communications. IEEE Journal*, *26*(1), 203–213.

Su, M., & van der Schaar, M. (2009). A new perspective on multi-user power control games in interference channels. *Wireless Communications. IEEE Transactions*, *8*(6), 2910–2919.

Su, M., & van der Schaar, M. (2009). From competition to coopetition: Stackelberg equilibrium in multiuser power control games. *Game Theory for Networks*, *2009*, 107–116.

Tran, N. H., Hong, C. S., Han, Z., & Lee, S. (2013). Optimal Pricing Effect on Equilibrium Behaviors of Delay-Sensitive Users in Cognitive Radio Networks. *Selected Areas in Communications. IEEE Journal*, *31*(11), 2566–2579.

Wang, B. B., Han, Z., & Liu, K. J. R. (2006). WLC41-4: Stackelberg Game for Distributed Resource Allocation over Multiuser Cooperative Communication Networks.*Global Telecommunications Conference, 2006. GLOBECOM '06. IEEE.*

Wang, B. B., Han, Z., & Liu, K. J. R. (2009). Distributed Relay Selection and Power Control for Multiuser Cooperative Communication Networks Using Stackelberg Game. *Mobile Computing. IEEE Transactions*, *8*(7), 975–990.

Wang, X., Ma, K., Han, Q., Liu, Z., & Guan, X. (2012). Pricing-based spectrum leasing in cognitive radio networks. *Networks, IET*, *1*(3), 116–125. doi:10.1049/iet-net.2012.0149

Wang, Y., Wang, X., & Wang, L. (2014). Low-complexity Stackelberg Game Approach for Energy-efficient Resource Allocation in Heterogeneous Networks. *Communications Letters, IEEE*, (99), 1-1.

Wang, Z., Jiang, L., & He, C. (2014). Optimal Price-based Power Control Algorithm in Cognitive Radio Networks. *Wireless Communications, IEEE Transactions*, (99), 1-1.

Wang, Z. Q., Jiang, L. G., & He, C. (2013). A Novel Price-Based Power Control Algorithm in Cognitive Radio Networks. *Communications Letters, IEEE*, *17*(1), 43–46. doi:10.1109/LCOMM.2012.120612.121587

Wu, Y., Zhang, T. Z., & Tsang, D. H. K. (2011). Joint Pricing and Power Allocation for Dynamic Spectrum Access Networks with Stackelberg Game Model. *Wireless Communications. IEEE Transactions, 10*(1), 12–19.

Wu, Y. L., & Liu, K. J. R. (2011). An Information Secrecy Game in Cognitive Radio Networks. *Information Forensics and Security. IEEE Transactions, 6*(3), 831–842.

Xiao, Y., Niyato, D., Han, Z., & Chen, K. C. (2014). Secondary Users Entering the Pool: A Joint Optimization Framework for Spectrum Pooling. *Selected Areas in Communications. IEEE Journal, 32*(3), 572–588.

Xiao, Y., Bi, G. A., Niyato, D., & DaSilva, L.A. (2012). A Hierarchical Game Theoretic Framework for Cognitive Radio Networks. *Selected Areas in Communications, IEEE Journal, 30*(10), 2053,-2069.

Xie, R. C., Yu, F. R., Ji, H., & Li, Y. (2012). Energy-Efficient Resource Allocation for Heterogeneous Cognitive Radio Networks with Femtocells. *Wireless Communications. IEEE Transactions, 11*(11), 3910–3920.

Xu, Y., & Mao, S. W. (2014). Stackelberg Game for Cognitive Radio Networks With MIMO and Distributed Interference Alignment. *Vehicular Technology. IEEE Transactions, 63*(2), 879–892.

Yang, D. J., Fang, X., & Xue, G. L. (2012). Game theory in cooperative communications. *Wireless Communications, IEEE, 19*(2), 44–49. doi:10.1109/MWC.2012.6189412

Yang, Y. F., & Quek, T. Q. S. (2014). Optimal Subsidies for Shared Small Cell Networks —A Social Network Perspective. *Selected Topics in Signal Processing. IEEE Journal, 8*(4), 690–702.

Yang, H., Zhang, X.N., & Meng, Q. (2007). *Stackelberg games and multiple equilibrium behaviors on networks*. Academic Press.

Yang, L., Kim, H., Zhang, J.S., Chiang, M., & Tan, C.W. (2013). Pricing-Based Decentralized Spectrum Access Control in Cognitive Radio Networks. *Networking, IEEE/ACM Transactions, 21*(2), 522-535.

Yi, Y. W., Zhang, J., Zhang, Q., & Jiang, T. (2011). Spectrum Leasing to Multiple Cooperating Secondary Cellular Networks. *Communications (ICC),2011 IEEE International Conference*.

Zhang, J., & Zhang, Q. (2009). Stackelberg Game for Utility-Based Cooperative Radio Networks. *Mobile and Ad Hoc Networking and Computing*, 23-32.

Zhang, N., Cheng, N., Lu, N., Zhou, H. B., Mark, J. W., & Shen, X. M. (2014). Risk-Aware Cooperative Spectrum Access for Multi-Channel Cognitive Radio Networks. *Selected Areas in Communications. IEEE Journal, 32*(3), 516–527.

Zhang, T., Chen, W., Han, Z., & Cao, Z. (2014). Hierarchic Power Allocation for Spectrum Sharing in OFDM-Based Cognitive Radio Networks. *Vehicular Technology. IEEE Transactions, 63*(8), 4077–4091.

Chapter 12
Pricing Methodology and Its Applications in Cognitive Radio and Multi-Tier Heterogeneous Cellular Networks

Chungang Yang
Xidian University, China

Lingxia Wang
Xidian University, China

Jia Xiao
Xidian University, China

Pengyu Huang
Xidian University, China

Jiandong Li
Xidian University, China

ABSTRACT

In this chapter, we concentrate on different technical applications and research directions of pricing theory and methodology, where we investigate following technical applications and functions of pricing including cooperative incentive mechanism design, Pareto- and social optimality improvement, distributed algorithm design with the low signaling overhead. We first clarify different concepts of pricing, summarize the motivation, present a taxonomy according to these different technical applications. Then, we survey applications of pricing theory and methodology with understandings and observations in cognitive radio and multi-tier heterogeneous cellular networks. We emphasize some of the recent critical problems, such as the cooperation incentive, resource and interference management and economics of small cells. Finally, we conclude this chapter with the possible research directions and more potential network applications of pricing theory and methodology.

INTRODUCTION

Wireless communications have experienced explosive growth, and will encounter the promising 1000x traffic enhancement challenge, which are further fueled by the popularity of smart devices, rich multimedia services, and high definition mobile Internet video. Three dimensions have been identified to achieve the required traffic increase in unit of bits per second per square kilometer (Soret et al., 2015),

DOI: 10.4018/978-1-5225-1712-2.ch012

(Bhushan et al., 2014). This is equivalent to bits per second per Hertz per cell (spectral efficiency) × Hertz (more spectrum) × cell per square kilometer (network densification). First, it is known that physical layer technology has approached its ceiling, and there exists limited space for spectrum efficiency enhancement, even though most of the recent transmission techniques, for instance, massive MIMO, really has really improved the spectrum efficiency. Second, we know that it is technically hard to explore and exploit more spectrum, and it is also expensive since most of valuable spectrum below 6GHz has been allocated to specific applications, in addition to the most latest research on the mmWave communications. These are why cognitive radio networks (Mitola, 2000), (Haykin, 2005), (Goldsmith et al., 2009) and the related dynamic spectrum access (Akyildiz et al., 2006) have found extensive studies from academia, industry, and standardization communities.

As the third dimension, dense deployment of small cells can both increase coverage and attain a capacity leap (Soret et al., 2015), (Bhushan et al., 2014), which is recognized as the most powerful dimension to address the 1000x capacity increase challenges. Certainly, it really introduces novel technical challenges, for instance, the mutual interference and serious traffic imbalance problems. Ultra-dense heterogeneous and small cell networks will contribute 56-fold traffic demand enhancement in a full frequency reuse way, thus improving the spectrum efficiency. At the same time, energy efficiency enhancement and green communications have been poured much research attention due to awareness of environment of common people and high energy consumption list of network operators. In summary, most of the current research has been focused on the spectrum efficiency optimization and the most recent energy efficiency optimization in cognitive radio and heterogeneous cellular networks.

However, wireless network operators encounter an embarrassing situation, which is the contradiction between the increasing network operation and maintenance cost and the disproportionately decreasing lower or flat monetary income. Such a situation is expected to be continuing for a long time due to the new proliferation of intelligent terminals with various resource-hungry service requirements and the promisingly challenged networking architecture and paradigm. According to the recent 5G activities, spectrum, energy, and cost efficiency are perceived as the three dimensional efficiency metrics (Bousia et al., 2014), (Leem et al., 2014). Cost efficiency optimization should be another critical performance metrics in 5G cognitive, ultra-dense green and cost effective networks, which has been the consensus of academia, industrial, and standardization communities. It is generally concluded that compared with 4G, 5G should have 5 to 15 times improvement on spectrum efficiency and more than 100 times improvement on energy and cost efficiency.

Joint consideration of the multiple dimensions of efficiency performance metrics involves the players of both operators and subscribers. Multiple network operators investigate various factors to better price resources or services, while multiple subscribers are always pursuing the better quality of experience (QoS) from network operators with less cost. These interactions and behaviors between operators and subscribers, or among themselves always involve both marketing and technical considerations. Therefore, it is necessary to jointly combine these economic factors into the technical design. Pricing techniques from economics provide researchers a new methodology to understand wireless networks (Huang & Gao, 2013), (Jiang et al., 2015), (Zhou et al., 2015). In (Jiang et al., 2014), the economic issues in cognitive cellular networks were studied from the perspectives of game theoretic modeling and mechanism design. Considering the selfishness of the owners of femtos, authors in (Zhou et al., 2014) formulated the two-tier interference coordination as pricing game.

Meanwhile, interactive dynamics and rational behaviors should be well investigated in cognitive radio and multi-tier heterogeneous cellular networks. Technical perspective and applications of pricing theory and methodology can well model, analyze, and design distributed schemes. Low cost and signaling overhead coupled with the backhaul capacity call for the distributed and self-organizing design in the 5G era. Currently pricing technology has extensive applications, and it has been also studied to improve the Pareto-optimal solution, thus achieving the social-welfare maximization. All these motivate us to complete this work. In this chapter, we view pricing methodology as the most powerful method to deal with these problems. It is necessary to first introduce basics, rationale and, technically challenging motivation of pricing theory and methodology in wireless communication networks.

RATIONALE AND MOTIVATION OF PRICING IN WIRELESS NETWORKS

The rationale behind combining economic factors into the technical design has been summarized in (Huang & Gao, 2013). Here, we use cognitive radio networks as an example to further explain the reasons. We know that there currently exist increasingly severe tensions between spectrum supply and spectrum demand, and the spectrum sharing between primary users and secondary users can be seen as a market, which is highly related to the pricing mechanism, and pricing is the core in the spectrum market between different operators or different spectrum-license holders. Authors in (Huang, 2013) summarized various market mechanisms to facilitate cooperative spectrum trading including pricing, bargaining, contract, and auction. This is an important factor to be considered during the technical design, for instance, the spectrum access and spectrum sharing.

We should jointly consider economic factors into the technical design. Previously, it is summarized from the resource utilization perspective. On the other hand, authors in (Huang & Gao, 2013) took the users' service differentiation perspective. It is known that different wireless networks are with different service features. For instance, WiFi networks provide high speed data service, while cellular networks provide wider coverage and better real-time voice service. Subscribers can choose different networks for different inelastic or elastic service applications, where subscribers are with different pricing models to be associated with different wireless networks. During this network selection process, wireless network operators always aim to maximize the monetary revenue with advanced pricing schemes. Various pricing models have received widespread attention including social optimal pricing, monopoly pricing, price differentiation, and oligopoly pricing (Huang & Gao, 2013). In addition, authors in (Courcoubetis & Weber, 2003) provided the basic terminology and models needed for readers to further understand these topics where technology, information and economics meet.

Summary: It is necessary to involve pricing theory and methodology in the technical design due to both the resource-limited and users' service differentiation.

NEW EMERGING APPLICATIONS OF PRICING METHODOLOGY

Pricing has been widely used to enable efficient resource management, for instance, rate control in wired networks (Kelly et al., 1998), bandwidth allocation in broadband networks (Yaiche et al., 2000), and the recent uplink power control in wireless networks (Yates, 1995), (Rasti et al., 2009). However, as mentioned above, nowadays we encounter novel network paradigms, serious technical challenges, rich resource hungry service and different pricing mechanism. Dynamic spectrum access and advanced spectrum sharing in cognitive radio networks, and interference management and backhaul design in heterogeneous cellular networks, and so on. Therefore, it is necessary to further understand and apply pricing theory and methodology to investigate these promising technical problems in these emerging wireless networks.

In this chapter, we take the cognitive radio, heterogeneous cellular networks as the most important network paradigms in future, although different network evolution perspectives towards 5G exist. In addition, spectrum, energy, and cost efficiency should be jointly considered in 5G cognitive, ultra-dense green and cost-effective networks.

Cognitive radio networks are an innovative approach, where radios are designed with a level of intelligence and agility. Cognitive radio technology enables radio devices to exploit, and explore spectrum space in an opportunistic way. Here, we know cognitive radios have the ability to monitor, sense, and detect the conditions of their operating environment, and dynamically reconfigure their own characteristics to best match those conditions. On the other hand, heterogeneous cellular networks are an attractive means of expanding mobile network capacity. A heterogeneous network is typically composed of multiple radio access technologies, architectures, transmission solutions, and different kinds of base stations of varying transmission power. Small-cell deployment in heterogeneous cellular networks increases coverage per square kilometer and is recognized as a promising technique to attain a capacity leap (Bhushan et al., 2014).

In fact, pricing as incentive mechanism and pricing as economic incentives for femtocells' openness have found a wide use in cognitive radio networks, cooperative relay networks, and heterogeneous cellular networks. Meanwhile, to the best of our knowledge, most of the pricing methodology is categorized as either resource-related one in the resource management mechanism design, or the service related one from the economic monetary revenue perspective. There exist some other pricing concepts have not been well understood. For instance, interference pricing appears as another dimension of pricing in cognitive radio networks, which is dependent on the predefined interference temperature of primary users. In addition, interference pricing has been extended to the heterogeneous cellular networks with multiple small cells under-laid with the macrocell.

Summary: Pricing theory and methodology have found extensive applications in cognitive radio and heterogeneous cellular network, and we will survey these pricing application with some deep understanding and explore it effectively to investigate there promising technical problems in these emerging wireless networks. Different from the previous survey works, we take the view point of more concentration on the technical applications and perspective of pricing theory and methodology.

CONTRIBUTIONS AND ORGANIZATION

Although several surveys on pricing have been published, most of them concentrated on economic perspective; or in some specific field or problem, for instance, (Dasilva, 2000) for Internet, (Almanthari et al., 2011) for congestion control, in (Fulp et al., 1998) for distributed algorithm for pricing network resources, and pricing in wireless networks (Gizelis & Vergados, 2011). Our work will provide a comprehensive survey of pricing methodology in the promising cognitive radio and heterogeneous cellular networks, and in particular, we provide a taxonomy of various pricing application from the technical perspective. Meanwhile, we look forward new directions, applications of pricing in the future.

The chapter is organized as follows: In Section II, we introduce the state-of-the-art survey of pricing and clarify the motivation of this work. In Section III, we summarize the basics of pricing techniques including the concepts, various technical applications, and taxonomy. In Section IV and Section V, we survey the applications of pricing theory and methodology in cognitive radio and multi-tier heterogeneous cellular wireless networks with the emphasis on cooperation incentive, interference management and economics of small cells. In Section VI, we summarize the future application and research directions of pricing theory and methodology. Finally, we conclude this chapter with Section VII.

STATE-OF-THE-ART SURVEY OF PRICING

Pricing methodology and techniques have found extensive applications in the promising communication networks. In general, there exist two research directions: the first one is the economic pricing techniques that relate to the monetary revenue of the wireless network operators; and the other one is the recent pricing scheme design for social optimality improvement, distributed algorithm design and low signaling overhead, which is from the technical perspective.

Authors in (Li & Huang, 2010) summarized two kinds of pricing designs including the optimization-oriented pricing and the economics-based pricing. First, the optimization-oriented pricing schemes always represent the Lagrangian multipliers of various constraints. It is known that the most important functions of this kind of pricing are used to distributively coordinate different network entities to maximize social-welfare performance. Second, the economics-based pricing schemes have been extensively applied to optimize various objectives, including the revenue maximization of network service providers or operators.

Most of the recent survey works of pricing methodology deal with the service pricing problems, which addresses the relationship between prices and traffic management. For instance, congestion control can well enhance the performance of wireless cellular networks and improve the the satisfaction of the users, thus introducing more revenues. Several congestion control techniques have been proposed for wireless cellular networks (Dasilva, 2000). These techniques, however, do not provide incentives to the users to use the wireless network rationally. Monetary incentives to the users through congestion pricing can motivate users to use the wireless network rationally and efficiently. Therefore, congestion pricing helps alleviate the problem of congestion. Congestion pricing is based on the idea of dynamically increasing or decreasing the prices of network services according to the congestion level in the whole network. Authors in (Almanthari et al., 2011) surveyed recent research work on congestion pricing.

Authors in (Fulp et al., 1998) presented a distributed flow control technique, where the networks are modeled as the competitive markets. In the markets, switches will set a price on their link bandwidth, which is based on supply and demand relationship. Users will purchase bandwidth to maximize their individual quality of service. This yields a decentralized flow control method that provides a Pareto optimal bandwidth distribution and high utilization. In (Gizelis & Vergados, 2011), the pricing schemes are classified, based on their ability to adapt to the needs of the wireless service providers and their subscribers during the entire service period, into static-based pricing and dynamic-based pricing Schemes. These schemes may improve the flexibility, the credibility, the adaptability, and the resource allocation mechanisms, and fulfill the today's subscriber needs. Consequently, they will allow wireless service providers to obtain a network optimization.

BASICS OF PRICING METHODOLOGY: A TECHNICAL PERSPECTIVE AND A TAXONOMY

On mention of pricing, we should first understand the concept of utility, which has been commonly used in game theory. Utility based network control and management algorithms have extensively been studied recently. They are not new concepts and have been commonly used in microeconomics. The utility refers to the level of satisfaction of the decision maker when it acquires certain amount of the resources or services with respect to the price paid for the achieve resources or services. Formally, a utility function that describes a particular set of preference rules is not unique. Actually utility and pricing based algorithms have been well studied for distributed flow control of best effort services in wired networks (Kelly et al., 1998), (Yaiche et al., 2000). Appropriate pricing techniques can be deployed among multiple selfish users to implement various resource allocation policies, including, but not limited to, revenue maximization, social-welfare improvement, user fairness guarantee and system-wide optimization.

GENERAL CLASSIFICATION AND DEFINITIONS

In general, there are two kinds of pricing methodology according to different perspectives of economics and techniques, as shown in Figure 1.

In this chapter, we define them as economic pricing and technical pricing, respectively. Here, economic pricing is always related to the revenue of the network operators, which studies the pricing problems from the economic viewpoint. Technical pricing always determines some technical functions introduced by pricing design, for instance, improving the Pareto optimality of inefficient equilibrium solution in non-cooperative game formulations. In detail, they are defined as follows.

Definition 1: *Economic Pricing* is the process of determining what a network operator will receive in exchange for its service to the subscribers, which is mainly related to the monetary revenue of the network operators.

By adopting an appropriate pricing policy, a service provider will be able to offer the necessary incentives for each subscriber to choose the service that best matches their needs, thereby discouraging over-allocation of resources and maximizing revenue. Therefore, economic pricing should be determined by both the service quality of the operators and service quantity required by the subscribers.

Figure 1. Classification of pricing methodology according to different perspectives of economics and techniques

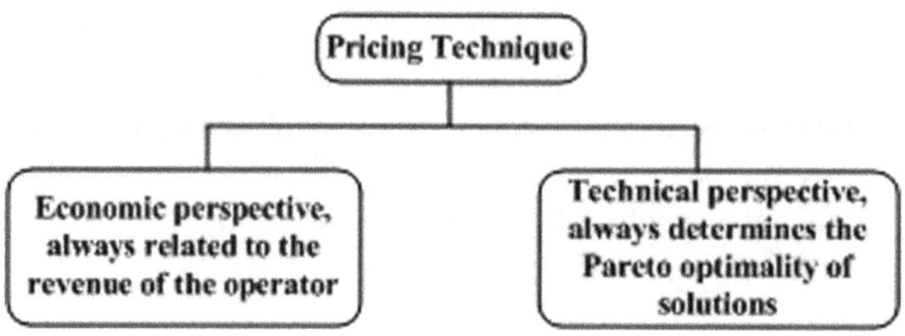

Definition 2: *Technical Pricing* is defined from the perspective of technical functions of introducing pricing, where these pricing factors always implement social optimality improvement, distributed algorithm design, or signaling overhead reduction, and so on.

For instance, the basic approach of social optimal pricing is to formulate the problem as an optimization problem, and design a dual-based distributed algorithm for the distributed resource allocation. Here the dual variables have the interpretation of pricing factors.

In this chapter, we have the following definitions according to what should be priced in the specific applications of pricing.

Definition 3: *Service Pricing* is highly related to the traffic forwarding and service pricing. It is normally from the economic perspective.

Definition 4: *Resource Pricing* is commonly known that pricing is highly related to the resource consumption in case of exorbitant consumption of resources, for instance, power resource or spectrum resource.

Here, power control was investigated in (Saraydar et al., 2001), where the consumed power was introduced as price. By now, we think both the service pricing and resource pricing have been extensively investigated and easily to be understood. Here, in this chapter we also concentrate on some other pricing concepts, including interference pricing, which is defined as follows.

Definition 5: *Interference Pricing* is the marginal decrease in utility for that user due to a marginal increase in interference.

Interference pricing can be viewed as a form of Pigovian tax, which was pointed in (Huang, 2009), where interference price was derived in the use case of distributed power control.

In fact, interference pricing has found several applications. Authors in (Huang et al., 2006) investigated joint power and price adaptation algorithm, where the price represents the interference among the users. The price measures the interference influence of other users to the investigated user. Authors in (Schmidt et al., 2009) discussed distributed resource allocation schemes based on the exchange of interference prices. An attractive property of these schemes is that they are scalable with limited information exchange. Authors in (Shi et al., 2009) defined the interference pricing in the MIMO channel case to avoid introducing significantly interference power to other players. Furthermore, the authors in (Man & Wong, 2013) studied random access with interference pricing based on the signal-to-interference plus noise ratio (SINR) model. Meanwhile, interference pricing was extended to cognitive radio networks in (Hong & Garcia, 2011), where interference pricing was introduced with the channel gain function as the time-varying price.

Above we basically clarify all the pricing concepts appeared in the current pricing-application literature. In this chapter, we concentrate on both technical and economic pricing, and our main contribution is to shed light on the technical perspective of pricing theory and methodology. Therefore, we first clarify these technical functions of pricing in the current extensive applications.

SKETCH OF APPLICATIONS OF PRICING: A TECHNICAL PERSPECTIVE

There exist several seminal work, which have been highly cited and mainly followed by the follow-up works. For instance, Kelly et al. in (Kelly et al., 1998) achieved the social optimal solution to maximize a social welfare utility function by allocating the resources according to the notion of proportional fairness per unit charge. Yaiche et al. in (Yaiche et al., 2000) obtained a Pareto-optimal and proportionally fair solution for elastic services in high-speed networks. It concluded with the pricing of elastic connections based on both users' bandwidth requirements and budget. The utility and pricing based designs have been applied to the power control problem in wireless networks (Yates, 1995). As follows, we summary these technical functions of pricing in detail.

1. **Pricing Improves Social and Pareto Optimality:** Roughly speaking, social welfare is defined as the sum of utilities, and social optimality means the optimization of all the utilities of players. A pricing policy is called incentive compatible if pricing enforces a equilibrium that improves social welfare. The basic idea of improving the Pareto optimality of non-cooperative behaviors is to control a player's behavior through the price of resources to obtain the desired results.

Authors in (Saraydar et al., 2001) investigated the distributed power control problem in CDMA system via non-cooperative game, where resource usage linear pricing function was formulated to improve the Pareto-optimality of Nash equilibrium, where the authors mathematically proved the opportunity of using pricing theory and methodology to improve efficiency of the equilibria. However, pricing scheme can not achieve the social optimality, and it only can improve the Pareto optimality. Similarly in (Rasti et al., 2009), a Pareto-efficient, goal-driven, and distributed power control was proposed, where the biggest difference from (Saraydar et al., 2001) was the proposed pricing scheme was based on signal-to-interference ratio (SIR). Both the consumed resource or the achieve SIR related pricing function should be adjusted to have the same units with the utility function, where a pricing factor was introduced. In (Wang et al., 2007), the pricing factor was optimally designed since it was determined from the Karush-Kuhn-Tucker (KKT) conditions, where the authors derived both the individual utility function with pricing design and the social-welfare utility maximization.

Summary: The first technical function of pricing theory and methodology is to improve social and Pareto-optimality.

2. **Pricing Facilitates Distributed Algorithm Design:** To improve the social and Pareto-optimality, there always involves social welfare optimization problem with coupled constraints and coupled objective functions. This is the biggest obstacle to design the distributed algorithm, and pricing theory and methodology can help decouple and facilitate the design of various distributed algorithm. For instance, different techniques to tackle coupling issues have been summarized in (Chiang et al., 2007), which enable developments of distributed algorithms and guarantee global optimality. The coupled objective functions are more difficult to deal with and thus less studied. One recent innovative approach to deal with coupled objective functions is to use the consistency pricing (Tan et al., 2006), which establishes a new framework of designing distributed algorithm.

On the limited resource couples, a linear pricing technique was employed to handle spectrum competition coupling (Xu et al., 2014) to maximize the energy efficiency requirement. A distributed algorithm was devised to harvest the multi-resource and multiuser gains, and the multi- resource allocation was transformed to the pricing-factor profile searching (Xu et al., 2014).

In most instances, the devised distributed algorithms converge to the Nash equilibrium of the corresponding power-control games. In addition, various pricing schemes are developed to achieve a balance between maximizing the total network utility and minimizing the power consumption, and to improve the efficiency of the equilibrium solutions. In (Maharjan et al., 2011), concentrating on spectrum trading problems in cognitive radio networks, it is important to understand the feasibility of using economic approaches as well as to realize the technical challenges associated with them. With this motivation, an extensive summary of the related work was provided in (Maharjan et al., 2011), while these studies used economic approaches such as game theory, price theory, and market theory to model the behavior of primary and secondary users for spectrum sharing and discuss the associated issues.

In (Yang & Li, 2010), a novel framework was constructed to facilitate the distributed spectrum resource matching between service requester pool and the service provider pool in the coexistence context of multiple heterogeneous cognitive radios. Both the optimal pricing schemes of the provider for the specific shared spectrum and optimal transmission strategies of the requester for rational spectrum requirement were designed and distributed algorithms were designed with low implementation complexity and limited iterations.

Summary: The second technical function of pricing theory and methodology is to help decouple and facilitate the design of various distributed algorithm.

3. **Pricing Encourages Cooperation:** The last but not the lease technical function of pricing theory and methodology is to motivate the cooperation between different players in the networks. For instance, an ad-hoc wireless network can be formulated as a free market economic model. Each node sets its own price for relaying services, with the goal of earning revenue that exceeds its time average reception and transmission expenses. A greedy pricing strategy that maximizes social welfare while ensuring all participants make non-negative profit was proposed in (Neely, 2009). Ren et al., in (Ren & Schaar, 2011) proposed a pricing framework that enabled the relay to set prices to maximize either its revenue or any desirable system utility. A pricing mechanism was proposed which gives the relay incentives to forward the users' signals to the destinations. In particular, the price is determined by the relay such that its revenue or system utility is maximized. A control signal motivates users to adopt a social behavior. An efficient pricing mechanism makes decentralized decisions compatible with overall system efficiency by encouraging efficient sharing of resources rather than the aggressive competition of the purely non-cooperative game.

Summary: The third technical function of pricing theory and methodology is to motivate the cooperation between different players in the networks.

TAXONOMY OF PRICING SCHEMES

After clarify concepts of pricing, we summarize the different technical application of pricing, and at this time, we present a taxonomy of pricing according to various networks, problems and technical applications. Different technical problems in various network architectures are investigated via pricing and related analysis method. Here, we provide a comprehensive survey and understanding of pricing in cognitive radio networks, where pricing theory and methodology are applied to these typical dynamic spectrum access, advance spectrum sharing

In fact, these technical applications also can be easily extend to the heterogeneous cellular networks due to the similar hierarchical network architecture between cognitive radio networks and heterogeneous cellular networks. That is why here we focus on the specific application of pricing with typical characteristics of heterogeneous cellular networks, where we provide the economics and pricing incentives of the wireless operators for introducing the small cells, for instance, the femtocells. Meanwhile, interference mitigation and capacity offload are critical to support the guaranteed performance.

In the third part, we concentrate on the future 5G networks, where we think it will be cognitive, green, and densely deployed of small cells to construct the heterogeneous networks. Here, we provide two typical applications of pricing in the field of green communications and the multi-resource management problems, details of which are illustrated in Figure 2.

For each application, we will concentrate on the technical perspective of them, where we Summarize like these: Pricing for improving Pareto-optimality, and this is the most technical application of pricing in the non-cooperative game formulations. Pricing can help to improve the optimality of the equilibrium solutions. Second, pricing for motivating cooperation be- tween difference selfish players; Third, pricing for facilitating distributed algorithm, reducing signaling overhead.

Figure 2. Taxonomy of pricing theory and methodology according to various networks, problems and technical applications

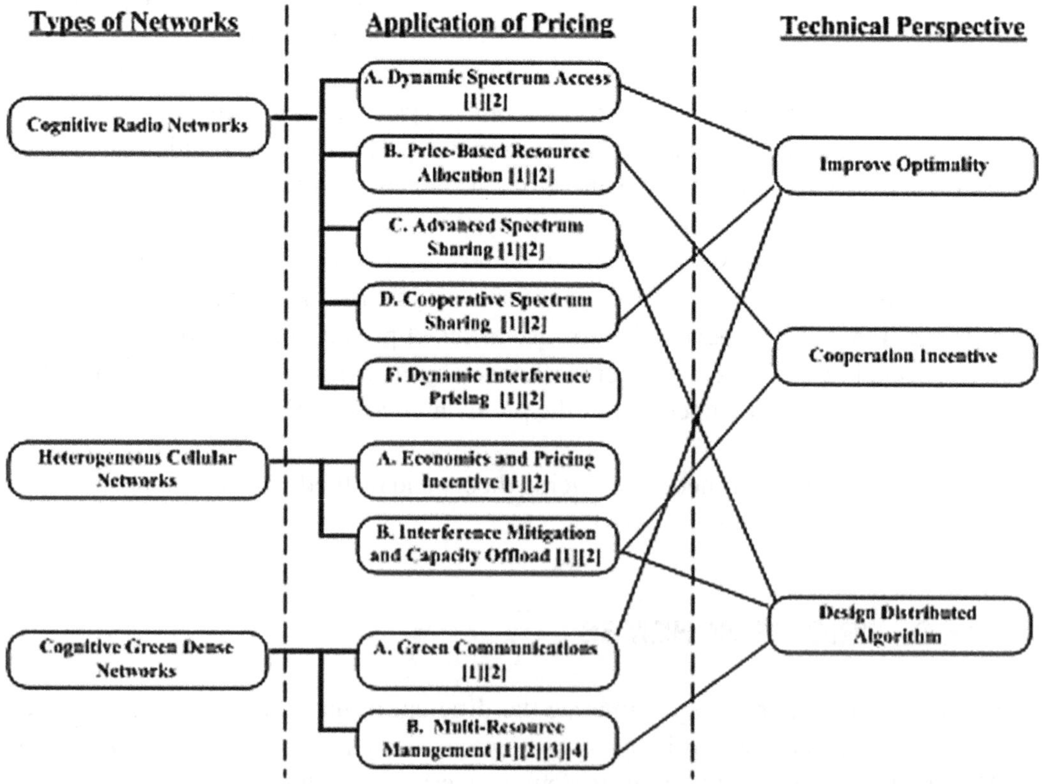

COGNITIVE RADIO NETWORKS

To release the spectrum scarcity, cognitive radio networks have been poured extensive research attention from academia, industry, and standardization communities. Cognitive radio networks can enhance spectrum utilization and thus improving the spectrum efficiency by opportunistically permitting secondary users access. To implement this process, secondary users should first sense the spectrum spaces and then determine the dynamic spectrum access; then secondary users should be allocated suitable resource including available channels and transmit power, to maximize the spectrum utilization performance. This process requires that the transmission of secondary users should not introduce uncontrollable interference power to the primary users. Primary users could coexist with secondary users in a spectrum underlay way, or in the overlay, or interweave way. On the other hand, shown in Figure 3 both centralized and distributed cognitive radio networks have been poured extensive attention on various kinds of problems, including spectrum sensing, spectrum access, and resource management and so on. On the centralized cognitive radio networks, we know cognitive CDMA and OFDMA cellular networks, and cognitive WiFi 2.0 networks have been studied for years. Certainly, the distributed cognitive radio networks are with more flexibilities and scalability.

We summarize different pricing schemes in cognitive radio networks, which are with the typical characteristics. Different applications of dynamic spectrum access, pricing-based resource allocation, and advanced spectrum sharing and so on, as shown in Figure 4.

DYNAMIC SPECTRUM ACCESS

Three aspects of dynamic spectrum access here are recognized that are highly related to the technical pricing, which are decentralized implementation, revenue of operators, and dynamics and behaviors of secondary users.

1. **Decentralized Implementation with Low Complexity:** Pricing is an efficient way to provide economic incentives for operators during dynamic spectrum access. For instance, authors in (Yang et al., 2013) investigated the pricing-based spectrum access control, where primary users can sell their temporarily unused spectrum to multiple secondary users; while secondary users compete via

Figure 3. Centralized and distributed cognitive radio networks

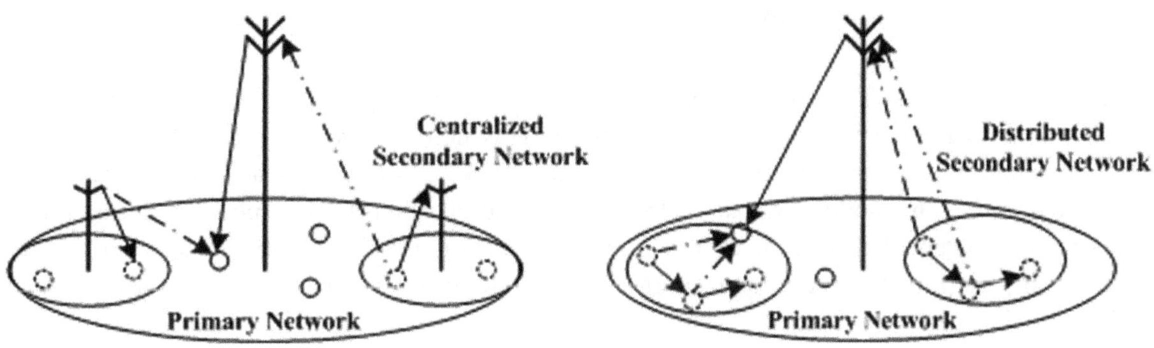

Figure 4. Different pricing-related applications in cognitive radio networks

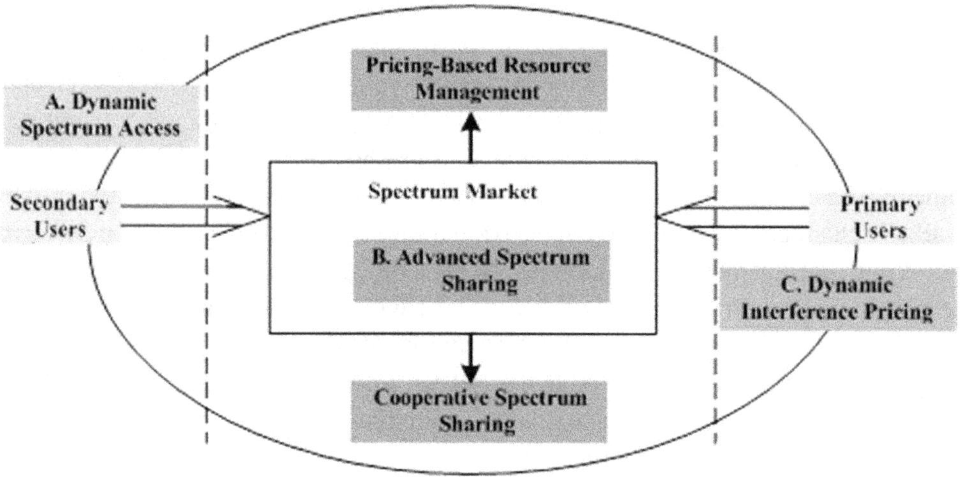

random access for such spectrum opportunities. Pricing-based spectrum management with random access provides a platform for secondary users contending for spectrum access and is amenable to decentralized implementation due to its low complexity. In (Wang et al., 2007), a novel joint power and channel allocation scheme that improves the performance through a distributed pricing approach. In this scheme, the spectrum allocation problem is modeled as a non-cooperative game, with each CR pair acting as a player. A price-based iterative water-filling algorithm is proposed, which enables CR users to reach a better Nash equilibrium.

2. **Maximize Operators' Revenue:** In dynamic spectrum access networks, to maximize a primary network operatorars revenue, a revenue sharing scheme was provided to allow secondary users to access in (Wu et al., 2014), where a two-layered game was formulated to investigate the coupling effect between the revenue sharing and resource allocation. In addition, an operator' revenue is highly related to the bearing traffic load and the serviced subscribers. To maximize the operators' revenue, load balance and maximization of the number of secondary users should be investigated. In (Tran et al., 2014), the authors observed the load balancing yields congestion avoidance incentives for secondary users. Here, they dealt with the problems with these constraints of the couplings among pricing, load balancing and secondary users' spectrum access decision, and the heterogeneity of primary users' traffic and secondary user types. Finally, a joint load balancing and pricing problem was to maximize operators' revenue in two cognitive radio markets including monopoly and duopoly (Tran et al., 2014).

To find the largest feasible set of secondary users that can be supported in the network, a vector-cardinality optimization problem (Zhai et al., 2014). This non-convex problem is however hard to solve, and we propose a convex relaxation heuristic based on the sum-of-in-feasibility in optimization theory. The presented methodology leads to the notion of admission price for spectrum access that can characterize the tradeoff between the total energy consumption and the system capacity. Price-driven algorithms for joint power and admission control are then proposed that quantify the benefits of energy-infeasibility balance.

3. **Economic Behaviors and Dynamics:** The collusion among selfish users may deteriorate the efficiency of dynamic spectrum allocation. The network users' behaviors and dynamics need to be taken into consideration for efficient and robust spectrum allocation. In (Ji & Liu, 2008), the spectrum allocation with multiple selfish legacy spectrum holders and unlicensed users was modeled as a multi-stage dynamic game, where a pricing-based collusion-resistant dynamic spectrum allocation approach to optimize overall spectrum efficiency, while not only keeping the participating incentives of the selfish users but also combating possible user collusion.

This buying decision of spectrum buyers depends on spectrum pricing, seller reputation, and duration of the contract and spectrum quality. However, existing pricing models for dynamic spectrum access only concern bandwidth making them unsuitable for real-world trading. In (Hassan et al., 2014), detailed issues related to the spectrum pricing were investigated with update of the spectrum sellers's reputation, modeling of trustworthiness of recommenders, incorporation of contract duration and spectrum quality, and dynamic pricing of the market. In (Do et al., 2014), the authors studies the pricing effect on the equilibrium behaviors of selfish secondary users' data packets. From the secondary users' point of view, a spectrum access decision on whether to join the queue of the BS or not is characterized through an individual optimal strategy that is joining the queue with a joining probability. This strategy also requires each secondary user to know the average queuing delay, which is a non-trivial problem.

Primary users sell the temporarily unused spectrum and secondary users compete via random access for such spectrum opportunities. Compared to existing market-based approaches with centralized scheduling, pricing-based spectrum management with random access provides a platform for secondary users contending for spectrum access and is amenable to decentralized implementation due to its low complexity. Focusing on two market models, one with a monopoly PU market and the other with a multiple-PU market. (Tran et al., 2013) characterized network operators' service provisions to delay-sensitive secondary users via pricing strategies. Based on the two paradigms of shared-use and exclusive-use dynamic spectrum access, we examine three network scenarios corresponding to three types of secondary markets.

Figure 5. Oligopoly, monopoly and duopoly market-based dynamic spectrum access

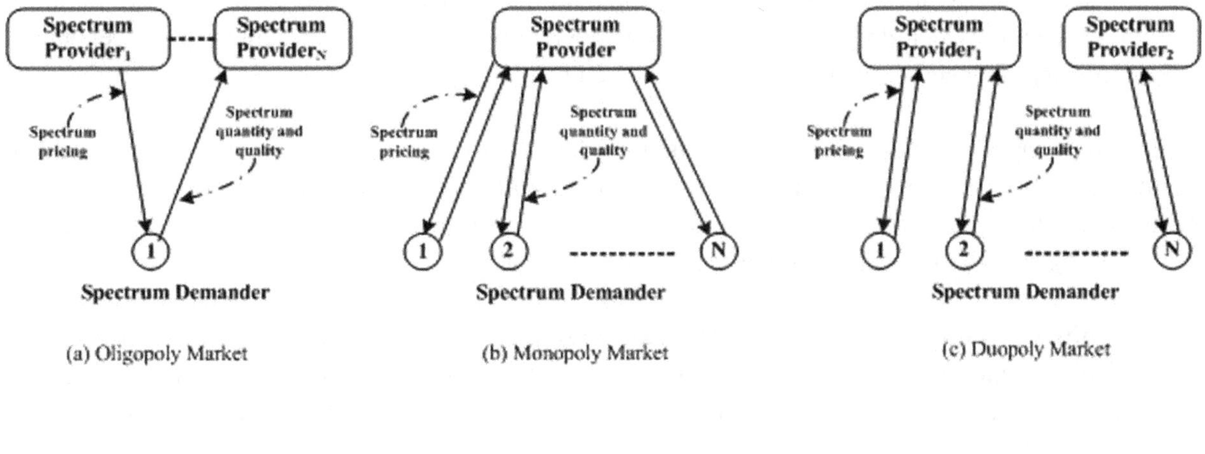

ADVANCED SPECTRUM SHARING

In this chapter, we refer to cognitive radio-inspired spectrum trading, leasing and auction as the advanced spectrum sharing schemes. In our previous work (Yang, 2015), (Yang & Li, 2013), we summarize these advanced and flexible spectrum management schemes and analyze their advantages and disadvantages. We take the viewpoints of both the spectrum marketing perspective and spectrum technical perspective. Meanwhile, in (Yang & Li, 2013), we proposed a pricing-based dynamic spectrum leasing in the framework of a dynamic hierarchical game. We found that spectrum pricing was critical for these advanced spectrum sharing. In fact, spectrum pricing during spectrum sharing process depends on the objective of spectrum trading, leasing and auction, and is also highly related to the behaviors of spectrum sellers and spectrum buyers. These spectrum sellers could be the primary service providers, primary users, or mobile virtual network operators. The spectrum pricing is implemented by the primary players, who can be primary service providers, primary users, or virtual network operators.

1. **Primary Service Providers:** Pricing can regulate secondary user churning behaviors among different primary service providers, and different pricing models affect both the revenue of operators and the experience quality of users. Authors in (Niyato et al., 2008) investigated spectrum pricing problem, where primary service providers compete to offer spectrum access opportunities to the secondary users. Primary service providers aim to maximize their revenue with an equilibrium pricing scheme. In (Niyato et al., 2008), authors provided different pricing models including market-equilibrium, competitive, and cooperative pricing models for spectrum trading. In these pricing models, the primary service providers have different behaviors to achieve different objectives of spectrum trading.
2. **Primary Users:** In addition to the primary service providers' pricing, primary users also can lease its spectrum to the secondary users, where it is necessary to characterize the outcomes of secondary spectrum markets when multiple primary users compete to provide secondary users' access. Price competition among primary users was investigated in (Kasbekar & Sarkar, 2012), where there were a random and unknown number of secondary users. The time slotted spectrum leasing process can be summarized as follows: with some probability primary users have unused spectrum to lease to secondary users, at this time, primary users must set a price with respect to the spectrum demands and spectrum quality. Authors in (Kavurmacioglu et al., 2014) studied a competition model in which each provider aims to enhance its revenue by opportunistically serving a price-dependent secondary demand, while also serving dedicated primary demand.

Summary: Optimal pricing is related to multiple factors to enhance spectrum revenue. Authors in (Duan et al., 2011) presented a comprehensive analytical study of two competitive secondary operators' spectrum leasing and pricing strategies, taking into account operators' heterogeneity in leasing costs and users' heterogeneity in transmission power and channel conditions.

3. **Cognitive Mobile Virtual Network Operators:** Authors in (Duan et al., 2011) studied the optimal investment and pricing decisions of a cognitive mobile virtual network operator under spectrum supply uncertainty. Compared with a traditional mobile virtual network operator who often leases spectrum via long-term contracts, a cognitive mobile virtual network operator can acquire spectrum dynamically in short-term by both sensing the empty "spectrum holes" of licensed bands

and dynamically leasing from the spectrum owner. As a result, a cognitive mobile virtual network operator can make flexible investment and pricing decisions to match demands of the secondary unlicensed users. The cognitive mobile virtual network operator determines the optimal amounts of spectrum sensing and leasing by evaluating the trade-off between cost and uncertainty. Authors in (Li et al., 2012) consider a downlink OFDM communication system with various network dynamics, including dynamic user demands, uncertain sensing spectrum resources, dynamic spectrum prices, and time-varying channel conditions. In addition, heterogeneous users and imperfect sensing technology are incorporated to make the network model more realistic. By exploring the special structure of the problem, low- complexity online control policies were implemented to determine pricing and resource scheduling without the statistics of dynamic network parameters.

Summary: Recent years market-driven advanced spectrum sharing schemes have been extensively exploited, and there are virtual money-inspired and resource exchange-based schemes. On the one hand, primary users temporarily sell the spectrum to secondary users in exchange of either a monetary reward or a performance improvement. On the other hand, cooperative spectrum sharing, where secondary users relay traffics for primary users to get their own share of spectrum.

By motivating the buyers truthfully reveal their private valuations, the seller can optimally allocate the resource. Contract theory is effective in designing incentive compatible mechanisms in a monopoly market under incomplete information, where the employer needs to sign a contract with employees before fully knowing their private information. The key idea is to offer the right contract items so that all agents have the incentive to truthfully reveal their private information (Duan et al., 2014).

DYNAMIC INTERFERENCE PRICING

Most of the pricing schemes are resource utilization-based pricing, for instance, authors in (Xu et al., 2010) studied the pricing issue where the secondary users strategically adjust their uplink transmission power to maximize the utility, and the primary service provider charges the secondary users on their transmitted power levels to enhance its own revenue. In this subsection, we summarize applications of interference pricing, where concentrate on power control and dynamic spectrum access control as two important use cases to study the interference pricing.

1. **Dynamic Interference Pricing for Power Control:** Authors in (Wang et al., 2014) investigated the price-based power control of spectrum sharing, where primary users will allow secondary users to access via pricing the interference power under the interference power constraint. In order to access the spectrum, the secondary users need to pay for their interference power. The primary user first decides the price for each secondary user to maximize its revenue. Then, each secondary user controls its transmit power to maximize its revenue based on non-cooperative game.

Authors in (Huang et al., 2005) studied the performance of a distributed and asynchronous power control scheme to improve the total network utility, and users need to adjust their transmission powers in an interference-aware fashion. The users exchange prices that reflect their loss in utility due to interference. The prices are then used to determine optimal power levels for each user. With logarithmic utilities, the pricing algorithm exhibits rapid convergence to the unique optimal power allocation.

It is a complex task for a primary user to allow secondary users to use their allocated spectrum. It requires the interference generated by the secondary users is well controlled. Authors in (Hong & Garcia, 2011) addressed this problem for a particular type of interference constraints, in which the aggregated powers of the interference generated by the unlicensed network are required not to exceed certain thresholds. A natural extension to these algorithms involves penalizing each unlicensed user with a set of time-varying prices based upon its contribution to the total interference.

2. **Dynamic Interference Pricing for Dynamic Spectrum Access:** Authors in (Man & Wong, 2013) studied the problem of random access with interference pricing. First, under the setting with fixed interference linear pricing, the existence of the Nash equilibrium is characterized in the random access game. In particular, when the utility functions of all the players satisfy a risk aversion condition, it is shown that the game is a S-modular game and characterize the convergence of the strategy profile to the Nash equilibrium. Then, under the setting with adaptive interference linear pricing, an iterative algorithm was proposed to solve the network utility maximization problem.

Summary: Interference pricing can be used to model the interactive behavior and constrain the rational behavior toward exorbitant resource utilization and resource collision. Interference pricing motivates the potential cooperations, and creates a distributively social-optimal environment from individual player's point of view to a certain extent. More complicated and practical interference pricing functions should be designed to achieve more efficient resource utilization and to avoid resource collision in addition to the current linear pricing functions.

MULTI-TIER HETEROGENEOUS CELLULAR NETWORKS

Ultra-dense deployment of various small cells in the coverage of macrocells makes up multi-tier heterogeneous cellular networks. Small cells are with several good properties compared to macrocells such as spectrum and energy efficient with high cost efficiency. In addition, different small cells suit for the traffic space and time characteristics of different scenarios, and small cells are regarded a promising technology to handle exponentially increasing wireless data traffic. For instance, the femtocell technique can address the poor in-building coverage problem and increase network capacity cost efficiently.

Extensive attention has been paid to resource management including power control, spectrum utilization, and load balancing in multi-tier heterogeneous cellular networks, and their success largely depends on whether operators and users accept this technology or not. Different wireless service operators and providers are always different pricing policies, which involves research of the economics and incentive mechanism design of the small cells.

At present, some wireless service providers have launched various services based on small cell deployment, although there are still plenty of challenges unsettled. Here, in this section, in addition to the economics of introducing small cells, we generally investigate three technical problems including inter-tier and intra-tier interference management, traffic offload, and the limited backhaul networking problems as shown in Figure 6. These are the typical and critical technical challenges in multi-tier heterogeneous cellular networks, where we concentrate on the service pricing, resource pricing, interference pricing, and traffic pricing.

Figure 6. Key applications of pricing methodology in multi-tier heterogeneous cellular networks

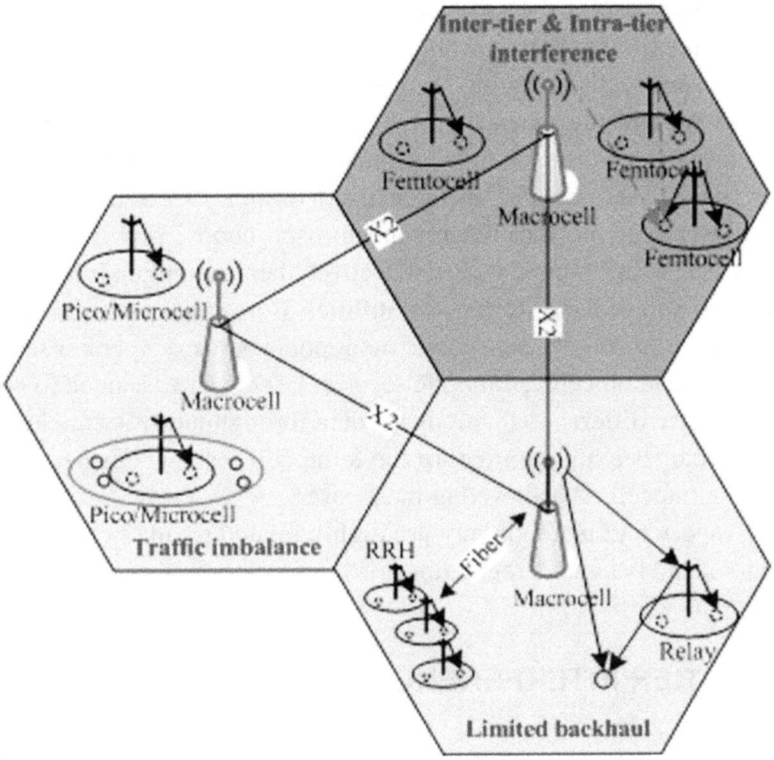

ECONOMICS AND PRICING INCENTIVE

Extensive research works should concentrate on these novel promising problems of inter- tier and intra-tier interference management, traffic offload, and the limited backhaul networking problems; however, we should first answer whether operators and users accept small cell technology or not. Authors in (Duan et al., 2012) pointed out one of the biggest challenges for femtocell deployment was lack of a clear business model, where the economic incentive for the cellular operator was investigated to enable femtocell service by leasing spectrum resource to an independent femtocell operator.

Small cells can be deployed by the same operator or different operators, or service providers. These kind of players' economic entry and interactive behaviors can be investigated via game theory.

Authors in (Kim et al., 2014) studied pricing schemes for femtocell service, where the interaction between a monopolist provider and users was formulated as a leader-follower game. Meanwhile, the competition between two providers in a duopoly market was modeled as a non-cooperative simultaneous game. In (Guijarro et al., 2012), a three-level multi-leader-follower game was formulated to analyze the entry effect of a femtocell operator into the market of macrocell operators, with the aims to answer the question of which benefit users can get from the entry of a femtocell operator into the market.

Economic aspects of femtocell services were also studied in (Yun et al., 2012) for the case of monopoly market, where flat pricing and partial volume pricing were exploited. Here, the basic conclusions are summarized as: first, open access femtocell service is beneficial to both users and providers; Second, flat pricing has a minor impact on operator revenue; Third, compared with partial volume pricing, flat pricing is advantageous to the operator when users are sensitive to price. To motivate the small cell service providers to open portion of the access opportunities to macro users, Lin et al. in (Lin et al., 2011) discussed three business-related frameworks in macro-femto heterogeneous networks according to different deployments of femtocells including joint deployment, service provider deployment, and user deployment frameworks. Meanwhile, incentive mechanism was studied in (Zhu et al., 2014), and authors in (Duan et al., 2013) investigated the economic incentive for a cellular operator to add femtocell service on top of its existing macrocell service. The operator first determines spectrum allocations and pricing of femtocell and macrocell services, and then heterogeneous users choose between the two services and the amount of resource to request. In (Yang et al., 2014) a novel linear refunding framework was proposed to both improves the operators' revenue and the users' utilities. Both split spectrum and common spectrum were studied in (Shetty et al., 2009), where a single monopolist network operator sets prices for both the mobile-only service and the femtocell and mobile service. Users were assumed to possess linear utility for data throughput and have different valuations for data throughput. It is concluded that the optimal pricing scheme always charges a higher price for the femtocell service. Further, at the optimal prices, almost full adoption of femtocells is achieved in most cases.

Summary: Pricing theory and methodology are highly related to and benefits the economic entry, motivations of operators, and behaviors regulations.

INTER-AND INTRA-TIER INTERFERENCE

Due to large-scale and self-deployed of small cells overlaid on traditional cellular networks, both inter- and intra-tier interference may arise and become serious due to full frequency reuse. We concentrate on resource pricing and interference-aware pricing schemes.

1. **Resource Pricing:** To mitigate the interference, authors in (Li et al., 2013) proposed a distributed power control scheme for the uplink transmission of two-tier femtocell networks. Both single-channel case and multiple-channel case were investigated in (Li et al., 2013). For single-channel case, each user announced a price that reflects compensation paid by other users. Both the distributed power levels and prices updating algorithm and cellular link protection scheme were proposed for protecting cellular users by increasing their prices.

For the downlink of dense femtocell networks, authors (Li et al., 2013) presented a distributed power self-optimization scheme. There are two steps, first, the authors established a non-cooperative power self-optimization game framework. Then, a novel convex pricing mechanism was presented to price the transmit power of femtocells in (Li et al., 2013), where the price coefficients were shown to be relative importanceto the profit function.

2. **Interference Pricing:** Authors in (Kang et al., 2011) investigated the price-based resource allocation in two-tier femtocell networks, where a central macrocell is underlaid with multiple distributed femtocells fully reusing the frequency band. Macrocell base station will protect themselves by pricing the interference from other femtocell users. Here, the authors in (Kang et al., 2011) proposed uniform pricing and non-uniform pricing for two practical femtocell network models including the sparsely deployed scenario for rural areas and densely deployed scenario for urban areas. Authors

in (Liu et al., 2013) introduced an energy-efficient power control in a convex pricing way with the considerations of both circuit and transmit powers. To achieve a targeted performance, the macrocells will protect themselves by its interference cap, which is based on the total interference macrocells will largely tolerate, thus pricing the interference from the femtocell users.

Summary: Certainly, there exist other interference mitigation techniques, interference can- cellation and interference alignment, and so on. Pricing theory and methodology can extend its applications in wider fields. Meanwhile, rich pricing schemes designed for cognitive radio networks can be expanded to the heterogeneous cellular networks. This is largely due to the similar hierarchy and network architecture between them, for instance, primary users and secondary users corresponding to macrocell and various small cells. Also, there exists natural hierarchy properties between service provider and the users, and macrocell and femtocell service providers in heterogeneous cellular networks. That is why the hierarchical game, also termed as the Stackelberg game has been extensively used in both cognitive radio and heterogeneous cellular networks.

TRAFFIC OFFLOAD

To counter the capacity crunch, mobile data offloading through the WiFi or small cell access points can significantly enhance user quality of service and alleviate the cellular congestion. This benefits both the mobile network operators and the mobile users, but comes with unique technical and economic challenges that must be jointly addressed. Authors in (Iosifidis et al., 2015) and (Gao et al., 2014) proposed a double auction mechanism and cooperative bargaining based offloading schemes, and (Cheung & Huang, 2015) achieved a good tradeoff between the user's payment and its QoS characterized by the file transfer deadline. Jointly considering the network congestion, switching penalty, and pricing, authors in (Yu et al., 2014), (Cheung et al., 2014) studied the network selection and data offloading problem in an integrated cellular Wi-Fi system. On the other hand, small cells can increase the capacity by offloading traffic from macrocells, where various low-power nodes can be distributed throughout the macrocells as a more targeted underlay to further enhance the area spectral efficiency, alleviate traffic imbalance, and thus improve the end-user experience. In this chapter, we summarize three ways to implement traffic offload in the pricing framework, which are resource-aware resource exchange and balance, mobile association, and D2D communication offload.

1. **Resource Exchange and Balance:** These small cells can be deployed by the same or the different wireless operators or the service providers with different pricing schemes. Therefore, different price decision and resource allocation schemes affect the revenue maximization of the service providers, and an economic framework was proposed in (Chen et al., 2011), where users choose either macrocell or femtocell service to optimize their own utility and the monopolistic wireless service providers tries to maximize its revenue via pricing and spectrum allocation strategy.

For instance, to avoid unnecessary high transmission power, a pricing-based power optimization scheme was proposed in (Liet al., 2013), where macrocell was set as the leader, and the leader will sell transmission power quota to femtocells with the non-uniform pricing scheme. Similarly, authors in (Zhang et al., 2013) devised a distributed power allocation algorithm with differentiated pricing update. In addition, a Stackelberg coordination game was formulated for macrocell and femtocell service providers to analyze a joint spectrum leasing, pricing and interference coordination process with the novel energy-aware utility functions (Yang & Li, 2014). To motivate the small cell service providers to open portion of the access opportunities to macrocell users, an pricing-related incentive mechanism was proposed in (Zhu et al., 2014), where the macrocell service provider could pay to the small cell service provider.

2. **Mobile Association:** Mobile association schemes were developed to balance the traffic among different nodes in wireless downlink heterogeneous cellular networks. Authors in (Gajic et al., 2014) studied the problem where wireless service providers compete for heterogeneous wireless users. In(Shen & Yu, 2014), a user and base-station association problem was studied with proportional fairness consideration, metric of which was based on the value of a designed utility function minus a price. The proposed price update algorithm was based on a coordinate descent method to solve the dual of the network utility maximization. Authors in (Li et al., 2012) investigated the cooperation between base stations and multiple low-power relay nodes on the downlink communications towards the users. A mobile association framework was proposed with intra-cell node cooperation to maximize the network capacity and balance the traffic load among the network nodes. Furthermore, a pricing mechanism was proposed to enable the distributed implementation. Meanwhile, different from conventional range expansion, authors in (Han et al., 2014) proposed a fair range expansion with interference mitigation, where not only the bias value of range expansion but also the number of users attached to the picocell was considered to improve the fairness of user association.
3. **D2D Communication Offload:** Introducing kinds of small cells can offload the traffic of the macrocell in heterogeneous cellular networks, and device-to-device (D2D) communications are regarded as one of the most important way to offload traffic. A D2D communication-based load balancing algorithm was proposed in (Liu et al., 2014), which utilizes D2D communications as bridges to offload traffic among different cells according to traffic distributions. A decentralized framework of joint spectrum allocation and power control was proposed in (Yin et al., 2015), where the pricing based method was used to mitigate interference from D2D pairs to the BS.

LIMITED BACKHAUL NETWORKS

Wireless heterogeneous networks are a cost and energy efficient alternative to provide high capacity. However, the cost and energy consumption of backhaul network may limit the benefits brought by the use of small cell, in particular, in the ultra-dense deployment of a large number of small cells. Authors in (Mahloo et al., 2014) presented a cost evaluation methodology to compute the total cost of ownership, and it can identify the most critical cost drivers in the backhaul networks and to have a better understanding of the backhaul dynamics when small cells are deployed (Mahloo et al., 2014).

Meanwhile, limited backhaul networks also restrict distributed control, management and maintenance among ultra-dense small cells. For instance, a signal-to-interference-plus-noise ratio (SINR)-pricing was proposed in (Xu et al., 2013) to limit the interference and avoid using a central processor, thus implementing a distributed iteration framework. Furthermore, it always involves a centralized management in the sum-rate maximization problem. Authors in (Carfagna & Barbarossa, 2012) admits a decentralized solution based on pricing mechanism, where the exchange of some control prices was required over a common signaling channel. These control prices are designed by using only statistical information about the cross-channels. A pricing mechanism was proposed in (Lorenzo et al., 2011) to maximize the sum-rate of a femtocell network in a distributed manner with a limited exchange of information. Another alternative game-theoretic techniques was proposed in (Barbarossa et al., 2011) to exploit the limited backhaul network, where an alternative pricing mechanism was proposed to allocate power in the joint time-frequency plane.

FUTURE APPLICATIONS AND RESEARCH DIRECTIONS

Both the orthogonal frequency-division multiple access and cognitive radio technologies offer great flexibility and feasibility for future green wireless communications. Meanwhile, cognitive femtocells have been envisioned as a promising technology for covering indoor environment and assisting heavy-loaded macrocell network. Although lots of technical issues of cognitive femtocell network have been studied, e.g., spectrum sharing, interference mitigation, etc., the economic issues that are very important for practical femtocell deployment have not been well investigated in the literature. However, most of previous works are focused on spectrum sharing and interference avoidance, and the energy efficiency aspect is largely ignored. Motivated by interference mitigation and energy-saving challenges of the heterogeneous networks and the promising cognitive radio techniques, more advanced energy- saving and interference control techniques based on cognitive radio should be developed for better performance. In (Yang & Li, 2014), we survey the interference mitigation and energy management from a cognitive radio perspective.

GREEN COMMUNICATIONS AND ENERGY EFFICIENCY

Energy efficiency enhancement and green communications should be poured more research attentions due to awareness of environment of common people and high energy consumption list of network operators. In summary, most of the current research has been focused on the spectrum efficiency optimization and the most recent energy efficiency optimization in cognitive radio and heterogeneous cellular networks. Authors in (Dixit et al., 2013) presented a techno-economic analysis for regulated secondary user access based on a novel base station-wide framework, where secondary users coexist with primary users. A dynamic incentive-based pricing model proposed in (Dixit et al., 2013) has the inherent capability of call admission control. It is useful to attract secondary users to obtain temporary wireless access during periods of low primary user demand, thus improving spectrum usage in the temporal domain. Concentrating on the pricing issues in the cognitive femtocell network, in (Jiang et al., 2014), a two-tier pricing game theoretic framework is formulated with two models including the static and the dynamic pricing model. In (Zhou et al., 2014), the pricing and resource allocation strategies were designed for the two-tier sensing-based cognitive femtocell networks, where the macrocell and femtocells are operating over the same frequency band. The macrocell base station protects itself by setting the maximum aggregate interference constraint and makes profit by pricing the interference from femtocell users. Different from the conventional underlay-based networks where there is one price only, in the proposed sensing-based networks two prices are adopted corresponding to the idle and busy states of the macrocell.

MULTI-RESOURCE MANAGEMENT

In 5G era, three efficiency aspects should be investigated including the spectrum efficiency, energy efficiency, and cost efficiency. This calls for fully exploring and exploiting multiple kinds of resources. This requires novel pricing methodology for the promising technical problems. For instance, to handle the limited resource couples, a linear pricing technique was employed to handle spectrum competition coupling (Xu et al., 2014) to maximize the energy efficiency requirement. A distributed algorithm was devised to harvest the multi-resource and multiuser gains, and the multi-resource allocation was transformed to the pricing-factor profile searching (Xu et al., 2014).

Pricing is highly related to cost and energy efficiency. In (Xie et al., 2012), the energy efficiency aspect of spectrum sharing and power allocation were investigated in heterogeneous cognitive radio networks with femtocells.

Cognitive and ultra-dense deployment of small cells and machine-to-machine communications are promisingly developing to provides an effective way to further improve the spectrum spatial and frequency reuse. In OFDMA-based hybrid overlay and underlay cognitive femto-cell networks, authors in (Ma et al., 2015) studied the subchannel allocation to fully exploit the potential of spatial and frequency reuse. In the future, pricing techniques will find new applications in ultra-dense deployment of small cells and machine-to-machine communications. Pricing theory and methodology will provide more and more new schemes to facilitate the self- organizing design of management and maintenance.

On the other hand, there has been a surging interest in dynamic pricing in smart grids. For instance, various demand-responsive pricing schemes have been proposed to coordinate the real-time energy consumption. Authors in (Qian et al., 2013) developed an optimal day-ahead pricing scheme for the electricity provider to minimize its cost. Meanwhile, pricing-based demand coordination in smart grids were summarized in (Ren et al., 2014), where the interaction of dynamic pricing and service request scheduling can enhance the wireless service provider's long-term profitability. Authors in (Qian et al., 2013) proposed a real-time pricing scheme that reduces the peak-to-average load ratio through demand response management in smart grid systems.

Last but not the least, the emergence of cloud computing provides a cost effective approach to deliver video streams to a large number of end users with the desired user quality of experience (QoE). Authors in (He et al., 2014) investigated the tradeoff between the cost incurred by virtual machine instance procurement and the achieved QoE of end users under Amazon EC2's pricing models. For cloud computing, authors in (Menache et al., 2014) proposed a pricing algorithm to regulate admission and perform resource allocation to maximize social welfare, and authors in (Wang et al., 2014) developed a computationally efficient pricing mechanism to enable fair competition for cloud resource. Finally, big data will help better pricing for differentiated services from the wireless operators' perspective.

CONCLUSION

Basics of pricing theory and methodology are first presented and summarized including rationale and motivation, concepts, various technical applications, and taxonomy. We survey the applications of pricing theory and methodology in cognitive radio and multi-tier heterogeneous cellular wireless networks with the emphasis on cooperative incentive mechanism design, Pareto- and social- optimality improvement, distributed algorithm design with the low signaling overhead. We look forwards the future application and research directions of the pricing theory and methodology in the fields of cloud computing, smart grids, cognitive and ultra-dense networks.

ACKNOWLEDGMENT

This work was supported in part by the National Science Foundation of China under Grant 61201139, 61231008; by the Open Research Fund of National Mobile Communications Research Laboratory, Southeast University under Grant No.2014D10; by the Fundamental Research Funds for the Central Universities under Grant JB150111; by the ISN02080001; by the 111 Project under Grant B08038; by the EU FP7 under Grant Number PIRSES-GA-2013-610524; and by the Shaanxi Province Science and Technology Research and Development Program (2011KJXX-40).

REFERENCES

Akyildiz, I. F., Lee, W. Y., Vuran, M. C., & Mohanty, S. (2006). Next generation/dynamic spectrum access/cognitive radio wireless networks: A survey. *Computer Networks*, *50*(13), 2127–2159. doi:10.1016/j.comnet.2006.05.001

Almanthari, B., Nasser, N., & Hassanein, H. (2011). Congestion pricing in wireless cellular networks. *IEEE Communications Surveys and Tutorials*, *13*(3), 358–371. doi:10.1109/SURV.2011.090710.00042

Barbarossa, S., Sardellitti, S., & Carfagna, A. (2011). Pricing mechanisms for interference management games in femtocell networks based on Markov modeling. *Future Network And Mobile Summit 2011* (Vol. 48, pp. 1 - 8).

Bhushan, N., Li, J., Malladi, D., & Gilmore, R. (2014). Network densification: The dominant theme for wireless evolution into 5g. *Communications Magazine IEEE*, *52*(2), 82–89. doi:10.1109/MCOM.2014.6736747

Bousia, A., Kartsakli, E., Antonopoulos, A., & Alonso, L. (2014). Sharing the small cells for energy efficient networking: How much does it cost? *Global Communications Conference (GLOBECOM), 2014 IEEE*. IEEE. doi:10.1109/GLOCOM.2014.7037207

Carfagna, A., & Barbarossa, S. (2012). *Distributed Bayesian Pricing for Sum-Rate Maximization in Small-Cell Networks. In Globecom Workshops (GC Wkshps), 2012 IEEE* (pp. 598–602). IEEE.

Chen, Y., Zhang, J., Lin, P., & Zhang, Q. (2011). Optimal Pricing and Spectrum Allocation for Wireless Service Provider on Femtocell Deployment.*IEEE International Conference on Communications* (Vol. 41, pp. 1 - 5). doi:10.1109/icc.2011.5963321

Cheung, M. H., & Huang, J. (2015). Dawn: Delay-aware wi-fi offloading and network selection. *IEEE Journal on Selected Areas in Communications*, *33*(6), 1214–1223. doi:10.1109/JSAC.2015.2416989

Chiang, M., Low, S. H., Calderbank, A. R., & Doyle, J. C. (2007). Layering as optimization decomposition: A mathematical theory of network architectures. *Proceedings of the IEEE, 95*(1), 255–312. doi:10.1109/JPROC.2006.887322

Courcoubetis, C., & Weber, R. (2003). *Pricing Communication Networks: Economic, Technology and Modeling*. John Wiley & Sons. doi:10.1002/0470867175

Dasilva, L. A. (2000). Pricing for qos-enabled networks: A survey. *IEEE Communications Surveys and Tutorials*, *3*(2), 2–8. doi:10.1109/COMST.2000.5340797

Dixit, S., Periyalwar, S., & Yanikomeroglu, H. (2013). Secondary user access in lte architecture based on a base-station-centric framework with dynamic pricing. *IEEE Transactions on Vehicular Technology*, *62*(1), 284–296. doi:10.1109/TVT.2012.2221753

Do, C. T., Tran, N. H., Han, Z., & Le, L. B. (2014). Optimal pricing for duopoly in cognitive radio networks: Cooperate or not cooperate? *Wireless Communications IEEE Transactions on*, *13*(5), 2574–2587. doi:10.1109/TWC.2014.031914.131363

Duan, L., Gao, L., & Huang, J. (2014). Cooperative spectrum sharing: A contract-based approach. *Mobile Computing IEEE Transactions on*, *13*(1), 1.

Duan, L., Huang, J., & Shou, B. (2011). Duopoly competition in dynamic spectrum leasing and pricing. *IEEE Transactions on Mobile Computing*, *11*(11), 1706–1719. doi:10.1109/TMC.2011.213

Duan, L., Huang, J., & Shou, B. (2011). Investment and pricing with spectrum uncertainty: A cognitive operator's perspective. *Mobile Computing IEEE Transactions on*, *10*(11), 1590–1604. doi:10.1109/TMC.2011.78

Duan, L., Huang, J., & Shou, B. (2013). Economics of femtocell service provision. *IEEE Transactions on Mobile Computing*, *12*(12), 2261–2273. doi:10.1109/TMC.2012.193

Duan, L., Shou, B., & Huang, J. (2012). Capacity Allocation and Pricing Strategies for Wireless Femtocell Services. *Production and Operations Management*.

Fulp, E. W., Ott, M., Reininger, D., & Reeves, D. S. (1998). Paying for QoS: an optimal distributed algorithm for pricing network resources. *Quality of Service, 1998. (IWQoS 98) 1998 Sixth International Workshop on* (pp.75-84).

Gajic, V., Huang, J., & Rimoldi, B. (2014). Competition of wireless providers for atomic users. *Proceedings of the Society for Photo-Instrumentation Engineers*, *22*(2), 512–525.

Gao, L., Iosifidis, G., Huang, J., Tassiulas, L., & Li, D. (2014). Bargaining-based mobile data offloading. *IEEE Journal on Selected Areas in Communications*, *32*(6), 1114–1125. doi:10.1109/JSAC.2014.2328393

Gizelis, C. A., & Vergados, D. D. (2011). A survey of pricing schemes in wireless networks. *IEEE Communications Surveys and Tutorials*, *13*(1), 126–145. doi:10.1109/SURV.2011.060710.00028

Goldsmith, A., Jafar, S. A., Marić, I., & Srinivasa, S. (2009). Breaking spectrum gridlock with cognitive radios: An information theoretic perspective. *Proceedings of the IEEE*, *97*(5), 894–914. doi:10.1109/JPROC.2009.2015717

Guijarro, L., Pla, V., Vidal, J. R., & Martinez--Bauset, J. (2012). Femtocell operator entry decision with spectrum bargaining and service competition. *IEEE Communications Letters*, *16*(12), 1976–1979. doi:10.1109/LCOMM.2012.101712.121645

Han, R., Feng, C., Xia, H., & Zhang, T. (2014). On the fairness of range expansion with interference mitigation in heterogeneous networks. *IEEE Communications Letters*, *18*(6), 1051–1054. doi:10.1109/LCOMM.2014.2320920

Hassan, M. R., Karmakar, G., & Kamruzzaman, J. (2014). Reputation and user requirement based price modeling for dynamic spectrum access. *IEEE Transactions on Mobile Computing*, *13*(9), 1–1. doi:10.1109/TMC.2013.78

Haykin, S. (2005). Cognitive radio: Brain-empowered wireless communications. *IEEE Journal on Selected Areas in Communications*, *23*(2), 201–220. doi:10.1109/JSAC.2004.839380

He, J., Wen, Y., Huang, J., & Wu, D. (2014). On the cost–qoe tradeoff for cloud-based video streaming under amazon ec2's pricing models. *IEEE Transactions on Circuits and Systems for Video Technology*, *24*(4), 669–680. doi:10.1109/TCSVT.2013.2283430

Hong, M., & Garcia, A. (2011). Equilibrium pricing of interference in cognitive radio networks. *IEEE Transactions on Signal Processing*, *59*(12), 6058–6072. doi:10.1109/TSP.2011.2165059

Huang, J. (2009). *Distributed algorithm design for network optimization problems with coupled objectives. In TENCON 2009 - 2009 IEEE Region 10 Conference* (pp. 1–6). IEEE.

Huang, J. (2013). Market mechanisms for cooperative spectrum trading with incomplete network information. *IEEE Communications Magazine*, *51*(51), 201–207. doi:10.1109/MCOM.2013.6619584

Huang, J., Berry, R., & Honig, M. L. (2005). Performance of distributed utility-based power control for wireless ad hoc networks.*Proceedings of IEEE Milcom.*

Huang, J., Berry, R. A., & Honig, M. L. (2006). Distributed interference compensation for wireless networks. *Selected Areas in Communications IEEE Journal on*, *24*(5), 1074–1084. doi:10.1109/JSAC.2006.872889

Huang, J., & Gao, L. (2013). Wireless network pricing. *Synthesis Lectures on Communication Networks*, *6*(2), 1–176. doi:10.2200/S00505ED1V01Y201305CNT013

Iosifidis, G., Gao, L., Huang, J., & Tassiulas, L. (2015). A double-auction mechanism for mobile data-offloading markets. *IEEE/ACM Transactions on Networking*, *23*(5), 1634–1647. doi:10.1109/TNET.2014.2345875

Ji, Z., & Liu, K. J. R. (2008). Multi-stage pricing game for collusion-resistant dynamic spectrum allocation. *IEEE Journal on Selected Areas in Communications*, *26*(1), 182–191. doi:10.1109/JSAC.2008.080116

Jiang, C., Chen, Y., Liu, K. J. R., & Ren, Y. (2014). Optimal pricing strategy for operators in cognitive femtocell networks. *IEEE Transactions on Wireless Communications*, *13*(9), 5288–5301. doi:10.1109/TWC.2014.2327970

Jiang, C., Chen, Y., Liu, K. J. R., & Ren, Y. (2015). Network economics in cognitive networks. *IEEE Communications Magazine*, *53*(5), 75–81. doi:10.1109/MCOM.2015.7105644

Joe-Wong, C., Sen, S., Ha, S., & Chiang, M. (2012). Optimized day-ahead pricing for smart grids with device-specific scheduling flexibility. *IEEE Journal on Selected Areas in Communications*, *30*(6), 1075–1085. doi:10.1109/JSAC.2012.120706

Kang, X., Zhang, R., & Motani, M. (2011). Price-Based Resource Allocation for Spectrum-Sharing Femtocell Networks: A Stackelberg Game Approach.*Global Telecommunications Conference (GLOBECOM 2011)*, (Vol. 30, pp.1 - 5). IEEE.

Kasbekar, G. S., & Sarkar, S. (2012). Spectrum pricing games with random valuations of secondary users. *IEEE Journal on Selected Areas in Communications*, *30*(30), 2262–2273. doi:10.1109/JSAC.2012.121218

Kavurmacioglu, E., Alanyali, M., & Starobinski, D. (2014). Competition in private commons: Price war or market sharing? *IEEE/ACM Transactions on Networking*.

Kelly, F. P., Maulloo, A. K., & Tan, D. K. H. (1998). Rate control for communication networks: Shadow price, proportional fairness and stability. *The Journal of the Operational Research Society*, *49*(3), 237–252. doi:10.1057/palgrave.jors.2600523

Kim, D., Kim, S., Kim, H., & Choi, S. (2014). Pricing policies for femtocell service adoption in two-tier cellular networks. *Communications (ICC), 2014 IEEE International Conference on* (pp. 2701-2706). IEEE.

Leem, H., Kim, J. Y., Jung, B. H., & Dan, K. S. (2014). Spectral-, energy-, & cost-efficient deployment of small cells in a HetNet topology. *ICC 2014 - 2014 IEEE International Conference on Communications*(Vol. 33, pp. 2709-2717).

Li, Q., Xu, Y., Hu, R. Q., & Wu, G. (2012). Pricing-based distributed mobile association for heterogeneous networks with cooperative relays. *Communications (ICC), 2012 IEEE International Conference on* (Vol. 11, pp. 5326-5331). IEEE.

Li, S., & Huang, J. (2010). Price differentiation for communication networks. *Networking IEEE/ACM Transactions on, 22*(3), 703-716.

Li, S., Huang, J., & Li, S. Y. R. (2012). Dynamic profit maximization of cognitive mobile virtual network operator. *IEEE Transactions on Mobile Computing*, *13*(3), 1.

Li, W., Zheng, W., Su, T., & Wen, X. (2013). Distributed power control and pricing for two-tier OFMDA femtocell networks using fictitious game. *Wireless Communications and Networking Conference (WCNC), 2013 IEEE* (pp.470-475). IEEE. doi:10.1109/WCNC.2013.6554609

Li, W., Zheng, W., Su, T., & Wen, X. (2013). Dual-utility based green power game in two-tier OFDMA femtocell networks with firefly algorithm. *Wireless Communications and Networking Conference (WCNC)*.

Li, X., Zhu, X., Wu, L., & Sandrasegaran, K. (2013). A distributed non-uniform pricing approach for power optimization in spectrum-sharing femtocell network. *IEEE Wireless Communications and Networking Conference (WCNC)*.

Lin, P., Zhang, J., Chen, Y., & Zhang, Q. (2011). Macro-femto heterogeneous network deployment and management: From business models to technical solutions. *IEEE Wireless Communications*, *18*(3), 64–70. doi:10.1109/MWC.2011.5876502

Liu, J., Kawamoto, Y., Nishiyama, H., Kato, N., & Kadowaki, N. (2014). Device-to-device communications achieve efficient load balancing in lte-advanced networks. *Wireless Communications IEEE*, *21*(2), 57–65. doi:10.1109/MWC.2014.6812292

Liu, J., Zheng, W., Li, W., Wang, X., Xie, Y., & Wen, X. (2013). Distributed uplink power control for two-tier femtocell networks via convex pricing.*Wireless Communications and Networking Conference (WCNC)*, (pp.458-463). IEEE.

Lorenzo, P. D., Omilipo, M., & Barbarossa, S. (2011). Distributed stochastic pricing for sum-rate maximization in femtocell networks with random graph and quantized communications. *Computational Advances in Multi-Sensor Adaptive Processing (CAMSAP), 2011 4th IEEE International Workshop on* (pp. 165-168). IEEE.

Ma, B., Man, H. C., Wong, V. W. S., & Huang, J. (2015). Hybrid overlay/underlay cognitive femtocell networks: A game theoretic approach. *IEEE Transactions on Wireless Communications*, *14*(6), 1–1. doi:10.1109/TWC.2015.2403363

Maharjan, S., Zhang, Y., & Gjessing, S. (2011). Economic approaches for cognitive radio networks: A survey. *Wireless Personal Communications*, *57*(1), 33–51. doi:10.1007/s11277-010-0005-9

Mahloo, M., Monti, P., Chen, J., & Wosinska, L. (2014). Cost modeling of backhaul for mobile networks. *Communications Workshops (ICC), 2014 IEEE International Conference on* (pp.397-402). IEEE.

Man, H. C., Southwell, R., & Huang, J. (2014). Congestion-aware network selection and data offloading. *2014 48th Annual Conference on Information Sciences and Systems (CISS)* (pp.1 - 6).

Man, H. C., & Wong, V. W. S. (2013). Interference pricing for sinr-based random access game. *IEEE Transactions on Wireless Communications*, *12*(5), 2292–2301. doi:10.1109/TWC.2013.031313.120938

Menache, I., Ozdaglar, A., & Shimkin, N. (2014). Socially optimal pricing of cloud computing resources. Institute for Computer Sciences, 322-331.

Mitola, J. (2000). *Cognitive radio: An integrated agent architecture for software defined radio*. Stockholm, Sweden: Royal Institute of Technology.

Neely, M. J. (2009). Optimal pricing in a free market wireless network. *Wireless Networks*, *15*(7), 901–915. doi:10.1007/s11276-007-0083-0

Niyato, D., & Hossain, E. (2008). Competitive pricing for spectrum sharing in cognitive radio networks: Dynamic game, inefficiency of nash equilibrium, and collusion. *Selected Areas in Communications IEEE Journal on*, *26*(1), 192–202. doi:10.1109/JSAC.2008.080117

Niyato, D., & Hossain, E. (2008). Market-equilibrium, competitive, and cooperative pricing for spectrum sharing in cognitive radio networks: Analysis and comparison. *IEEE Transactions on Wireless Communications*, *7*(11), 4273–4283. doi:10.1109/T-WC.2008.070546

Niyato, D., Hossain, E., & Han, Z. (2009). Dynamics of multiple-seller and multiple-buyer spectrum trading in cognitive radio networks: A game-theoretic modeling approach. *IEEE Transactions on Mobile Computing*, *8*(8), 1009–1022. doi:10.1109/TMC.2008.157

Qian, L. P., Zhang, Y. J. A., Huang, J., & Wu, Y. (2013). Demand response management via real-time electricity price control in smart grids. *IEEE Journal on Selected Areas in Communications*, *31*(7), 1268–1280. doi:10.1109/JSAC.2013.130710

Rasti, M., Sharafat, A. R., & Seyfe, B. (2009). Pareto-efficient and goal-driven power control in wireless networks: A game-theoretic approach with a novel pricing scheme. *IEEE/ACM Transactions on Networking*, *17*(2), 556–569. doi:10.1109/TNET.2009.2014655

Ren, S. (2014). Dynamic scheduling and pricing in wireless cloud computing. *Skull Base Surgery*, *13*(10), 2283–2292.

Ren, S., & Schaar, M. V. D. (2011). Pricing and distributed power control in wireless relay networks. *Signal Processing IEEE Transactions on*, *59*(6), 2913–2926. doi:10.1109/TSP.2011.2121064

Saraydar, C. U., Mandayam, N. B., & Goodman, D. J. (2001). Efficient power control via pricing in wireless data networks. *IEEE Transactions on Communications*, *50*(2), 291–303. doi:10.1109/26.983324

Schmidt, D. A., Shi, C., Berry, R. A., Honig, M. L., & Utschick, W. (2009). *Distributed Resource Allocation Schemes-Pricing algorithms for power control and beamformer design in interference networks.* Retrieved from http:www.ece.northwestern.edurberryPubsSPmag_09.pdf

Shen, K., & Yu, W. (2014). Distributed pricing-based user association for downlink heterogeneous cellular networks. *Selected Areas in Communications IEEE Journal on*, *32*(6), 1100–1113. doi:10.1109/JSAC.2014.2328143

Shetty, N., Parekh, S., & Walrand, J. (2009). Economics of femtocells. *IEEE Conference on Global Telecommunications* (pp.1-6). IEEE Press.

Shi, C., Schmidt, D. A., Berry, R. A., Honig, M. L., & Utschick, W. (2009). Distributed interference pricing for the MIMO interference channel. *Proceedings of the 2009 IEEE international conference on Communications* (pp.1-5). IEEE Press. doi:10.1109/ICC.2009.5198927

Soret, B., Pedersen, K. I., Jorgensen, N. T. K., & Fernandez-Lopez, V. (2015). Interference coordination for dense wireless networks. *IEEE Communications Magazine*, *53*(1), 102–109. doi:10.1109/MCOM.2015.7010522

Tan, C. W., Palomar, D. P., & Chiang, M. (2006). Distributed optimization of coupled systems with applications to network utility maximization. In *Proc. IEEE International Conference on Acoustics, Speech, and Signal Processing (ICASSP)*.

Tran, N. H., Hong, C. S., Han, Z., & Lee, S. (2013). Optimal pricing effect on equilibrium behaviors of delay-sensitive users in cognitive radio networks. *IEEE Journal on Selected Areas in Communications*, *31*(11), 2566–2579. doi:10.1109/JSAC.2013.131134

Tran, N. H., Le, L. B., Ren, S., Han, Z., & Hong, C. S. (2014). Joint pricing and load balancing for cognitive spectrum access: Non-cooperation vs cooperation. *IEEE Journal on Selected Areas in Communications*.

Wang, F., Krunz, M., & Cui, S. (2007). Price-based spectrum management in cognitive radio networks. *IEEE Journal of Selected Topics in Signal Processing*, *2*(1), 70–78.

Wang, Q., Ren, K., & Meng, X. (2012). When cloud meets eBay: Towards effective pricing for cloud computing. *Proceedings of the IEEE, 131*, 936–944.

Wang, X., Zheng, W., Liu, J., Li, W., & Wen, X. (2013). Distributed power self-optimization with convex pricing in dense femtocell networks via an exact potential game. *Personal Indoor and Mobile Radio Communications (PIMRC), 2013 IEEE 24th International Symposium on* (pp. 1919-1923). IEEE.

Wang, Z., Jiang, L., & He, C. (2013). A novel price-based power control algorithm in cognitive radio networks. *IEEE Communications Letters, 17*(1), 43–46. doi:10.1109/LCOMM.2012.120612.121587

Wang, Z., Jiang, L., & He, C. (2014). Optimal price-based power control algorithm in cognitive radio networks. *IEEE Transactions on Wireless Communications, 13*(11), 5909–5920. doi:10.1109/TWC.2014.2318708

Wu, Y., & Song, W. Z. (2014). Cooperative resource sharing and pricing for proactive dynamic spectrum access via nash bargaining solution. *IEEE Transactions on Parallel and Distributed Systems, 25*(11), 2804–2817. doi:10.1109/TPDS.2013.285

Wu, Y., Zhu, Q., Huang, J., & Tsang, D. H. K. (2014). Revenue sharing based resource allocation for dynamic spectrum access networks. *IEEE Journal on Selected Areas in Communications, 32*(11), 2280–2296. doi:10.1109/JSAC.2014.141125

Xie, R., Yu, F. R., Ji, H., & Li, Y. (2012). Energy-efficient resource allocation for heterogeneous cognitive radio networks with femtocells. *IEEE Transactions on Wireless Communications, 11*(11), 3910–3920. doi:10.1109/TWC.2012.092112.111510

Xu, C., Sheng, M., Yang, C., Wang, X., & Wang, L. (2014). Pricing-based multiresource allocation in ofdma cognitive radio networks: An energy efficiency perspective. *IEEE Transactions on Vehicular Technology, 63*(5), 2336–2348. doi:10.1109/TVT.2013.2280617

Xu, Y., He, C., Jiang, L., & Li, J. (2013). Distributed dynamic SINR pricing for multi-cell beamforming with limited backhaul signaling. *Global Communications Conference (GLOBECOM)*, 2013IEEE (pp.3620-3624).

Yaïche, H., Mazumdar, R. R., & Rosenberg, C. (2000). A game theoretic framework for bandwidth allocation and pricing in broadband networks. *IEEE/ACM Transactions on Networking, 8*(5), 667–678. doi:10.1109/90.879352

Yang, C. (2015). *Advanced Cognitive Radio-Enabled Spectrum Management*. IGI Global. doi:10.4018/978-1-4666-6571-2.ch018

Yang, C., & Li, J. (2010). *Joint Economical and Technical Consideration of Dynamic Spectrum Sharing: A Multi-Stage Stackelberg Game Perspective. In Vehicular Technology Conference Fall (VTC 2010-Fall), 2010 IEEE 72nd* (pp. 1–5). IEEE.

Yang, C., & Li, J. (2013). Pricing-based dynamic spectrum leasing: A hierarchical multi-stage stackelberg game perspective. *IEICE Transactions on Communications, 96*(6), 1511–1521. doi:10.1587/transcom.E96.B.1511

Yang, C., & Li, J. (2014). Cooperative Spectrum Leasing to Femtocells with Interference Compensation.*Vehicular Technology Conference (VTC Spring), 2014 IEEE 79th* (pp.1-6). IEEE. doi:10.1109/VTCSpring.2014.7022865

Yang, C., & Li, J. (2014). Interference mitigation and energy management in heterogeneous networks: a cognitive radio perspective. *Convergence of Broadband Broadcast & Cellular Network Technologies.*

Yang, L., Kim, H., Zhang, J., Chiang, M., & Tan, C. W. (2013). Pricing-based decentralized spectrum access control in cognitive radio networks. *IEEE/ACM Transactions on Networking*, *21*(2), 522–535. doi:10.1109/TNET.2012.2203827

Yang, L., Kim, H., Zhang, J., Chiang, M., & Tan, C. W. (2013). Pricing-based decentralized spectrum access control in cognitive radio networks. *IEEE/ACM Transactions on Networking*, *21*(2), 522–535. doi:10.1109/TNET.2012.2203827

Yang, Y., Quek, T. Q. S., & Duan, L. (2014). Backhaul-constrained small cell networks: Refunding and qos provisioning. *IEEE Transactions on Wireless Communications*, *13*(9), 5148–5161. doi:10.1109/TWC.2014.2326401

R. D. Yates. (1995). A framework for uplink power control in cellular radio systems. *Selected Areas in Communications IEEE Journal on, 13.*

Yin, R., Yu, G., Zhang, H., Zhang, Z., & Li, G. Y. (2015). Pricing-based interference coordination for d2d communications in cellular networks. *IEEE Transactions on Wireless Communications*, *14*(3), 1519–1532. doi:10.1109/TWC.2014.2368151

Yu, H., Gao, L., Li, Z., Wang, X., & Hossain, E. (2010). Pricing for uplink power control in cognitive radio networks. *IEEE Transactions on Vehicular Technology*, *59*(4), 1769–1778. doi:10.1109/TVT.2010.2040492

Yu, H., Man, H. C., Huang, L., & Huang, J. (2014). Predictive delay-aware network selection in data offloading.*Global Communications Conference (GLOBECOM)*. IEEE.

Yun, S. Y., Yi, Y., Cho, D. H., & Mo, J. (2012). The economic effects of sharing femtocells. *IEEE Journal on Selected Areas in Communications*, *30*(3), 595–606. doi:10.1109/JSAC.2012.120409

Zhai, X., Zheng, L., & Tan, C. W. (2014). Energy-infeasibility tradeoff in cognitive radio networks: Price-driven spectrum access algorithms. *Selected Areas in Communications IEEE Journal on*, *32*(3), 528–538. doi:10.1109/JSAC.2014.140313

Zhang, H., Chu, X., & Wen, X. (2013). Differentiated-pricing based power allocation in dense femtocell networks. *Springerbriefs in Computer Science*, 599-603.

Zhou, Q., Chen, Z., Gao, F., Li, J. C. F., & Lei, M. (2014). Pricing and power allocation in sensing-based cognitive femtocell networks. *Communications (ICC), 2014 IEEE International Conference on* (Vol. 99, pp. 5342-5347). IEEE.

Zhou, T., Chen, Y., Jiang, C., & Liu, K. J. R. (2014). Pricing game for time mute in femto-macro coexistent networks. *IEEE Transactions on Wireless Communications*, *14*, 1–1.

Zhu, K., Hossain, E., & Niyato, D. (2014). Pricing, spectrum sharing, and service selection in two-tier small cell networks: A hierarchical dynamic game approach. *IEEE Transactions on Mobile Computing*, *13*(8), 1843–1856. doi:10.1109/TMC.2013.96

Zhu, K., Hossain, E., & Niyato, D. (2014). Pricing, spectrum sharing, and service selection in two-tier small cell networks: A hierarchical dynamic game approach. *IEEE Transactions on Mobile Computing*, *13*(8), 1843–1856. doi:10.1109/TMC.2013.96

Compilation of References

3G Americas. (2009). *The benefits of SON in LTE: Self optimizing and self organizing networks.* Whitepaper.

4G Americas. (2011). *Self optimizing networks: The benefits of SON in LTE.* Whitepaper.

Abdelnasser, A., Hossain, E., & Kim, D. I. (2014). Clustering and resource allocation for dense femtocells in a two-tier cellular OFDMA network. *IEEE Transactions on Communications*, *13*(3), 1628–1641.

Abdullah, L. M., Baba, M. D., & Ali, S. G. A. (2014). Parameters optimization for handover between femtocell and macrocell in LTE-based network.*2014 IEEE International Conference on Control System, Computing and Engineering (ICCSCE)*. doi:10.1109/ICCSCE.2014.7072796

Academic group of Baidu. (n.d.). *ICIC*. Retrieved from http://baike.baidu.com

Adhikary, A., Nam, J., Ahn, J.-Y., & Caire, G. (2013). Joint spatial division and multiplexing: The large-scale array regime. *IEEE Transactions on Information Theory*, *59*(10), 6441–6463. doi:10.1109/TIT.2013.2269476

Aguado, M., Onandi, O., Agustin, P. S., Higuero, M., & Taquet, E. J. (2008). WiMax on Rails. *IEEE Vehicular Technology Magazine*, *3*(3), 47–56. doi:10.1109/MVT.2008.927489

Aguilar-Garcia, A., Barco, R., Fortes, S., & Munoz, P. (2015). Load balancing mechanisms for indoor temporarily overloaded heterogeneous femtocell networks. *EURASIP Journal on Wireless Communications and Networking*, *2015*(29).

Ahmed, M., Yoon, S.-G., Jeong, S., & Bahk, S. (2013). Geo-location based self-organization scheme for femtocell networks. *2013International Conference on ICT Convergence (ICTC)*. doi:10.1109/ICTC.2013.6675415

Akin, S., & Gursoy, M. C. (2013). On the throughput and energy efficiency of cognitive MIMO transmissions. *IEEE Transactions on Vehicle Technology*, *62*(7), 3245–3260. doi:10.1109/TVT.2013.2254142

Akkarajitsakul, K., Hossain, E., Niyato, D., & Dong, I. K. (2011). Game theoretic approaches for multiple access in wireless networks: A survey. *IEEE Communications Surveys and Tutorials*, *13*(3), 372–395. doi:10.1109/SURV.2011.122310.000119

Akyildiz, I. F., Lee, W. Y., Vuran, M. C., & Mohanty, S. (2006). Next generation/dynamic spectrum access/cognitive radio wireless networks: A survey. *Computer Networks*, *50*(13), 2127–2159. doi:10.1016/j.comnet.2006.05.001

Ali, M., & Tse, D. (2010). *On the degrees of freedom of miso broadcast channels with delayed feedback.* Available from http://www.researchgate.net/profile/Mohammad_Maddah-Ali/publication/264869062_On_the_Degrees_of_Freedom_of_MISO_Broadcast_Channels_with_Delayed_Feedback/links/54c7b3800cf238bb7d0b0a81.pdf

Ali, S. H., & Leung, V. C. M. (2009). Dynamic frequency allocation in fractional frequency reused OFDMA networks. *IEEE Transactions on Wireless Communications*, *8*(8), 4286–4295. doi:10.1109/TWC.2009.081146

Almanthari, B., Nasser, N., & Hassanein, H. (2011). Congestion pricing in wireless cellular networks. *IEEE Communications Surveys and Tutorials*, *13*(3), 358–371. doi:10.1109/SURV.2011.090710.00042

Al-Tous, H., & Barhumi, I. (2013). Joint Power and Bandwidth Allocation for Amplify-and-Forward Cooperative Communications Using Stackelberg Game. *Vehicular Technology. IEEE Transactions*, *62*(4), 1678–1691.

Al-Zahrani, A. Y., & Yu, F. R. (2011). A game theory approach for inter-cell interference management in OFDM networks.*IEEE International Conference on Communications (ICC)*. doi:10.1109/icc.2011.5962512

Andrews, J. G. (2005). Interference cancellation for cellular systems: A contemporary overview. *IEEE Wireless Commun.*, *2*(3), 19–29. doi:10.1109/MWC.2005.1421925

Andrews, J. G., Claussen, H., Dohler, M., Rangan, S., & Reed, M. (2012). Femtocells: Past, present, and future. *IEEE Journal on Selected Areas in Communications*, *30*(3), 497–508. doi:10.1109/JSAC.2012.120401

Andrews, J. G., Ghosh, A., & Muhamed, R. (2007). *Fundamentals of wimax: Understanding broadband wireless networking*.Prentice Hall Communications Engineering & Emerging Technologies.

Ao, X., Yu, F. R., Yuan, H., & Jiang, S. (2014). Traffic-aware link scheduling with interference alignment for multi-user MIMO networks. In *Proceedings of IEEE International Conference on Communications*. Sydney: IEEE. doi:10.1109/ICC.2014.6883352

Arbi, A., & O'Farrell, T. (2015). *Energy efficiency in 5G access networks: Small cell densification and high order sectorisation.* Paper presented at the Communication Workshop (ICCW), 2015 IEEE International Conference on.

Arraycomm.com. (2015). *Cooper's Law*. Retrieved 17 December, 2015, from http://www.arraycomm.com/technology/coopers-law/

Arslan, M. Y., Yoon, J., Sundaresan, K., Krishnamurthy, S. V., & Banerjee, S. (2011). FERMI: A FEmtocell Resource Management System for Interference Mitigation in OFDMA networks. In *Proceedings of International Conference on Mobile Computing and Networking*. Las Vegas, NV: ACM. doi:10.1145/2030613.2030617

Arslan, M. Y., Yoon, J., Sundaresan, K., Krishnamurthy, S. V., & Banerjee, S. (2013). A resource management system for interference mitigation in enterprise OFDMA femtocells. *IEEE/ACM Transactions on Networking*, *21*(5), 1447–1460. doi:10.1109/TNET.2012.2226245

Athley, F., Johansson, M. N., & Nilsson, A. (2013). *Increased Sectorization: Horizontal or Vertical?* Paper presented at the Vehicular Technology Conference (VTC Fall), 2013 IEEE 78th.

Auer, G., Gódor, I., Hévizi, L., Imran, M. A., Malmodin, J., Fazekas, P., & Tafazolli, R. (2010). *Enablers for Energy Efficient Wireless Networks. In Vehicular Technology Conference Fall (VTC 2010-Fall)* (pp. 1–5). Ottawa, ON: IEEE.

Ayach, O. E., & Heath, R. W. (2012). Interference alignment with analog channel state feedback. *Wireless Communications IEEE Transactions on*, *11*(2), 626–636. doi:10.1109/TWC.2011.122211.101814

Badic, B., O'Farrrell, T., Loskot, P., & He, J. (2009). *Energy Efficient Radio Access Architectures for Green Radio: Large versus Small Cell Size Deployment*. Paper presented at the Vehicular Technology Conference Fall (VTC 2009-Fall), 2009 IEEE 70th.

Bai, F., Sadagopan, N., & Helmy, A. (2003, March). *Important: A framework to systematically analyze the impact of mobility on performance of routing protocols in adhoc networks*. Paper presented at the meeting of IEEE International Conference on Computer Communications, San Francisco, CA. doi:10.1109/INFCOM.2003.1208920

Balakrishnan, R., Yang, X., Venkatachalam, M., & Akyildiz, I. F. (2011, March). *Mobile relay and group mobility for 4G WiMAX networks*. Paper presented at the meeting of the IEEE Wireless Communications and Networking Conference, Cancun, Mexico. doi:10.1109/WCNC.2011.5779334

Bao, Y., Wang, X., Zhou, S., & Niu, Z. (2012, October). *An energy-efficient client pre-caching scheme with wireless multicast for video-on-demand services*. Paper presented at the meeting of IEEE Asia-Pacific Conference on Communications, Jeju Island, South Korea.

Barbarossa, S., Sardellitti, S., & Carfagna, A. (2011). Pricing mechanisms for interference management games in femtocell networks based on Markov modeling. *Future Network And Mobile Summit 2011* (Vol. 48, pp. 1 - 8).

Barbieri, A., Damnjanovic, A., Tingfang, J., Montojo, J., Yongbin, W., Malladi, D., & Horn, G. et al. (2012). LTE femtocells: System design and performance analysis. *IEEE Journal on Selected Areas in Communications*, *30*(3), 586–594. doi:10.1109/JSAC.2012.120408

Barth, U., & Kuehn, E. (2010). Self-organization in 4G mobile networks: motivation and vision. *2010 7th International Symposium on Wireless Communication Systems (ISWCS)*.

Bartoli, G., Fantacci, R., Marabissi, D., & Pucci, M. (2014). Angular interference suppression in cognitive LTE-A femtocells.*2014 International Wireless Communications and Mobile Computing Conference (IWCMC)*. doi:10.1109/IWCMC.2014.6906488

Bastug, E., Bennis, M., & Debbah, M. (2014). Living on the edge: The role of proactive caching in 5G wireless networks. *IEEE Communications Magazine*, *52*(8), 82–89. doi:10.1109/MCOM.2014.6871674

Bello, O., & Zeadally, S. (2014). Intelligent device-to-device communications in the internet of things. *IEEE Systems Journal,* (99), 1–11.

Bennis, M., & Niyato, D. A. (2010) Q-learning based approach to interference avoidance in self-organized femtocell networks. *IEEE GLOBECOM Workshops (GC Wkshps)*.

Bennis, M., Perlaza, S. M., Blasco, P., & Han, Z. (2013). Self-organization in small cell networks: A reinforcement learning approach. *IEEE Transactions on Wireless Communications*, *1*(1), 1–11.

Bernardo, F. F., Agusti, R. R., Perez-Romero, J. J., & Sallent, O. (2011). Intercell interference management in OFDMA networks: A decentralized approach based on reinforcement learning. *IEEE Transactions on Systems, Man and Cybernetics. Part C, Applications and Reviews*, *41*(6), 968–976. doi:10.1109/TSMCC.2010.2099654

Bhatnagar, R. K., & Panigrahi, P. K. (2013). *Energy Efficiency, Clean Power as Engine for Sustainable Growth*. Retrieved from http://www.itu.int/en/ITU-T/climatechange/symposia/201305/Documents/info/Presentations/s1p3-India.pdf

Bhat, P., Nagata, S., Campoy, L., Berberana, I., Derham, T., & Liu, G. et al. (2012). LTE-advanced: An operator perspective. *IEEE Communications Magazine*, *50*(2), 104–114. doi:10.1109/MCOM.2012.6146489

Bhushan, N., Li, J., Malladi, D., & Gilmore, R. (2014). Network densification: The dominant theme for wireless evolution into 5g. *Communications Magazine IEEE*, *52*(2), 82–89. doi:10.1109/MCOM.2014.6736747

Birk, Y., & Kol, T. (1998). Informed-source coding-on-demand (ISCOD) over broadcast channels. In *Proceedings of the conference on computer communications: seventeenth annual joint conference of the IEEE Computer and Communications Societies, Gateway to the 21st century.* IEEE.

Birk, Y., & Kol, T. (2006). Coding on demand by an informed source (iscod) for efficient broadcast of different supplemental data to caching clients. *IEEE Transactions on Information Theory*, *52*(6), 2825–2830. doi:10.1109/TIT.2006.874540

Bjornson, E., Hammarwall, D., & Ottersten, B. (2009). Exploiting quantized channel norm feedback through conditional statistics in arbitrarily correlated MIMO systems. *IEEE Transactions on Signal Processing*, *57*(10), 4027–4041. doi:10.1109/TSP.2009.2024266

Bjornson, E., Zetterberg, P., Bengtsson, M., & Ottersten, B. (2013). Capacity limits and multiplexing gains of MIMO channels with transceiver impairments. *IEEE Communications Letters*, *17*(1), 91–94. doi:10.1109/LCOMM.2012.112012.122003

Blair, J. R. S., & Peyton, B. (1993). *An introduction to chordal graphs and clique trees, in graph theory and sparse matrix computation*. New York: Springer.

Blasco, P., & Gunduz, D. (2014, June). *Learning-based optimization of cache content in a small cell base station*. Paper presented at the meeting of IEEE International Conference on Communications, Sydney, Australia. doi:10.1109/ICC.2014.6883600

Blume, O., Ambrosy, A., Wilhelm, M., & Barth, U. (2013). *Energy Efficiency of LTE networks under traffic loads of 2020*. Paper presented at the Wireless Communication Systems (ISWCS 2013).

Blume, O., Zeller, D., & Barth, U. (2010). *Approaches to energy efficient wireless access networks. In Communications, Control and Signal Processing (ISCCSP)* (pp. 1–5). Limassol: IEEE.

Boccardi, F., & Braun, V. (2014). Scenarios for 5G mobile and wireless communications: The vision of the METIS project. *IEEE Communications Magazine*, *52*(5), 26–35. doi:10.1109/MCOM.2014.6815890

Bodanese, E. (2011). *A Brief Introduction to Heterogeneous Networks (HetNets) and its Challenges. In Conmmunication Technology and Application, ICCTA* (pp. 605–609). Beijing: IEEE.

Boucher, N. J. (2001). *The cellular radio handbook: a reference for cellular system operation*. Retrieved from http://www.gbv.de/dms/ilmenau/toc/321833627.PDF

Boudreau, G., Panicker, J., Guo, N., Chang, R., Wang, N., & Vrzic, S. (2009). Interference coordination and cancellation for 4g networks. *IEEE Communications Magazine*, *47*(4), 74–81. doi:10.1109/MCOM.2009.4907410

Bouras, C., Kavourgias, G., Kokkinos, V., & Papazois, A. (2012). Interference management in LTE femtocell systems using an adaptive frequency reuse scheme. *Wireless Telecommunications Symposium (WTS)*. doi:10.1109/WTS.2012.6266120

Bousia, A., Kartsakli, E., Antonopoulos, A., & Alonso, L. (2014). Sharing the small cells for energy efficient networking: How much does it cost? *Global Communications Conference (GLOBECOM), 2014 IEEE*. IEEE. doi:10.1109/GLOCOM.2014.7037207

Boyd, S., Kim, S.-J., Vandenberghe, L., & Hassibi, A. (2007). A tutorial on geometric programming. *Optimization and Engineering*, *8*(1), 67–127. doi:10.1007/s11081-007-9001-7

Boyd, S., & Vandenberghe, L. (2004). *Convex Optimization*. Cambridge, UK: Cambridge University Press. doi:10.1017/CBO9780511804441

Braun, L., Klein, A., Carle, G., Reiser, H., & Eisl, J. (2012, April). *Analyzing caching benefits for You Tube traffic in edge networks*. Paper presented at the meeting of IEEE Network Operations and Management Symposium, Maui, HI.

Breslau, L., Cao, P., Fan, L., Phillips, G., & Shenker, S. (1999, March). *Web caching and Zipf-like distributions: evidence and implications*. Paper presented at the meeting of IEEE International Conference on Computer Communications, New York, NY. doi:10.1109/INFCOM.1999.749260

Bresler, G., Parekh, A., & Tse, D. N. C. (2010). The approximate capacity of the many-to-one and one-to-many gaussian interference channels. *IEEE Transactions on Information Theory*, *56*(9), 4566–4592. doi:10.1109/TIT.2010.2054590

Bu, S. R., & Yu, F. R. (2012). Dynamic energy-efficient resource allocation in cognitive heterogeneous wireless networks with the smart grid.*Global Communications Conference*. doi:10.1109/GLOCOM.2012.6503579

Bu, S. R., & Yu, F. R. (2014). Green Cognitive Mobile Networks With Small Cells for Multimedia Communications in the Smart Grid Environment. *Vehicular Technology. IEEE Transactions*, *63*(5), 2115–2126.

Bu, S. R., Yu, F. R., & Qian, Y. (2013). Energy-efficient cognitive heterogeneous networks powered by the smart grid. *INFOCOM, 2013. Proceedings of the IEEE*, 980–988.

Bu, S. R., Yu, F. R., & Senarath, G. (2013). Interference-aware energy-efficient resource allocation for heterogeneous networks with incomplete channel state information. *Communications (ICC),2013 IEEE International Conference.*

Bu, S., Yu, F. R., Cai, Y., & Liu, X. (2012). When the smart grid meets energy-efficient communications: Green wireless cellular networks powered by the smart grid. *IEEE Transactions on Wireless Communications*, *11*(8), 3014–3024.

C2Power. (2016, March 16). Retrieved from Cognitive Radio and Cooperative Strategies for POWER saving in multi-standard wireless devices: http://ict-power.eu

Cadambe, V. R., & Jafar, S. A. (2008). Interference alignment and degrees of freedom of the K-user interference channel. *IEEE Transactions on Information Theory*, *54*(8), 3425–3441. doi:10.1109/TIT.2008.926344

Canzian, L., Badia, L., & Zorzi, M. (2013). Promoting Cooperation in Wireless Relay Networks Through Stackelberg Dynamic Scheduling. *Communications. IEEE Transactions*, *61*(2), 700–711. doi:10.1109/TCOMM.2012.120512.110654

Cao, Q., Zhao, H. V., & Jing, Y. D. (2012). Power Allocation and Pricing in Multiuser Relay Networks Using Stackelberg and Bargaining Games. *Vehicular Technology. IEEE Transactions*, *61*(7), 3177–3190.

Carfagna, A., & Barbarossa, S. (2012). *Distributed Bayesian Pricing for Sum-Rate Maximization in Small-Cell Networks. In Globecom Workshops (GC Wkshps), 2012 IEEE* (pp. 598–602). IEEE.

Chaganti, R., Sathya, V., Ahammed, S. A., Rex, R. A., & Tamma, B. R. (2013). Efficient SON handover scheme for enterprise Femtocell networks.*2013 IEEE International Conference on Advanced Networks and Telecommunications Systems (ANTS)*. doi:10.1109/ANTS.2013.6802870

Chakaravarthy, V. T., & Roy, S. (2009). Approximating maximum weight K-colorable subgraphs in chordal graphs. *Information Processing Letters*, *109*(7), 365–368. doi:10.1016/j.ipl.2008.12.007

Chandrasekhar, V., & Andrews, J. G. (2007). Uplink capacity and interference avoidance for two-tier femtocell networks. *IEEE Transactions on Wireless Communications*, *8*(7), 3498–3509. doi:10.1109/TWC.2009.070475

Chandrasekhar, V., & Andrews, J. G. (2009). Spectrum allocation in tiered cellular networks. *IEEE Transactions on Communications*, *57*(10), 3059–3068. doi:10.1109/TCOMM.2009.10.080529

Chandrasekhar, V., Andrews, J. G., & Gatherer, A. (2008). Femtocell networks: A survey. *IEEE Communications Magazine*, *46*(9), 59–67. doi:10.1109/MCOM.2008.4623708

Chang, R. Y., Tao, Z., Zhang, J., & Kuo, C. C. J. (2009). Multicell ofdma downlink resource allocation using a graphic framework. *IEEE Transactions on Vehicular Technology*, *58*(7), 3494–3507. doi:10.1109/TVT.2009.2014384

Chan, Y., Zhang, S., & Xu, S. (2011). Fundamental trade-offs on green wireless networks. *IEEE Communications Magazine*, *49*(6), 30–37. doi:10.1109/MCOM.2011.5783982

Chen, X. F., Zhang, H. G., & Lasanen, M. (2013). Combined learning for energy efficiency in heterogeneous cellular networks. *Personal, Indoor and Mobile Radio Communications (PIMRC Workshops), 2013 IEEE 24th International Symposium.*

Chen, Z., & Kountouris, M. (2015, June). *Cache-enabled small cell networks with local user interest correlation.* Paper presented at the meeting of IEEE International Workshop on Signal Processing Advances in Wireless Communications, Stockholm, Sweden.

Cheng, Z., & Devroye, N. (2014). The degrees of freedom of the K-pair-user full-duplex two-way interference channel with a MIMO relay. In *Proc. IEEE Int. Symp. Inf. Theory (ISIT)*, (pp. 2714-2718).

Chen, L., Libman, L., & Leneutre, J. (2011). Conflicts and Incentives in Wireless Cooperative Relaying: A Distributed Market Pricing Framework. *Parallel and Distributed Systems. IEEE Transactions*, *22*(5), 758–772.

Chen, P., Qiu, L. D., & Wang, Y. (2012). Uplink power control for CDMA Satellite cognitive radio. *Systems Engineering and Electronics*, *34*(11), 2354–2360.

Chen, S., & Cheng, R. S. (2014). Clustering for interference alignment in multiuser interference network. *IEEE Transactions on Vehicular Technology*, *63*(6), 2613–2624. doi:10.1109/TVT.2013.2292897

Chen, Y. D., Hoo Teo, K., Kishore, S., & Zhang, J. Y. (2008). Inter-Cell Interference Management in WiMAX Downlinks by A Stackelberg Game between BSs. *ICC*, *08*, 3442–3446.

Chen, Y. J., Zhang, J., & Zhang, Q. (2012). Utility-Aware Refunding Framework for Hybrid Access Femtocell Network. *Wireless Communications. IEEE Transactions*, *11*(5), 1688–1697.

Chen, Y., Lin, P., & Zhang, Q. (2014). Lotus: Location-aware online truthful double auction for dynamic spectrum access. *IEEE Transactions on Wireless Communications*, *14*(2), 510–518.

Chen, Y., Zhang, J., Lin, P., & Zhang, Q. (2011). Optimal Pricing and Spectrum Allocation for Wireless Service Provider on Femtocell Deployment.*IEEE International Conference on Communications* (Vol. 41, pp. 1 - 5). doi:10.1109/icc.2011.5963321

Cheung, M. H., & Huang, J. (2015). Dawn: Delay-aware wi-fi offloading and network selection. *IEEE Journal on Selected Areas in Communications*, *33*(6), 1214–1223. doi:10.1109/JSAC.2015.2416989

Chiang, M., Low, S. H., Calderbank, A. R., & Doyle, J. C. (2007). Layering as optimization decomposition: A mathematical theory of network architectures. *Proceedings of the IEEE*, *95*(1), 255–312. doi:10.1109/JPROC.2006.887322

Chiang, M., Tan, C. W., Palomar, D. P., O'neill, D., & Julian, D. (2007). Power control by geometric programming. *IEEE Transactions on Wireless Communications*, *6*(7), 2640–2651. doi:10.1109/TWC.2007.05960

China Mobile Research Institute. (2011). *C-RAN- The Road Towards Green RAN White Paper (Version 2.5)* [PDF document]. Retrieved from Lecture Notes Online Web site: http://labs.chinamobile.com/report/view_59826

Choi, B.-G., Cho, E. S., Chung, M. Y., Cheon, K.-Y., & Park, A.-S. (2011). A femtocell power control scheme to mitigate interference using listening TDD frame.*2011 International Conference on Information Networking (ICOIN)*. doi:10.1109/ICOIN.2011.5723186

Cierny, M., Wang, H., Wichman, R., Ding, Z., & Wijting, C. (2013). On number of almost blank subframes in heterogeneous cellular networks. *IEEE Transactions on Wireless Communications*, *12*(10), 5061–5073. doi:10.1109/TWC.2013.090513.121756

Cisco Visual Networking. (2015). *Global Mobile Data Traffic Forecast Update 2014–2019*. White Paper c11-520862. Available on http://www.cisco.com/c/en/us/solutions/collateral/serviceprovider/visual-networking-index-vni/white_paper_c11-520862. html

Claussen, H., Ho, L. T. W., & Samuel, L. G. (2008). An overview of femtocell concept. *Bell Labs Technical Journal.*, *13*(1), 221–246. doi:10.1002/bltj.20292

Cohen, K., Leshem, A., & Zehavi, E. (2013). Game Theoretic Aspects of the Multi-Channel ALOHA Protocol in Cognitive Radio Networks. *Selected Areas in Communications. IEEE Journal on*, *31*(11), 2276–2288.

Congzheng, H., Harrold, T., Armour, S., Krikidis, I., Vide, S., Grant, P. M., & Hanzo, L. et al. (2011). Green Radio: Radio Techniques to Enable Energy-Efficient Wireless Networks. *IEEE Communications Magazine*, 46–54.

Cost effective Enterprise small cell deployment with UltraSON. (2016). Qualcomm Technologies Inc. Retrieved June 26, 2016, from https://www.qualcomm.com/documents/whitepaper-cost-effective-enterprise-small-cell-deployment-ultrason

Courcoubetis, C., & Weber, R. (2003). *Pricing Communication Networks: Economic, Technology and Modeling*. John Wiley & Sons. doi:10.1002/0470867175

Cui, H., Song, L., & Jiao, B. (2014). Multi-pair two-way amplify-and-forward relaying with very large number of relay antennas. *IEEE Transactions on Wireless Communications*, *13*(5), 2636–2645. doi:10.1109/TWC.2014.032514.130885

Cumanan, K., Musavian, L., Lambotharab, S., & Gershman, A. B. (2010). SINR balancing technique for downlink beamforming cognitive radio networks. *IEEE Signal Processing Letters*, *17*(2), 133–136. doi:10.1109/LSP.2009.2033650

D. Bharadia, E. McMilin, & S. Katti. (2013). Full duplex radios. *ACM Sigcomm*, 1-5.

Da, B., & Zhang, R. (2011). Exploiting interference alignment in multi-cell cooperative OFDMA resource allocation. In *Proceedings of Global Telecommunications Conference*. Houston, TX: IEEE.

Dai, J., Chen, M., & Chung, P. (2013). Uplink capacity analysis of single-user SA-MIMO system. *Journal of Southeast University*, *29*(1), 1–6.

Damnjanovic, A., Montojo, J., Wei, Y., Ji, T., Luo, T., Vajapeyam, M., & Malladi, D. et al. (2011). A survey on 3GPP heterogenous networks. *IEEE Wireless Communications.*, *18*(3), 10–21. doi:10.1109/MWC.2011.5876496

Dasilva, L. A. (2000). Pricing for qos-enabled networks: A survey. *IEEE Communications Surveys and Tutorials*, *3*(2), 2–8. doi:10.1109/COMST.2000.5340797

Day, B. P., Margetts, A. R., Bliss, D. W., & Schniter, P. (2012). Full-duplex MIMO relaying: Achievable rates under limited dynamic range. *IEEE Journal on Selected Areas in Communications*, *30*(9), 1541–1553. doi:10.1109/JSAC.2012.120921

Debaillie, B., Desset, C., & Louagie, F. (2015). *A Flexible and Future-Proof Power Model for Cellular Base Stations*. Paper presented at the Vehicular Technology Conference (VTC Spring), 2015 IEEE 81st.

Desset, C., Debaillie, B., Giannini, V., Fehske, A., Auer, G., Holtkamp, H., et al. (2012). *Flexible power modeling of LTE base stations*. Paper presented at the Wireless Communications and Networking Conference (WCNC).

Di Taranto, R., Popovski, P., Simeone, O., & Yomo, H. (2010). Efficient Spectrum Leasing via Randomized Silencing of Secondary Users. *Wireless Communications. IEEE Transactions*, *9*(12), 3739–3749.

Dinkelbach, W. (1967). On nonlinear fractional programming. *Management Science*, *13*(7), 492–498. doi:10.1287/mnsc.13.7.492

Dirani, M., & Altman, Z. (2010). *A cooperative reinforcement learning approach for inter-cell interference coordination in ofdma cellular networks*. Avignon, France: WiOpt.

Dixit, S., Periyalwar, S., & Yanikomeroglu, H. (2013). Secondary user access in lte architecture based on a base-station-centric framework with dynamic pricing. *IEEE Transactions on Vehicular Technology*, *62*(1), 284–296. doi:10.1109/TVT.2012.2221753

Do, C. T., Tran, N. H., Han, Z., & Le, L. B. (2014). Optimal pricing for duopoly in cognitive radio networks: Cooperate or not cooperate? *Wireless Communications IEEE Transactions on*, *13*(5), 2574–2587. doi:10.1109/TWC.2014.031914.131363

Do, C. T., Tran, N. H., Han, Z., Le, L. B., Lee, S., & Hong, C. S. (2014). Optimal Pricing for Duopoly in Cognitive Radio Networks: Cooperate or not Cooperate? *Wireless Communications. IEEE Transactions*, *13*(5), 2574–2587.

Duan, L., Shou, B., & Huang, J. (2012). Capacity Allocation and Pricing Strategies for Wireless Femtocell Services. *Production and Operations Management.*

Duan, L. J., Huang, J. W., & Shou, B. Y. (2013). Economics of Femtocell Service Provision. *Mobile Computing. IEEE Transactions*, *12*(11), 2261–2273.

Duan, L., Gao, L., & Huang, J. (2014). Cooperative spectrum sharing: A contract-based approach. *Mobile Computing IEEE Transactions on*, *13*(1), 1.

Duan, L., Huang, J., & Shou, B. (2011). Duopoly competition in dynamic spectrum leasing and pricing. *IEEE Transactions on Mobile Computing*, *11*(11), 1706–1719. doi:10.1109/TMC.2011.213

Duan, L., Huang, J., & Shou, B. (2011). Investment and pricing with spectrum uncertainty: A cognitive operator's perspective. *Mobile Computing IEEE Transactions on*, *10*(11), 1590–1604. doi:10.1109/TMC.2011.78

Duan, L., Huang, J., & Shou, B. (2013). Economics of femtocell service provision. *IEEE Transactions on Mobile Computing*, *12*(12), 2261–2273. doi:10.1109/TMC.2012.193

Dudnikova, A., & Panno, D. (2013). An energy saving approach for femtocell coverage function. *2013 22nd ITC Specialist Seminar on Energy Efficient and Green Networking (SSEEGN).*

Duel-Hallen, A., Holtzman, J., & Zvonar, Z. (1995). Multiuser detection for cdma systems. *Personal Communications IEEE*, *2*(2), 46–58. doi:10.1109/98.382531

EARTH. (2014, April 1). Retrieved from EARTH: http://ict-earth.eu

El-Hajj, A. M., Dawy, Z., & Saad, W. (2012, June). *A stable matching game for joint uplink/downlink resource allocation in OFDMA wireless networks.* Paper presented at the meeting of IEEE International Conference on Communications, Ottawa, Canada.

El-Absi, M., Shaat, M., Bader, F., & Kaiser, T. (2014). Interference alignment with frequency-clustering for efficient resource allocation in cognitive radio networks. In *Proceedings of Global Communications Conference*. Austin, TX: IEEE. doi:10.1109/GLOCOM.2014.7036936

Elayoubi, S. E., & Fourestié, B. (2006). On frequency allocation in 3G LTE systems. In *Proceedings of the IEEE 17th International Symposium on Personal, Indoor and Mobile Radio Communications (PIMRC '2006).* Helsiniki, Finland: IEEE. doi:10.1109/PIMRC.2006.254118

ElBamby, M. S., Bennis, M., Saad, W., & Latvaaho, M. (2014, August). *Content-aware user clustering and caching in wireless small cell networks.* Paper presented at the meeting of IEEE International Symposium on Wireless Communications Systems, Barcelona, Spain. doi:10.1109/ISWCS.2014.6933489

Elhakeem, A. K., Girolamo, R. D., Bdira, I. B., & Talla, M. (1993). Delay and throughput characteristics of TH, CDMA, TDMA, and hybrids for multipath faded data transmission channels. *Universal Personal Communications, 1993. Personal Communications: Gateway to the 21st Century. Conference Record., 2nd International Conference on.*

Elias, J., Martignon, F., Chen, L., & Altman, E. (2013). Joint Operator Pricing and Network Selection Game in Cognitive Radio Networks: Equilibrium, System Dynamics and Price of Anarchy. *Vehicular Technology. IEEE Transactions, 62*(9), 4576–4589.

Ericsson SON Optimization Manager. (2012). Retrieved June 26, 2016, from http://archive.ericsson.net/service/internet/picov/get?DocNo=28701-FGC1011634&Lang=EN&HighestFree=Y

EU FP7 Project CoRaSat. (n.d.). Retrieved from http://www.ict-corasat.eu

Everett, E., Sahai, A., & Sabharwal, A. (2014). Passive self-interference suppression for full-duplex infrastructure nodes. *IEEE Transactions on Wireless Communications, 24*(2), 680–694. doi:10.1109/TWC.2013.010214.130226

Fan, S., & Tian, H. (2015). Cooperative resource allocation for self healing in small cell networks. *IEEE Communications Letters, 19*(7), 1221–1224. doi:10.1109/LCOMM.2015.2433268

Fehske, A. J., Richter, F., & Fettweis, G. P. (2009). *Energy Efficiency Improvements through Micro Sites in Cellular Mobile Radio Networks.* Paper presented at the GLOBECOM Workshops, 2009 IEEE.

Felegyhazi, M., & Hubaux, J.-P. (2006). *Game Theory in Wireless Networks: A Tutorial.* EPFL Technical Report, LCA-REPORT-2006-002.

Feng, S., & Seidel, E. (2008). *Self organizing networks (SON) in 3GPP Long Term Evolution.* Nomor Research GmbH. White Paper.

Feng, D., Jiang, C., Lim, G., Cimini, L. J. Jr, & Feng, G. (2013). A Survey of Energy-Efficient Wireless Communications. *IEEE Communications Surveys and Tutorials, 15*(1), 167–177. doi:10.1109/SURV.2012.020212.00049

Feng, L., & Li, W. (2013). A Dynamic Game Algorithm for Power Allocation in Cognitive Relay Transmission *Communications. IEEE Transactions, 61*(11), 4440–4447. doi:10.1109/TCOMM.2013.101113.130034

Feng, X., Sun, G., Gan, X., Yang, F., Tian, X., Wang, X., & Guizani, M. (2014). Cooperative spectrum sharing in cognitive radio networks: A distributed matching approach. *IEEE Transactions on Communications, 62*(8), 2651–2664. doi:10.1109/TCOMM.2014.2322352

Forster, C., Dickie, I., Maile, G., Smith, H., & Crisp, M. (2009). *Understanding the environmental impact of communication systems.* Ofcom Final Report.

Foschini, G. J., & Gans, M. J. (1998). On limits of wireless communications in a fading environment when using multiple antennas. *Wireless Personal Communications, 6*(3), 311–335. doi:10.1023/A:1008889222784

Frenger, P. K., Orten, P., & Ottosson, T. (2000). Code-spread cdma with interference cancellation. *IEEE Journal on Selected Areas in Communications, 17*(12), 2090–2095. doi:10.1109/49.814815

Frenkiel, R., & Schwartz, M. (2010). Creating cellular: A history of the amps project (1971-1983). *Communications Magazine IEEE, 48*(9), 14–24. doi:10.1109/MCOM.2010.5560579

Fu, T. H., Xu, S. Y., & Kwak, K. S. (2014). *Geometric Programming Based Resource Allocation for Mobile Relays on Vehicles.* Paper presented at the meeting of the IEEE 79st Vehicular Technology Conference, Seoul, South Korea. doi:10.1109/VTCSpring.2014.7022930

Fulp, E. W., Ott, M., Reininger, D., & Reeves, D. S. (1998). Paying for QoS: an optimal distributed algorithm for pricing network resources. *Quality of Service, 1998. (IWQoS 98) 1998 Sixth International Workshop on* (pp.75-84).

Furht, B., & Ahson, S. A. (2007). *Long Term Evolution: 3GPP LTE Radio and Cellular Technology*. Prentice-Hall.

Gabry, F., Li, N., Schrammar, N., Girnyk, M., Rasmussen, L. K., & Skoglund, M. (2014). On the Optimization of the Secondary Transmitter's Strategy in Cognitive Radio Channels with Secrecy. *Selected Areas in Communications. IEEE Journal*, *32*(3), 451–463.

Gajic, V., Huang, J., & Rimoldi, B. (2014). Competition of wireless providers for atomic users. *Proceedings of the Society for Photo-Instrumentation Engineers*, *22*(2), 512–525.

Gale, D., & Shapley, L. S. (1962). College admissions and the stability of marriage. *The American Mathematical Monthly*, *69*(1), 9–15. doi:10.2307/2312726

Gao, G., Zhang, W., Wen, Y., Wang, Z., & Zhu, W. (2015). Cost-efficient video transcoding in media cloud by leveraging user viewing pattern. *IEEE Transactions on Multimedia,* (99), 1–10.

Gao, L., Iosifidis, G., Huang, J., Tassiulas, L., & Li, D. (2014). Bargaining-based mobile data offloading. *IEEE Journal on Selected Areas in Communications*, *32*(6), 1114–1125. doi:10.1109/JSAC.2014.2328393

Gerami, M., Ming, X., & Skoglund, M. (2014). Partial repair for wireless caching networks with broadcast channels. *IEEE Wireless Communications Letters*, *4*(2), 145–148. doi:10.1109/LWC.2014.2386879

Gizelis, C. A., & Vergados, D. D. (2011). A survey of pricing schemes in wireless networks. *IEEE Communications Surveys and Tutorials*, *13*(1), 126–145. doi:10.1109/SURV.2011.060710.00028

Goldsmith, A. (2005). *Wireless Communications*. Cambridge, UK: Cambridge University Press. doi:10.1017/CBO9780511841224

Goldsmith, A., Jafar, S. A., Marić, I., & Srinivasa, S. (2009). Breaking spectrum gridlock with cognitive radios: An information theoretic perspective. *Proceedings of the IEEE*, *97*(5), 894–914. doi:10.1109/JPROC.2009.2015717

Golrezaei, N., Molisch, M., Ji, A. F., Dimakis, A. G., & Caire, G. (2012. November). *Device-to-device communications for wireless video delivery*. Paper presented at the meeting of IEEE Asilomar Conference on Signals, Systems and Computers, Pacific Grove, CA.

Golrezaei, N., Molisch, A. F., Dimakis, A. G., & Caire, G. (2013). Femtocaching and device-to-device collaboration: A new architecture for wireless video distribution. *IEEE Communications Magazine*, *51*(4), 142–149. doi:10.1109/MCOM.2013.6495773

Gomadam, K., Cadambe, V. R., & Jafar, S. A. (2011). A distributed numerical approach to interference alignment and applications to wireless interference networks. *IEEE Transactions on Information Theory*, *57*(6), 3309–3322. doi:10.1109/TIT.2011.2142270

Green Radio. (2016, March 21). Retrieved from Mobile vce Green Radio: www.mobilevce.com

GreenTouch. (2016, March 21). Retrieved from GreenTouch Web site: http://greentouch.org

Gu, Y., Zhang, Y., Pan, M., & Han, Z. (2014, December). *Cheating in matching of device to device pairs in cellular networks*. Paper presented at the meeting of IEEE Global Communications Conference, Austin, TX. doi:10.1109/GLOCOM.2014.7037583

Gu, Y., Zhang, Y., Pan, M., & Han, Z. (2015, March). *Student admission matching based content-cache allocation*. Paper presented at the meeting of IEEE Wireless Communications and Networking Conference, New Orleans, LA.

Guan, Y., Xiao, Y., Feng, H., Shen, C. C., & Cimini, L. J. (2014, December). *MobiCacher: mobility-aware content caching in small-cell networks.* Paper presented at the meeting of IEEE Global Communiocations Conference, Austin, TX. doi:10.1109/GLOCOM.2014.7037523

Guelzim, T., & Obaidat, M. S. (2013). Green Computing and Communication Architecture. In M. S. Obaidat, A. Anpalagan, & I. Woungang (Eds.), *Handbook of green information and communication systems* (pp. 209–227). Academic Press. doi:10.1016/B978-0-12-415844-3.00008-5

Guijarro, L., Pla, V., Vidal, J. R., & Martinez--Bauset, J. (2012). Femtocell operator entry decision with spectrum bargaining and service competition. *IEEE Communications Letters*, *16*(12), 1976–1979. doi:10.1109/LCOMM.2012.101712.121645

Guler, B., & Yener, A. (2014). Selective interference alignment for mimo cognitive femtocell networks. *IEEE Journal on Selected Areas in Communications*, *32*(3), 439–450. doi:10.1109/JSAC.2014.140306

Guler, B., & Yener, A. (2014). Uplink interference management for coexisting MIMO femtocell and macrocell networks: An interference alignment approach. *IEEE Transactions on Wireless Communications*, *13*(4), 2246–2257. doi:10.1109/TWC.2014.030314.131408

Guo, W., & O'Farrell, T. (2012). *Capacity-Energy-Cost Tradeoff in Small Cell Networks.* Paper presented at the Vehicular Technology Conference (VTC Spring), 2012 IEEE 75th.

Guo, W., Rigelsford, J. M., Ford, L., & O'Farrell, T. (2012). Dynamic basestation antenna design for low energy networks. *Progress in Electromagnetics Research, C*, 31, 153–168.

Guo, W., Wang, S., Chu, X., Zhang, X., Chen, J., & Song, H. (2013). Automated small-cell deployment for heterogeneous cellular networks. *IEEE Communications Magazine*, *51*(5), 46–53. doi:10.1109/MCOM.2013.6515046

Gupta, A., & Jha, R. K. (2015). A survey of 5G network: Architecture and emerging technologies. *Access IEEE*, *3*, 1206–1232. doi:10.1109/ACCESS.2015.2461602

Guruacharya, S., Niyato, D., Kim, & Hossain, E. (2013). Hierarchical Competition for Downlink Power Allocation in OFDMA Femtocell Networks. *Wireless Communications, IEEE Transactions*, *12*(4), 1543-1553.

Guruacharya, S., Niyato, D., Kim, D. I., & Hossain, E. (2013). Hierarchical competition for downlink power allocation in OFDMA femtocell networks. *IEEE Transactions on Wireless Communications*, *12*(4), 1543–1553. doi:10.1109/TWC.2013.022213.120016

Guvenc, I. (2011). Capacity and fairness analysis of heterogeneous networks with range expansion and interference coordination. *IEEE Communications Letters*, *15*(10), 1084–1087. doi:10.1109/LCOMM.2011.082611.111387

Gu, Y., Saad, W., Bennis, M., Debbah, M., & Han, Z. (2015). Matching theory for future wireless networks: Fundamentals and applications. *IEEE Communications Magazine*, *53*(5), 52–59. doi:10.1109/MCOM.2015.7105641

Haider, F., Dianati, M., & Tafazolli, R. (2011, July). *A simulation based study of Mobile Femtocell assisted LTE networks.* Paper presented at the meeting of the 7th International Wireless Communications and Mobile Computing Conference, Istanbul, Turkey. doi:10.1109/IWCMC.2011.5982771

Haile, B. B., Mutafungwa, E., & Hämäläinen, J. (2015). Coordinated multi-point transmission for relaxation of self-backhauling bottlenecks in heterogeneous networks. *EURASIP Journal on Wireless Communications and Networking*, (1): 1–17.

Haldar, K. L., Li, H., & Agrawal, D. P. (2013). A cluster-aware soft frequency reuse scheme for inter-cell interference mitigation in LTE based femtocell networks. *2013IEEE 14th International Symposium and Workshops on World of Wireless, Mobile and Multimedia Networks (WoWMoM).* doi:10.1109/WoWMoM.2013.6583436

Halonen, T., Romero, J., & Melero, J. (2002). *GSM, GPRS and EDGE performance: evolution towards 3G/UMTS*. New York: Wiley.

Hamdoun, H., Loskot, P., O'Farrell, T., & He, J. (2012). Survey and applications of standardized energy metrics to mobile networks. *Annals of Telecommunications, 67*(3-4), 113-123.

Hamidouche, K., Saad, W., & Debbah, M. (2014, May). *Many-to-many matching games for proactive social-caching in wireless small cell networks*. Paper presented at the meeting of IEEE International Symposium on Modeling and Optimization in Mobile, Ad Hoc, and Wireless Networks, Hammamet, Tunisia. doi:10.1109/WIOPT.2014.6850348

Hamouda, S., Zitoun, M., & Tabbane, S. (2014). Win–win relationship between macrocell and femtocells for spectrum sharing in LTE-A. *Communications, IET, 8*(7), 1109–1116. doi:10.1049/iet-com.2013.0845

Han, K., Choi, Y., Kim, D., Na, M., Choi, S., & Han, K. (2009). Optimization of femtocell network configuration under interference constraints.*Proceedings of the 7th international conference on Modeling and Optimization in Mobile, Ad Hoc, and Wireless Networks*. IEEE Press.

Han, R., Feng, C., Xia, H., & Zhang, T. (2014). On the fairness of range expansion with interference mitigation in heterogeneous networks. *IEEE Communications Letters, 18*(6), 1051–1054. doi:10.1109/LCOMM.2014.2320920

Hao, L., & Zhuang, W. (2012). Efficient on-demand data service delivery to high-speed trains in cellular/infostation integrated networks. *IEEE Journal on Selected Areas in Communications, 30*(4), 780–791. doi:10.1109/JSAC.2012.120512

Harte, L. (2004). *Introduction to Wcdma, Physical Channels, Logical Channels, Network, and Operation*. Retrieved from https://www.land.ufrj.br/laboratory/repository/upfiles/book/introduction_to_evdo_book.pdf

Hasan, Z., Boostanimehr, H., & Bhargava, V. K. (2011). Green Cellular Networks: A Survey, Some Research Issues and Challenges. *IEEE Communications Surveys and Tutorials, 13*(4), 524–540. doi:10.1109/SURV.2011.092311.00031

Hassan, M. R., Karmakar, G., & Kamruzzaman, J. (2014). Reputation and user requirement based price modeling for dynamic spectrum access. *IEEE Transactions on Mobile Computing, 13*(9), 1–1. doi:10.1109/TMC.2013.78

Hassibi, B., & Hochwald, B. M. (2003). How much training is needed in multiple-antenna wireless links? *IEEE Transactions on Information Theory, 49*(4), 951–963. doi:10.1109/TIT.2003.809594

Haykin, S. (2005). Cognitive radio: Brain-empowered wireless communications. *IEEE Journal on Selected Areas in Communications, 23*(2), 201–220. doi:10.1109/JSAC.2004.839380

He, J., Loskot, P., O'Farrell, T., Friderikos, V., Armour, S., & Thompson, J. (2010). *Energy efficient architectures and techniques for Green Radio access networks*. Paper presented at the Communications and Networking in China (CHINACOM), 2010 5th International ICST Conference on. doi:10.4108/chinacom.2010.136

He, A., Amanna, A., Tsou, T., Chen, X., Dinesh, D., Gaeddert, J., & Bose, T. et al. (2011). Green Communications: A Call for Power Efficient Wireless Systems. *Journal of Communication, 6*(4), 340–351.

He, J., Wen, Y., Huang, J., & Wu, D. (2014). On the cost–qoe tradeoff for cloud-based video streaming under amazon ec2's pricing models. *IEEE Transactions on Circuits and Systems for Video Technology, 24*(4), 669–680. doi:10.1109/TCSVT.2013.2283430

Hevizi, L. G., & Godor, I. (2011). *Power savings in mobile networks by dynamic base station sectorization*. Paper presented at the Personal Indoor and Mobile Radio Communications (PIMRC), 2011 IEEE 22nd International Symposium on.

Himayat, N., Yeh, S.-p., Panah, A. Y., Talwar, S., Gerasimenko, M., Andreev, S., & Koucheryavy, Y. (2014). Multi-Radio Heterogeneous Networks: Architectures and Performance.*2014 Internacional Conference on Computing Networking and Communicatio*. doi:10.1109/ICCNC.2014.6785341

Holtkamp, H., Auer, G., Giannini, V., & Haas, H. (2013). A Parameterized Base Station Power Model. *Communications Letters, IEEE, 17*(11), 2033–2035. doi:10.1109/LCOMM.2013.091213.131042

Hong, M., & Garcia, A. (2011). Equilibrium pricing of interference in cognitive radio networks. *IEEE Transactions on Signal Processing, 59*(12), 6058–6072. doi:10.1109/TSP.2011.2165059

Honig, M. L. (2009). *Advances in Multiuser Detection.* Available from http://www.ece.northwestern.edu/~mh/Advances/Peacock_Collings_Honig.pdf

Honig, M., Madhow, U., & Verdu, S. (2010). Blind adaptive multiuser detection. *Information Theory IEEE Transactions on, 6*(4).

Hossain, E., Rasti, M., Tabassum, H., & Abdelnasser, A. (2014). Evolution toward 5G multi-tier cellular wireless networks: An interference management perspective. *IEEE Wireless Communications, 21*(3), 118–127. doi:10.1109/MWC.2014.6845056

Hossain, M. F., Munasinghe, K. S., & Jamalipour, A. (2013). Energy-Aware Dynamic Sectorization of Base Stations in Multi-Cell OFDMA Networks. *Wireless Communications Letters, IEEE, 2*(6), 587–590. doi:10.1109/WCL.2013.080813.130481

Hou, I.-H., & Chen, C. S. (2013). An energy-aware protocol for self-organizing heterogeneous LTE systems. *IEEE Journal on Selected Areas in Communications, 31*(5), 937–946. doi:10.1109/JSAC.2013.130512

Hoydis, J., Brink, S., & Debbah, M. (2013). Massive MIMO in UL/DL of cellular networks: How many antennas do we need? *IEEE Journal on Selected Areas in Communications, 31*(2), 160–171. doi:10.1109/JSAC.2013.130205

Hu, Z., Zheng, Z., Wang, T., & Song, L. (2015, April). *Small-cell caching mechanism for multi-service providers*. Paper presented at the meeting of IEEE Conference on Computer Communications Workshops, Hong Kong, China.

Huang, G., & Li, J. (2015). Interference mitigation for femtocell networks via adaptive frequency reuse. *IEEE Transactions on Vehicular Technology,* (99). 1-1.

Huang, H., Alrabadi, O., Daly, J., Samardzija, D., Tran, C., Valenzuela, R., & Walker, S. (2010). *Increasing throughput in cellular networks with higher-order sectorization*. Paper presented at the Signals, Systems and Computers (ASILOMAR), 2010 Conference Record of the Forty Fourth Asilomar Conference on. doi:10.1109/ACSSC.2010.5757638

Huang, J. (2009). *Distributed algorithm design for network optimization problems with coupled objectives. In TENCON 2009 - 2009 IEEE Region 10 Conference* (pp. 1–6). IEEE.

Huang, J. (2013). Market mechanisms for cooperative spectrum trading with incomplete network information. *IEEE Communications Magazine, 51*(51), 201–207. doi:10.1109/MCOM.2013.6619584

Huang, J., Berry, R. A., & Honig, M. L. (2006). Distributed interference compensation for wireless networks. *Selected Areas in Communications IEEE Journal on, 24*(5), 1074–1084. doi:10.1109/JSAC.2006.872889

Huang, J., Berry, R., & Honig, M. L. (2005). Performance of distributed utility-based power control for wireless ad hoc networks.*Proceedings of IEEE Milcom.*

Huang, J., & Gao, L. (2013). Wireless network pricing. *Synthesis Lectures on Communication Networks, 6*(2), 1–176. doi:10.2200/S00505ED1V01Y201305CNT013

Hua, S., Liu, H., Zhuo, X. J., Wu, M. Q., & Panwar, S. S. (2014). Exploiting Multiple Antennas in Cooperative Cognitive Radio Networks. *Vehicular Technology. IEEE Transactions*, *63*(7), 3318–3330.

Hu, R., & Qian, Y. (2014). An energy efficient and spectrum efficient wireless heterogeneous network framework for 5G systems. *IEEE Communications Magazine*, *52*(5), 94–101. doi:10.1109/MCOM.2014.6815898

Hwang, I., Song, B., & Soliman, S. S. (2013). A holistic view on hyper-dense heterogeneous and small cell networks. *IEEE Communications Magazine*, *51*(6), 20–27. doi:10.1109/MCOM.2013.6525591

IMT-2020. (5G) Promotion Group. (2014). *White paper on 5G vision and requirements*. Retrieved February 20, 2015, from http://www.imt-2020.cn/zh

Iosifidis, G., Gao, L., Huang, J., & Tassiulas, L. (2015). A double-auction mechanism for mobile data-offloading markets. *IEEE/ACM Transactions on Networking*, *23*(5), 1634–1647. doi:10.1109/TNET.2014.2345875

Jafar, S. A. (2010). *Exploiting channel correlations - simple interference alignment schemes with no csit*. Retrieved from http://newport.eecs.uci.edu/~syed/papers/nocsitalignconf.pdf

Jafar, S. A. (2010). Interference alignment — a new look at signal dimensions in a communication network. *Foundations & Trends® in Communications & Information Theory, 1*(1), 1-136.

Jafar, S. (2011). Interference alignment: A new look at signal dimensions in a communication network. *Foundations and Trends in Communication and Information Theory*, *7*(1), 1–136. doi:10.1561/0100000047

Jafar, S. A., & Shamai, S. (2006). Degrees of freedom region for the mimo x channel. *Eprint Arxiv Cs*, *54*(1), 151–170.

Jain, R., Chiu, D., & Hawe, W. (1998). A quantitative measure of fairness and discrimination for resource allocation in shared computer systems. *Computer Science*. Retrieved from cs.ni/9809099

Jaiswal, S., Ganz, A., & Mettu, R. (2009). An optimization framework for demand-based fair stream allocation in MIMO ad hoc Networks. *Mobile Networks and Applications*, *14*(4), 451–469. doi:10.1007/s11036-009-0161-x

Jiang, C., Chen, Y., & Liu, K. J. R. (2013, November). Distributed adaptive networks: A graphical evolutionary game-theoretic view. *IEEE Transactions on Signal Processing*, *61*(22), 5675–5688. doi:10.1109/TSP.2013.2280444

Jiang, C., Chen, Y., Liu, K. J. R., & Ren, Y. (2014). Optimal pricing strategy for operators in cognitive femtocell networks. *IEEE Transactions on Wireless Communications*, *13*(9), 5288–5301. doi:10.1109/TWC.2014.2327970

Jiang, C., Chen, Y., Liu, K. J. R., & Ren, Y. (2015). Network economics in cognitive networks. *IEEE Communications Magazine*, *53*(5), 75–81. doi:10.1109/MCOM.2015.7105644

Jin, Z., & Vandendorpe, L. *Resource allocation in multi-cellular DF relayed OFDMA systems*. (2011, December). Paper presented at the meeting of the IEEE Global Telecommunications Conference, Houston, TX.

Ji, Z., & Liu, K. J. R. (2008). Multi-stage pricing game for collusion-resistant dynamic spectrum allocation. *IEEE Journal on Selected Areas in Communications*, *26*(1), 182–191. doi:10.1109/JSAC.2008.080116

Joe-Wong, C., Sen, S., Ha, S., & Chiang, M. (2012). Optimized day-ahead pricing for smart grids with device-specific scheduling flexibility. *IEEE Journal on Selected Areas in Communications*, *30*(6), 1075–1085. doi:10.1109/JSAC.2012.120706

Jo, H. S., Mun, C., Moon, J., & Yook, J. G. (2009). Interference mitigation using uplink power control for two-tier femtocell networks. *IEEE Transactions on Wireless Communications*, *8*(10), 4906–4910. doi:10.1109/TWC.2009.080457

Jo, H.-S., Xia, P., & Andrews, J. G. (2012). Open, closed, and shared access femtocells in the downlink. *EURASIP Journal on Wireless Communications and Networking*, *2012*(1), 363. doi:10.1186/1687-1499-2012-363

Jo, M., Maksymyuk, T., Batista, R. L., Maciel, T. F., De Almeida, A. L. F., & Klymash, M. (2014). A survey of converging solutions for heterogeneous mobile networks. *IEEE Wireless Communications*, *21*(8), 54–62.

Joyce, R., Morris, D., Brown, S., Vyas, D., & Zhang, L. (2015). *Higher Order Horizontal Sectorisation Gains for 6, 9, 12 and 15 Sectored Cell Sites in a 3GPP/HSPA+ Network*. Academic Press.

Kamel, M. I., & Elsayed, K. M. F. (2012). Performance evaluation of a coordinated time-domain eICIC framework based on ABSF in heterogeneous LTE-Advanced networks.*2012 IEEE Global Communications Conference (GLOBECOM)*. doi:10.1109/GLOCOM.2012.6503967

Kang, X., Zhang, R., & Motani, M. (2011). Price-Based Resource Allocation for Spectrum-Sharing Femtocell Networks: A Stackelberg Game Approach.*Global Telecommunications Conference (GLOBECOM 2011)*, (Vol. 30, pp.1 - 5). IEEE.

Kang, X., Zhang, R., & Motani, M. (2012). Price-Based Resource Allocation for Spectrum-Sharing Femtocell Networks: A Stackelberg Game Approach. *Selected Areas in Communications. IEEE Journal*, *30*(3), 538–549.

Kang, X., Zhang, R., & Motani, M. (2012). Price-based resource allocation for spectrum-sharing femtocell networks: A Stackelberg game approach. *IEEE Journal on Selected Areas in Communications*, *30*(3), 538–549. doi:10.1109/JSAC.2012.120404

Kantarci, M. E. (2014, April). *Multicast-aware caching for small cell networks*. Paper presented at the meeting of IEEE Wireless Communications and Networking Conference, Istanbul, Turkey.

Kantarci, M. E. (2015, March). *Content caching in small cells with optimized uplink and caching power*. Paper presented at the meeting of IEEE Wireless Communications and Networking Conference, New Orleans, LA.

Karlsson, R. S. (1999). Radio resource sharing and capacity of some multiple access methods in hierarchical cell structures.*Vehicular Technology Conference*. *5*, 2825-2829. doi:10.1109/VETECF.1999.800302

Kasbekar, G. S., & Sarkar, S. (2012). Spectrum pricing games with random valuations of secondary users. *IEEE Journal on Selected Areas in Communications*, *30*(30), 2262–2273. doi:10.1109/JSAC.2012.121218

Katkovnik, V., Lee, M., & Kim, Y. (2006). Performance study of the minimax robust phased array for wireless communications. *IEEE Transactions on Communications*, *54*(4), 608–613. doi:10.1109/TCOMM.2006.873068

Kavurmacioglu, E., Alanyali, M., & Starobinski, D. (2014). Competition in private commons: Price war or market sharing? *IEEE/ACM Transactions on Networking*.

Kelly, F. P., Maulloo, A. K., & Tan, D. K. H. (1998). Rate control for communication networks: Shadow price, proportional fairness and stability. *The Journal of the Operational Research Society*, *49*(3), 237–252. doi:10.1057/palgrave.jors.2600523

Khandekar, A., Blushan, N., Tingfang, J., & Vanghi, V. (2010). LTE-Advanced: Heterogeneous Networks.*2010 European Wireless Conference* (pp. 978-982). Lucca, Italy: IEEE. doi:10.1109/EW.2010.5483516

Kim, D., Kim, S., Kim, H., & Choi, S. (2014). Pricing policies for femtocell service adoption in two-tier cellular networks. *Communications (ICC), 2014 IEEE International Conference on* (pp. 2701-2706). IEEE.

Kim, K. H., Lee, J. H., Lee, C. H., Jeon, N. R., & Kim, S. C. (2009). Coordinated Beamforming with Limited BS Cooperation for Multicell Multiuser MIMO Broadcast Channel. In *Proceedings of Vehicular Technology Conference (VTC '2009 Spring)*.Barcelona, Spain: IEEE. doi:10.1109/VETECS.2009.5073473

Kolios, P., Friderikos, V., & Papadaki, K. (2011). Future Wireless Mobile Networks. *IEEE Vehicular Technology Magazine*, *6*(1), 24–30. doi:10.1109/MVT.2010.939905

Kotecha, J. H., & Sayeed, A. M. (2004). Transmit signal design for optimal estimation of correlated MIMO channels. *IEEE Transactions on Signal Processing*, *52*(2), 546–557. doi:10.1109/TSP.2003.821104

Koutlia, K., Perez-Romero, J., & Agusti, R. (2015). On Enhancing Almost Blank Subframes Management for Efficient eICIC in HetNets. *Vehicular Technology Conference (VTC '2015 Spring)*. Glasgow, UK: IEEE. doi:10.1109/VTC-Spring.2015.7145960

Kulkarni, P., Chin, W. H., & Farnham, T. (2010). Radio resource management considerations for lte femto cells. *Computer Communication Review*, *40*(1), 26–30. doi:10.1145/1672308.1672314

Kumar, A., & Saad, W. (2015, June). *On the tradeoff between energy harvesting and caching in wireless networks.* Paper presented at the meeting of IEEE Communicaitons Conference, London, UK. doi:10.1109/ICCW.2015.7247470

Kurda, R., Boukhatem, L., & Kaneko, M. (2015). Femtocell power control methods based on users' context information in two-tier heterogeneous networks. *EURASIP Journal on Wireless Communications and Networking*, *2015*(1), 129. doi:10.1186/s13638-015-0328-z

Kurda, R., Boukhatem, L., & Yahiya, T. A. (2014). Interference mitigation in mobile environment through power adjustment in macro and femto cell systems.*2014 IEEE Wireless Communications and Networking Conference (WCNC).* doi:10.1109/WCNC.2014.6952491

Kwon, Y. M., Oh, S. M., Shin, J., & Chung, M. Y. (2015, November). *Radio resource management for moving small-cells with proactive content cache*. Paper presented at the meeting of IEEE Region 10 Conference, Macao, China.

Kwon, H., & Birdsall, T. (1986). Channel capacity in bits per joule. *IEEE Journal of Oceanic Engineering*, *11*(1), 97–99. doi:10.1109/JOE.1986.1145138

Larsson, E. G., Edfors, O., Tufvesson, F., & Marzetta, T. L. (2014). Massive MIMO for next generation wireless systems. *IEEE Communications Magazine*, *52*(2), 186–195. doi:10.1109/MCOM.2014.6736761

Lasaulce, S., Hayel, Y., El Azouzi, R., & Debbah, M. (2009). Introducing hierarchy in energy games. *Wireless Communications. IEEE Transactions*, *8*(7), 3833–3843.

Le Treust, M., Lasaulce, S., Hayel, Y., & He, G. N. (2013). Green Power Control in Cognitive Wireless Networks. *Vehicular Technology, IEEE Transactions*, *62*(4), 1741-1754.

Lee, H.-C., Oh, D.-C., & Lee, Y.-H. (2010). *Mitigation of Inter-Femtocell Interference with Adaptive Fractional Frequency Reuse.* Paper presented at the ICC Conference, Cape Town, South Africa. doi:10.1109/ICC.2010.5502298

Lee, D., Seo, H., Clerckx, B., Hardouin, E., Mazzarese, D., Nagata, S., & Syana, K. (2012). Coordinated multipoint transmission and reception in lte-advanced: Deployment scenarios and operational challenges. *IEEE Communications Magazine*, *50*(2), 148–155. doi:10.1109/MCOM.2012.6146494

Leem, H., Kim, J. Y., Jung, B. H., & Dan, K. S. (2014). Spectral-, energy-, & cost-efficient deployment of small cells in a HetNet topology. *ICC 2014 - 2014 IEEE International Conference on Communications*(Vol. 33, pp. 2709-2717).

Lee, N., & Heath, R. W. Jr. (2014). Advanced Interference Management Technique: Potential and Limitations. *IEEE Wireless Communications*, *23*(3), 30–38. doi:10.1109/MWC.2016.7498072

Lertwiram, N., Popovski, P., & Sakaguchi, K. (2012). A study of trade-off between opportunistic resource allocation and interference alignment in femtocell scenarios. *IEEE Wireless Communications Letters*, *1*(4), 356–359. doi:10.1109/WCL.2012.051712.120268

Li, H., Wang, Z., & Hu, D. (2015, August). *Joint wireless and backhaul load balancing in cooperative caches enabled small-cell networks*. Paper presented at the meeting of IEEE Personal, Indoor and Mobile Radio Communications, Hong Kong, China. doi:10.1109/PIMRC.2015.7343607

Li, H., Xu, X., Hu, D., & Qu, X. (2010). *Graph Method Based Clustering Strategy for Femtocell Interference Management and Spectrum Efficiency Improvement*. Paper presented at the WiCOM Conference, Chengdu, China. doi:10.1109/WICOM.2010.5601236

Li, J., Chen, W., Xiao, M., Shu, F., & Liu, X. (2015). Efficient video pricing and caching in heterogeneous networks. *IEEE Transactions on Vehicular Technology*, (99), 1-8.

Li, Q., Xu, Y., Hu, R. Q., & Wu, G. (2012). Pricing-based distributed mobile association for heterogeneous networks with cooperative relays. *Communications (ICC), 2012 IEEE International Conference on* (Vol. 11, pp. 5326-5331). IEEE.

Li, S., & Huang, J. (2010). Price differentiation for communication networks. *Networking IEEE/ACM Transactions on*, *22*(3), 703-716.

Liang, Y. S., Chung, W. H., Zhang, H., & Kuo, S. Y. (2011). Interference mitigation through self-organization in OFDMA femtocells. *IEEE 22nd International Symposium on Personal Indoor and Mobile Radio Communications (PIMRC)*.

Liang, Y. S., Chung, W. H., Ni, G. K., Chen, I. Y., Zhang, H., & Kuo, S. Y. (2012). Resource allocation with interference avoidance in ofdma femtocell networks. *IEEE Transactions on Vehicular Technology*, *61*(5), 2243–2255. doi:10.1109/TVT.2012.2191164

Lien, S. Y., Tseng, C. C., Chen, K. C., & Su, C. W. (2010). *Cognitive radio resource management for qos guarantees in autonomous femtocell networks*. Paper presented at the ICC Conference, Cape Town, South Africa. doi:10.1109/ICC.2010.5502784

Li, J. D., Meng, Y., Li, H. Y., & Suo, L. (2015). The Graph-based Fair Resource Allocation Scheme Combing Interference Alignment in Femtocell Networks. *IET Communications*, *9*(2), 211–218. doi:10.1049/iet-com.2014.0611

Lim, S. H., Ko, Y. B., Jung, G. H., Kim, J., & Jang, M. W. (2014). Inter-chunk popularity-based edge-first caching in content-centric networking. *IEEE Communications Letters*, *18*(8), 1331–1334. doi:10.1109/LCOMM.2014.2329482

Lin, P., Zhang, J., Chen, Y., & Zhang, Q. (2011). Macro-femto heterogeneous network deployment and management: From business models to technical solutions. *IEEE Wireless Communications*, *18*(3), 64–70. doi:10.1109/MWC.2011.5876502

Liolos, K., Schlueter, G., Krause, J., Zimmer, F., Combelles, L., Grotz, J., … Vanelli, C. A. (2013). Cognitive radio scenarios for satellite communications: The CoRaSat approach. In Proceedings of Future Network and Mobile Summit. Lisbon: IEEE.

Li, Q., Niu, H., Papathanassiou, A., & Wu, G. (2014). 5G network capacity: Key elements and technologies. *IEEE Vehicular Technology Magazine*, *9*(1), 71–76. doi:10.1109/MVT.2013.2295070

Li, S. H. (2003). *The standard of the third generation wireless communication. TD-SCDMA*. Beijing, China: Posts & Telecom Press.

Li, S., Huang, J., & Li, S. Y. R. (2012). Dynamic profit maximization of cognitive mobile virtual network operator. *IEEE Transactions on Mobile Computing*, *13*(3), 1.

Lister, D. (2009). *An operator's view on green radio*. Keynote Speech, GreenComm.

Liu, D. T., Chen, Y., Zhang, T. K., Keong Chai, K., Loo, J., & Vinel, A. (2013). Stackelberg Game Based Cooperative User Relay Assisted Load Balancing in Cellular Networks. *Communications Letters, IEEE*, *17*(2), 424–427. doi:10.1109/LCOMM.2012.122012.122377

Liu, H. X., & Zhu, J. K. (2008). Dynamic spectrum allocation of little satellite group communication based on cognitive radio. *Communications Technology*, *41*(7), 1–3.

Liu, J., Kawamoto, Y., Nishiyama, H., Kato, N., & Kadowaki, N. (2014). Device-to-device communications achieve efficient load balancing in lte-advanced networks. *Wireless Communications IEEE*, *21*(2), 57–65. doi:10.1109/MWC.2014.6812292

Liu, J., Zhao, T., Zhou, S., Cheng, Y., & Niu, Z. (2014). CONCERT: A cloud-based architecture for next-generation cellular systems. *IEEE Wireless Communications*, *21*(6), 14–22. doi:10.1109/MWC.2014.7000967

Liu, J., Zheng, W., Li, W., Wang, X., Xie, Y., & Wen, X. (2013). Distributed uplink power control for two-tier femtocell networks via convex pricing.*Wireless Communications and Networking Conference (WCNC)*, (pp.458-463). IEEE.

Liu, Q. J., Bi, S. J., Sun, J., & Liu, T. X. (2012). Potential applications of cognitive radio technology in satellite navigation system. In *Proceedings of 3rd Annual Conference of China's Satellite Navigation*. Guangzhou: CSNC.

Liu, Y., Zhang, Y., & Xie, S. (2015). Integrated Energy and Spectrum Harvesting for 5G Wireless Communications. *IEEE Network*, *29*(3), 75–81. doi:10.1109/MNET.2015.7113229

Li, W., Zheng, W., Su, T., & Wen, X. (2013). Distributed power control and pricing for two-tier OFMDA femtocell networks using fictitious game. *Wireless Communications and Networking Conference (WCNC), 2013 IEEE* (pp.470-475). IEEE. doi:10.1109/WCNC.2013.6554609

Li, W., Zheng, W., Su, T., & Wen, X. (2013). Dual-utility based green power game in two-tier OFDMA femtocell networks with firefly algorithm. *Wireless Communications and Networking Conference (WCNC).*

Li, X. H. (2007). Space information acquisition based on software-defined radio and cognitive radio. *Aerospace Electronic Warfare*, *23*(6), 33–35.

Li, X., Qian, L., & Kataria, D. (2009). Downlink power control in co-channel macrocell femtocell overlay. In *Proceedings of 43rd Annual Conference on Information Sciences and Systems, CISS 2009*. IEEE. doi:10.1109/CISS.2009.5054750

Li, X., Zhu, X., Wu, L., & Sandrasegaran, K. (2013). A distributed non-uniform pricing approach for power optimization in spectrum-sharing femtocell network. *IEEE Wireless Communications and Networking Conference (WCNC).*

Li, Y., Sheng, M., Yang, C., & Wang, X. (2013). Energy Efficiency and Spectral Efficiency Tradeoff in Interference-Limited Wireless Networks. *IEEE Communications Letters*, *17*(10), 1924–1927. doi:10.1109/LCOMM.2013.082613.131286

Li, Y., Vucetic, B., Zhou, Z., & Dohler, M. (2007). Distributed adaptive power allocation for wireless relay networks. *IEEE Transactions on Wireless Communications*, *6*(3), 948–958. doi:10.1109/TWC.2007.05256

Li, Z., Li, J. D., Liu, Q., & Shen, B. (2014). Interference alignment based spectrum sharing and user scheduling for cognitive radio MIMO system. *Journal of Communication*, *35*(1), 167–172.

López-Pérez, D., Guvenc, I., De la Roche, G., Kountouris, M., & Zhang, J. (2011). Enhanced Intecell Interference Coordination Challenges in Heterogeneous Networks. *IEEE Wireless Communications*, 22-30.

Lopez-Perez, D., Chu, X., Vasilakos, A. V., & Claussen, H. (2013). On distributed and coordinated resource allocation for interference mitigation in self-organizing LTE networks. *IEEE/ACM Transactions on Networking*, *21*(4), 1145–1158. doi:10.1109/TNET.2012.2218124

López-Pérez, D., Guvenc, I., de la Roche, G. d., Kountouris, M., Quek, T. Q., & Zhang, J. (2011). *Enhanced Intercell Interference Coordination Challenges in Heterogeneous Networks*. IEEE Wireless Communication.

Lopez-Perez, D., Ming, D., Claussen, H., & Jafari, A. H. (2015). Towards 1 Gbps/UE in Cellular Systems: Understanding Ultra-Dense Small Cell Deployments. *IEEE Communications Surveys and Tutorials*, *17*(4), 2078–2101. doi:10.1109/COMST.2015.2439636

Lorenzo, P. D., Omilipo, M., & Barbarossa, S. (2011). Distributed stochastic pricing for sum-rate maximization in femtocell networks with random graph and quantized communications. *Computational Advances in Multi-Sensor Adaptive Processing (CAMSAP), 2011 4th IEEE International Workshop on* (pp. 165-168). IEEE.

Lorenz, R. G., & Boyd, S. P. (2005). Robust minimum variance beamforming. *IEEE Transactions on Signal Processing*, *53*(5), 1684–1696. doi:10.1109/TSP.2005.845436

Loyka, S. (2001). Channel capacity of MIMO architecture using the exponential correlation matrix. *IEEE Communications Letters*, *5*(9), 369–371. doi:10.1109/4234.951380

Luan, T., Gao, F. F., Zhang, X. D., Li, C. F., & Lei, M. (2012). Rate maximization and beamforming design for relay-aided multiuser cognitive networks. *IEEE Transactions on Vehicle Technology, 61*(4), 1940-1945.

Luo, S., Zhang, R., & Lim, T. J. (2014). Downlink and uplink energy minimization through user association and beamforming in C-RAN. *IEEE Transactions on Wireless Communications*, *14*(1), 494–508. doi:10.1109/TWC.2014.2352619

Ma, B., Man, H. C., Wong, V. W. S., & Huang, J. (2015). Hybrid overlay/underlay cognitive femtocell networks: A game theoretic approach. *IEEE Transactions on Wireless Communications*, *14*(6), 1–1. doi:10.1109/TWC.2015.2403363

Mackenzie, A., & Dasilva, L. (2006). Game theory for wireless engineers. *Synthesis Lectures on Communications,* (1), 1-86.

Maddah-Ali, M. A., Motahari, A. S., & Khandani, A. K. (2008). Communication over mimo x channels: Interference alignment, decomposition, and performance analysis. *Information Theory IEEE Transactions on*, *54*(8), 3457–3470. doi:10.1109/TIT.2008.926460

Maharjan, S., Zhang, Y., & Gjessing, S. (2011). Economic approaches for cognitive radio networks: A survey. *Wireless Personal Communications*, *57*(1), 33–51. doi:10.1007/s11277-010-0005-9

Mahloo, M., Monti, P., Chen, J., & Wosinska, L. (2014). Cost modeling of backhaul for mobile networks. *Communications Workshops (ICC), 2014 IEEE International Conference on* (pp.397-402). IEEE.

Ma, L., Chen, X. T., Liu, H. J., & Liang, X. W. (2010). The application of cognitive radio technology in LEO satellite system. In *Proceedings of 6th Annual Conference of Satellite Communications New Business New Technology*. Beijing: ChinaInstitute of Communications.

Man, H. C., Southwell, R., & Huang, J. (2014). Congestion-aware network selection and data offloading. *2014 48th Annual Conference on Information Sciences and Systems (CISS)* (pp.1 - 6).

Man, H. C., & Wong, V. W. S. (2013). Interference pricing for sinr-based random access game. *IEEE Transactions on Wireless Communications*, *12*(5), 2292–2301. doi:10.1109/TWC.2013.031313.120938

Mao, L., Xu, S. Y., Fu, T. H., & Huang, Q. *Game theory based power allocation algorithm in high-speed mobile environment*. (2012, September). Paper presented at the meeting of the IEEE 72nd Vehicular Technology Conference, Ottawa, Canada. doi:10.1109/VTCFall.2012.6399136

Marden, J. R., Arslan, G., & Shamma, J. S. (2009). Joint strategy fictitious play with inertia for potential games. *IEEE Transactions on Automatic Control*, *54*(2), 208–220. doi:10.1109/TAC.2008.2010885

Marzetta, T. L. (2010). Noncooperative cellular wireless with unlimited numbers of base station antennas. *IEEE Transactions on Wireless Communications*, *9*(11), 3590–3600. doi:10.1109/TWC.2010.092810.091092

Maslennikov, R., Trushanin, A., Testov, O., Vechkanov, M., Antipova, A., Thomas, T. A., et al. (2013). *Azimuth and elevation sectorization for the stadium environment*. Paper presented at the Global Communications Conference (GLOBECOM). doi:10.1109/GLOCOM.2013.6831694

Menache, I., Ozdaglar, A., & Shimkin, N. (2014). Socially optimal pricing of cloud computing resources. Institute for Computer Sciences, 322-331.

Meng, Y., Li, J. D., Li, H. Y., & Pan, M. (2015). A transformed conflict graph based resource allocation scheme combining interference alignment in OFDMA femtocell networks. *IEEE Transactions on Vehicular Technology*, *64*(10), 4728–4737. doi:10.1109/TVT.2014.2366747

Mezghani, A., Damak, N., & Nossek, J. A. (2010). Circuit aware design of power-efficient short range communication systems.*In Proc. IEEE Int. Symp. Wireless Commun. Syst. (ISWCS)*, (pp. 869-873). doi:10.1109/ISWCS.2010.5624403

Miao, G., Himayat, N., & Li, Y. (2010). Energy-efficient link adaptation in frequency-selective channels. *IEEE Transactions on Communications*, *58*(2), 545–554. doi:10.1109/TCOMM.2010.02.080587

Micallef, G., Mogensen, P., Scheck, H. O., & Lang, E. (2011). *Energy Efficient Evolution of Mobile Networks: Macro-Only Upgrades vs. a Joint-Pico Deployment Strategy*. Paper presented at the Vehicular Technology Conference (VTC Spring), 2011 IEEE 73rd.

Ming, C. N., (2012). Some suggestions on demand for research and construction of China satellite mobile communication system. *Space International*, 1-9.

Mishra, A. R. (2006). *Advanced Cellular Network Planning and Optimisation*. Available from http://download.e-bookshelf.de/download/0000/5683/51/L-G-0000568351-0002356592.pdf

Mitola, J. (2000). *Cognitive radio: An integrated agent architecture for software defined radio*. Stockholm, Sweden: Royal Institute of Technology.

Mohammad, S. O., & Alagan Anpalagan, I. W. (2013). Handbook of Green Information and Communications Systems. Academic Press.

Mohammed, S., & Larsson, E. (2013). Per-antenna constant envelope precoding for large multi-user MIMO systems. *IEEE Transactions on Communications*, *61*(3), 1059–1071. doi:10.1109/TCOMM.2013.012913.110827

Monderer, D., & Shapley, L. S. (1996). Potential games. *Games and Economic Behavior*, *14*(1), 124–143. doi:10.1006/game.1996.0044

Monti, P., Tombaz, S., Wosinska, L., & Zander, J. (2012). *Mobile backhaul in heterogeneous network deployments: Technology options and power consumption*. Paper presented at the Transparent Optical Networks (ICTON), 2012 14th International Conference on.

Mueller, C. M., Bakker, H., & Ewe, L. (2011). Evaluation of the automatic neighbor relation function in a dense urban scenario. *IEEE 73rd Vehicular Technology Conference (VTC Spring)*.

Namgoong, W. (2005). Modeling and analysis of nonlinearities and mismatches in AC-coupled direct-conversion receiver. *IEEE Transactions on Wireless Communications*, *4*(1), 163–173. doi:10.1109/TWC.2004.840239

Nam, W., Bai, D., Lee, J., & Kang, I. (2014). Advanced interference management for 5g cellular networks. *IEEE Communications Magazine*, *52*(5), 52–60. doi:10.1109/MCOM.2014.6815893

Nazer, B., Gastpar, M., Jafar, S. A., & Vishwanath, S. (2011). Ergodic interference alignment. *IEEE Transactions on Wireless Communications*, *11*(2), 842–851.

Nazir, M., Bennis, M., Ghaboosi, K., MacKenzie, A. B., & Latva-aho, M. (2010). Learning based mechanisms for interference mitigation in self-organized femtocell networks.*Conference Record of the Forty Fourth Asilomar Conference on Signals, Systems and Computers (ASILOMAR)*. doi:10.1109/ACSSC.2010.5757866

Neely, M. J. (2009). Optimal pricing in a free market wireless network. *Wireless Networks*, *15*(7), 901–915. doi:10.1007/s11276-007-0083-0

Nesterov, Y., Nemirovskii, A. S., & Ye, Y. (Eds.). (1994). *Interior-point polynomial algorithms in convex programming*. Philadelphia: Society for Industrial and Applied Mathematics. doi:10.1137/1.9781611970791

Ngo, H. Q., Larsson, E. G., & Marzetta, T. L. (2013). Energy and spectral efficiency of very large multiuser MIMO systems. *IEEE Transactions on Communications*, *61*(4), 1436–1449. doi:10.1109/TCOMM.2013.020413.110848

Ngo, H. Q., Suraweera, H. A., Matthaiou, M., & Larsson, E. G. (2014). Multipair full-duplex relaying with massive arrays and linear processing. *IEEE Journal on Selected Areas in Communications*, *32*(9), 1721–1737. doi:10.1109/JSAC.2014.2330091

Ngo, H. Q., Suraweerat, H. A., & Larsson, E. G. (2013). Multipair massive MIMO full-duplex relaying with MRC/MRT processing. *Proceedings of the IEEE, ICC*, 4807–4813.

Nguyen, D., Tran, L.-N., Pirinen, P., & Latva-aho, M. (2014). Precoding for full duplex multiuser MIMO systems: Spectral and energy efficiency maximization. *IEEE Transactions on Signal Processing*, *61*(16), 4038–4050. doi:10.1109/TSP.2013.2267738

Nguyen, T. M., Jeong, Y., Quek, T. Q. S., Tay, W. P., & Shin, H. (2013). Interference alignment in a Poisson field of MIMO femtocells. *IEEE Transactions on Wireless Communications*, *12*(6), 2633–2645. doi:10.1109/TWC.2013.040413.120024

Nie, W., Yi, Z., Zheng, F., Zhang, W., & Farrell, T. O. (2015). HetNets with Random DTX Scheme: Local Delay and Energy Efficiency. *IEEE Transactions on Vehicular Technology*, (99), 1-1. doi: 10.1109/TVT.2015.2477374

Nielsen, T. T., & Wigard, J. (2000). *Performance Enhancements in a Frequency Hopping GSM Network*. Available from http://www.openisbn.org/download/0792378199.pdf

Nikolaou, S., Renesse, R. V. & Schiper, N. (2015). Proactive cache placement on cooperative client caches for online social networks and applications. *IEEE Transactions on Parallel Distribution Systems,* (99), 1–10.

Niyato, D., Wang, P., Pink, T. H., Saad, W., & Kim, D. I. (2014). Cooperation in delay tolerant networks with wireless energy transfer: Performance analysis and optimization. *IEEE Transactions on Vehicular Technology,* (99), 3740-3754.

Niyato, D., & Hossain, E. (2008). Competitive pricing for spectrum sharing in cognitive radio networks: Dynamic game, inefficiency of nash equilibrium, and collusion. *Selected Areas in Communications IEEE Journal on*, *26*(1), 192–202. doi:10.1109/JSAC.2008.080117

Niyato, D., & Hossain, E. (2008). Market-equilibrium, competitive, and cooperative pricing for spectrum sharing in cognitive radio networks: Analysis and comparison. *IEEE Transactions on Wireless Communications*, *7*(11), 4273–4283. doi:10.1109/T-WC.2008.070546

Niyato, D., & Hossain, E. (2009). Dynamics of network selection in heterogeneous wireless networks: An evolutionary game approach. *IEEE Transactions on Vehicular Technology*, *58*(4), 2008–2017. doi:10.1109/TVT.2008.2004588

Niyato, D., Hossain, E., & Han, Z. (2009). Dynamics of multiple-seller and multiple-buyer spectrum trading in cognitive radio networks: A game-theoretic modeling approach. *IEEE Transactions on Mobile Computing*, *8*(8), 1009–1022. doi:10.1109/TMC.2008.157

O'Farrell, T., & Fletcher, S. (2015). Green Communication Concepts, Energy Metrics and Throughput Efficiency for Wireless Systems. In K. Samdanis, P. Rost, A. Maeder, M. Meo, & C. Verikoukis (Eds.), *Green Communications: Principles, Concepts and Practice* (pp. 19–42). Chichester, UK: John Wiley & Sons, Ltd. doi:10.1002/9781118759257.ch2

O'Malley, G. (2007). *Algorithmic aspects of stable matching problems* (Unpublished doctoral dissertation). University of Glasgow, Glasgow, UK.

Oh, E., Krishnamachari, B., Liu, X., & Niu, Z. (2011). Toward dynamic energy-efficient operation of cellular network infrastructure. *IEEE Communications Magazine*, *49*(6), 56–61. doi:10.1109/MCOM.2011.5783985

OPERA-Net 2. (2016, March 16). Retrieved from Celtic-Plus Smart Connected World: http://projects.celticplus.eu

Oussama, H., El-Azouzi, R., & Yezekael, H. (2013). A Stackelberg Model for Opportunistic Sensing in Cognitive Radio Networks. *Wireless Communications, IEEE Transactions*, *12*(5), 2148-2159.

Pantiano, F., Bennis, M., Saad, W., & Debbah, M. (2015, June). *Match to cache: joint user association and backhaul allocation in cache-aware small cell networks*. Paper presented at the meeting of IEEE International Conference on Communications, London, UK.

Pantisano, F., Bennis, M., Saad, W., & Debbah, M. (2014a, May). *Cache-aware user association in backhaul-constrained small cell networks*. Paper presented at the meeting of International Symposium on Modeling and Optimization in Mobile, Ad Hoc, and Wireless Networks, Hammamet, Tunisia. doi:10.1109/WIOPT.2014.6850276

Pantisano, F., Bennis, M., Saad, W., & Debbah, M. (2014b, November). *In-network caching and content placement in cooperative small cell networks*. Paper presented at the meeting of International Conference on 5G Ubiquitous Connectivity, Levi, Finland. doi:10.4108/icst.5gu.2014.258230

Pantisano, F., Bennis, M., Saad, W., Debbah, M., & Latva-aho, M. (2013). Interference alignment for cooperative femtocell networks: A game-theoretic approach. *IEEE Transactions on Mobile Computing*, *12*(11), 2233–2246. doi:10.1109/TMC.2012.196

Park, S., Seo, W., Choi, S., & Hong, D. (n.d.). A beamforming codebook restriction for cross-tier interference coordination in two-tier femtocell networks. *IEEE Transaction on Vehicular Technology, 60*(4), 1651-1663.

Park, J., & Van der Schaar, M. (2009). *Stackelberg Contention Games in Multiuser Networks*. Computer Science and Game Theory.

Park, S. H., Kim, M. J., & Ko, Y. C. (2014). Resource allocation for spatial-frequency domain based interference alignment scheme in MIMO-OFDM systems. In *Proceedings of International Conference on Communications*. Sydney: IEEE. doi:10.1109/ICC.2014.6884239

Patel, P., & Holtzman, J. (2010). Analysis of a simple successive interference cancellation scheme in a ds/cdma system. *IEEE Journal on Selected Areas in Communications*, *12*(5), 796–807. doi:10.1109/49.298053

Pauli, V., Naranjo, J. D., & Seidel, E. (2010). Heterogeneous LTE Networks and Inter-Cell Interference Coordination. *Nobel Mobile Radio*, 1-9.

Pejanovic-Djurisic, M. E. S., & Ilic-Delibasic, M. (2012). Fundamental Optimization Criteria for Green Wireless Communications. *MIPRO*, *2012*, 734–736.

Pejanovic-Djurisic, M., Stovrag, E., & Ilic-Delibasic, M. (2012). *Fundamental optimization criteria for Green wireless communications. In MIPRO* (pp. 733–736). Opatija: IEEE.

Peng, M., Wang, C., Li, J., Xiang, H., & Lau, V. (2015). Recent advances in underlay heterogeneous networks: Interference control, resource allocation, and self-organization. *IEEE Communications Surveys and Tutorials*, *17*(2), 700–729. doi:10.1109/COMST.2015.2416772

Peng, M., & Wang, W. (2010). Technologies and standards fortd-scdma evolutions to imt-advanced. *IEEE Communications Magazine*, *47*(12), 50–58. doi:10.1109/MCOM.2009.5350368

Perabathini, B., Bastug, E., Kountouris, M., Debbah, M., & Conte, A. (2015, June). *Caching at the edge: a green perspective for 5G networks*. Paper presented at the meeting of IEEE International Conference on Communications, London, UK. doi:10.1109/ICCW.2015.7247608

Persson, D., Eriksson, T., & Larsson, E. G. (2013). Amplifier-Aware Multiple-Input Multiple-Output Power Allocation. *Communications Letters, IEEE*, *17*(6), 1112–1115. doi:10.1109/LCOMM.2013.043013.130050

Pervaiz, H., Musavian, L., Ni, Q., & Ding, Z. (2015). Energy and spectrum efficient transmission techniques under qos constraints towards green heterogeneous networks. *IEEE Access*, *3*, 1655–1671. doi:10.1109/ACCESS.2015.2479406

Peters, S. W., & Heath, R. W. (2009). Interference alignment via alternating minimization. In *Proceedings of International Conference on Acoustics, Speech and Signal Processing*. Taipei: IEEE.

Peters, S. W., & Heath, R. W. (2012). User partitioning for less overhead in MIMO interference channels. *IEEE Transactions on Wireless Communications*, *11*(2), 592–603. doi:10.1109/TWC.2011.122211.101188

Pingyod, A., & Somchit, Y. (2015, May). *Content updating method in femtocaching*. Paper presented at the meeting of IEEE International Joint Conference Computer Science and Software Engineering, Chonburi, Thailand.

Pollin, S., Mangharam, R., Bougard, B., Van den Perre, L., Moerman, I., & Rajkumar, R. (2007). MEERA: Cross-Layer Methodology for Energy Efficient Resource Allocation in Wirelles Networks. *IEEE Transactions on Wireless Communications*, *6*(2), 617–628. doi:10.1109/TWC.2007.05356

Pollin, S., Mangharam, R., Bourgard, B., Moerman, I., Rajkumar, R., & Catthoor, F. (2008). MEERA: Cross-Layer Methodology for Energy Efficient Resource Allocation in Wireless Network. *IEEE Transactions on Wireless Communications*, *7*(1), 98–109. doi:10.1109/TWC.2008.05356

Poularakis, K., Iosifidis, G., Sourlas, V., & Tassiulas, L. (2016). Exploiting caching and multicast for 5G wireless networks. *IEEE Transactions on Wireless Communications,* (99), 1–13.

Poularakis, K., Iosifidis, G., Sourlas, V., & Tassiulas, L. (2014). Approximation algorithms for mobile data caching in small cell networks. *IEEE Transactions on Communications*, *62*(10), 3665–3677. doi:10.1109/TCOMM.2014.2351796

Qian, L. P., Zhang, Y. J. A., Huang, J., & Wu, Y. (2013). Demand response management via real-time electricity price control in smart grids. *IEEE Journal on Selected Areas in Communications*, *31*(7), 1268–1280. doi:10.1109/JSAC.2013.130710

Qian, L., Li, K. H., & The, K. C. (2014). Achieving Optimal Diversity-Multiplexing Tradeoff for Full-Duplex MIMO Multihop Relay Networks. *IEEE Transactions on Information Theory*, *57*(1), 303–316.

Qualcomm. (2011). *LTE advanced: heterogenous networks*. Author.

Quek, T. Q. S., Cheung, W. C., & Kountouris, M. (2011). *Energy Efficiency Analysis of Two-Tier Heterogeneous Networks*. Paper presented at the Wireless Conference 2011 - Sustainable Wireless Technologies (European Wireless), 11th European.

R. D. Yates. (1995). A framework for uplink power control in cellular radio systems. *Selected Areas in Communications IEEE Journal on, 13*.

Rahman, M., Yanikomeroglu, H., & Wong, W. (2009). Interference Avoidance with Dynamic Inter-Cell Coordination for Downlink LTE System. In *Proceedings of IEEE Wireless Communications & Networking Conference*. Budapest, Hungary: IEEE. doi:10.1109/WCNC.2009.4917761

Rao, X., Ruan, L., & Lau, V. K. N. (2013a). Csi feedback reduction for mimo interference alignment. *IEEE Transactions on Signal Processing*, *61*(18), 4428–4437. doi:10.1109/TSP.2013.2269902

Rao, X., Ruan, L., & Lau, V. K. N. (2013b). Limited feedback design for interference alignment on mimo interference networks with heterogeneous path loss and spatial correlations. *IEEE Transactions on Signal Processing*, *61*(10), 2598–2607. doi:10.1109/TSP.2013.2252168

Rappaport, T. S., Sun, S., Mayzus, R., Zhao, H., Azar, Y., & Wang, K. et al. (2013). Millimeter Wave Mobile Communications for 5G Cellular: It Will Work! *IEEE Access*, *1*, 335–349. doi:10.1109/ACCESS.2013.2260813

Rasti, M., Sharafat, A. R., & Seyfe, B. (2009). Pareto-efficient and goal-driven power control in wireless networks: A game-theoretic approach with a novel pricing scheme. *IEEE/ACM Transactions on Networking*, *17*(2), 556–569. doi:10.1109/TNET.2009.2014655

Ren, S. (2014). Dynamic scheduling and pricing in wireless cloud computing. *Skull Base Surgery*, *13*(10), 2283–2292.

Ren, S., & Schaar, M. V. D. (2011). Pricing and distributed power control in wireless relay networks. *Signal Processing IEEE Transactions on*, *59*(6), 2913–2926. doi:10.1109/TSP.2011.2121064

Riedel, I., & Fettweis, G. (2011). *Increasing throughput and fairness in the downlink of cellular systems with N-fold sectorization.* Paper presented at the GLOBECOM Workshops (GC Wkshps), 2011 IEEE. doi:10.1109/GLOCOMW.2011.6162429

Rittenhouse, G., Goya, S. N., & Samuel, S. (2011). Sustainable Telecommunications.*Tecnical Symposium at ITU* (pp. 19-23). Geneva: IEEE.

Ronasi, K., Niu, B., Wong, V. W. S., Gopalakrishman, S., & Schober, R. (2014). Throughput-Efficient Scheduling and Interference Alignment for MIMO Wireless Systems. *IEEE Transactions on Wireless Communications*, *13*(4), 1779–1789. doi:10.1109/TWC.2014.031314.122040

Rose, L., Belmega, E. V., Saad, W., & Debbah, M. (2014). Pricing in Heterogeneous Wireless Networks: Hierarchical Games and Dynamics. *Wireless Communications. IEEE Transactions*, *13*(9), 4985–5001.

Roth, A. E., & Sotomayor, M. (1991). *Two sided matching: A study in game-theoretic modeling and analysis*. Cambridge, UK: Cambridge University Press.

Rusek, F., Persson, D., Lau, B. K., & Larsson, E. G. et al.. (2013). Scaling up MIMO: Opportunities and challenges with very large arrays. *IEEE Signal Processing Magazine*, *30*(1), 40–60. doi:10.1109/MSP.2011.2178495

Sabharwal, A., Schniter, P., Guo, D., Bliss, D. W., Rangarajan, S., & Wichman, R. (2014). In-band full-duplex wireless: Challenges and opportunities. *IEEE Journal on Selected Areas in Communications*, *32*(9), 1637–1652. doi:10.1109/JSAC.2014.2330193

Sallakh, U., Mwanje, S. S., & Mitschele-Thiel, A. (2014). Multi-parameter Q-Learning for downlink inter-cell interference coordination in LTE SON.*IEEE Symposium on Computers and Communication (ISCC)*. doi:10.1109/ISCC.2014.6912599

Santella, G., & Mazzenga, F. (1998). A hybrid analytical-simulation procedure for performance evaluation in M-QAM-OFDM schemes in presence of nonlinear distortions. *IEEE Trans. Veh. Tech.*, *47*(2), 142–151. doi:10.1109/25.661041

Santhi, K. R., Srivastava, V. K., Senthilkumaran, G., & Butare, A. (2003). Goals Of True Broad band's Wireless Next Wave (4G-5G). In *Proceedings of IEEE Vehicular Technology Conference*. Orlando, FL: IEEE.

Saquib, N., Hossain, E., Le, L. B., & Kim, D. I. (2012). Interference management in ofdma femtocell networks: Issues and approaches. *IEEE Wireless Communications*, *19*(3), 86–95. doi:10.1109/MWC.2012.6231163

Saraydar, C. U., Mandayam, N. B., & Goodman, D. J. (2001). Efficient power control via pricing in wireless data networks. *IEEE Transactions on Communications*, *50*(2), 291–303. doi:10.1109/26.983324

Sathya, V., Gudivada, H. V., Narayanam, H., Krishna, B. M., & Tamma, B. R. (2013). Enhanced distributed resource allocation and interference management in LTE femtocell networks. *2013 IEEE 9th International Conference on Wireless and Mobile Computing, Networking and Communications (WiMob).*

Schmidt, D. A., Shi, C., Berry, R. A., Honig, M. L., & Utschick, W. (2009). *Distributed Resource Allocation Schemes-Pricing algorithms for power control and beamformer design in interference networks.* Retrieved from http:www.ece.northwestern.edurberryPubsSPmag_09.pdf

Self Optimizing Networks. (n.d.). Retrieved June 26, 2016, from http://www.amdocs.com/products/centralized-management-optimization-solution/pages/self-optimizing-networks.aspx

Semasinghe, P., Zhu, K., & Hossain, E. (2013). Distributed resource allocation for self-organizing small cell networks: An evolutionary game approach. *IEEE Globecom Workshops (GC Wkshps).*

Semasinghe, P., Hossain, E., & Zhu, K. (2015). An evolutionary game for distributed resource allocation in self-organizing small-cells. *IEEE Transactions on Mobile Computing*, *14*(2), 274–287. doi:10.1109/TMC.2014.2318700

Sesia, S., Toufik, I., & Baker, M. (2009). *LTE: The UMTS Long Term Evolution: From Theory to Practice*. New York: Wiley Publishing. doi:10.1002/9780470742891

Shahid, A., Aslam, S., Sohaib, S., Kim, H. S., & Lee, K.-G. (2014). A self-organized metaheuristic approach towards inter-cell interference management for LTE-Advanced. *EURASIP Journal on Wireless Communications and Networking*, 2–15.

Shang, C. Y. A., Smith, P. J., Woodward, G. K., & Suraweera, H. A. (2014). Linear transceivers for full duplex MIMO relays. *in Proc.Australian Communications Theory Workshop (AusCTW 2014).* doi:10.1109/AusCTW.2014.6766420

Shanmugam, K., Golrezaei, N., Dimakis, A. G., Molisch, A. F., & Caire, G. (2013). Femtocaching: Wireless content delivery through distributed caching helpers. *IEEE Transactions on Information Theory*, *59*(12), 8402–8413. doi:10.1109/TIT.2013.2281606

Shen, K., & Yu, W. (2014). Distributed pricing-based user association for downlink heterogeneous cellular networks. *Selected Areas in Communications IEEE Journal on*, *32*(6), 1100–1113. doi:10.1109/JSAC.2014.2328143

Shetty, N., Parekh, S., & Walrand, J. (2009). Economics of femtocells. *IEEE Conference on Global Telecommunications* (pp.1-6). IEEE Press.

Shi, C., Wang, Y., Wang, T., & Zhang, T. (2011). *Cross-layer design for interference-limited spectrum sharing systems with heterogeneous QoS*. Paper presented at the meeting of the IEEE Global Telecommunications Conference, Houston, TX.

Shi, C., Schmidt, D. A., Berry, R. A., Honig, M. L., & Utschick, W. (2009). Distributed interference pricing for the MIMO interference channel. *Proceedings of the 2009 IEEE international conference on Communications* (pp.1-5). IEEE Press. doi:10.1109/ICC.2009.5198927

Shim, J. N., Min, B. Y., Kim, K., Jang, J., & Kim, D. K. (2014, May). *Advanced femto-caching file placement technique for overlapped helper coverage*. Paper presented at the meeting of IEEE Vehicular Technology Conference, Seoul, South Korea.

Shin, W., Noh, W., Jang, K., & Choi, H. H. (2012). Hierarchical interference alignment for downlink heterogeneous networks. *Wireless Communications IEEE Transactions on*, *11*(12), 4549–4559. doi:10.1109/TWC.2012.101912.120421

Shomorony, I., & Avestimehr, A. S. (2014). Degrees of Freedom of Two-Hop Wireless Networks: Everyone Gets the Entire Cake. *IEEE Transactions on Information Theory*, *60*(5), 2417–2431. doi:10.1109/TIT.2014.2308172

Silverstein, J. W., & Bai, Z. D. (1995). On the empirical distribution of eigenvalues of a class of large dimensional random matrices. *Journal of Multivariate Analysis*, *54*(2), 175–192. doi:10.1006/jmva.1995.1051

Simeone, O., Stanojev, I., Savazzi, S., Bar-Ness, Y., Spagnolini, U., & Pickholtz, R. (2008). Spectrum Leasing to Cooperating Secondary Ad Hoc Networks. *Selected Areas in Communications. IEEE Journal*, *26*(1), 203–213.

Simsek, M., & Czylwik, A. (2012). Improved decentralized fuzzy Q-learning for interference reduction in heterogeneous LTE-networks. *17th Proceedings of International OFDM Workshop 2012 (InOWo'12).*

Singh, R., O'Farrell, T., & David, J. P. R. (2013). *Performance evaluation of IEEE 802.15.7 CSK physical layer*. Paper presented at the 2013 IEEE Globecom Workshops (GC Wkshps).

Singh, R., O'Farrell, T., & David, J. P. R. (2014). An Enhanced Color Shift Keying Modulation Scheme for High-Speed Wireless Visible Light Communications. *Journal of Lightwave Technology*, *32*(14), 2582–2592. doi:10.1109/JLT.2014.2328866

Single, S. O. N. (n.d.). Retrieved June 26, 2016, from http://www.huawei.com/ilink/en/solutions/broader-smarter/HW_104774

Sithamparanathan, K., Nardis, L. D., Benedetto, M. G., Guidotti, A., Corazza, G., & Corazza, G. E. (2010). Cognitive satellite terrestrial radio. In *2010 IEEE Global Telecommunications Conference* (pp. 1-6).

Skubic, B., De Betou, E. I., Ayhan, T., & Dahlfort, S. (2012). Energy-efficient next-generation optical access networks. *Communications Magazine, IEEE*, *50*(1), 122–127. doi:10.1109/MCOM.2012.6122542

Soh, Y. S., Quek, T. Q. S., Kountouris, M., & Shin, H. (2013). Energy Efficient Heterogeneous Cellular Networks. *IEEE Journal on Selected Areas in Communications*, *31*(5), 840–850. doi:10.1109/JSAC.2013.130503

SON (Self-Organizing Networks) Ecosystem. (2015). *2015 – 2030 – Opportunities, Challenges, Strategies & Forecasts. Market Research Report. Signals and Systems Telecom.* Retrieved June 26, 2016, from http://www.snstelecom.com/son

Song, J., Song, H., & Choi, W. (2015, June). *Optimal caching placement of caching system with helpers*. Paper presented at the meeting of IEEE International Conference on Communications, London, UK. doi:10.1109/ICC.2015.7248590

Soret, B., Pedersen, K. I., Jorgensen, N. T. K., & Fernandez-Lopez, V. (2015). Interference coordination for dense wireless networks. *IEEE Communications Magazine*, *53*(1), 102–109. doi:10.1109/MCOM.2015.7010522

Studer, C., & Larsson, E. (2013). PAR-aware large-scale multi-user MIMO OFDM downlink. *IEEE Journal on Selected Areas in Communications*, *31*(2), 303–313. doi:10.1109/JSAC.2013.130217

Su, M., & van der Schaar, M. (2009). A new perspective on multi-user power control games in interference channels. *Wireless Communications. IEEE Transactions*, *8*(6), 2910–2919.

Su, M., & van der Schaar, M. (2009). From competition to coopetition: Stackelberg equilibrium in multi-user power control games. *Game Theory for Networks*, *2009*, 107–116.

Sundaresan, K., & Rangarajan, S. (2009). Efficient resource management in OFDMA Femto cells.*ACM Interational Symposium on Mobile Ad Hoc NETWORKING and Computing*, (pp. 33-42). doi:10.1145/1530748.1530754

Sun, W., Zhang, P., Chen, Y., Qin, Z., & Teng, D. (2013). A tight coupling cooperation scheme in WiFi/WiMAX heterogeneous mesh networks. *International Journal of Computers, Communications & Control*, *8*(3), 460–468. doi:10.15837/ijccc.2013.3.10

Suraweera, H. A., Hien Quoc, N., & Duong, T. Q. et al. (2013). Multi-pair amplify-and-forward relaying with very large antenna arrays. *Proceedings of the IEEE, ICC*, 4635–4640.

Suraweera, H. A., Krikidis, I., Zheng, G., Yuen, C., & Smith, P. J. (2014). Low-complexity end-to-end performance optimization in MIMO full-duplex relay systems. *IEEE Transactions on Wireless Communications*, *13*(2), 913–927. doi:10.1109/TWC.2013.122313.130608

Su, Z., & Xu, Q. (2015). Content distribution over content centric mobile social networks in 5G. *IEEE Communications Magazine*, *53*(6), 66–72. doi:10.1109/MCOM.2015.7120047

Suzuki, H., Tran, T., & Collings, I. B. et al. (2008). Transmitter noise effect on the performance of a MIMO-OFDM hardware implementation achieving improved coverage. *IEEE Journal on Selected Areas in Communications*, *26*(8), 867–876. doi:10.1109/JSAC.2008.080804

Tall, A., Altman, Z., & Altman, E. (2014). Self organizing strategies for enhanced ICIC (eICIC).*12th International Symposium on Modeling and Optimization in Mobile, Ad Hoc, and Wireless Networks (WiOpt)*. doi:10.1109/WIOPT.2014.6850315

Tan, C. W., Palomar, D. P., & Chiang, M. (2006). Distributed optimization of coupled systems with applications to network utility maximization. In *Proc. IEEE International Conference on Acoustics, Speech, and Signal Processing (ICASSP)*.

Tanbourgi, R., Singh, S., Andrews, J. G., & Jondral, F. K. (2014). Analysis of Non-Coherent Joint-Transmission Cooperation in Heterogeneous Cellular Networks. In P*roceedings of IEEE International Conference on Communications (ICC '2014)*. Sydney, Australia: IEEE. doi:10.1109/ICC.2014.6884140

Tawil, C., & Tawil, S. (2009). *Apparatus and method for transmitting terrestrial signals on a common frequency with satellite transmissions*. USA: 6892050 B1.

Telatar, E. (1999). Capacity of multi-antenna Gaussian channels. *European Trans. Telecom.*, *10*(6), 585–595. doi:10.1002/ett.4460100604

Tembine, H. (2012). *Distributed strategic learning for wireless engineers*. CRC Press. doi:10.1201/b11896

Thompson, J. S., Grant, P. M., Fletcher, S., & O'Farrell, T. (2012). Minimising Power Consumption to Achieve More Efficient Green Cellular Radio Base Station Designs. In H. Zhang (Ed.), *Green Communications: Theoretical Fundamentals, Algorithms and Applications* (pp. 555–578). CRC Press. doi:10.1201/b13083-22

Tombaz, S., Monti, P., Farias, F., Fiorani, M., Wosinska, L., & Zander, J. (2014). *Is backhaul becoming a bottleneck for green wireless access networks?* Paper presented at the Communications (ICC), 2014 IEEE International Conference on.

Tombaz, S., Usman, M., & Zander, J. (2011). *Energy efficiency improvements through heterogeneous networks in diverse traffic distribution scenarios*. Paper presented at the Communications and Networking in China (CHINACOM), 2011 6th International ICST Conference on.

Tran, N. H., Hong, C. S., Han, Z., & Lee, S. (2013). Optimal Pricing Effect on Equilibrium Behaviors of Delay-Sensitive Users in Cognitive Radio Networks. *Selected Areas in Communications. IEEE Journal*, *31*(11), 2566–2579.

Tran, N. H., Hong, C. S., Han, Z., & Lee, S. (2013). Optimal pricing effect on equilibrium behaviors of delay-sensitive users in cognitive radio networks. *IEEE Journal on Selected Areas in Communications*, *31*(11), 2566–2579. doi:10.1109/JSAC.2013.131134

Tran, N. H., Le, L. B., Ren, S., Han, Z., & Hong, C. S. (2014). Joint pricing and load balancing for cognitive spectrum access: Non-cooperation vs cooperation. *IEEE Journal on Selected Areas in Communications*.

Trend. (2016, March 19). Retrieved from Trend Toward Real Energy-efficient Network Design: www.fp7-trend.eu

Tresch, R., Guillaud, M., & Riegler, E. (2009). A clustered alignment-based interference management approach for large OFDMA cellular networks. In *Joint NEWCOM++/COST2100 Workshop on Radio Resource Allocation for LTE.*

Tresch, R., & Guillaud, M. (2009). Clustered interference alignment in large cellular networks. In *Proceedings of International Symposium on Personal, Indoor and Mobile Radio Communications*. Tokyo: IEEE. doi:10.1109/PIMRC.2009.5450107

Tresch, R., & Guillaud, M. (2010). Performance of interference alignment in clustered wireless ad hoc networks. In *Proceedings of International Symposium on Information Theory*. Austin, TX: IEEE. doi:10.1109/ISIT.2010.5513243

Van, W. F., Janssen, G. M. J., & Prasad, R. (1995). Groupwise successive interference cancellation in a DS/CDMA system. In *Proceedings of Personal, Indoor and Mobile Radio Communications (PIMRC '95),Sixth IEEE International Symposium on Wireless: Merging onto the Information Superhighway*. Toronto, Canada: IEEE.

Vanghi, V., Damnjanovic, A., Harte, L., & Vojcic, B. (2004). *The cdma2000 System for Mobile Communications: 3G Wireless Evolution.* Retrieved from http://www.openisbn.org/download/0131416014.pdf?name=The_Cdma2000_System_For_Mobile_Communications_3G_W

Wagner, S., Couillet, R., Debbah, M., & Slock, D. T. M. (2012). Large system analysis of linear precoding in correlated MISO broadcast channels under limited feedback. *IEEE Transactions on Information Theory*, *58*(7), 4509–4537. doi:10.1109/TIT.2012.2191700

Wang, T., Song, L., & Han, Z. (2015, March). *Dynamic femtocaching for mobile users*. Paper presented at the meeting of IEEE Wireless Communications and Networking Conference, New Orleans, LA.

Wang, X., Zheng, W., Liu, J., Li, W., & Wen, X. (2013). Distributed power self-optimization with convex pricing in dense femtocell networks via an exact potential game. *Personal Indoor and Mobile Radio Communications (PIMRC), 2013 IEEE 24th International Symposium on* (pp. 1919-1923). IEEE.

Wang, Y., Wang, X., & Wang, L. (2014). Low-complexity Stackelberg Game Approach for Energy-efficient Resource Allocation in Heterogeneous Networks. *Communications Letters, IEEE*, (99), 1-1.

Wang, Z., Jiang, L., & He, C. (2014). Optimal Price-based Power Control Algorithm in Cognitive Radio Networks. *Wireless Communications, IEEE Transactions*, (99), 1-1.

Wang, B. B., Han, Z., & Liu, K. J. R. (2006). WLC41-4: Stackelberg Game for Distributed Resource Allocation over Multiuser Cooperative Communication Networks. *Global Telecommunications Conference, 2006. GLOBECOM '06. IEEE.*

Wang, B. B., Han, Z., & Liu, K. J. R. (2009). Distributed Relay Selection and Power Control for Multiuser Cooperative Communication Networks Using Stackelberg Game. *Mobile Computing. IEEE Transactions*, *8*(7), 975–990.

Wang, C. X., Haider, F., Gao, X., You, X. H., Yang, Y., Yuan, D., & Hepsaydir, E. et al. (2014). Cellular architecture and key technologies for 5G wireless communication networks. *IEEE Communications Magazine*, *52*(2), 122–130. doi:10.1109/MCOM.2014.6736752

Wang, F., Krunz, M., & Cui, S. (2007). Price-based spectrum management in cognitive radio networks. *IEEE Journal of Selected Topics in Signal Processing*, *2*(1), 70–78.

Wang, H. X., Pan, C. S., & Song, J. H. (2013). *Spaceborne smart antenna beamforming technology*. Beijing, China: Defense Industry Press.

Wang, K., Chen, Z., & Liu, H. (2014). Push-based wireless converged networks for massive multimedia content delivery. *IEEE Transactions on Wireless Communications*, *13*(5), 2894–2905. doi:10.1109/TWC.2014.040914.131422

Wang, Q., Ren, K., & Meng, X. (2012). When cloud meets eBay: Towards effective pricing for cloud computing. *Proceedings of the IEEE*, *131*, 936–944.

Wang, W., Liao, Q., & Zhang, Q. (2014). COD: A cooperative cell outage detection architecture for self organizing femtocell networks. *IEEE Transactions on Wireless Communications*, *13*(11), 6007–6014. doi:10.1109/TWC.2014.2360865

Wang, W., & Zhang, Q. (2014). Local cooperation architecture for self healing femtocell networks. *IEEE Wireless Communications.*, *21*(2), 42–49. doi:10.1109/MWC.2014.6812290

Wang, X., Ma, K., Han, Q., Liu, Z., & Guan, X. (2012). Pricing-based spectrum leasing in cognitive radio networks. *Networks, IET*, *1*(3), 116–125. doi:10.1049/iet-net.2012.0149

Wang, Z. Q., Jiang, L. G., & He, C. (2013). A Novel Price-Based Power Control Algorithm in Cognitive Radio Networks. *Communications Letters, IEEE*, *17*(1), 43–46. doi:10.1109/LCOMM.2012.120612.121587

Wang, Z., Jiang, L., & He, C. (2014). Optimal price-based power control algorithm in cognitive radio networks. *IEEE Transactions on Wireless Communications*, *13*(11), 5909–5920. doi:10.1109/TWC.2014.2318708

Wang, Z., Xiong, W., Dong, C., Wang, J., & Li, S. (2011). A novel downlink power control scheme in LTE heterogeneous network.*International Conference on Computational Problem-Solving (ICCP)*. doi:10.1109/ICCPS.2011.6089792

Weber, S., Andrews, J. G., & Jindal, N. (2007). The effect of fading, channel inversion, and threshold scheduling on ad hoc networks. *IEEE Transactions on Information Theory*, *53*(11), 4127–4149. doi:10.1109/TIT.2007.907482

Weeraddana, P. C., Codreanu, M., Latva-aho, M., & Ephremides, A. (2011). Resource allocation for cross-layer utility maximization in wireless networks. *IEEE Trans. Veh. Tech.*, *60*(10), 2790–2809. doi:10.1109/TVT.2011.2157544

Wei, N., & Collings, I. B. (2013). A new adaptive small-cell architecture. *IEEE Journal on Selected Areas in Communications*, *31*(5), 829–839. doi:10.1109/JSAC.2013.130502

West, D. (2001). *Introduction to Graph Theory* (2nd ed.). Upper Saddle River, NJ: Prentice Hall.

Wie, Y., Song, M., Liu, B., Wang, X., & Li, Y. (2011). Energy Efficient Cooperative Relaying and Cognitive Radio Technologies to Deliver Green Communication. In *Pervasive Computing and Applications (ICPCA), 2011 6th International Conference* (pp. 105-109). Port Elizabeth: IEEE.

Williams, E., Kahhat, R., Allenby, B., Kavazanjian, E., Kim, J., & Xu, M. (2008). Environmental, Social, and Economic Implications of Global Reuse and Recycling of Personal Computers. *Environmental Science & Technology*, *42*(17), 6446–6454. doi:10.1021/es702255z PMID:18800513

wireless@kth. (2016, March 16). Retrieved from https://wireless.kth

Wu, D., Zhang, L., & Zhou, J. (2015). *Self-organized spectrum access in small-cell networks: A noncooperation interference minimization game solution*. Paper presented at the WCSP Conference, Nanjing, China. doi:10.1109/WCSP.2015.7341204

Wu, Y., Jiang, H., Wu, H., & Zhang, D. (2010). Physical cell identity self-organization for home eNodeB deployment in LTE. *2010 6th International Conference on Wireless Communications Networking and Mobile Computing (WiCOM).*

Wu, D., Wu, Q., & Xu, Y. (2016). Self-organized Spectrum Access in Small-cell Networks with Dynamic Loads. *Transactions on Internet and Information Systems (Seoul), 10*(5), 1976–1997.

Wu, J. (2012). Green Wireless Communications: From concept to reality (Industry Perspectives). *IEEE Wireless Communications, 19*(4), 4–5. doi:10.1109/MWC.2012.6272415

Wu, J., Zhang, Y., Zukerman, M., & Yung, E. K.-N. (2015). Energy-Efficient Base-Station Sleep-Mode Techniques in Green Cellular Networks: A Survey. *IEEE Communications Surveys and Tutorials, 17*(2), 803–826. doi:10.1109/COMST.2015.2403395

Wu, W. L., & Niu, K. (2005). *Principle of Mobile communication.* Beijing, China: Electronics Industry Press.

Wu, Y. L., & Liu, K. J. R. (2011). An Information Secrecy Game in Cognitive Radio Networks. *Information Forensics and Security. IEEE Transactions, 6*(3), 831–842.

Wu, Y., & Song, W. Z. (2014). Cooperative resource sharing and pricing for proactive dynamic spectrum access via nash bargaining solution. *IEEE Transactions on Parallel and Distributed Systems, 25*(11), 2804–2817. doi:10.1109/TPDS.2013.285

Wu, Y., Zhang, T. Z., & Tsang, D. H. K. (2011). Joint Pricing and Power Allocation for Dynamic Spectrum Access Networks with Stackelberg Game Model. *Wireless Communications. IEEE Transactions, 10*(1), 12–19.

Wu, Y., Zhu, Q., Huang, J., & Tsang, D. H. K. (2014). Revenue sharing based resource allocation for dynamic spectrum access networks. *IEEE Journal on Selected Areas in Communications, 32*(11), 2280–2296. doi:10.1109/JSAC.2014.141125

Xiangfang, L., Lijun, Q., & Kataria, D. (2009). Downlink power control in co-channel macrocell femtocell overlay.*Annual Conference on Information Sciences and Systems (CISS).*

Xiao, Y., Bi, G. A., Niyato, D., & DaSilva, L.A. (2012). A Hierarchical Game Theoretic Framework for Cognitive Radio Networks. *Selected Areas in Communications, IEEE Journal, 30*(10), 2053,-2069.

Xiao, L., Fuja, T. E., & Costello, D. J. (2010). Mobile Relaying: Coverage Extension and Throughput Enhancement. *IEEE Transactions on Communications, 58*(9), 2709–271. doi:10.1109/TCOMM.2010.080310.090199

Xiao, Y., Niyato, D., Han, Z., & Chen, K. C. (2014). Secondary Users Entering the Pool: A Joint Optimization Framework for Spectrum Pooling. *Selected Areas in Communications. IEEE Journal, 32*(3), 572–588.

Xia, X., Xie, W., & Zhang, D. et al. (2015). Multi-pair full-duplex amplify-and-forward relaying with very large antenna arrays. In *Proc. IEEE WCNC.*

Xie, R. C., Yu, F. R., Ji, H., & Li, Y. (2012). Energy-Efficient Resource Allocation for Heterogeneous Cognitive Radio Networks with Femtocells. *Wireless Communications. IEEE Transactions, 11*(11), 3910–3920.

Xie, R., Yu, F. R., Ji, H., & Li, Y. (2012). Energy-efficient resource allocation for heterogeneous cognitive radio networks with femtocells. *IEEE Transactions on Wireless Communications, 11*(11), 3910–3920. doi:10.1109/TWC.2012.092112.111510

Xu, S. Y., Han, J., & Chen, T. (May, 2012). *Enhanced Inter-Cell Interference Coordination in Heterogeneous Networks for LTE-Advanced.* Paper presented at the meeting of the Vehicular Technology Conference, Yokohama, Japan. doi:10.1109/VETECS.2012.6240203

Xu, C., Sheng, M., Yang, C., Wang, X., & Wang, L. (2014). Pricing-based multiresource allocation in ofdma cognitive radio networks: An energy efficiency perspective. *IEEE Transactions on Vehicular Technology*, *63*(5), 2336–2348. doi:10.1109/TVT.2013.2280617

Xu, J., Wang, J., Zhu, Y., Yang, Y., Zheng, X., & Wang, S. et al.. (2014). Cooperative distributed optimization for the hyper-dense small-cell deployment. *IEEE Communications Magazine*, *52*(5), 61–67. doi:10.1109/MCOM.2014.6815894

Xu, Y., He, C., Jiang, L., & Li, J. (2013). Distributed dynamic SINR pricing for multi-cell beamforming with limited backhaul signaling. *Global Communications Conference (GLOBECOM)*, 2013IEEE (pp.3620-3624).

Xu, Y., & Mao, S. W. (2014). Stackelberg Game for Cognitive Radio Networks With MIMO and Distributed Interference Alignment. *Vehicular Technology. IEEE Transactions*, *63*(2), 879–892.

Xu, Y., Wang, J., Wu, Q., Zheng, J., Shen, L., & Anpalagan, A. (2015). Dynamic spectrum access with statistical QoS provisioning: A distributed learning approach beyond expectation optimization. *Computer Science*, (2): 174–175.

Xu, Y., Wu, Q., Shen, L., Wang, J., & Anpalagan, A. (2013). Opportunistic spectrum access with spatial reuse: Graphical game and uncoupled learning solutions. *IEEE Transactions on Wireless Communications*, *12*(10), 4814–4826. doi:10.1109/TWC.2013.092013.120862

Xylomenos, G., Vasilakos, X., Tsilopoulos, C., Siris, V. A., & Polyzos, G. C. (2012). Caching and mobility support in a publish-subscribe internet architecture. *IEEE Communications Magazine*, *50*(7), 52–58. doi:10.1109/MCOM.2012.6231279

Yaacoub, E. (2015). Interference mitigation in femtocell networks with joint channel sensing and resource allocation.*2015 IEEE Wireless Communications and Networking Conference (WCNC).* doi:10.1109/WCNC.2015.7127569

Yaïche, H., Mazumdar, R. R., & Rosenberg, C. (2000). A game theoretic framework for bandwidth allocation and pricing in broadband networks. *IEEE/ACM Transactions on Networking*, *8*(5), 667–678. doi:10.1109/90.879352

Yamada, K., Sakai, Y., Suzuki, T., Kawahara, Y., Asami, T., & Aida, H. (2010, September). *A Communication System with a Fast Handover under a High Speed Mobile Environment.* Paper presented at the meeting of the IEEE 72nd Vehicular Technology Conference, Ottawa, Canada. doi:10.1109/VETECF.2010.5594584

Yang, C., & Li, J. (2014). Interference mitigation and energy management in heterogeneous networks: a cognitive radio perspective. *Convergence of Broadband Broadcast & Cellular Network Technologies.*

Yang, C., Guo, K., Sheng, M., Li, J., & Yue, J. (2014). *Energy-efficient capacity offload to smallcells with interference compensation.* Paper presented at the 2014 IEEE Wireless Communications and Networking Conference (WCNC). doi:10.1109/WCNC.2014.6952513

Yang, H., Zhang, X.N., & Meng, Q. (2007). *Stackelberg games and multiple equilibrium behaviors on networks.* Academic Press.

Yang, L., Kim, H., Zhang, J.S., Chiang, M., & Tan, C.W. (2013). Pricing-Based Decentralized Spectrum Access Control in Cognitive Radio Networks. *Networking, IEEE/ACM Transactions*, *21*(2), 522-535.

Yang, C. (2015). *Advanced Cognitive Radio-Enabled Spectrum Management.* IGI Global. doi:10.4018/978-1-4666-6571-2.ch018

Yang, C., & Li, J. (2010). *Joint Economical and Technical Consideration of Dynamic Spectrum Sharing: A Multi-Stage Stackelberg Game Perspective. In Vehicular Technology Conference Fall (VTC 2010-Fall), 2010 IEEE 72nd* (pp. 1–5). IEEE.

Yang, C., & Li, J. (2013). Pricing-based dynamic spectrum leasing: A hierarchical multi-stage stackelberg game perspective. *IEICE Transactions on Communications*, *96*(6), 1511–1521. doi:10.1587/transcom.E96.B.1511

Yang, C., & Li, J. (2014). Cooperative Spectrum Leasing to Femtocells with Interference Compensation.*Vehicular Technology Conference (VTC Spring), 2014 IEEE 79th* (pp.1-6). IEEE. doi:10.1109/VTCSpring.2014.7022865

Yang, D. J., Fang, X., & Xue, G. L. (2012). Game theory in cooperative communications. *Wireless Communications, IEEE*, *19*(2), 44–49. doi:10.1109/MWC.2012.6189412

Yang, L., Kim, H., Zhang, J., Chiang, M., & Tan, C. W. (2013). Pricing-based decentralized spectrum access control in cognitive radio networks. *IEEE/ACM Transactions on Networking*, *21*(2), 522–535. doi:10.1109/TNET.2012.2203827

Yang, Y. F., & Quek, T. Q. S. (2014). Optimal Subsidies for Shared Small Cell Networks —A Social Network Perspective. *Selected Topics in Signal Processing. IEEE Journal*, *8*(4), 690–702.

Yang, Y., Quek, T. Q. S., & Duan, L. (2014). Backhaul-constrained small cell networks: Refunding and qos provisioning. *IEEE Transactions on Wireless Communications*, *13*(9), 5148–5161. doi:10.1109/TWC.2014.2326401

Yao, H., Zhang, Q., & Niu, Z. (2010, March). *Resource allocation in multi-cell OFDMA-based relay networks*. Paper presented at the meeting of the IEEE INFOCOM, San Diego, CA.

Yinan, Q., Imran, M. A., & Tafazolli, R. (2011). *Energy-aware adaptive sectorisation in LTE systems*. Paper presented at the Personal Indoor and Mobile Radio Communications (PIMRC), 2011 IEEE 22nd International Symposium on.

Yin, H., Gesbert, D., & Cottatellucci, L. (2014). Dealing with interference in distributed large-scale MIMO systems: A statistical approach. *IEEE Journal of Selected Topics in Signal Processing*, *8*(5), 942–953. doi:10.1109/JSTSP.2014.2322583

Yin, R., Yu, G., Zhang, H., Zhang, Z., & Li, G. Y. (2015). Pricing-based interference coordination for d2d communications in cellular networks. *IEEE Transactions on Wireless Communications*, *14*(3), 1519–1532. doi:10.1109/TWC.2014.2368151

Yiu, S., Vu, M., & Tarokh, V. (2009). Interference and noise reduction by beamforming cognitive networks. *IEEE Transactions on Communications*, *57*(10), 3144–3153. doi:10.1109/TCOMM.2009.10.080501

Yi, Y. W., Zhang, J., Zhang, Q., & Jiang, T. (2011). Spectrum Leasing to Multiple Cooperating Secondary Cellular Networks. *Communications (ICC),2011 IEEE International Conference*.

Yuan, F., Yang, C., Wang, G., & Lei, M. (2013). Adaptive channel feedback for coordinated beamforming in heterogeneous networks. *IEEE Transactions on Wireless Communications*, *12*(8), 3980–3994. doi:10.1109/TWC.2013.071913.121553

Yu, H., Gao, L., Li, Z., Wang, X., & Hossain, E. (2010). Pricing for uplink power control in cognitive radio networks. *IEEE Transactions on Vehicular Technology*, *59*(4), 1769–1778. doi:10.1109/TVT.2010.2040492

Yu, H., Man, H. C., Huang, L., & Huang, J. (2014). Predictive delay-aware network selection in data offloading.*Global Communications Conference (GLOBECOM)*. IEEE.

Yun, S. Y., Yi, Y., Cho, D. H., & Mo, J. (2012). The economic effects of sharing femtocells. *IEEE Journal on Selected Areas in Communications*, *30*(3), 595–606. doi:10.1109/JSAC.2012.120409

Zahir, T., Arshad, K., Nakata, A., & Moessner, K. (2013). Interference management in femtocells. *IEEE Communications Surveys and Tutorials*, *15*(1), 293–311. doi:10.1109/SURV.2012.020212.00101

Zhai, X., Zheng, L., & Tan, C. W. (2014). Energy-infeasibility tradeoff in cognitive radio networks: Price-driven spectrum access algorithms. *Selected Areas in Communications IEEE Journal on*, *32*(3), 528–538. doi:10.1109/JSAC.2014.140313

Zhang, D., & Cuthbert, L. (2011). *Dynamic sectorisation based on genetic algorithms for OFDMA networks*. Academic Press.

Zhang, F., Xu, C., & Zhang, Y. (2015, June). *EdgeBuffer: caching and prefetching content at the edge in the mobility first future Internet architecture*. Paper presented at the meeting of IEEE Symposium on a World of Wireless, Mobile and Multimedia Networks, Boston, MA.

Zhang, G. X., Gan, Z. M., & Li, G. X. (2010). Study on the development of China satellite mobile communication. *Satellite & Networks, 94*(3), 32-43.

Zhang, H., Chu, X., & Wen, X. (2013). Differentiated-pricing based power allocation in dense femtocell networks. *Springerbriefs in Computer Science*, 599-603.

Zhang, J., & Zhang, Q. (2009). Stackelberg Game for Utility-Based Cooperative Radio Networks. *Mobile and Ad Hoc Networking and Computing*, 23-32.

Zhang, G. X., & Gan, Z. M. (2005). Study on the development ideas and strategies of China satellite mobile communication system. *Digital Communication World, 95*(7), 24–27.

Zhang, J., Jiang, B. Q., & Zheng, L. (2013). The application of cognitive radio network in the satellite communication. *Journal of Guilin University of Electronic Technology, 33*(4), 284–287.

Zhang, J., Wen, C. K., Jin, S., Gao, X., & Wong, K. (2013). On capacity of large-scale MIMO multiple access channels with distributed sets of correlated antennas. *IEEE Journal on Selected Areas in Communications, 31*(2), 133–148. doi:10.1109/JSAC.2013.130203

Zhang, N., Cheng, N., Lu, N., Zhou, H. B., Mark, J. W., & Shen, X. M. (2014). Risk-Aware Cooperative Spectrum Access for Multi-Channel Cognitive Radio Networks. *Selected Areas in Communications. IEEE Journal, 32*(3), 516–527.

Zhang, Q., Zhu, X., Wu, L., & Sandrasegaran, K. (2013). A coloring-based resource allocation for OFDMA femtocell networks.*2013 IEEE Wireless Communications and Networking Conference (WCNC)*. doi:10.1109/WCNC.2013.6554644

Zhang, T., Chen, W., Han, Z., & Cao, Z. (2014). Hierarchic Power Allocation for Spectrum Sharing in OFDM-Based Cognitive Radio Networks. *Vehicular Technology. IEEE Transactions, 63*(8), 4077–4091.

Zhang, X., & Shin, K. G. (2013). Cooperative carrier signaling: Harmonizing coexisting WPAN and WLAN devices. *IEEE/ACM Transactions on Networking, 21*(2), 426–439. doi:10.1109/TNET.2012.2200499

Zhao, L., Zhao, G., & O'Farrell, T. (2013). *Efficiency metrics for wireless communications*. Paper presented at the 2013 IEEE 24th Annual International Symposium on Personal, Indoor, and Mobile Radio Communications (PIMRC).

Zhao, X., & Chen, P. (2014). Improving UE SINR and networks energy efficiency based on femtocell self-optimization capability.*2014 IEEE Wireless Communications and Networking Conference Workshops (WCNCW)*.

Zhao, Z., Peng, M., Ding, Z., Wang, W., & Chen, H.-H. (2014). Denoise-and-forward network coding for two-way relay mimo systems. *IEEE Transactions on Vehicular Technology, 63*(63), 775–788. doi:10.1109/TVT.2013.2281330

Zheng, L., Cai, Y., & Xu, Y. (2014). *MAC-layer interference mitigation in dynamic and distributed environment: dynamic graphic game with stochastic learning*. Paper presented at the GAMENETS Conference, Beijing, China. doi:10.1109/GAMENETS.2014.7043692

Zheng, G., Krikidis, I., & Ottersten, B. (2013). Full-duplex cooperative cognitive radio with transmit imperfections. *IEEE Transactions on Wireless Communications, 12*(8), 2498–2511. doi:10.1109/TWC.2013.032513.121464

Zheng, W., Zhang, H., Chu, X., & Wen, X. (2013). Mobility robustness optimization in self-organizing LTE femtocell networks. *EURASIP Journal on Wireless Communications and Networking*, *2013*(27).

Zhou, Q., Chen, Z., Gao, F., Li, J. C. F., & Lei, M. (2014). Pricing and power allocation in sensing-based cognitive femtocell networks. *Communications (ICC), 2014 IEEE International Conference on* (Vol. 99, pp. 5342-5347). IEEE.

Zhou, Y., Pan, Z., Hu, J., Shi, J., & Mo, X. (2011, April). *Broadband wireless communications on high speed trains.* Paper presented at the meeting of the 20th Annual Wireless and Optical Communications Conference, Newark, NJ. doi:10.1109/WOCC.2011.5872303

Zhou, S., Gong, J., Zhou, Z., Chen, W., & Niu, Z. (2015). GreenDelivery: Proactive content caching and push with energy harvesting-based small cells. *IEEE Communications Magazine*, *53*(4), 142–149. doi:10.1109/MCOM.2015.7081087

Zhou, T., Chen, Y., Jiang, C., & Liu, K. J. R. (2014). Pricing game for time mute in femto-macro co-existent networks. *IEEE Transactions on Wireless Communications*, *14*, 1–1.

Zhou, X., Feng, S., Ding, Y., Liu, Y., Wang, W., & Zhang, Y. (2013). Game-theoretical frequency reuse method for complex cognitive femto-cell network.*8th International ICST Conference on Communications and Networking in China (CHINACOM).*

Zhou, Z., Dong, M., Ota, K., & Chang, Z. (2015). Energy-efficient context-aware matching for resource allocation in ultra-dense small cells. *IEEE Access*, *3*, 1849–1860. doi:10.1109/ACCESS.2015.2478863

Zhu, H. (2012). Radio resource allocation for OFDMA systems in high speed environments. *IEEE Journal on Selected Areas in Communications*, *30*(4), 748–759. doi:10.1109/JSAC.2012.120509

Zhu, J., & Yang, H.-C. (2011). Interference control with beamforming coordination for two-tier femtocell networks and its performance analysis.*IEEE International Conference on Communications (ICC)*. doi:10.1109/icc.2011.5963130

Zhu, K., Hossain, E., & Niyato, D. (2014). Pricing, spectrum sharing, and service selection in two-tier small cell networks: A hierarchical dynamic game approach. *IEEE Transactions on Mobile Computing*, *13*(8), 1843–1856. doi:10.1109/TMC.2013.96

About the Contributors

Chungang Yang is currently an associated professor in Xidian University, where Dr. Yang leads the research team of "GUIDE, Game, Utility, Intelligent computing Design for Emerging communications". Dr. Yang received his Bachelor and Doctoral Degree in Xidian University, Xi'an, China, in 2006 and 2011, respectively. Between Sep. 2010 and March 2011, he held the visiting scholar position in the department of electrical and computer engineering at Michigan Technological University. Since July 2011, he has been with Xidian University initially as an assistant professor, and now an associated professor since July 2013. His research interests include resource and interference management, network optimization, and mechanism design for cognitive radio networks, heterogeneous cellular networks, and game theory for wireless communication, computing networks, and network economics.

Jiandong Li graduated from Xidian University with Bachelor Degree, Master Degree and Ph. D in Communications and Electronic System respectively in 1982, 1985 and 1991. He is with Xidian University from 1985, an associate professor from 1990 to 1994, professor from 1994, Ph.D student supervisor from 1995, and Dean of School of Telecommunication Engineering, Xidian University from 1997 respectively. Now, he also serves as Executive Vice Dean Graduate School, Xidian University. Dr. Li is a senior member of IEEE and China Institute of Electronics (CIE) and the fellow of China Institute of Communication (CIC). He was the member of PCN specialist group for China 863 Communication high technology program between Jan. 1993 and Oct. 1994 and from 1999 to 2000. He is also the member of Communication specialist group for The Ministry of Information Industry. His current research interest and projects are funded by the 863 High Tech Project, NSFC, National Science Fund for Distinguished Young Scholars, TRAPOYT, MOE and MOI.

* * *

Abdelrahman Arbi received his B.Sc in Electrical and Electronic Engineering from the University of Garyounis in Libya, and his M.Sc in Computer Science from the Ecole Nationale Supérieure des Télécommunications (ENST) in France. Mr. Abdelrahman has worked in telecommunication sector with the national telecom operator (GPTC) in Libya, and in the IT sector with CGI and Babel media in Canada. He is currently a PhD candidate in the Department of Electronic & Electrical Engineering, University of Sheffield, UK. His research interests include next generation wireless networks, energy and spectrum efficient wireless networks, radio access network planning and optimisation and resource allocation and scheduling in cellular networks.

Norma Alicia Barboza-Tello is professor of Bioelectronics at the School of Sciences of Engineering and Technology in Autonomous University of Baja California, since 2011. She teaches mostly subjects from physics areas as electric circuits, optics, optoelectronics and signals processing. She obtained her Bachelor's degree in Electronic Engineering in 2001 from the Technological Institute of Mexicali. In 2005 she earned her Master of Science degree in Optics, and then in 2010 she obtained her Ph.D. in Optics degree, both titles granted by the Center of Scientific Research and Higher Education of Ensenada (CICESE). Her research areas of interest includes Biosignals processing, development of tunable compact light sources for application in Bioengineering, development of optical biosensors and analysis and also the creation of techniques to improve the experience on the teaching-learning process. Her work aims to propose less invasive medical diagnosis and treatment methods.

Zheng Chang received the B.Eng. degree from Jilin University, Changchun, China in 2007, M.Sc. (Tech.) degree from Helsinki University of Technology (Now Aalto University), Espoo, Finland in 2009 and Ph.D degree from the University of Jyvaskyla, Jyvaskyla, Finland in 2013. Since 2008, he has held various research positions at Helsinki University of Technology, University of Jyvaskyla and Magister Solutions Ltd in Finland. He was a visiting researcher at Tsinghua University, China, from June to August in 2013, and at University of Houston, TX, during from April to May in 2015. He has been awarded by the Ulla Tuominen Foundation, the Nokia Foundation and the Riitta and Jorma J. Takanen Foundation for his research work. Currently he is working with University of Jyvaskyla and his research interests include signal processing, radio resource allocation, cross-layer optimizations for wireless networks, and green communications.

Ling Chen is a graduate student in the Institute of Communications of Chongqing University.

Yuan Dong received her B.E. degree in Electronic and Information Engineering from Northwestern Polytechnical University in 2009, and she successively received her Master Degree and Ph.D. in Computer Science in 2013 from University of Technology of Troyes, France. Since 2014, she has been doing her research and teaching at the Chang'an University. Her main research interests include machine learning, pattern recognition and wireless hetergeneous network.

Simon Fletcher is Chief Technology Officer of Real Wireless and takes responsibility for the overall technology offering and strategy of Real Wireless. Prior to Real Wireless, Simon worked at NEC Corporation leading developments of emerging global market product strategies, portfolio alignment, and innovation programmes including early 5G, Cities and IoT business. With a primary focus on infrastructure and network edge technologies he represented NEC in joint ventures developing 3G and 4G global market products; working in various Systems, Consulting and development project management roles. Through his engagements in Industry Open Innovation activities he has been Industrial Chair of the Mobile VCE Green Radio Programme, and currently serves as a Director of mVCE. He has also served as a Steering Board member of the LTE SAE Trials Initiative and a Director of the ICT-KTN, and is currently the Conference Chair of the Cambridge Wireless Future of Wireless International Conference.

Tianhang Fu is a master candidate student in Beijing Jiaotong University. Her research interset includes radio resource allocation in high-speed railway systems.

Yi Hu is a postgraduate student in Chongqing University.

Pengyu Huang received the B.S. and M.S. degrees from Xidian University, Xi'an, China, in 1998 and 2004 respectively. He is currently a Lecturer with the Broadband Wireless Communication Laboratory, School of Telecommunication Engineering, Xidian University. His research interests include cognitive radio networks and wireless sensor networks.

Yufeng Li is a postgraduate student in Chongqing University.

Yanxia Liang is a Lecturer in school of Telecommunication and Information Engineering in Xi'an University of Posts and Telecommunications(XUPT) (since 2011), China. She received her Bachelor of Engineering in XUPT in 2004, and completed her Master of Engineering in 2007 and Doctor of Military Communication in 2011 in Xidian University, China. Her research interests include Interference management, radio resource management in heterogeneous networks, Source Coding, Speech communication and quantum compression.

Yong Liao, PhD, research associate, deputy director of Key Laboratory of Aircraft TT&C and Communication (Chongqing University), Ministry of Education, China. His research interests include high-speed mobile communication and aircraft TT&C and communication . He hosted one National Natural Science Foundation of China, one of Chongqing Frontier and Basic Research Projects, two projects of State Key Laboratory of Integrated Services Networks Open Fund (Xidian University), one project of Shaanxi Key Laboratory of Network and System Security Open Fund, one of the major projects of Sichuan Key Laboratory of Artificial Intelligence (AI) Open Fund, one of key projects and two projects of Fundamental Research Funds for the Central Universities; research participation in one of major research projects of National Natural Science Foundation of China, three projects of National Natural Science Foundation of China, one project of National Development and Reform Commission demonstration project CNGI 2005, one of 863 Program. More than 40 papers have been published or accepted by internationally renowned journals and domestic authoritative journals, such as International Journal of Adaptive Control and Signal Processing, IEEE Geoscience and Remote Sensing Letters, IEEE Communications Letters, Physics Letters A, Acta Electronica Sinica and Journal on Communications, as the first or corresponding author, and there are 30 SCI/EI papers among them. Meanwhile, he has obtained or applied for more than 30 national invention patents. He holds the post of Invited Review Journal staff of Communications of the CCF (China Computer Federation), an Editorial Board Member of Chinese Journal of Engineering, multiple journals review experts at home and abroad such as IEEE Transactions on Cybernetics, IEEE Geoscience and Remote Sensing Letters, International Journal of Electronics, China Communications, Security and Communication Networks, J Acta Electronica Sinica, Journal of Computer Research and Development, Journal of Radars, Journal of Computer Science, Journal of Computer Applications, Application Research of Computers and Telecommunication Engineering. He is a Senior member of CCF, Public Policy Committee member of CCF, Network and Data Communications Professional Committee member of CCF, Program Committee member of the 12th National Network and Data Communications Conference (NDCC2014), member of Chongqing of CCF Young Computer Scientists & Engineers Forum (YOCSEF), member of IEEE, Chinese Institute of Electronics and Chinese Association of Automation.

Paul Medina-Castro received the B.Eng. degree from Instituto de Tecnologico de Sonora (ITSON), Cd. Obregon, Mexico, and the M.Sc. degree from Centro de Investigación Científica y de Educación Superior de Ensenada (CICESE), Ensenada, Mexico, both in electrical engineering. From July to September 2005, he worked as a research associate at the Broadband Wireless and Internetworking Research Laboratory of the University of Ottawa, Canada. From October 2005 to January 2007, he worked as a lead engineer in projects related to routing and access control in wireless sensor networks as well as IP telephony over wireless LANs. From August 2007 to July 2008, he was a visiting student at the University of Western Ontario, Canada. In September 2011 he obtained a Ph.D. degree at CICESE. Currently he works as full time professor at the Universidad Autónoma de Baja California (UABC). His main interest areas of study are related to Wireless Access Links and Bioinstrumentation.

Yun Meng received her B.S. degree and the Ph.D. in Communication Engineering in Xi'dian University in 2009 and 2015 respectively. She is currently working in Chang'an University. Her research interests include radio resource management in heterogeneous networks and graph theory for wireless networks.

Aradhana Misra is currently working as Assistant Professor in the Department of Electronics and Communication Engineering, Gauhati University. Completed B.Tech from Sikkim Manipal Institute of Technology, Gangtok, India and M.Tech in Electronics and Communication from Gauhati University and Pursuing Ph.D in Gauhati University. Field of research interest includes signal processing and wireless communications.

Timothy O'Farrell is Chair Professor of Wireless Communications at the University of Sheffield, UK. He is an expert in wireless communication systems specialising in physical layer signal processing, radio resource management and wireless network planning. He has pioneered research on energy efficient mobile cellular communications, the mathematical modeling of CSMA based MAC protocols for WiFi, iterative block coding for wireless communication systems and spreading sequence design for CDMA wireless networks. He is an entrepreneur, being the cofounder and CTO of Supergold Communication (2000-2007), a start-up that participated in the standardisation of IEEE 802.11g with the MBCK proposal. In the framework of Mobile VCE (mVCE), Professor O'Farrell was the Academic Coordinator of the Core 5 Green Radio project (2009-2012) and a leader in establishing energy efficiency as a global research field in wireless communication systems. He has managed 24 major research projects as principal investigator; published over 250 journal and conference papers; published 15 patents; and has participated in standards, consultancies and expert witness activities within the wireless sector. Currently, Professor O'Farrell leads the UK Research Strategy Community Organisation in Communications, Mobile Computing and Networking (www.commnet.ac.uk) funded by EPSRC. He also leads the EPSRC funded project on Frequency Agile Radio (FARAD). Professor O'Farrell is a Chartered Engineer and a member of the IET and IEEE.

Kandarpa Kumar Sarma is currently an Associate Professor in Department of Electronics and Communication Technology, Gauhati University, Guwahati, Assam, India. His areas of interest are Soft-Computation and its Applications, Mobile Communication, Antenna Design, Speech Processing, Document Image Analysis and Signal Processing Applications in High Energy Physics, Neuro-computing and Computational Models for Social-Science Applications.

Xuanfan Shen is a postgraduate student in Chongqing University.

Song Shi received his B.S. degree in Electrical Engineering and Automation in Chang'an University in 2003, and M.S. degree Traffic Engineering and Control in 2013 respectively. He is currently working for his Ph.D. in Traffic Engineering and Control in Chang'an University. His research focus on the network planning.

Irma Uriarte-Ramírez is a full professor at the School of Sciences of Engineering and Technology in Autonomous University of Baja California, since 2013. She obtained her Bachelor's degree in Electronic Engineering in 2002 and the Master of Engineering Degree in Electronics in 2008 from the Engineering Faculty Mexicali from the Autonomous University of Baja California. Her research areas of interest includes 5G wireless communication and Green Communications. In Education Area she is interested in Accreditation of University Programs and techniques to improve the teaching-learning process in Basic Sciences.

Ducheng Wu received his B.S. degree in electronic and information engineering from Wuhan University of Technology, Wuhan, China, in 2010 and M.S. degree from College of Communications Engineering, PLA University of Science and Technology, Nanjing, China, in 2013, respectively. He is currently a Ph.D. candidate in College of Communications Engineering, PLA University of Science and Technology. His research interests include resource allocation in small cell networks, game theory and statistical learning.

Qihui Wu received his B.S. degree in communications engineering, M.S. degree and Ph.D. degree in communications and information systems from Institute of Communications Engineering, Nanjing, China, in 1994, 1997 and 2000, respectively. From 2003 to 2005, he was a Postdoctoral Research Associate at Southeast University, Nanjing, China. From 2005 to 2007, he was an Associate Professor with the Institute of Communications Engineering, PLA University of Science and Technology, Nanjing, China, where he is currently a Full Professor. From March 2011 to September 2011, he was an Advanced Visiting Scholar in Stevens Institute of Technology, Hoboken, USA. Dr. Wu's current research interests span the areas of wireless communications and statistical signal processing, with emphasis on system design of software defined radio, cognitive radio, and smart radio.

Xiaochen Xia was born in China, 1987. He received his B.E. degree in electronic science and technology from Tianjin University (TJU) in 2010, and M.S. degree in communication and information system from PLA University of Science and Technology (PLAUST), in 2013. He is currently working toward the Ph.D. degree in institution of communications engineering, PLAUST. His research interests include relaying network, full-duplex communication, Network coding, MIMO techniques. He received the 2013 excellent master degree dissertation award of Jiangsu Province, China.

Jia Xiao is with the State Key Laboratory on Integrated Services Networks, Xidian University, Xi'an, China.

Kui Xu was born in 1982. He received the B.S. degree in wireless communications from the PLA University of Science and Technology, Nanjing, China in 2004, and the Ph.D. degree in software defined radio from the PLA University of Science and Technology, Nanjing, China, in 2009. He is currently an associate professor in the College of Communications Engineering, PLA University of Science and Technology. From Dec. 2013, he was a Postdoctoral Fellow with PLA University of Science and Technology. His research interests include broadband wireless communications, signal processing for communications, network coding, wireless communication networks. He is the author of more than papers in refereed journals and conference proceedings and holds 6 patents in China. Prof. Xu currently served on the technical program committee of the IEEE WCSP 2014, WCSP 2015, ICSPDM 2015, SIRS 2015, ISSPIT 2015. He received the URSI Young Scientists Award in 2014 and the 2010 ten excellent doctor degree dissertation award of PLAUST. He serves as the reviewer for IEEE Trans. Wireless Communication, IEEE Transaction Vehicle Technology, IEEE Communications Letter, IEEE Signal Processing Letters.

Shaoyi Xu received her Ph. D degree from Inha University, South Korea in 2007. After that, she worked in Nokia Research Center in Nokia China Ltd. as a senior researcher and a postdoctor. From 2009 to now, she worked in Beijing Jiaotong University. From Feb. 2013 to Feb. 2014, she is a visiting scholar in Department of Electrical and Computer Engineering, University of California, USA. She has obtained more than 20 international and Chinese patents and published more than 70 journal papers and conference papers. Her research interests include D2D communication, cooperative communication in high-speed environments, cognitive radio and UWB, etc.

Youyun Xu was born in 1966. He was graduated from Shanghai Jiao Tong University with Ph.D. Degree in Information and Communication Engineering in 1999. He is currently a professor with Nanjing Institute of Communication Engineering, China. He is also a part-time professor with the Institute of Wireless Communication Technology of Shanghai Jiao Tong University (SJTU), China. Professor Xu has more than 20 years professional experience of teaching and researching in communication theory and engineering. Now, his research interests are focusing on New Generation Wireless Mobile Communication System (IMT-Advanced and Related), Advanced Channel Coding and Modulation Techniques, Multi-user Information Theory and Radio Resource Management, Wireless Sensor Networks, Cognitive Radio Networks, etc. He is a senior member of IEEE, a senior member of Chinese Institute of Electronics.

Yuhua Xu received his B.S. degree in Communications Engineering, and Ph.D. degree in Communications and Information Systems from College of Communications Engineering, PLA University of Science and Technology, in 2006 and 2014 respectively. He has been with College of Communications Engineering, PLA University of Science and Technology since 2012, and currently as an Assistant Professor. His research interests focus on opportunistic spectrum access, learning theory, game theory, and distributed optimization techniques for wireless communications. He has published several papers in international conferences and reputed journals in his research area. He served as Associate Editor for Wiley Transactions on Emerging Telecommunications Technologies and KSII Transactions on Internet and Information Systems, and the Guest Editor of Special Issue on "The Evolution and the Revolution of 5G Wireless Communication Systems" for IET Communications. In 2011 and 2012, he was awarded Certificate of Appreciation as Exemplary Reviewer for the IEEE Communications Letters. He was selected to receive the IEEE Signal Processing Society's (SPS) 2015 Young Author Best Paper Award, and the Funds for Distinguished Young Scholars of Jiangsu Province in 2015.

Dongmei Zhang was born in China, 1972. She received the B.S. degree in institute of communications engineering, Nanjing, China in 1994, and the M.S. degree in communications and information system from the institute of communications engineering, Nanjing, China in 1997. She is currently an associate professor with the PLA University of Science and Technology, Nanjing, China. Her research interests include broadband wireless communications, cognitive radio, wireless resource allocation.

Shumin Zhang is a postgraduate student in Chongqing University.

Ming Zhao is a postgraduate student in Chongqing University.

Fu-Chun Zheng obtained the BEng (1985) and MEng (1988) degrees in radio engineering from Harbin Institute of Technology, China, and the PhD degree (1992) in Electrical Engineering from the University of Edinburgh, UK. From 1992 to 1995, he was a post-doctoral research associate with the University of Bradford, UK, Between May 1995 and August 2007, he was with Victoria University, Melbourne, Australia, first as a lecturer and then as an associate professor in mobile communications. He joined the University of Reading, UK, in September 2007 as Professor (Chair) of Signal Processing. He has been awarded two UK EPSRC Visiting Fellowships - both hosted by the University of York (UK): first from August 2002 to July 2003 and then from August 2006 to July 2007. He has also been a "1000 talents" scholar with Southeast University, China, since 2010. Over the past 25 years, Dr Zheng has also carried out many industry-sponsored projects. He has been both a short term visiting fellow and a long term visiting research fellow with British Telecom, UK. Dr Zheng's current research interests include signal processing for communications, multiple antenna systems, and green communications. He has been an active IEEE member since 1995. He was an editor (2001 – 2004) of IEEE Transactions on Wireless Communications. In 2006, Dr Zheng served as the general chair of IEEE VTC 2006-Spring in Melbourne, Australia (www.ieeevtc.org/vtc2006spring) - the first ever VTC held in the southern hemisphere. He is the executive TPC chair for the recent IEEE VTC 2016-S in Nanjing, China (the first ever VTC in mainland China: www.ieeevtc.org/vtc2016spring).

Xin Zhou is a postgraduate student in Chongqing University.

Zhenyu Zhou is a professor in North China Electric Power University.

Index

H

I

J

K

L

M

N

O

P

R

S

www.ingramcontent.com/pod-product-compliance
Lightning Source LLC
LaVergne TN
LVHW081250011026
849162LV00006BA/1399